Rainwater Tank Systems for Urban Water Supply

Rainwater Tank Systems for Urban Water Supply

Design, Yield, Energy, Health Risks, Economics and Social Perceptions

Edited by
Ashok K. Sharma, Donald Begbie and Ted Gardner

IWA PUBLISHING

Published by **IWA Publishing**
Alliance House
12 Caxton Street
London SW1H 0QS, UK
Telephone: +44 (0)20 7654 5500
Fax: +44 (0)20 7654 5555
Email: publications@iwap.co.uk
Web: www.iwapublishing.com

First published 2015

British Library Cataloguing in Publication Data
A CIP catalogue record for this book is available from the British Library

ISBN: 9781780405353 (Paperback)
ISBN: 9781780405360 (eBook)

Contents

Editors xiii

List of Authors xv

Foreword xxi

Preface xxiii

Editorial xxv

Chapter 1
Rainwater harvesting systems for urban developments ***1***
Stephen Cook, Ashok K. Sharma and Ted Gardner

1.1 Introduction 1
1.2 International Experiences with Rainwater Harvesting Systems 3
1.3 The Australian Experience with Rainwater Tank Systems 5
1.4 Key Issues for Greater Uptake of Rainwater Tank Systems 7
1.4.1 Estimating rainwater system yield and mains water savings 7
1.4.2 Understanding the risks associated with rainwater quality 8
1.4.3 Guidelines for managing risks of rainwater use 9
1.4.4 Evaluating the cost-effectiveness of rainwater tanks 10
1.4.5 Understanding the indirect costs and benefits of rainwater tanks (externalities) . . 11
1.4.6 Impact of rainwater systems on stormwater flows and nutrient loads 12
1.4.7 Impact of rainwater tank systems on centralised water systems and water quality 13
1.4.8 Energy consumption in rainwater tank systems 13
1.5 Conclusions 14
1.6 References 14

Chapter 2
Rainwater tank modelling ... ***19***
Alison M. Vieritz, Luis E. Neumann and Stephen Cook

2.1 Introduction ... 19
2.2 General Concepts Underlying A Rain Water Tank Model ... 20
2.3 Aspects of Some Existing Rain Water Tank Models ... 24
2.3.1 Time-step and climate data inputs ... 24
2.3.2 Water demand data inputs ... 25
2.3.3 Roof runoff modelling ... 27
2.3.4 Tank water balance configurations ... 27
2.3.5 Model outputs ... 28
2.4 Influence of Different Variables in the Tank Simulation ... 29
2.4.1 Time-step and spill rule ... 30
2.4.2 Roof area ... 33
2.4.3 Tank size ... 33
2.4.4 Demand. ... 34
2.4.5 Initial and continuing losses ... 35
2.4.6 Simulation length ... 36
2.4.7 Validation ... 37
2.5 Upscaling of Rainwater Tank Behaviour to Multiple Tanks ... 39
2.6 Conclusions ... 41
2.7 References ... 42

Chapter 3
Quantifying mains water savings from residential rainwater tanks ... ***47***
Cara D. Beal, Meng Nan Chong, Julian Fyfe, Andrea Turner and Ted Gardner

3.1 Introduction ... 48
3.1.1 Why quantify mains water savings? ... 48
3.1.2 Previous studies on mains water savings ... 48
3.1.3 Chapter objectives and scope ... 50
3.2 Case Study 1 – Desktop Analysis of Mains Water Savings ... 50
3.2.1 Background ... 50
3.2.2 Methods ... 51
3.2.3 Results ... 55
3.2.4 Discussion and implications ... 57
3.2.5 Limitations of Case Study 1 ... 57
3.2.6 Concluding remarks ... 58
3.3 Case Study 2 – Benchmark Analysis of Mains Water Savings ... 58
3.3.1 Background ... 58
3.3.2 Research aims ... 58
3.3.3 Methods ... 59
3.3.4 Results and discussion ... 60

3.3.5 Challenges and limitations 62
3.3.6 Concluding remarks 62
3.4 Case Study 3 – Water Savings from Rebated Rainwater Tanks 63
3.4.1 Background 63
3.4.2 Methods 63
3.4.3 Mains water savings results 65
3.4.4 Interpretation and implications 66
3.4.5 Challenges and limitations 67
3.4.6 Concluding remarks 67
3.5 Key Considerations in Quantifying Mains Savings 68
3.5.1 Quality of the datasets 68
3.5.2 Mixed method and analyses 70
3.5.3 Sample size v quality of datasets 70
3.6 Summary and Conclusions 70
3.7 References 71

Chapter 4
Monitoring of household rainwater tank systems for rainwater usage ***73***
Shivanita Umapathi, Reid Butler, Matthew Ferguson, David Pezzaniti and Ashok K. Sharma

4.1 Introduction 73
4.2 Monitoring of Water Usage in Rainwater Tank Systems 74
4.2.1 Drivers for monitoring 75
4.2.2 Case studies in Australia 75
4.3 Rainwater System Components, Accessories and Configurations 76
4.4 Experimental Approaches 78
4.4.1 Monitoring methods 78
4.4.2 Instrumentation 79
4.4.3 Sample number and ethics approval 86
4.4.4 Site inspections and audits 86
4.4.5 Correlation with weather and seasonal patterns 86
4.4.6 Monitoring duration and intervals 87
4.5 Data Management 88
4.5.1 Data cleansing and validation 88
4.5.2 Data analysis 89
4.6 Case Study 1: Rainwater Tank Monitoring, South East Queensland, Australia 89
4.6.1 Data analysis and results 90
4.7 Case Study 2: Sydney Water, Sydney, New South Wales, Australia 93
4.7.1 Data analysis and results 94
4.8 Other Cost Considerations Associated with Monitoring 95
4.9 Conclusion 96
4.10 References 97

Chapter 5
Physical verification of household rainwater tank systems ... 101
Sharon Biermann and Reid Butler

5.1 Introduction ... 101
5.2 Rainwater Tank Installation Compliance Monitoring Framework ... 103
5.2.1 Sydney BASIX assessment ... 103
5.2.2 Queensland Development Code Performance criteria ... 104
5.2.3 Rainwater tank installation audit protocol ... 105
5.3 Application of the Method to the SEQ Case Study Area ... 110
5.3.1 Context ... 110
5.3.2 Data collection ... 111
5.4 Results ... 112
5.4.1 General rainwater tank characteristics ... 112
5.4.2 Rainwater tank storage volume ... 112
5.4.3 Connected roof area ... 115
5.4.4 Connection to toilets, washing machines and external use ... 119
5.4.5 Continuous supply ... 119
5.4.6 Water quality protection ... 119
5.5 Summary and Conclusion ... 121
5.6 References ... 123

Chapter 6
Understanding energy usage in rainwater tank systems through laboratory and household monitoring ... 127
Grace Tjandraatmadja, Monique Retamal, Shivanita Umapathi and Guenter Hauber-Davidson

6.1 Energy Usage in Rainwater Systems ... 127
6.2 Rainwater System Set-Up ... 129
6.3 Energy Use Estimation ... 133
6.3.1 Factors impacting energy consumption ... 133
6.3.2 Pump performance ... 133
6.3.3 Pump design and characteristics ... 133
6.3.4 End use water requirements ... 136
6.3.5 Dwelling occupancy and characteristics ... 140
6.3.6 Friction losses ... 141
6.3.7 Other components ... 141
6.4 Reducing Energy Use for Rainwater Systems – Lessons from Australia ... 145
6.4.1 Policy considerations for rainwater system energy use ... 146
6.5 Conclusions ... 146
6.6 References ... 147

Chapter 7
Management and operational needs for urban rainwater tanks ***151***
Magnus Moglia, Andrea Walton, John Gardner and Grace Tjandraatmadja

7.1 The Need for Managing Rainwater Tanks 151
7.2 Issues to Consider in the Management of Urban Rainwater Tanks 152
7.3 Private Ownership vs Public Benefit 155
7.4 Factors that Influence Condition 156
7.4.1 Maintenance 156
7.4.2 Design 156
7.4.3 Installation 157
7.5 Local Context 158
7.5.1 Local regulatory environment 158
7.5.2 Understanding behaviour 159
7.6 Research Framework 161
7.7 Application of the Research Framework: The South East Queensland Case Study 163
7.7.1 South East Queensland context 163
7.7.2 Definition of strategies 166
7.7.3 Assessing strategies 168
7.8 Discussion 175
7.9 Conclusions 177
7.10 References 178

Chapter 8
Public perceptions, motivational drivers, and maintenance behaviour for urban rainwater tanks ***181***
Aditi Mankad, Kelly Fielding and Sorada Tapsuwan

8.1 Introduction 181
8.2 Past Social Research on Rainwater Use in Urban Australia 183
8.3 Theoretical Frameworks for Understanding Public Acceptance, Adoption and Maintenance of Rainwater Tanks 184
8.4 Investigating Public Acceptance of Rainwater Tanks: Qualitative Methods 186
8.4.1 Identifying facilitators and barriers to rainwater tank adoption in South East Queensland 186
8.4.2 Case study findings 188
8.5 Investigating Public Acceptance of Rainwater Tanks: Quantitative Methods 190
8.5.1 Predictors of rainwater tank adoption in South East Queensland 190
8.5.2 Case study findings 192
8.6 Identifying Motivational Drivers of Rainwater Tank Maintenance: Quantitative Methods 194
8.6.1 Drivers of rainwater tank maintenance behaviour for mandated tank owners 195
8.6.2 Case study results 196
8.6.3 Drivers of rainwater tank maintenance behaviour for retrofitted versus mandated tanks 198
8.6.4 Case study results 199

8.7 Conclusion ... 200
8.7.1 How can we influence public acceptance and adoption of rainwater tanks? ... 201
8.7.2 How can we encourage more effective maintenance of domestic rainwater tanks? ... 202
8.8 References ... 203

Chapter 9
Chemical quality of rainwater in rain tanks ... **207**
Mirela I. Magyar and Anthony R. Ladson

9.1 Introduction ... 207
9.1.1 The uses of tank water ... 208
9.1.2 Water quality guidelines ... 209
9.2 Water Quality from Rainfall to Roof to Tank ... 210
9.2.1 Rainwater and atmospheric influences on inflow quality ... 210
9.2.2 The effect of roof material on the quality of tank inflows ... 211
9.2.3 A statistical overview of roof runoff quality ... 212
9.2.4 Improving tank inflow quality ... 215
9.2.5 The rain tank as a water treatment device ... 216
9.3 Measurements of Rain Tank Water Quality ... 217
9.3.1 Introduction ... 217
9.3.2 Studies of tank water quality: A summary ... 217
9.3.3 Meta-analysis: Proportion of tanks with high lead concentrations ... 219
9.3.4 pH of tank water and relationship with lead concentration ... 221
9.3.5 Quality aspects of tank supply to hot water systems ... 221
9.4 Improving the Quality of Water Supplied by Rain Tanks ... 223
9.5 Conclusion ... 223
9.6 References ... 224

Chapter 10
Microbiological quality and associated health risks with the use of roof-captured rainwater ... **229**
Warish Ahmed and Simon Toze

10.1 Introduction ... 229
10.2 Faecal Indicators and Pathogens in Roof-Captured Rainwater ... 230
10.2.1 Faecal indicators ... 230
10.2.2 Bacterial pathogens ... 232
10.2.3 Opportunistic bacterial pathogens ... 233
10.2.4 Protozoa pathogens ... 234
10.2.5 Likely sources of *Escherichia coli* harboring toxin genes in rainwater tanks ... 235
10.2.6 Presence and source of faecal indicators and zoonotic pathogens in household drinking water taps fed from rainwater tanks in South East Queensland ... 237
10.2.7 Inactivation of faecal indicator bacteria in a roof-captured rainwater system ... 238
10.3 Health Risks Associated with Roof Captured Rainwater ... 242

10.3.1 Quantitative Microbial Risk Assessment to determine health risk from the use of roof-captured rainwater . . . 245
10.4 Concluding Remarks and Recommendations . . . 248
10.5 References . . . 249

Chapter 11
Cluster-scale rainwater harvesting . . . 253
Stephen Cook, Ashok K. Sharma, Thulo Ram Gurung, Luis E. Neumann, Magnus Moglia and Priya Chacko

11.1 Introduction . . . 253
11.2 Literature Review . . . 255
11.2.1 Examples of cluster-scale rainwater harvesting . . . 255
11.2.2 Impediments and benefits . . . 256
11.3 Case Studies of Cluster-Scale Harvesting . . . 257
11.3.1 Capo di Monte . . . 257
11.3.2 Wannon water's roof water harvesting project . . . 259
11.3.3 Christie walk . . . 260
11.3.4 Fitzgibbon chase potable roof water (PotaRoo) scheme . . . 261
11.4 Modelling the Performance of Individual vs. Cluster Rainwater Harvesting . . . 263
11.5 Method for the Design of CSRH Systems . . . 264
11.5.1 Designing the CSRH system . . . 264
11.6 Water Treatment . . . 268
11.6.1 Preliminary treatment and filtration . . . 268
11.6.2 Disinfection . . . 269
11.6.3 Post-treatment methods . . . 269
11.7 Management of CSRH Systems . . . 269
11.8 Conclusions . . . 271
11.9 References . . . 271

Chapter 12
Economics of individual and communal rainwater tank systems . . . 275
Murray R. Hall, Thulo Ram Gurung and Kym Whiteoak

12.1 Introduction . . . 275
12.2 Cost-Effectiveness of Individual Rainwater Tanks . . . 276
12.2.1 Method . . . 276
12.2.2 Case study – mandated rainwater tanks in South East Queensland, Australia . . . 277
12.3 Cost-Effectiveness of Communal Rainwater Tanks . . . 285
12.3.1 Communal rainwater tank systems . . . 285
12.3.2 Method for cost-effectiveness estimation . . . 286
12.4 Cost Benefit Analysis of Rainwater Tanks . . . 290
12.4.1 The role of cost-effectiveness assessment . . . 290
12.4.2 Cost-benefit analysis . . . 291
12.4.3 Proposed economic framework . . . 291
12.4.4 Case study – Toowoomba . . . 296

12.5 Discussion 297
12.6 Conclusions 298
12.7 References 299

Chapter 13
Impact of rainwater tanks on urban hydrology and stormwater quality* *301
Matthew J. Burns, Anthony R. Ladson and Tim D. Fletcher

13.1 Introduction 301
13.2 Impacts of Urbanization on Stream Hydrology and Water Quality 302
13.2.1 Stream hydrology 302
13.2.2 Water quality 302
13.2.3 Stream ecological consequences 306
13.3 Conventional Approaches to Urban Stormwater Management in Australia 306
13.4 How Urban Stormwater can be Managed for Urban Stream Protection or Restoration 308
13.4.1 Increase volumetric losses 308
13.4.2 Increase infiltration (filtered-flow) 308
13.4.3 Increase equivalent initial loss 308
13.5 The Role of Rainwater Tanks in Flow-Regime Management 309
13.5.1 Increase volumetric losses 309
13.5.2 Increase infiltration (filtered-flow) 309
13.5.3 Increase equivalent initial loss 313
13.5.4 Water quality and tanks 313
13.6 Other Benefits of Rainwater Tanks 313
13.7 Conclusion 314
13.8 References 314

Chapter 14
Rainwater tanks in Australia: Their social/political context, a research overview, policy implications, future research needs, and application of findings to other countries* *319
Ted Gardner, Don Begbie, Ashok K. Sharma and Grace Tjandraatmadja

14.1 Introduction 320
14.2 Drivers for Rainwater Tanks Implementation 321
14.2.1 Australian context 321
14.2.2 International context 322
14.3 Current Availability of Water Resources in Australia 324
14.3.1 Changes in approach to alternative water supplies in Australia 324
14.4 Policy Implications from the Chapters 326
14.5 Lessons for Other Countries 331
14.6 Conclusions 333
14.7 References 333

Index 339

Editors

Dr. Ashok K. Sharma is an Adjunct Professor at Institute of Sustainability and Innovation, Victoria University, Melbourne, Australia and Principal Research Engineer at CSIRO Land and Water Flagship, Australia. He has 30 years of research, teaching and industrial experience on planning and design of centralised and decentralised water, wastewater and stormwater systems; integrated urban water management and water sensitive urban design. At CSIRO, he has led research on alternative water, wastewater and stormwater systems to address knowledge gaps in their mainstream uptake. He also worked as a Planning Engineer at Department of Natural Resources and Mines, Queensland, Australia and as an Engineer at Uttar Pradesh State Water Corporation, and Assistant Professor at Delhi College of Engineering in India. He has co-authored 2 books, 6 book chapters, 58 journal and 64 conference publications, and 40 technical reports. He completed his B Tech (Civil Eng.) at G B P Agriculture and Technology University, Pantnagar, India; ME (Environmental Eng.) and PhD (Civil Eng.) at Indian Institute of Technology, Roorkee, India. He is a Fellow of the Institution of Engineers (Australia) and CP Eng. (Australia).

Don Begbie is the Program Manager, Research and Development, Australian Water Recycling Centre of Excellence. Don was the Director of the Urban Water Security Research Alliance in South East Queensland until it was disbanded in February 2013. He managed and coordinated the delivery of research for urban water security with a focus on alternative water sources such as rainwater tanks and stormwater harvesting. He was previously Director Water Science, Queensland Department of Natural Resources and Water, where he managed the delivery of research into urban water systems, groundwater and surface water modelling, and freshwater quality and aquatic ecosystem health. Don completed both his Bachelor of Agricultural Science and Masters of Agricultural Studies at The University of Queensland.

Adjunct Professor Ted Gardner (University Sunshine Coast) is a retired scientist with a number of adjunct appointments with regional universities in Queensland, where he enjoys interacting with students. Prior to his retirement in 2010, Ted was Principal Research Scientist in the Integrated Urban Water Systems group of CSIRO Land and Water. He led the Urban Water Security Research Alliance funded projects into decentralised technologies, and storm water harvesting and reuse in South East Queensland. He was also the Principal Scientist with the Queensland Department of Environment and Resource Management, leading the Urban Water Cycle group which focused on soil physics, irrigation and catchment hydrology/salinity, effluent reuse, and urban water sustainability. Ted has won several achievement awards, including the 2005 Australia Day Award Public Service Medal for his work on water recycling and urban water supply. In 2014, he was award the biennial McLean-Idema award from Irrigation Australia for his career work on irrigation using recycled water. He has an extensive publication record including over 200 peer reviewed journal and conference papers, 3 book chapters, editor of a scientific monograph on purified recycled water, and has made numerous presentations to technical and community groups. Ted completed his Bachelor of Agricultural Science and Masters of Agricultural Science at the University of Queensland.

List of Authors

Dr. Warish Ahmed
Land and Water Flagship
Commonwealth Scientific and Industrial Research Organisation, Australia,
41 Boggo Road, Dutton Park, QLD 4102 Australia.
E-mail: warish.ahmed@csiro.au

Dr. Cara D. Beal
Smart Water Research Centre and School of Engineering,
Griffith University,
Southport, QLD 4222 Australia.
E-mail: c.beal@griffith.edu.au

Donald Begbie
Land and Water Flagship
Commonwealth Scientific and Industrial Research Organisation, Australia,
41 Boggo Road, Dutton Park, QLD 4102 Australia.
Australian Water Recycling Centre of Excellence
Level 5, 200 Creek St, Brisbane, QLD 4000 Australia.
E-mail: donaldbegbie@bigpond.com; don.begbie@australianwaterrecycling.com.au; don.begbie@csiro.au

Prof. Sharon Biermann
School of Earth and Environment, The University of Western Australia,
M004, 35 Stirling Highway, Crawley, WA 6009 Australia.
E-mail: sharon.biermann@uwa.edu.au

Dr. Matthew J. Burns
Department of Resource Management and Geography, The University of Melbourne,
221 Bouverie St, Parkville, VIC 3010 Australia.
E-mail: matthew.burns@unimelb.edu.au

Reid Butler
REID Environmental,
48 Piper St., Lilyfield, NSW 2040 Australia.
E-mail: reid@reidenvironmental.com

Priya Chacko
Land and Water Flagship
Commonwealth Scientific and Industrial Research Organisation, Australia,
Waite Road, Glen Osmond, SA 5064 Australia.
Email: priya.chacko@csiro.au; priyachacko79@gmail.com

Dr. Meng Nan Chong
School of Engineering (Chemical Engineering Discipline),
Monash University Malaysia Campus,
Bandar Sunway, Selangor, DE 46150 Malaysia.
E-mail: chong.meng.nan@monash.edu

Stephen Cook
Land and Water Flagship
Commonwealth Scientific and Industrial Research Organisation, Australia,
37 Graham Road Highett, VIC 3149 Australia.
E-mail: Stephen.cook@csiro.au

Matthew Ferguson
Sydney Water,
10 Valentine Avenue Parramatta, NSW 2150 Australia.
E-mail: matthew.ferguson@sydneywater.com.au

Dr. Kelly Fielding
The University of Queensland, Level 4, GPN3 (Building 39A),
Campbell Road, St. Lucia, QLD 4072 Australia.
E-mail: k.fielding@uq.edu.au

Prof. Tim D Fletcher
Department of Resource Management and Geography, The University of Melbourne,
221 Bouverie St, Parkville, VIC 3010 Australia.
E-mail: tim.fletcher@unimelb.edu.au

Julian Fyfe
Institute for Sustainable Futures,
University of Technology Sydney,
235 Jones St, Ultimo, NSW 2007 Australia.
E-mail: Julian.Fyfe@uts.edu.au

Dr. John Gardner
Land and Water Flagship
Commonwealth Scientific and Industrial Research Organisation, Australia,
41 Boggo Road, Dutton Park, QLD 4102 Australia.
E-mail: john.gardner@csiro.au

Prof. Ted Gardner
School of Science, Education and Engineering,
University of the Sunshine Coast,
Sippy Downs, QLD 4558 Australia.
E-mail: egardner@usc.edu.au; tedandkayegardner@bigpond.com

Thulo Ram Gurung
Centre for Infrastructure Engineering and Management,
Griffith School of Engineering, Griffith University,
Parklands Drive, Southport, QLD 4222 Australia.
E-mail: ram.gurung@griffithuni.edu.au

Murray R. Hall
Land and Water Flagship
Commonwealth Scientific and Industrial Research Organisation, Australia,
41 Boggo Road, Dutton Park, QLD 4102 Australia.
E-mail: murray.hall@csiro.au

Guenter Hauber-Davidson
Water Conservation Group,
2/246 Albert St, East Melbourne, Victoria 3002 Australia.
E-mail: guenterhd@watergroup.com

Dr. Anthony R. Ladson
School of Civil, Environmental and Mining Engineering,
The University of Adelaide, SA 5005 Australia.
Department of Resource Management and Geography, The University of Melbourne,
221 Bouverie St, Parkville, VIC 3010 Australia.
E-mail: anthony.ladson@adelaide.edu.au; tony.ladson@gmail.com

Dr. Mirela I. Magyar
Mirela Magyar Consulting
E-mail: mirela_magyar@hotmail.com

Dr. Aditi Mankad
Land and Water Flagship
Commonwealth Scientific and Industrial Research Organisation, Australia,
41 Boggo Road, Dutton Park, QLD 4102 Australia.
E-mail: aditi.mankad@csiro.au

Dr. Magnus Moglia
Land and Water Flagship
Commonwealth Scientific and Industrial Research Organisation, Australia,
37 Graham Road Highett, VIC 3149 Australia.
E-mail: magnus.moglia@csiro.au

Dr. Luis E. Neumann
Land and Water Flagship
Commonwealth Scientific and Industrial Research Organisation, Australia,
37 Graham Road Highett, VIC 3149 Australia.
E-mail: luis.neumann@csiro.au

David Pezzaniti
Centre for Water Reuse and Management,
University of South Australia, Mawson Lakes, SA 5095 Australia.
E-mail: david.pezzaniti@unisa.edu.au

Monique Retamal
Institute for Sustainable Futures,
University of Technology Sydney,
PO Box 123, Broadway, NSW 2007 Australia.
E-mail: Monique.Retamal@uts.edu.au

Prof. Ashok K. Sharma
Land and Water Flagship
Commonwealth Scientific and Industrial Research Organisation, Australia,
37 Graham Road Highett, VIC 3149 Australia.
Institute for Sustainability and Innovation, Victoria University,
Hoppers Lane, Werribee, VIC 3030 Australia.
E-mail: asharma2006@gmail.com; ashok.sharma@csiro.au; ashok.sharma@vu.edu.au

Dr. Sorada Tapsuwan
Land and Water Flagship
Commonwealth Scientific and Industrial Research Organisation, Australia,
Private Bag 5, Wembley, WA 6913 Australia.
E-mail: sorada.tapsuwan@csiro.au

Grace Tjandraatmadja
Land and Water Flagship
Commonwealth Scientific and Industrial Research Organisation, Australia,
37 Graham Road Highett, VIC 3149 Australia.
E-mail: grace.tjandra@csiro.au; gracetjand@gmail.com

Dr. Simon Toze
Land and Water Flagship
Commonwealth Scientific and Industrial Research Organisation, Australia,

41 Boggo Road, Dutton Park, QLD 4102 Australia.
E-mail: simon.toze@csiro.au

Dr. Andrea Turner
Institute for Sustainable Futures,
University of Technology Sydney,
235 Jones St, Ultimo, NSW 2007 Australia.
E-mail: Andrea.Turner@uts.edu.au

Shivanita Umapathi
Land and Water Flagship
Commonwealth Scientific and Industrial Research Organisation, Australia,
Centre for Water Reuse and Management,
University of South Australia,
Mawson Lakes, SA 5095 Australia.
E-mail: Shivanita.Umapathi@mymail.unisa.edu.au; shiv.umapathi@csiro.au

Alison Vieritz
Department of Science, Information Technology, Innovation and the Arts, Science Delivery,
41 Boggo Road, Dutton Park, QLD 4102 Australia.
E-mail: alison.vieritz@dsitia.qld.gov.au

Dr. Andrea Walton
Land and Water Flagship
Commonwealth Scientific and Industrial Research Organisation, Australia,
41 Boggo Road, Dutton Park, QLD 4102 Australia.
E-mail: andrea.walton@csiro.au

Kym Whiteoak
RM Consulting Group,
Suite 1, 357 Camberwell Road, Camberwell, VIC 3124 Australia.
E-mail: kymw@rmcg.com.au

Foreword

A book on rainwater tanks! Really? What is there to write about concerning those concrete, plastic or corrugated iron structures that many will associate as the source of water on rural or remote properties where the water distribution system has not reached? Haven't they been around for ages? Do you not just install them and allow them to fill from rainwater runoff from a roof or from pumped groundwater? Then doesn't gravity cause the water to flow to taps, hot water systems, toilets and gardens? So simple, so established, so what?

What, however, if we wish to include rainwater tanks as part of a regional water supply system, whether rural, urban or in-between? Are rainwater tanks an efficient means of reducing demand from other water supplies? Is the water in urban rainwater tanks suitable for drinking? Does plumbing the tanks into a household for non-drinking purposes make sense? How energy efficient are rainwater tanks? What are the maintenance requirements? What is the public acceptability of such tanks in an urban setting where there are other options for supplying water? Are there alternatives to small-scale rainwater harvesting? What is the trade-off between keeping spare capacity in a rainwater tank to capture the first flush of stormwater, thereby reducing the environmental load, as opposed to keeping the tank as full as possible to maximize available supply?

While individual articles address these and many other questions, this book is unique in bringing together the many aspects of this deceptively simple device – the rainwater tank. The chapters are informed both by the past findings of others and the latest research results. The motivation for the book was the extended extremely low rainfall period endured by south-east Queensland in the early part of this century and the impact that this was having on the communities that lived there. As one response, the Queensland Government, CSIRO, Griffith University and The University of Queensland created and resourced the Urban Water Security Research Alliance (UWSRA), to address a range of water supply options – one of which was the systematic and comprehensive introduction of rainwater harvesting.

As chair of the Research Advisory Committee (RAC) of UWSRA, I am delighted that Ashok Sharma, Don Begbie and Ted Gardner, and ultimately, IWA Publishing took up the suggestion that there was a need for a 'rainwater tank book', so that others throughout the world might gain from our experiences in Australia.

So simple? – perhaps;
So established? – undoubtedly;
So what? – read the book.

Paul Greenfield
Chair International Water Centre,
Brisbane, Australia & Chair,
Local Organising Committee,
IWA Congress,
Brisbane, 2016.

Preface

Historically, household rainwater tank systems have been implemented in rural and peri-urban areas for potable water supply where municipal reticulated water supply systems were not feasible due to economic and/or technical considerations. They have also been implemented in urban areas in a number of countries to provide a local, decentralised water source. However, their widespread implementation in urban areas with a centralised water supply is comparatively novel. The importance of these systems in cities has grown as water managers seek to address capacity constraints of current water supply systems as well as increasing resilience to drought and the adverse impacts of climate change. Rainwater tank systems are now implemented under integrated urban water management (IUWM) and water sensitive urban design (WSUD) approaches, which take a holistic view of the urban water system for water supply and stormwater management. These approaches include stormwater quality management and flood mitigation.

Dwellings with rainwater tanks increased from 15% to 28% of Australian capital city households (about 5.5 million dwellings) over the last six years; however, the increase in rainwater tanks outside capital city areas was much smaller, increasing from 38% of households to 44% over the same time period. Taken overall, 34% of all Australian households (8.9 million) had installed a rainwater tank by 2013. This rapid uptake of rainwater tanks in Australian cities was encouraged through financial incentives, including rebates for homeowners, but more importantly, from changes to residential building codes that effectively mandated the installation of rainwater tanks in response to the 'millennium drought' (2003 to 2008), as well as environmental sustainability objectives. These actions were taken by water policy makers to diversify the urban water supply source mix in order to address drought, which was particularly acute in many of the urban water supply catchments. Similar trends for increased uptake of rainwater tanks in urban areas can also be expected in other parts of the world impacted by potable water shortage to meet present and future demand, reduced availability of good quality water, or where there is an environmental driver to reduce the adverse impacts of urban runoff on receiving waters.

Much of the research to date has been focussed on tanks in developing countries for potable supply applications, or for flood mitigation in highly-urbanised, developed countries. This book is based on a comprehensive research program on rainwater tanks in South East Queensland, Australia, undertaken with funding from the Urban Water Security Research Alliance (UWSRA). The UWSRA, a partnership between the State government and selected research institutions, was motivated by the millennium drought, whilst the rainwater program's brief was to provide an evidence base for expanding scientific knowledge about rainwater tank systems, and their role as an auxiliary water supply source in modern urban water systems.

This book addresses many of the significant knowledge gaps for the successful implementation of rainwater tanks systems as part of an integrated urban water management approach, including: actual harvested yield and the corresponding mains water savings, optimal sizing for rainwater storages and roof collection systems, modelling tools for sizing tanks and estimating long-term yields, expected chemical and microbiological water quality and implications for managing public health risks, energy consumption of rainwater systems, successful approaches for operation and management of rainwater tanks, the sociology of community acceptance of tanks and their maintenance, impact of rainwater tanks on stormwater quality and quantity, economics of distributed and communal rainwater systems, and regulatory and policy implications in adopting rainwater tanks.

We believe this book will provide students and researchers around the world with a valuable resource on the biophysical aspects and social implications of rainwater tanks. The book will also be a valuable resource for developers, civil designers, architects and plumbers seeking to implement sustainable water servicing approaches using rainwater tanks for residential, industrial and commercial developments. And last but not least, the book should also be useful for water professionals who are involved in the strategic water planning for a town or at a larger scale. Although this book is based on research projects conducted in South East Queensland, Australia, we believe the generalised approaches and methodologies described in the book can be applied to rainwater tank implementation programs in most geographic and social contexts of the international community. Furthermore, the book provides insights on the expected performance and potential pitfalls of the adoption of rainwater tanks systems as part of an integrated approach in managing urban water systems. The book also identifies a number of remaining knowledge gaps that need to be addressed to better inform the policy development and management decisions needed for encouraging rainwater tank systems as one of the mainstream solutions for augmenting centralised water supply systems. We believe their widespread adoption will move society closer to the goal of ecologically sustainable development.

We posit that the future of tanks seem bright in both developed and developing countries. In the former, communal systems in dense urban areas give the opportunity to provide water at a quality assured potable standard, as well as reducing peak stormwater discharge into combined sewers, drains and creeks. In developing countries, the suite of available construction materials and treatment devices make distributed rainwater systems a very attractive source of safe household water. But in all cases, the engagement of the community is essential to ensure that the system operates to design specifications over the long term.

Good information allows good policy development and good decision making. The information presented in this book should empower designers, planners and regulators to better incorporate rainwater systems into the urban fabric of the world, after allowing for some local context specific modifications.

We hope you enjoy reading it.

Ashok K. Sharma
Don Begbie
Ted Gardner

Editorial

There is growing pressure on urban fresh water resources due to an ever increasing population and climate change. Implementing rainwater tanks is one option to supplement freshwater resources based on the *fit for purpose* concept. A rainwater tank system looks apparently simple, but various complexities occur in their design and implementation to achieve long-term economic, social and environmental benefits. We have championed this book to provide knowledge to water professionals, managers and regulators to help them better realise the complexities in rainwater tank systems implementation so as to achieve the desired benefits. We are proud to have been able to facilitate contributions from Australian authors who have worked on various aspects of rainwater tank systems covering modelling, auditing, monitoring, management, social perception, optimal system configuration design, economics, and chemical and biological water quality.

In Chapter 1, Stephen Cook, Ashok Sharma and Ted Gardner describe the history of rainwater tanks implementation and reviewed rainwater harvesting practices around the world. They have indicated that rainwater harvesting systems have been used for local water supply as far back as the ancient civilisations of Greece, Jordan and Persia. Rainwater harvesting can provide a source of better quality drinking water in developing countries where surface water becomes contaminated, and good quality ground water is not available. In peri-urban and rural areas of developed countries, rainwater is often the only potable source, as reticulated water supply is not available due to economic considerations. However, this chapter mainly focusses on the provision of rainwater systems in urban areas as a supplementary source to mains water supply, to address resource constraints due to population growth and climate change, with a highlight on Australian experiences.

Numerical models are now readily available on personal computers and can simulate performance of rainwater tank systems over long time periods. Model outputs are used across the globe for supporting decisions on the suitability of rainwater tank systems to meet household water demand. In Chapter 2, Alison Vieritz, Luis Neumann and Stephen Cook present a generalised model of a rainwater tank/roof catchment system, discuss the various hydraulic and computational processes, and explore the impact of key model parameters, the choice of time-step, order of calculation and simulation length on predicted rainwater yield. The authors also discuss the modelling issues in representing the collective behaviour of a large number of identical tank systems in an urban area (also known as spatial lumping) for estimating the average reliability and yield of a large rainwater tank population.

Chapter 3 covers the various methods to quantity mains water savings from installing residential rainwater tanks to (partially) substitute for potable water in urban settings. It is important for water planners to quantify mains water savings to ensure rainwater tanks contribute their component to the strategic water plan for a city or a region. The rainwater usage estimated with modelling tools or theoretical approaches can be significantly different from the actual rainwater usage due to the variation in various modelling parameters as described in Chapter 2. In this Chapter 3, Cara Beal and colleagues describe three different statistical methods for assessing mains water savings, using examples from Australia where rainwater tanks were installed on a suburb wide basis. It is anticipated that these methods can be applied in any part of the globe for quantifying mains water savings where rainwater tanks are mandated or encouraged in urban households. The factors that mainly influence the mains water savings are also described in this chapter.

Mathematical models and statistical analysis are described in Chapters 2 and 3 respectively to estimate mains water savings. However direct measurement of rainwater use is the *gold standard* of quantifying potable water savings. In Chapter 4, Shivanita Umapathi and colleagues describe a method for real-time monitoring of rainwater usage from household raintanks and the associated instrumentation required. Two Australian case studies are described in this chapter, where real-time monitoring of water flows and energy usage was conducted and monitoring outcomes were compared with mathematical and other mains water saving approaches. Limitations of small sample size are also discussed.

Local guidelines or development codes are prescribed for the installation of rainwater tank systems to achieve certain rainwater yield from the raintanks. These guidelines usually specify minimum raintank size (kL), connected roof area (m^2), and water quality improvement devices. However, compliance with these guidelines or development codes is not know until a post installation physical audit of these systems is undertaken. Non-compliance of local development codes could seriously impact the mains water saving assumed to occur from installing rainwater tanks. Sharon Biermann and Reid Butler, in Chapter 5, describe a generic rainwater tank installation compliance audit protocol, and demonstrate its application in South East Queensland (SEQ), Australia, where Queensland Development Code MP 4.2 specified minimum tank size, connected roof area, water quality improvement devices, and connected water end uses for non-potable application in new detached dwellings.

Pumps are an integral part of most rainwater systems, so understanding energy consumption and exploring ways to minimise energy use are important from economic and environmental sustainability considerations. A large number of these systems have been installed in urban developments, and thus, any attempt to improve energy efficiency should have significant overall economic and environmental impact. In Chapter 6, Grace Tjandraatmadja and colleagues examine the factors that influence energy use in commonly used rainwater pump systems. These factors are pump size, flow rate, system and infrastructure design, and indirectly, water policy. The authors also describe how the system configuration could be improved to reduce energy use. A description of laboratory setup for validating various system configurations is provided for wider professional interest, as well as providing guidance to allow similar studies to be undertaken in other parts of the globe.

Supply from rainwater tank systems at desired quantity and quality can only be achieved if the systems are regularly maintained, usually by the tank owners. There are also substantial public health risks from mosquito borne arboviruses if mesh screens fitted to raintanks are poorly maintained. Magnus Moglia and colleagues, in Chapter 7, describe strategies to ensure that privately owned rainwater tanks remain both functional and safe. These strategies are not straight forward as, in many cases, the private owners are either not motivated or not technically competent to undertake maintenance. Whilst it is relatively simple to maintain a single rainwater tank, the task becomes much more complex to ensure that the entire stock of rainwater tanks is maintained. Problem definition and communication strategies described in this chapter will help water professionals to achieve the desired maintenance objectives for rainwater tanks.

The adoption of rainwater tanks, including their regular maintenance, is very much influenced by community perception and attitude, which in turn are impacted by psychological factors. In Chapter 8, Aditi Mankad, Kelly Fielding and Sorada Tapsuwan describe methodologies to investigate these factors, and demonstrate their application using case studies from various regions of Australian. The authors explore attitudes of users and non-users of rainwater tanks, psychological and behavioural data comparing mandated installation of rainwater tanks with voluntary adoption through rebate programs, and the main psychological variables likely to influence rainwater use and tank maintenance beyond public acceptance.

In urban areas where mains water is available, health agencies usually do not recommend rainwater for potable applications for microbiological reasons. However, it is also important to understand the health implications of the chemical water quality of rainwater if it is used for potable purposes. In Chapter 9, Mirela Magyar and Anthony Ladson review the chemical water quality aspects of rainwater tanks and compared them with drinking water guidelines. The authors indicate that high lead concentrations, elevated metal concentration and low pH are common issues in rainwater. High concentration of lead was reported in 31 of the 32 studies. The authors suggest that the risk of lead contamination in urban areas should be taken seriously. Most aspects of water quality can be improved by suitable treatment, on-going maintenance and improvement in tank design, especially the pump intake hose.

As mentioned above, rainwater is not recommended for potable applications in urban areas with reticulated water supply. However, such use can't be completely stopped, whilst potable use of rainwater in peri-urban and rural communities is common. Thus, the understanding of microbiological quality of rainwater is essential. Warish Ahmed and Simon Toze, in Chapter 10, describe the microbiological quality of rainwater and associated health risks from zoonotic pathogens in roof captured rainwater. The conclusions provided in this chapter are based on water samples from residential rainwater tanks in South East Queensland, Australia. However, the information provided, especially on the kinetics of natural die off, will be of general use for all professionals involved with rainwater supply in any residential urban or rural development.

Cluster-scale (communal) rainwater tanks are an alternative to individual householder tanks and are discussed in Chapter 11. In communal systems, roof water is collected through a common gravity conveyance (collection) system from a group of houses and stored in a communal tank. After appropriate treatment, the water is supplied back to the homes through a dedicated, pressurised water reticulated system. Depending on the level of treatment, the water supply from cluster-scale systems can be used for potable applications rather than just non-potable uses such as toilet flushing, laundry and garden irrigation, as occurs for example for individual tanks. Moreover, these communal systems can be well managed through a formalised, quality assured management and maintenance arrangement, thereby overcoming the serious maintenance limitations of household-scale rainwater harvesting systems. In this chapter, Stephen Cook and colleagues have described the multiple benefits that cluster-scale rainwater tank systems can offer, including economies of scale based on life cycle costs, reduced land footprint per allotment, centralised treatment and disinfection, and matching overall demand for different households.

Understanding the economics of rainwater tank systems is essential to compare them with centralised and other alternative water supply systems. The comparison is generally based on the cost of water ($/kL) supplied by an asset over its useful life. In Chapter 12, Murray Hall, Thulo Ram Gurung and Kym Whiteoak describe the economics of both individual household and cluster-scale rainwater tank systems. They examine the levelised cost and cost benefit approaches for economic assessment using data from case studies in South East Queensland, Australia. The chapter also covers a method for evaluating the economies of scale of cluster-scale rainwater tanks system. The authors also describe other benefits from rainwater tank systems (externalities), such as delaying infrastructure upgrades and reduced stormwater flows to receiving waters, using cost benefit analysis.

Rainwater tanks in urban areas also impact on urban hydrology and stormwater quality. Urbanisation increases runoff, reduces water quality and adversely impacts the receiving water environment; all responses which can be counter balanced to some extent by installing rainwater tanks. In Chapter 13, Matthew Burns, Anthony Ladson and Tim Fletcher describe the role of rainwater tanks in stormwater flow management and water quality improvement, with a focus on restoring natural stream flows and water quality regimes. They also describe other benefits such as urban cooling and flood mitigation due to the implementation of rainwater tanks in the urban environment.

Chapter 14 summarises the findings of the preceding 13 chapters and highlights the policy implications of these findings for Australian states, and lessons for the international community. In this chapter, Ted Gardner and colleagues also provide a social/technical context of the development of rainwater tanks in Australia. The authors highlight that the major Australian water authorities have invested huge amounts of money in seawater desalination plants as a key strategy in developing climate resilient potable water supplies. Thus, the future of rainwater tanks and dual reticulation recycled water systems as alternative supplies have become uncertain in urban areas. However, the authors also highlight that Australian water professionals consider rainwater tanks and recycled water will continue to play a major role in mitigating the adverse impacts of climate change on the urban environment. The authors also conclude that rainwater tanks can supply superior water quality in developing countries, and the body of work included in this book can be used across the globe either directly, or with some modifications based on the local social/technical/political contexts.

Ashok K. Sharma
Don Begbie
Ted Gardner

Chapter 1

Rainwater harvesting systems for urban developments

Stephen Cook, Ashok K. Sharma and Ted Gardner

ABSTRACT

Rainwater harvesting systems have been used as a local water supply source since the first human settlements. In recent times, rainwater harvesting systems have become an important water supply source in rural and remote areas where reticulated water supply systems are not available. Harvested rainwater can also provide an 'improved' drinking water source in rural and peri-urban areas of developing countries where surface water can be contaminated by faecal pathogens, and/or good quality groundwater is not readily available. However, this chapter is focussed on the potential of rainwater harvesting as a secondary water source in modern cities that have a centralised reticulated water supply system. Rainwater tanks are now being implemented under integrated urban water management concepts, to reduce the use of mains water for non-potable household uses. This substitution concept is based on a 'fit for purpose' water quality to help address the increase in demand for freshwater resources due to rapid population growth and urbanisation.

This chapter reviews rainwater harvesting practices around the world with a particular emphasis on the drivers behind the adoption of rainwater harvesting. There is a particular focus on the Australian experience due in part to the rapid uptake of rainwater systems over the last decade or so. This uptake was in response to pressures on mains water supply due to an extended drought and growing population.

The chapter also outlines some of the issues that are confronted when planning rainwater harvesting systems as a source of non-potable water in modern cities. These issues include likely yield from rainwater systems (kL/household/year), managing public health risks, cost-effectiveness, energy demand and environmental benefits. This chapter provides foundation knowledge on the issues, which are then explored in more detail in subsequent chapters of this book.

Keywords: rainwater tanks; modelling; strategic water planning; water quality.

1.1 INTRODUCTION

Rainwater harvesting systems have been an important water source for human settlements since the earliest examples of civilised society. There are examples of rainwater harvesting in the ancient civilisations of the

Middle East (AbdelKhaleq & Alhaj Ahmed, 2007; Evenari *et al.* 1971). Mays *et al.* (2007) noted that in ancient Greece cities used rainwater collection to augment supply from aqueduct systems fed by natural springs when population growth resulted in demand exceeding supply from existing sources. In modern cities of developed countries, there has been a renewed interest in the potential benefits of rainwater to complement drinking water supply systems. Whilst in rural areas, as well as in developing countries, rainwater harvesting can provide an important source of drinking water.

Whilst rainwater harvesting was an important water source in early forms of human settlement as cities populations grew, the main approach to water supply was sourcing water from catchments outside of cities, which is then supplied via large-scale reticulated networks. These centrally managed and operated systems have usually provided safe and reliable safe services in developed countries since the mid-1850s. However, over the last decade or so, there has been increased adoption of rainwater tanks for augmenting urban water supply in developed countries. For example in Australian cities, the number of household rainwater tanks more than doubled over the period 1994 to 2010, with more than one million households in Australian cities now having a rainwater tank (ABS, 2010). In developed countries, rainwater tanks were historically used in areas on the urban fringe as an interim measure before reticulated potable water became available. An exception of course is rural areas where mains water supply is usually limited to towns. While the focus of this chapter is on the experiences of rainwater tanks in cities of the developed world, there is some attention paid to the use of rainwater harvesting in developing countries. In these countries, rainwater can supply an 'improved' drinking water source. Rainwater systems can be particularly useful in conditions where surface water is contaminated by faecal pathogens, and groundwater is either not readily available or contaminated by chemicals, such as naturally occurring arsenic in Bangladesh (Chakraborti *et al.* 2010).

The basic design of household rainwater systems has not changed significantly over the centuries in that runoff from an imperious area, such as a roof, is directed by gravity into to storage vessels where it can then be drawn upon to meet household water demands. In modern settlements, building roofs provide the impervious runoff area with collection via roof gutters and down pipes to storage tanks. By adding an electric pump, water can be supplied to any elevation in the dwelling at a flow rate and pressure acceptable for most domestic uses. There is also the potential to expand the use of harvested rainwater through treatments such as filtration and disinfection for potable application. Whilst the basic elements of rainwater tanks are well understood, the re-emergence of rainwater tanks as an important water source in modern cities has exposed a number of knowledge gaps. Current knowledge gaps on rainwater harvesting systems include:

- The likely yield and reliability of rainwater harvesting systems in different development contexts and system configurations;
- The public health risks associated with the use of rainwater tanks;
- The treatment options available to supply rainwater that is fit for the intended purpose;
- The life cycle costs associated with constructing and operating a rainwater system;
- The impact on flows in the centralised water supply and stormwater networks;
- The environmental benefits of rainwater harvesting systems due to avoided stormwater discharge;
- The energy demand associated with rainwater system pumps;
- The community acceptance of rainwater systems and their associated maintenance tasks; and
- The appropriate management models to best mitigate risks of decentralised rainwater systems.

Addressing these knowledge gaps is important for encouraging greater adoption of rainwater harvesting systems, which we believe should contribute to a more sustainable urban water system.

This chapter provides an overview of rainwater harvesting practices across the world, with a particular focus on the drivers for their adoption. The Australian experience with rainwater harvesting is explored in more detail. The chapter then introduces the knowledge gaps that we believe are impeding greater adoption of rainwater harvesting in modern cities. This chapter sets the context, and introduces the issues, that are then explored in more detail in subsequent chapters of this book.

1.2 INTERNATIONAL EXPERIENCES WITH RAINWATER HARVESTING SYSTEMS

In this section we review the literature on experiences with rainwater harvesting systems across the world. In particular, we examine the drivers for their adoption in different urban contexts and the level of uptake across countries.

Rainwater harvesting has undergone a surge in popularity in the United States. The uptake of rainwater tanks has mostly been in areas where there is a lack of high-quality freshwater supplies or where households are motivated by environmental concerns (Mendez *et al.* 2011). Thomas *et al.* (2014) undertook a survey to better understand rainwater harvesting practices and the motivation for harvesting rainwater in the United States. This survey found the main use of harvested rainwater was irrigation. However, 25% of the 222 respondents indicated that they used rainwater for potable purposes. Of the households using rainwater for potable uses, 70% employed ultraviolet light as their main method of disinfecting the water. Moreover, 21% of respondents who used rainwater tanks for potable uses undertook no water quality testing (Thomas *et al.* 2014). The significant number of households not monitoring the quality of the harvested water has implications for the risk management and design of an appropriate treatment train. Mendez *et al.* (2011) investigated the effect of roofing material on the quality of harvested rainwater in the United States. This study recommended that first-flush diversion, filtration and disinfection be used to meet United States Environmental Protection Agency (USEPA) water reuse guidelines (Mendez *et al.* 2011).

In Canada, harvesting roof runoff for domestic purposes has mostly been practiced in rural communities, but there has been renewed interest in urban areas (Farahbakhsh *et al.* 2009). The increased interest has been motivated by the potential of harvested rainwater to contribute to sustainability objectives, including water conservation and stormwater quality (Farahbakhsh *et al.* 2009). Farahbakhsh *et al.* (2009) noted that while the benefits of rainwater harvesting are well understood, there are a number of impediments for greater uptake in Canada, including the lack of a clear policy for rainwater harvesting, and uncertainty on costs and risks. Despins *et al.* (2009) undertook an assessment of rainwater quality from systems in Ontario, Canada. Their results indicated that microbiological quality improved during colder weather, whilst the physiochemical quality of harvested rainwater was most influenced by roofing and storage materials, as well as local environmental conditions (Despins *et al.* 2009). They found that the quality of harvested rainwater could be managed to acceptable standards via the selection of appropriate materials, and the implementation of post-tank treatment (Despins *et al.* 2009). Although Canada has abundant freshwater resources, there are many rural communities that rely upon water from local tanks for potable water supply (Baird *et al.* 2013). These tanks are filled by a range of water sources including trucked-in water and rainwater harvesting. It has been identified that there is the need for improved risk management policies and practices to reduce the potential for adverse human health impacts from water tanks. First Nations communities (i.e., the indigenous communities) are particularly vulnerable to water quality issues from storage tanks given they often lack access to alternative safe water supply (Baird *et al.* 2013).

In Southern Brazil, the projected increased demand for drinking water due to population growth has generated interest in the benefits of rainwater harvesting (Ghisi *et al.* 2007). An analysis of 62 cities in Southern Brazil indicated that rainwater harvesting could potentially reduce potable water demand by 34% to 92% (Ghisi *et al.* 2007). However, this analysis did not consider seasonality of demand, or the dynamics of roof runoff volume, available storage volume and demand.

In developed countries of Europe and Asia, the interest in rainwater harvesting systems is often driven by concerns for stormwater discharge into combined sewers (i.e., sewers that collect both sewage and stormwater flows), local flooding control, and lastly, water supply. In Germany for example, water scarcity is not a major issue yet the uptake of rainwater tanks has been widespread since the 1980s (Nolde, 2007). The rainwater industry is well established with over 1.6 million installations (4% of households) and approximately 50,000 rainwater systems being installed every year in new homes (Nolde, 2007). The major driver for government and water companies to encourage greater adoption of rainwater harvesting systems is to reduce the capital costs associated with providing additional flow capacity in the sewerage system to cope with increases in wastewater and stormwater discharges. Hence, rainwater tanks are part of a system to retain stormwater runoff on site for subsequent local infiltration (Hermann & Schmida, 1999). The 'disconnection of impervious areas' can lead to direct government subsidy and/or a reduction in household sewage charges from water companies (Hermann & Schmida, 1999). Despite the subsidies, the cost-benefit ratio for households to install a rainwater harvesting systems remains unfavourable. However, there is still a high incidence of voluntary installation of rainwater harvesting systems in new private dwellings in Germany (over 65%). This uptake is thought to be due to peoples' desire to contribute to sustainability (by reducing stormwater discharge into waterways) as well as the concept of self-reliance (Schuetze, 2013). Rainwater tanks in new dwellings can be used to meet water demand for toilet flushing, clothes washing, garden irrigation and other external uses. The harvested rainwater can also be used to recharge. In some cases, there is reluctance from local authorities and water companies to promote greater adoption of rainwater tanks to help substitute for mains water demand. This is due to the potential for the increased use of harvested rainwater to reduce mains water billing revenue that is needed to invest in, and maintain, the centralised infrastructure (Schuetze, 2013).

In France, adoption of rainwater harvesting systems has been limited, which is due in part to a French law that restricts the domestic use of rainwater for clothes laundry, showering and drinking (Vialle *et al.* 2011). This law, which constrains the domestic use of rainwater to external uses such as garden irrigation and cleaning, is based on concerns with water quality and potential public health impacts. This contrasts markedly with a European Union Directive that puts a priority on urban water savings including rainwater harvesting and reuse in buildings (Palla *et al.* 2011). In other parts of Europe, such as Sweden, the concept of ecologically sustainable development is motivating installation of rainwater harvesting systems in large (1100 units) community housing complexes (Villarreal & Dixon, 2005).

In southern Europe, the Mediterranean rainfall pattern of dry, hot summers means there is considerable interest in rainwater harvesting systems for public and private dwellings because of drought induced water supply restrictions (Domènech & Saurí, 2011). Domènech and Saurí (2011) reported on rainwater harvesting systems in a municipality of Barcelona, Spain, where rainwater tanks were mandated for individual and communal houses with gardens greater than 300 m^2. Their study, which largely focussed on socio-economic factors, reported a high user satisfaction level with rainwater tanks despite the long payback time, and a community pride in the water savings that benefits the whole society.

In the United Kingdom, there are government policies that encourage rainwater harvesting in private and public dwellings (DEFRA, 2008; Ward, 2010) as well as publications providing technical detail for its implementation (e.g., UK Environmental Agency, 2008). As with Germany, a driver for the adoption of rainwater harvesting is to reduce stormwater discharge into waterways, and mitigate local flooding issues (Farnsworth, 2012). Hence, rainwater tanks in individual and communal dwellings are seen as part of implementing a Sustainable Urban Drainage System (SUDS) (Farnsworth, 2012). However, as there is no rebate available for rainwater tank installation, adoption by households has been limited (Farnsworth, 2012; Ward, 2010).

Rainwater tanks can also be one measure to adapt to uncertainty in supply from traditional water sources due to inherent climate variability and the impacts of climate change. Pandey *et al.* (2003) articulated a theory that in a changing climate, people have resorted to local rainwater harvesting in order to adapt and ensure survival. This theory was explored by undertaking a detailed analysis of the relationship between periods of climate change (in particular droughts) and the adoption of local water harvesting over recent millennia. The study found a correlation between periods of drought and archaeological evidence of increased human construction of rainwater harvesting systems (Pandey *et al.* 2003). The analysis extended to modern times where in 1999, the hottest summer experienced in central India during the 20th century, resulted in a government program of investment in rainwater tanks for augmenting domestic water supply. Said (2014) reported on a study that assessed the potential for rainwater harvesting in New Delhi, India. This study found that the benefits of rainwater harvesting, as a measure to encourage more sustainable use of water resources, could be maximised when combined with a program encouraging water efficiency (Said, 2014). Islam *et al.* (2011) undertook an analysis of the potential for rainwater harvesting to provide an improved supply of safe water to slum dwellers in Dhaka City, Bangladesh. This analysis found that rainwater harvesting was a feasible option for improved water supply, which was accepted by the local community provided there was some Government assistance to offset the costs of installation (Islam *et al.* 2011).

In Japan, rainwater harvesting systems are promoted largely to reduce urban flooding, noting that roofs comprise about 50% of the total impervious area of cities (Ward, 2012). In China, rainwater harvesting has been explored as a measure to mitigate urban flooding in areas with monsoonal rainfall (Zhang & Hu, 2014). In the semiarid areas of China, rainwater harvesting is promoted to address water scarcity for domestic and agricultural uses (Li *et al.* 2004).

Korea is another developed Asian country with a strong interest in rainwater harvesting and reuse systems, which is motivated by the objectives of reducing urban flash floods and combined sewer overflows during the wet season. However, Kim and Yoo (2009) found that having 10% of the impervious urban area connected to rainwater harvesting systems only reduced the level of urban flooding by 1% during the wet season. This was due to the limitations of storage volume that could be provided relative to the runoff volume.

Experiences with rainwater tank systems for urban water supply can largely be partitioned into developed and developing countries. While this chapter has largely focussed on developed countries experiences, the following considers the importance of rainwater harvesting systems in developing countries. For many countries in Africa, and many parts of South East Asia, access to a safe drinking water supply is poor, especially in rural areas, which leads to increased rates of illness (Baguma *et al.* 2010; Kahinda *et al.* 2007). In addition, the source of water supply is often remote from village settlements and travel time and effort for its collection is onerous (Baguma *et al.* 2010). Although household rainwater systems generally does not have the same microbiological standards as potable water (Ahmed *et al.* 2011), it is usually of far superior quality to untreated water accessed from surface water bodies and some shallow wells. Consequently, there is a range of programs in Africa where non-government organisations encourage the installation of local rainwater harvesting systems to improve health outcomes and quality of life (e.g., RAIN, 2011).

1.3 THE AUSTRALIAN EXPERIENCE WITH RAINWATER TANK SYSTEMS

Australia has significant uptake of rainwater tank systems in cities relative to other developed countries (Ward, 2010). In 2013, 34% of households had a rainwater tank; an increase from 24% of households in 2007 (ABS, 2013). In this section, we explore the drivers for such high adoption of rainwater tanks.

The uptake of rainwater tanks as a non-potable water source in Australian cities accelerated during the Millennium drought, which occurred from 2000 to about 2008. The extended drought saw capital cities' water catchments fall to unprecedented low levels, with the real possibility of cities running out of water unless radical steps were taken (Apostolidis *et al.* 2011). It was this factor that 'focused the minds' of the water planners in all Australian states and lead to an explosion of investment in alternative water supply sources (seawater desalination by reverse osmosis, recycling of wastewater, stormwater harvesting and reuse) and rainwater tanks, as well as demand reduction programs.

In Australian cities, rainwater tank uptake was encouraged both by financial rebates for installation, and changes to building codes that encouraged households to install a secondary alternative water source for non-potable demands such as toilet flushing and garden irrigation (NSW Planning and Infrastructure, 2004; Queensland Department of Infrastructure and Planning, 2008). The installation of rainwater tanks was also a response to increased water restrictions during the drought (ABS, 2013). Also, rainwater tanks were promoted under Water Sensitive Urban Design (WSUD) and Integrated Urban Water Management (IUWM) approaches as a tool for reducing the impact of urban stormwater on the natural environment (Sharma *et al.* 2012; Sharma *et al.* 2013). Rainwater tank systems continue to be encouraged in Australia as a supplementary water source through financial incentives and regulations such as requirements in building codes for alternative water sources and/or water conservation measures. Rainwater tanks have featured as part of an overall strategy to secure long-term water supply demand balance in Australian cities (e.g., Queensland Water Commission, 2010). However, in Queensland, the mandatory requirement for rainwater tanks to be installed in new homes was removed due to concerns that it was increasing the construction costs of residential development (Mander, 2013).

Rainwater harvesting is a mainstream practice in rural areas of Australia, with 44% of households outside of capital cities having a rainwater tank (ABS, 2013). Also, rainwater tanks have also been used as an interim water supply source in peri-urban areas where the low population density does not yet support reticulated water supply systems. In urban areas that are serviced by reticulated water supply, rainwater has provided a non-potable water source for garden irrigation. The supply of an alternative water supply source for irrigation is particularly important during periods of drought where mains water use for this purpose is often restricted by regulations. However, there has been a shift to extend the use of harvested rainwater to uses inside the home, particularly for toilet flushing. More than half (51%) of Australian households with a rainwater tank have them plumbed for internal uses (ABS, 2013). The use of rainwater systems for toilet flushing requires management of the risks associated with the potential for backflow of rainwater to contaminate potable water supply, as well as ensuring continuity of supply to the appliance via a back-up water supply source. The primary reason householders installed a rainwater tank was to save water (Figure 1.1), and as mentioned previously, there has been a substantial rise in installed numbers since 2007 (Figure 1.2) (ABS, 2013). The significant proportion of households in Australia that use rainwater for domestic purposes demonstrates their importance in contributing to water supply security, especially in rural Australia.

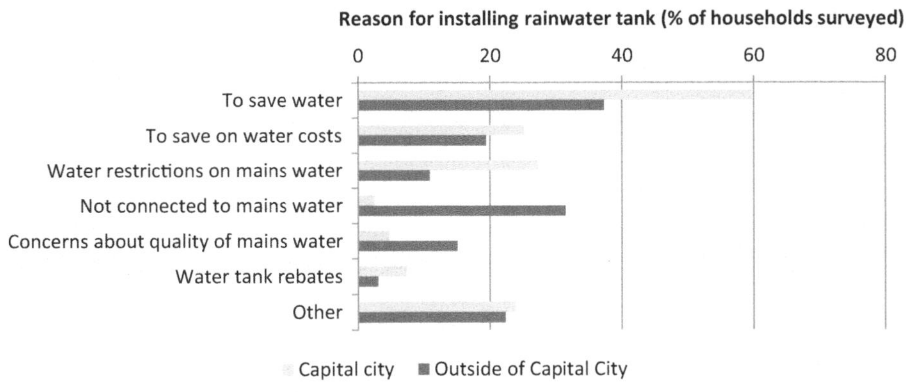

Figure 1.1 Reason for installing a rainwater tank in Australia (ABS, 2013).

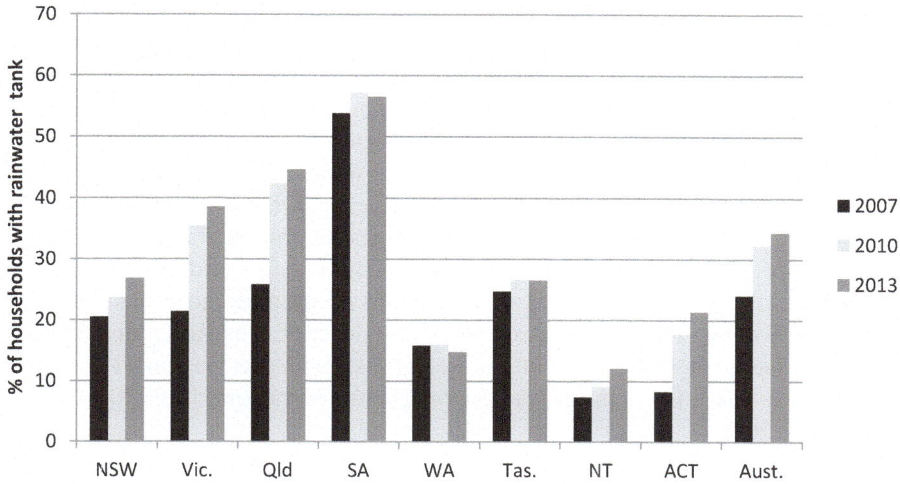

Figure 1.2 Percentage of households with rainwater tanks in Australia (ABS, 2013).

1.4 KEY ISSUES FOR GREATER UPTAKE OF RAINWATER TANK SYSTEMS

This section summarises a range of issues faced in the greater uptake of rainwater tank systems. Rainwater tanks have become very common in urban areas, which raises a number of issues that include: the integration of rainwater tanks with existing centralised water supply systems; impact on reducing demand for mains water; long-term reliability and yield; energy implications; supply cost relative to other supply and demand options; water quality considerations and public health risk; and the need for improved planning and design guidelines. Some of these issues are considered in more detail in the remainder of this book, whilst the following sections summarise some of the main challenges that need to be addressed in securing greater mainstream acceptance of rainwater tanks in modern cities.

1.4.1 Estimating rainwater system yield and mains water savings

A significant driver for the uptake of rainwater tanks was the need to reduce demand for potable water supply due to drought conditions. Rainwater tanks were included in strategic water supply-demand balance planning (Queensland Water Commission, 2010), so there was a need for rigour when estimating the likely yield and mains water savings due to the uptake of rainwater tanks.

Understanding the likely yield from a rainwater system is also important for aiding the system design, as it can inform the sizing of roof catchments and storages. Also, estimating the likely mains water savings provides an understanding of the reliability of the rainwater system for the designated end-uses.

The amount of mains water saved due to the installation of the rainwater tank is primarily a function of the following factors: catchment area; storage volume; end-uses; household demand profile; and rainfall pattern. Modelling tools have been developed to simulate the dynamic interactions amongst these factors, which can be used to assess likely mains water savings due to the installation of rainwater harvesting system (see Chapter 3). However, there is also the need for monitoring studies to validate results from simulation tools. The comparison of modelling results with the measured household potable water savings is essential to gauge the accuracy of these modelling predictions. The validation of the potable water savings is also essential to avoid systemic errors in the long-term water planning for a city or a region. Hence, modelling predictions *alone* should not be used for strategic water planning purposes.

Chong *et al.* (2011) assessed the potable water savings from rainwater tanks by comparing mains water consumption data from households with rainwater tanks against the regional average residential water consumption for the same period. Their study found the average mains water savings for households with a 5 kL rainwater tank was 58.8 kL per household per year (kL/hh/yr), but with significant variation between local government areas. Beal *et al.* (2012) estimated the mains water savings associated with rainwater tanks plumbed for indoor non-potable use. This estimate was based on comparing water billing data for households with internally plumbed rainwater tanks with those households with no rainwater tank plumbed for indoor use. This pairwise comparison found there was significant spatial variation in mains water savings due to internally plumbed rainwater tanks, with average annual mains water savings of 50 kL/hh/yr for the South East Queensland region, Australia, ranging from 20 to 95 kilolitres per household (Beal *et al.* 2012).

1.4.2 Understanding the risks associated with rainwater quality

There is a need to consider the likely quality of rainwater harvested and the associated risks that are posed to households. The inclusion of rainwater tanks in government policies has renewed interest in measuring the microbiological and chemical quality of captured rainwater (Magyar *et al.* 2007). Rainwater quality can be affected by the roof and storage tank materials, overhanging trees, and the rigor of maintenance of rainwater system (Rodrigo *et al.* 2010). Lee *et al.* (2012) investigated the quality of harvested rainwater in Korea as affected by roofing materials. Wooden, concrete, clay tiles and galvanised steel roofing materials were investigated, and galvanised steel was found to provide the best rainwater quality, which was attributed by the authors to disinfection provided by the high daily surface temperature of this material (Lee *et al.* 2012).

Huston *et al.* (2012) conducted a study to identify the contributors of heavy metals and ionic contaminants in rainwater tanks in Brisbane, Australia. They identified four source factors influencing the bulk deposition at various locations in Brisbane, which included crustal matter/sea salt, car exhaust/road dust, industrial dust and aged sea salt/secondary aerosols. These factors, on average, contributed 65% of the total contaminants. They also identified six collection system factors which included plumbing, building material, galvanizing, roofing, steel and lead flashing/paint. These factors contribute nearly 35% of the contaminants.

Magyar *et al.* (2007) reported that the concentration of lead exceeded the Australian Drinking Water Guidelines (ADWG) values in 5 of the 9 tanks investigated in metropolitan Melbourne. O'Connor *et al.* (2009) expanded this study to 52 tanks and identified relationships between lead concentration in the tank water and tank sediments, and various environmental variables. In 14 of the 52 tanks, lead concentrations exceeded ADWG health guidelines. Lead flashing, prevailing winds, proximity to roads and commercial zones had statistically significant relationships with lead concentration in rainwater. However, no single factor was identified as the major cause. Lead has the most potential to impact seriously on human health, particularly on infants and children (Goyer, 1993). These issues are discussed in more detail in Chapter 9.

The microbial quality of rainwater depends on number of factors. For example, climate can have a significant impact on the microbial composition of harvested rainwater. For example, Evans *et al.* (2006) found airborne microorganism contributed a significant non-pathogenic bacterial load to roof water, with the magnitude influenced by wind velocities and wind direction. Yaziz *et al.* (1989) found that the bacterial contamination of rainwater, as measured by total coliform concentration, increased the longer the period between rainfall events. This study also found that roof materials and the intensity of the rainfall event influenced total coliform concentration. In a study of 24 household rainwater tanks in South East Queensland, 63% of rainwater tanks contained *E.coli* exceeding the limit of ADWG (Ahmed *et al.* 2012), which was related to faecal matter from small animals such as possums and birds.

The studies reported above show that the chemical and microbiological quality of roof harvested rainwater systems varies significantly, with a number of studies finding that rainwater quality exceeded drinking water guidelines for both chemical and microbiological parameters (Magyar *et al.* 2007; O'Connor *et al.* 2009; Yaziz *et al.* 1989). However, epidemiological studies of children in South Australia could identify no adverse health link between ingestion of rainwater and gastroenteritis (Heyworth *et al.* 2006). Care should be taken in using rainwater based on a 'fit for purpose' concept, and proper treatment of rainwater will be required if used for drinking purposes. Limiting rainwater for irrigation and selected indoor non-potable uses may be the most appropriate uses as they can significantly reduce mains water demand, whilst requiring minimal or very little treatment. When rainwater systems are plumbed for indoor non-potable applications, the primary concerns of guidelines are to avoid the risks of cross contamination with drinking water supply (via backflow events) and to reduce human contact with pathogens that may be present in harvested rainwater.

1.4.3 Guidelines for managing risks of rainwater use

The widespread adoption of rainwater tanks as an alternative water source has been encouraged to reduce mains water use, and to provide environmental benefits such as reduced stormwater flows. However, the increased use of rainwater tanks in urban areas does raise the need for guidelines that manage the risks associated with their use. The quality of rainwater for human use (potable or non-potable) is not regulated by any standard that is internationally recognised (Birks *et al.* 2004). Therefore, many countries and regions have developed local guidelines for local needs (Schuetze, 2013). Guidelines for rainwater use need to consider the water quality requirements of the intended use of the rainwater tank, and the likely microbiological and chemical risks posed. The preceding section has (briefly) detailed the quality of rainwater, whilst this section assesses the guidelines that are available to manage risks associated with rainwater quality.

In Australia, there is a national guidance document, *Guidance on the Use of Rainwater,* which takes a systematic analysis of the potential hazards and risks associated with rainwater use (EnHealth Council, 2010). This document argues that the health risks associated with drinking rainwater from a properly maintained tank are low in most parts of Australia, however, the microbial and chemical quality of rainwater is likely to be lower than mains water supply (EnHealth Council, 2010). Therefore, state guidelines recommend drinking of mains water supply where it is available (NSW Health, 2007). While the potable use of rainwater is possible with disinfection treatment, this section focuses on the guidelines for non-potable use of rainwater. Where the rainwater collected in a tank is intended for ingestion, drinking water guidelines, such as the World Health Organisation (2011), should be applied. Chapman *et al.* (2008) recommend a comprehensive health risk assessment is undertaken if the harvested rainwater is intended for drinking and other potable uses. The health risk assessment should consider if there are potential sources of pollution, suitability of materials used for rainwater system, the likelihood of vulnerable people (such as the very young and the very old) drinking the water, and how ongoing maintenance is assured (Chapman *et al.* 2008).

The EnHealth Council (2010) identified practical preventative measures that can be used to minimise most of the health and aesthetic hazards associated with rainwater collected in tanks. These preventative measures include regular, simple maintenance practices. The maintenance practices include keeping intact insect screen on the tanks, cleaning the area around tank to prevent breeding ground for mosquitoes, regular cleaning of gutters, and trimming of overhanging trees to discourage possums. In Australia, there is a range of other guidelines and standards that provide comprehensive technical advice on all aspects of rainwater tank design, installation and maintenance. These include a handbook for rainwater tank design and installation (Standards Australia, 2008), which includes guidance on maintenance tasks for rainwater tanks, and the frequency these tasks should be carried out.

In the United States, there is a lack of uniform guidance for the use of rainwater. This has resulted in a range of guidelines for use and treatment amongst state and local governments (Kloss, 2008). This lack of specific guidance on the use of rainwater has meant that some jurisdictions have based requirements on guidelines used for recycled water sources, which resulted in overly stringent regulations (Kloss, 2008). Kloss (2008) recommended that development of national guidelines on rainwater use, needs to consider the likely harvested rainwater quality and yield with the demand requirements of targeted end uses.

Fewtrell and Kay (2007) found that there are no regulations on the microbial quality required for the non-potable use of rainwater in the United Kingdom. Furthermore, where water supplied from rainwater tanks is used solely for non-potable purposes, there is often no requirement for quality monitoring. Fewtrell and Kay (2007) considered that adaption of other microbial water guidelines to non-potable use of rainwater was unlikely to be successful. Rather, they argue that the development of a standard for the microbial quality of rainwater tank water should take a health impact assessment approach, and that microbial concentrations are at set a level to protect human health, based on the intended uses.

1.4.4 Evaluating the cost-effectiveness of rainwater tanks

Prior to adopting rainwater tanks as official government policy, there is a need to consider their cost-effectiveness relative to centralised mains water supply, or alternative non-potable local water sources. The cost-effectiveness of a rainwater tank will be a product of whole of life cost and the water yield delivered over time (Marsden Jacobs Associates, 2007). Cost-effectiveness can be considered from either the perspective of an individual householder (levelised cost) or that of the whole of community (Benefit Cost analysis, Marsden Jacobs Associates, 2007). The whole of community cost considers not only the costs and benefits borne by the households, but also accounts for avoided costs for the capital and operating expenses associated with reduced capacity of public water supply and stormwater systems (Marsden Jacobs Associates, 2007).

The direct capital costs for rainwater tanks include the storage tank, tank installation and fittings, concrete slab or tank stand, household plumbing and a pump. The ongoing operating costs for a household are: energy costs for pumping; and maintenance of the tank and pump (Marsden Jacobs Associates, 2007). Vivian *et al.* (2010) undertook an analysis of the installation costs for a rainwater tank in Australian cities. Their study found that for a 5 kL tank plumbed for indoor and outdoor uses, installation cost was approximately AU$3,900 for the Gold Coast. In most Australian cities, there are rebates available to offset the costs of a household installing of a rainwater tank. The amount of rebate varies among jurisdictions, along with the size of the tank and the connected end uses. Hall (2013) examined the cost-effectiveness of rainwater tanks in South East Queensland, Australia. This study found that the average levelised cost of rainwater tanks for the scenario was $9.22 per kilolitre (kL) with 95% confidence limits of $6.73 and $12.77/kL. The variation in yield, pump and tank life and maintenance had the largest effect on the variation in the cost-effectiveness. The results were also sensitive to discount rate assumptions (Hall, 2013). As with all models, the outcomes of economic studies are strongly dependent on the input assumptions. But the results indicate that the supply costs of household rainwater systems in South East Queensland cannot compete economically with other alternative water supply sources.

As the cost-effectiveness of rainwater tank is a function of both yield and whole of life costs, there is the need to consider factors that influence yield. The efficiency of a rainwater tank system is influenced by: tank size, roof catchment area, and the amount and regularity of the water demand (Vivian *et al.* 2010). Of course, the local rainfall amount and its temporal pattern strongly influences the yield from rainwater tank. The analysis by Marsden Jacobs Associates (2007) found that roof catchment area had the greatest impact on yield, followed by annual rainfall and tank size. This highlights the need to consider both the optimal configuration of rainwater systems, as well as the local climate in maximising cost-effective investment in rainwater systems.

The lifecycle costs and benefits for a household can be calculated using a discounted cash flow analysis, which accounts for the time value of money with future costs being discounted to their present value. This analysis uses estimated costs (initial and ongoing) of rainwater system, rainwater yield, and the avoided costs (e.g., avoided water tariffs) (Marsden Jacob Associates, 2007). If a householder is considering investing in a rainwater tank system and the present value life cycle costs exceeds the present value life cycle benefits, then it may be necessary to offer a rebate to cover the shortfall (Marsden Jacobs Associates, 2007).

In considering diversification of water supply sources, there is the need to consider cost-effectiveness as one of the criteria in developing a preferred portfolio of sources in an urban water supply system. PMSEIC (2007) undertook a comparison of the direct costs associated with different water supply/demand options. That study found that rainwater tanks can often be expensive in terms of cost per unit of water supplied. The results indicated that household scale rainwater systems have a higher unit cost for water supplied when compared to centralised water supply options. However, it is necessary to consider more than direct costs in developing a water supply source mix for cities. Other factors to be considered include reliability, environmental impact, community acceptance, freedom to water gardens and the suitability of current governance models (PMSEIC, 2007; Sharma *et al.* 2012).

In South East Queensland the mandatory requirements for rainwater tanks in new dwellings was removed in 2013 (Mander, 2013). The decision to remove the mandatory requirements for rainwater tanks was based on a report by the Queensland Competition Authority, which found that costs of the rainwater exceeded the benefits (Queensland Competition Authority, 2012). The capital and operating costs of rainwater tanks were found to be greater than anticipated benefits, which include deferred augmentation of mains water supply system, reduced operating costs for mains water supply and reduced bioretention costs (Queensland Competition Authority, 2012).

1.4.5 Understanding the indirect costs and benefits of rainwater tanks (externalities)

Outcomes from rainwater tanks that aren't captured in the direct costs and benefits to the affected parties can be considered as externalities. The parties that generally bear the direct costs or receive the direct benefits from a rainwater tank are the householder and the water utility. The latter may be impacted by reduced demand for mains water and/or offer a rebate for rainwater tank adoption. More specifically, an externality is any cost (or benefit) that is not incurred directly by the utility or the householder. Externalities are usually not adequately captured by the market price to reflect the full cost (or benefit) of the service or product, in this case rainwater tanks. There can be positive as well as negative externalities. This section discusses some of the externalities that are associated with the adoption of rainwater tanks as an alternative local water source. One of the most challenging aspects of evaluating the costs and benefits of alternative water policy options is economic quantification of the externalities.

The economic 'cost benefit ratio' may be more favourable where the rainwater tanks can be used to avoid or defer investment in stormwater infrastructure. For example, Coombes *et al.* (2000) found that a water sensitive design approach that included rainwater tanks could be 25% more cost effective than a conventional approach to providing stormwater infrastructure. Coombes and Kuczera (2003) found that investment in rainwater tanks provided greater economic benefits to the community than investment in traditional water supply and traditional stormwater management options. Marsden Jacob Associates (2007) made the point that the savings from deferring, downsizing or avoiding investment in traditional infrastructure due to rainwater tanks is only likely to be realised in *greenfield areas* prior to the construction of water supply and stormwater infrastructure.

Rainwater tanks uptake can provide some significant socio-economic benefits for the general public. For example, investments in rainwater tanks can leave money in local communities and create local jobs. Rainwater tanks installations require skilled labour and significant materials, while utilities starting a program to encourage rainwater tank may add staff and hire contractors. These jobs, and the materials purchased, can add significantly to the local economy and benefit the general public (Maddaus, 1999).

1.4.6 Impact of rainwater systems on stormwater flows and nutrient loads

A study by Parkinson (2005) investigated the impacts of domestic water conservation on urban sewerage systems in cities with combined sewer overflow (CSO), and concluded that rainwater harvesting was the most beneficial water conservation method. The justification for this was that rainwater tanks could assist in reducing discharge into the combined sewer system during high rainfall events (Parkinson, 2005). This finding is supported by Vaes and Berlamont (2001) who showed, based on modelling, that rainwater storage systems can significantly reduce rainfall to combined sewer systems, including reduced peak flow if installed on a sufficiently large scale. However, there is the need to account for tank storage levels at the start of the rainfall event to accurately estimate the impact of rainwater storages on downstream flows (Vaes & Berlamont, 2001).

In Australian cities (which have separate sewer and rainwater systems), the objectives of rainwater systems are primarily the augmentation of potable water supplies, and to some extent, restoration of pre-development flow regimes in the receiving waterways (Burns *et al.* 2012). The impact of rainwater tanks on the latter is not yet well understood nor quantified. Modelling undertaken by Burns *et al.* (2012) found that rainwater tanks can assist in restoring catchment retention capacity (units of mm) closer to that of the natural conditions. However, this will depend upon the match between roof size, storage volume and demand profile. A report to the National Water Commission considered the impact of rainwater tanks on the size of stormwater drains, and demonstrated that under certain conditions rainwater tanks can reduce peak flows to stormwater drains by up to 50% (Marsden Jacob Associates, 2007). This means that stormwater drain capacity could be reduced in new developments. However, there will be a limit to size reduction since drains are typically designed to manage peak flow events for a 1 in 20 year event (Marsden Jacob Associates, 2007).

The mandatory installation of rainwater tanks in all new homes of greenfield developments has implications for both receiving water quality and management of stormwater discharge. Analysis by Mackay Regional Council, located in tropical North Queensland, found that removing rainwater tanks from the stormwater treatment train increased the required wetland from 3.5% of the development catchment area to 5.8%, a 65% increase in the size of constructed wetlands needed to meet water quality targets (Galea, 2014).

The discharge of urban stormwater to receiving waters impacts on the ecological health of streams and other waterways (Walsh *et al.* 2009). Walsh *et al.* (2009) argued that if stream ecosystems are to be restored or protected, there is the need to reduce the hydraulic connectivity between impervious surfaces in urban areas and streams for the small, frequent rainfall events. Water harvesting by rainwater tanks will help reduce both the number and peak flow of runoff events which in turn seems to have a beneficial effect on the macro-invertebrate health of ephemeral streams around Melbourne. Reduction in nitrogen export has also been shown to be significant when rainwater tanks are installed (Sharma *et al.* 2008). Khastagir and Jayasuriya (2010) modelled the impact of rainwater tanks on runoff quality. They found that a 3 kL rainwater tank used for toilet flushing, laundry and garden watering could reduce nitrogen loads by 81%.

1.4.7 Impact of rainwater tank systems on centralised water systems and water quality

The widespread rainwater harvesting and use practices can also affect the mains water demand and quality. Grandet *et al.* (2010) studied the effect of rainwater harvesting on centralised urban water systems in Northern France. The study highlighted rainwater use resulted in a permanent decrease in mains water demand leading to an increase in water age, and hence quality deterioration in the distribution system. Water age was generally affected when rainwater supplied more than 30% of the overall water demand. However, the rainwater supply systems might be profitable for the community if rainwater use allowed the deferment of new water mains infrastructure. Lucas *et al.* (2010) investigated the impact of diurnal water use patterns, demand management and rainwater tanks on water supply network design. The study suggested that the rainwater tanks combined with mains water 'trickle top-up' produced diurnal 'mains water' use patterns different to that of 'household' water use patterns. When simulated correctly, this significantly reduced peak hour mains water demand. This outcome impacts upon water supply network design criteria and provides opportunities to offset water infrastructure costs.

1.4.8 Energy consumption in rainwater tank systems

The energy associated with household scale rainwater pumping is a function of the pump characteristics, the system set-up, and the household water use patterns (Retamal *et al.* 2009). The energy can be expressed as the 'total energy' consumed over a period of time for pumping (e.g., kWh per year) or as 'specific energy', that is, the energy required to pump a set amount of water (e.g., kWh per kL of rainwater).

The tank location relative to where water is used and the land topography can influence the head required and the associated energy usage. Gardner *et al.* (2006) and Beal *et al.* (2008) examined the energy for rainwater pumping for 4 to 6 properties located on a steep slope in Brisbane, Australia, where household rainwater tanks received back-up supply from a communal tank at the base of the slope. This study found that the energy required for water supply from the rainwater system was between 2.1 and 3.8 kWh/kL. Retamal *et al.* (2009) found the energy footprint of household-scale rainwater harvesting systems varied from 0.9 to 2.3 kWh/kL for 8 households equipped with various rainwater pump and mains backup systems, different numbers of occupants, and diverse rainwater end uses.

The large variability observed between dwellings is a reflection of the wide range of rainwater systems configurations (mainly pump type/size) and the demand characteristics of the different households (low flow or high flow appliances/end uses). It is also an indication that the energy requirements for rainwater pumping can be optimised for dwellings through appropriate system design and configuration set-up. Analysis by Umapathi *et al.* (2013), based on a monitoring study of 20 homes in South East Queensland, found that rainwater systems with automatic switching devices for topping up tanks with mains water had significantly less energy demand than systems using trickle top-up.

Small pumps used for rainwater supply are typically much less efficient than large pumps used for bulk water supply (Retamal *et al.* 2009). In Australian capital cities, pumping for centralised water supply typically has an average energy demand of 0.9 kWh/kL, with a range between 0.09 and 1.85 kWh/kL depending upon pumping distance and lift required for each city (Kenway *et al.* 2008). This places the median energy for household rainwater supply of 1.5 to 1.8 kWh/kL within the upper range of centralized gravity fed water supply. However, when compared to other alternative sources such as recycled water or desalinated water, which use on average 2.8 kWh/kL (Knights *et al.* 2007) and 3.5 kWh/kL (Apostolidis *et al.* 2011) respectively, rainwater supply is the least energy intensive option for alternative water supplies.

1.5 CONCLUSIONS

Rainwater harvesting for local water supply is not a new concept as there is well documented evidence of its practice since the first urban societies evolved. This chapter has reviewed rainwater harvesting practices around the world, with a particular emphasis on the drivers behind the adoption of rainwater harvesting. There is a particular focus on the Australian experience with rainwater systems, which is due to the rapid uptake of rainwater systems over the last decade. This uptake was in response to growing pressures on mains water supply systems due to an extended drought and growing population.

Rainwater tanks can provide reliable and affordable water supply in rural and remote areas where piped water supply systems are not available. In developing countries, rainwater can supply an 'improved' drinking water source where surface water is contaminated by faecal pathogens, or good quality groundwater is not readily available. In modern cities, rainwater tanks are now being implemented under integrated urban water management concepts, to reduce the volume of mains water consumed for non-potable household uses. This substitution concept is based on fit for purpose water quality to reduce unnecessary treatment costs.

This chapter has outlined some of the issues that are faced when considering the potential for rainwater harvesting systems as a secondary source of non-potable water in modern cities. These issues include understanding the likely yield from rainwater systems, both for optimising the design of the rainwater system as well as estimating the likely impact on reducing mains peak flows. There is also the need to manage the risks associated with the use of rainwater, which requires an understanding of its likely quality as well as improved guidelines on how to manage risks. The chapter has highlighted the challenges of evaluating the cost-effectiveness of rainwater harvestings relative to other alternative water sources, as well as centralised water supply systems. Whilst rainwater harvesting often has a higher cost per unit of water supplied than other sources (e.g., recycled water), consideration of externalities, such as mitigating the environmental impact of stormwater discharge to receiving waters, can increase the cost to benefit ratio. However, economic analysis has found that under most assumptions the operating and capital costs of installing a rainwater tank as a secondary water source will be greater than the benefits. The challenge is to identify the urban contexts and rainwater tanks configurations that are best suited to maximise the benefits relative to the costs. This challenge is explored in greater detail in the subsequent chapters of this book.

1.6 REFERENCES

AbdelKhaleq R. A. and Alhaj Ahmend I. (2007). Rainwater harvesting in ancient civilizations in Jordan. *Water Science and Technology: Water Supply*, **7**(1), 85–93.

ABS (2013). Environmental issues: Water use and Conservation March 2013–4602.0.55.003. Australian Bureau of Statistics, Canberra.

Ahmed W., Gardner T. and Toze S. (2011). Microbiological quality of roof-harvested rainwater and health risks: a review. *Journal of Environmental Quality*, **40**(1), 13–21.

Ahmed W., Hodgers L., Sidhu J. P. S. and Toze S. (2012). Fecal indicators and zoonotic pathogens in household drinking water taps fed from rainwater tanks in Southeast, Queensland, Australia. *Applied and Environmental Microbiology*, **78**, 219–226.

Apostolidis N., Hertle C. and Young R. (2011). Water recycling in Australia. *Water*, **3**, 869–881. http://www.yvw.com.au/yvw/groups/public/documents/document/yvw1003236.pdf (accessed June 2011).

Baguma D., Loiskandl H. and Jung H. (2010). Water management, rainwater harvesting and predictive variables in rural households. *Water Resource Management*, **24**, 3333–3348.

Baird J. M., Summers R. and Plummer R. (2013). Cisterns and safe drinking water in Canada. *Canadian Water Resources Journal*/Revue Canadienne des Ressources Hydriques, **38**(2), 121–134.

Beal C. D., Hood B., Gardner T., Lane J. and Christiansen C. (2008). Energy and water metabolism of a sustainable subdivision in South East Queensland: a reality check, *Enviro 2008*, Melbourne.

Beal C., Sharma A., Gardner T. and Chong M. (2012). A desktop analysis of potable water savings from internally plumbed rainwater tanks in South East Queensland. *Water Resource Management*, **26**, 1577–1590.

Birks R., Colbourne R. J., Hills S. and Hobson R. (2004). Microbiological water quality in a large in-building, water recycling facility, *Water Science and Technology*, **50**(2), 165–172.

Burns M. J., Fletcher T. D., Duncan H. P., Hatt B. E., Ladson A. R. and Walsh C. J. (2012). The stormwater retention performance of rainwater tanks at the landparcel scale. *7th international WSUD conference*, Melbourne, Australia, 21–23 February 2012.

Chakraborti D., Rahman M. M., Das B., Murrill M., Dey S., Mukherjee S. C., Dhar R. K., Biswas B. K., Chowdhury U. K., Roy S., Sorif S., Selim M., Rahman M. and Quamruzzaman Q. (2010). Status of groundwater arsenic contamination in Bangladesh: a 14 years study report. *Water Research*, **44**, 5789–5802.

Chapman H., Cartwright T. and Tripodi N. (2008). Guidance Manual for the Design and Installation of Urban Roofwater Harvesting Systems in Australia (Edition 1), CRC for Water Quality and Treatment – Research Report 39.

Chong M. N., Umapathi S., Mankad A., Gardner E. A., Sharma A. and Biermann S. (2011). Estimating water savings from mandated rainwater tanks in South East Queensland. Urban Water Security Research Alliance Technical Report No. 49. http://www.urbanwateralliance.org.au/publications/UWSRA-tr49.pdf (accessed 1 August 2014).

Coombes P. (2003). On-site water balance modelling using PURRS. Workshop on Rainwater Tank and Wastewater Reuse Modelling - Key Elements In Water Cycle Analysis and Sustainable Development, **19**–20 June, Melbourne.

Coombes P. J. and Kuczera G. (2003). A sensitivity analysis of an investment model used to determine the economic benefits of rainwater tanks [online]. In: 28th International Hydrology and Water Resources Symposium: About Water; Symposium Proceedings, M. J. Boyd, J. E. Ball, M. K. Babister and J. Green (eds), Institution of Engineers, Barton, ACT, Australia, pp. 2.243–2.250.

Coombes P. J., Argue J. and Kuczera G. (2000). Figtree Place: a case study in water sensitive urban development (WSUD). *Urban Water*, **1**(4), 335–343.

DEFRA (2008). Future Water: The Government's Water Strategy for England. https://www.gov.uk/government/uploads/system/uploads/attachment_data/file/69346/pb13562-future-water-080204.pdf (accessed May 2013).

Despins C., Farahbakhsh K. and Leidl C. (2009). Assessment of rainwater quality from rainwater harvesting systems in Ontario, Canada. *Journal of Water Supply: Research and Technology – AQUA*, **58**(2), 117–134.

Domènech L. and Saurí D. (2011). A comparative appraisal of the use of rainwater harvesting in single and multi-family buildings of the Metropolitan Area of Barcelona (Spain): social experience, drinking water savings and economic costs. *Journal of Cleaner Production*, **19**(6–7), 598–608.

EnHealth (2010). Guidance on the use of rainwater tanks, Department of Health and Aging - Commonwealth of Australia. http://www.health.gov.au/internet/publications/publishing.nsf/Content/ohp-enhealth-raintank-cnt-l (accessed May 2013).

Evans C. A., Coombes P. J. and Dunstan R. H. (2006). Wind, rain and bacteria: the effect of weather or the microbial composition of roof-harvested rainwater. *Water Research*, **40**, 37–44.

Evenari M., Shanan L. and Tadmor N. (1971). The Negev: The Challenges of a Desert. Harvard University Press, United States of America.

Farahbakhsh K., Despins C. and Leidl C. (2009). Developing capacity for large-scale rainwater harvesting in Canada. *Water Quality Resources Journal of Canada*, **44**(1), 92–102.

Farnsworth L. (2012). Sustainable water supplies and surface water management. *Environment Industries Magazine*, **20**, 118–120.

Fewtrell L. and Kay D. (2007). Microbial quality of rainwater supplies in developed countries: a review. Urban *Water Journal*, **4**(4), 253–260.

Galea L. (2014). Voluntary mechanism for stormwater quality management, 2014 Stormwater Queensland Conference, 6–8 August 201, Noosa Australia.

Gardner E. A., Millar G. E., Christiansen C., Vieritz A. M. and Chapman H. (2006). Energy and water use at a WSUD subdivision in Brisbane, Australia. *Australian Journal of Water Resources*, **10**(3), 283–291.

Ghisi E., Bressan D. L. and Martini M. (2007). Rainwater tank capacity and potential for potable water savings by using rainwater in the residential sector of southeastern Brazil. *Building and Environment*, **42**(4), 1654–1666.

Goyer R. A. (1993). Lead toxicity – current concerns. *Environmental Health Perspectives*, **100**, 177–187.

Grandet C., Binning P. J., Mikkelsen P. S. and Blanchet F (2010). Effect of rainwater harvesting on centralised urban water supply systems. *Water Science and Technology: Water Supply*, **10**(4), 570–576. IWA Publication.

Hall M. (2013). Review of rainwater tank cost-effectiveness in South East Queensland. Urban Water Security Research Alliance, Technical Report No. 105, 2nd edn. http://www.urbanwateralliance.org.au/publications/UWSRA-tr105.pdf (accessed 01 September 2014).

Herrmann T. and Schmida U. (1999). Rainwater utilisation in Germany: efficiency, dimensioning, hydraulic and environmental aspects. *Urban Water*, **1**, 307–316.

Heyworth J. S., Glonek G., Maynard E. J., Baghurst P. A. and Finlay-Jones, J. (2006). Consumption of untreated tank rainwater and gastroenteritis among young children in South Australia. *International Journal of Epidemiology*, **35**, 1051–1058. doi:10.1093/ije/dyl105.

Huston R., Chan Y. C., Chapman H., Gardner T. and Shaw G. (2012). Source apportionment of heavy metals and ionic contaminants in rainwater tanks in a subtropical urban area in Australia. *Water Research*, **46**(4), 1121–1132.

Islam M. M., Chou F. N. F. and Kabir M. R. (2011). Feasibility and acceptability study of rainwater use to the acute water shortage areas in Dhaka City, Bangladesh. *Natural Hazards*, **56**(1), 93–111.

Kahinda J., Taigbenu A. and Boroto, J. (2007). Domestic rainwater harvesting to improve water supply in rural South Africa. *Physics and Chemistry of the Earth*, **32**, 1050–1057.

Kenway S. (2008). Preliminary LCA of the SEQ Water Strategy. Urban Water Security Research Alliance, September 2008.

Khastagir A. and Jayasuriya L. N. (2010). Impacts of using rainwater tanks on stormwater harvesting and runoff quality. *Water Science and Technology*, **62**(9), 324–329.

Kim K. and Yoo, C. (2009). Hydrological modelling and evaluation of rainwater harvesting facilities: case study on several rainwater harvesting facilities in Korea. *Journal of Hydrologic Engineering*, **14**(6), 545–561.

Kloss C. (2008). Managing wet weather with green infrastructure municipal handbook – rainwater harvesting polices, US EPA, report EPA-833-F-08-010. http://water.epa.gov/infrastructure/greeninfrastructure/upload/gi_munichandbook_harvesting.pdf (accessed May 2013).

Knights D., MacGill I. and Passey R. (2007). The sustainability of desalination plants in Australia: is renewable energy the answer? Ozwater'07 conference proceedings, 5 March 2007, Australian Water Association, Sydney, Australia.

Lee J. Y., Bak G. and Han M. (2012). Quality of roof-harvested rainwater – Comparison of different roofing materials. *Environmental Pollution*, **162**, 422–429.

Li X.-Y., Xie Z.-K. and Yan X.-K. (2004). Runoff characteristics of artificial catchment materials for rainwater harvesting in the semiarid regions of China. *Agricultural Water Management*, **65**(3), 211–224.

Lucas S. A., Coombes P. J. and Sharma A. K. (2010). The impact of diurnal water reuse patterns, demand management and rainwater tanks on the supply network design. *Water Supply and Technology: Water Supply-WSTWS*, **10**(1), 69–80.

Maddaus W. (1999). Realising the benefits from water conservation. *Journal of Contemporary Water Research and Education*, **114**, 8–17.

Magyar M. I., Mitchell V. G., Ladson A. R. and Diaper C. (2009). An investigation of rainwater tanks quality and sediment dynamics. *Water Science and Technology*, **56**(9), 21–28, IWA Publishing.

Mander T. Minister for Housing and Public Works, Queensland (2013). Tanks no longer necessary, media release. Queensland Government.

Marsden Jacob Associates (2007). The cost-effectiveness of rainwater tanks in urban Australia, Waterlines Report No. 1 – March 2007.

Mays L. W., Koutsoyiannis D. and Angelakis A. N. (2007). A brief history of urban water supply in antiquity. *Water Science and Technology: Water Supply*, **7**(1), 1–12.

Melbourne Water (1996). The port phillip bay environmental study. Publication number 1324–7905.

Mendez C. B., Klenzendorf J. B., Afshar B. R., Simmons M. T., Barrett M. E., Kinney K. A. and Kirisits M. J. (2011). The effect of roofing material on the quality of harvested rainwater. *Water Research*, **45**(5), 2049–2059.

Nolde E. (2007). Possibilities of rainwater utilisation in densely populated area including precipitation runoffs from traffic surfaces. *Desalination*, **215**, 1–11.

NSW Health (2007). Use of rainwater tanks where a public water supply is available, Environmental Health Branch, NSW Health. http://www.health.nsw.gov.au/policies/gl/2007/GL2007_009.html (accessed May 2013).

NSW Planning and Infrastructure (2004). BASIX Building Sustainability Index. NSW Government, Department of Planning and Infrastructure. https://www.basix.nsw.gov.au/basixcms/ (accessed May 2013).

O'Connor J. B., Mitchell V. G., Magyar M. I., Ladson A. R. and Diaper C. (2009). Where is the lead in rainwater tanks coming from? 32nd Hydrology and Water Resources Symposium, Newcastle, Australia, Engineers Australia. 30 November – 3 December, 653–660. http://search.informit.com.au/documentSummary;dn=754058694996455;res=IELENG (accessed 01 September 2014).

Palla A., Gnecco I. and Lanza L. G. (2011). Non-dimensional design parameters and performance assessment of rainwater harvesting systems. *Journal of Hydrology*, **40**(1–2), 65–76.

Pandey D. N., Gupta A. K. and Anderson D. M. (2003). Rainwater harvesting as an adaptation to climate change. Current *Science*, **85**(1), 46–59.

Parkinson J. S. and Butler D. (2005). Modelling the impacts of domestic water conservation on the sustainability of the urban sewerage system. *Water and Environment Journal*, **19**(1), 49–56.

PMSEIC (2007). Water for Our Cities. Prime Minister's Science, Engineering and Innovation Council, Canberra, Australia. http://www.innovation.gov.au/Science/PMSEIC/Documents/WaterforOurCities.pdf (accessed May 2013).

Queensland Competition Authority (2012). Assessment of Proposed Repeal of Water Saving Regulations, Queensland Competition Authority.

Queensland Department of Infrastructure and Planning (2008). Queensland Government Water saving targets. A guide to the Queensland Development Code Mandatory Part 4.2. http://www.dlgp.qld.gov.au/resources/guidline/development-code/water-saving-targets-guidleines.pdf (accessed May 2013).

Queensland Water Commission (2010). South East Queensland Water Strategy. Queensland Water Commission, The State of Queensland, Brisbane.

RAIN (2012). Rainwater Harvesting Implementation Network. Annual Report 2011. http://www.rainfoundation.org/fileadmin/PublicSite/Annual_reports/Annual_report_11-ENG-1.pdf (accessed May 2013).

Retamal M., Glassmire J., Abeysuriya K., Turner A. and White S. (2009). The Water-Energy Nexus: Investigation into the Energy Implications of Household Rainwater Systems. CSIRO-Institute for Sustainable Futures, University of Technology, Sydney.

Rodrigo S., Sinclair M. and Leder K. (2010). A survey of the characteristics and maintenance of rainwater tanks in urban areas of South Australia. *Water Science and Technology*, **61**(6), 1569–1577.

Said S. (2014). Assessment of roof-top rain water harvesting potential in South Delhi, India: a case study. *International Journal of Environmental Research and Development*, **4**(2), 141–146.

Schuetze T. (2013). Rainwater harvesting and management–policy and regulations in Germany. *Water Science and Technology, Water Supply*, **13**, 376–385.

Sharma A., Gray S., Diaper C., Liston P. and Howe C. (2008). Assessing integrated water management options for urban developments - Canberra case study. *Urban Water Journal*, **5**(2), 147–159.

Sharma A., Cook S., Tjandraatmadja G. and Gregory A. (2012). Impediments and constraints in the uptake of water sensitive urban design measures in greenfield and infill developments. *Water Science and Technology*, **65**(2), 340–352.

Sharma A., Tjandraatmadja G., Cook S. and Gardner T. (2013). Decentralised systems: definition and drivers in the current context, *Water Science and Technology*, **67**(9), 2091–2101, IWA Publication.

Standards Australia (2008). HB230-2008, Rainwater tank design and installation handbook.

Thomas R. B., Kirisits M. J., Lye D. J. and Kinney K. A. (2014). Rainwater harvesting in the United States: a survey of common system practices. *Journal of Cleaner Production*, **75**(0), 166–173.

Umapathi S., Chong M. N. and Sharma A. K. (2013). Evaluation of plumbed rainwater tanks in households for sustainable water resource management: a real-time monitoring study. *Journal of Cleaner Production*, **42**(0), 204–214.

Vaes G. and Berlamont J. (2001). The effect of rainwater storage tanks on design storms. *Urban Water*, **3**(4), 303–307.

Vialle C., Sablayrolles C., Lovera M., Huau M. C. and Montréjaud-Vignoles M. (2011). Modelling of a roof runoff harvesting system: The use of rainwater for toilet flushing. *Water Science and Technology: Water Supply*, **11**(2), 151–158.

Villarreal E. L. and Dixon A. (2005). Analysis of a rainwater collection system for domestic water supply in Ringdansen, Norrkoping, Sweden. *Building and Environment*, **40**, 1174–1184.

Vivian W., Tam Y., Tam L. and Zeng S. X. (2010). Cost effectiveness and tradeoff on the use of rainwater tank: an empirical study in Australian residential decision-making. *Resources, Conservation and Recycling*, **54**(3), 178–186.

Walsh C., Fletcher T. and Ladson A. (2009). Retention capacity: a metric to link stream ecology and storm-water management. *Journal of Hydrologic Engineering*, **14**(4), 399–406.

Ward S. L. (2010). Rainwater harvesting in the UK: a strategic framework to enable transition from novel to mainstream. PhD thesis, University of Exeter UK.

World Health Organisation (2011). Guidelines for Drinking Water Quality, 4 edn. World Health Organisation, Geneva, Switzerland.

Yaziz M. I., Gunting H., Sapari N. and Ghazali A. W. (1989). Variations in rainwater quality from roof catchments. *Water Research*, **23**, 761–765.

Zhang X. and Hu M. (2014). Effectiveness of rainwater harvesting in runoff volume reduction in a planned Industrial Park, China. *Water Resources Management* **28**(3), 671–682.

Chapter 2

Rainwater tank modelling

Alison M. Vieritz, Luis E. Neumann and Stephen Cook

ABSTRACT

Mathematical models that simulate the performance of rainwater systems are important in supporting decisions on the suitability of rainwater systems to not only meet potable and non-potable water demands, but also reduce discharge of stormwater and associated pollutants to the environment. A variety of modelling tools have been developed over the last few decades to support the design of rainwater systems and optimise the combination of connected roof area and storage size to best meet demands. However, the selection of the rainwater system modelling approach can influence the predicted outcomes, with different choices of time-step, algorithms and simulation length significantly influencing the results.

This chapter presents a generalized model of a rainwater tank, and discusses its components in relation to the different rainwater tank models that have been developed around the world. We then explore the impact of key model parameters, including connected roof area, tank storage size, rainfall loss factor, water demand and climate inputs, as well as the choice of time-step and simulation length on the output, using the rainwater tank model described by Mitchell (2007). We demonstrate how each of these parameters can affect the simulated results of the tank behaviour. Finally, we show that using an 'average tank' behaviour to represent the collective behaviour of a large number of identical tanks in an urban area (also known as spatial lumping) can significantly overestimate the reliability and yield of the rainwater system. The errors introduced by such approaches are discussed using examples of yield and overflow estimation.

Keywords: Rainwater tank model; yield estimation; overflow estimation; model time-step; spatial averaging.

2.1 INTRODUCTION

It may be a little surprising to observe that rainwater tank systems have been adopted in many modern cities even though reticulated (mains) water supplies exist in these areas. There are a number of reasons for this. Firstly, they provide a secondary, non-potable water source to complement reticulated drinking water supply. Water scarcity and the need to reduce demand on the drinking water system has been the primary driver for the uptake of rainwater systems in Australia (Cook *et al.* 2013), Spain (Domènech & Saurí, 2011), south-eastern United States (Jones & Hunt, 2010) and Jordan (Abdulla & Al-Shareef, 2009). Secondly, in regions where water scarcity is not an issue, rainwater harvesting has been adopted to address broader urban sustainability issues, such as the mitigation of stormwater environmental impact. In Germany, rainwater harvesting has been adopted by environmentally conscious households (Herrmann & Schmida, 2000), whilst in Sweden, rainwater harvesting has been investigated as an approach to improve environmental sustainability of multi residential developments (Villarreal & Dixon, 2005). In developing countries, the context for adopting rainwater systems is that they can provide much improved quality of water relative to alternative local sources, are more affordable for low-income households, and can provide a water supply source where no piped system is provided (Campbell, 1986).

But the challenge in many places is how can these rainwater tank systems be designed for optimal performance? Indeed, how can we predict the performance of a rainwater tank system before it is even built? To support the design of rainwater systems, a variety of modelling tools have been developed over the last few decades. These tools have been developed and used to estimate the performance of rainwater tanks in various locations throughout the world including: Australia (Coombes & Barry, 2007; Mitchell, 2007), Taiwan (Su *et al.* 2009), United States (Jones & Hunt, 2010), Poland (Słyś, 2009), United Kingdom (Fewkes, 2000), Brazil (Ghisi *et al.* 2007), West Africa (Cowden *et al.* 2008), Italy (Palla *et al.* 2012) and Spain (Farreny *et al.* 2011) amongst many others.

Some of these tools are very simple, considering only the variability in annual rainfall (mm/year) when sizing rainwater tanks (e.g., Gould & Nissen-Petersen, 1999). However, this approach can obscure reliability estimates as two cities with similar annual rainfall can have significantly different seasonal rainfall patterns, and hence would be expected to show quite different tank water capture/yield behaviour (Basinger *et al.* 2010).

Another modelling approach is to develop analytical formulas that estimate the required storage volume based on the anticipated water use rate, local climate, and the required level of reliability of the tank to supply water to meet demand at any point in time. These models are very location-specific and use probabilistic relationships developed using local rainfall records (Basinger *et al.* 2010). But once developed, these models provide engineers and planners with a robust tool for rapid interpretation of likely tank performance. Further details of this approach can be found elsewhere (e.g., Guo & Baetz, 2007; Basinger *et al.* 2010) and will not be discussed in this chapter.

The most common approach for modelling rainwater system performance is to use a continuous water balance simulation that applies historical rainfall observations to generate inflow, with an assumed water demand as outflow, to calculate the volume of water within a rainwater tank as a function of time (Basinger *et al.* 2010). In this chapter, we will focus on this approach and provide examples of models that are used around the world, including countries with severe data constraints.

Finally, when modelling the behaviour of a rainwater tank for a household, or the impact of rainwater tanks in a multi residential development, a number of important factors need to be considered including time-step, simulation length, processes to include in the model, order of calculations used within the model, and output required. These factors, and their influence on the model predictions, are explored in this chapter to guide the selection of the modelling approach for a given application.

2.2 GENERAL CONCEPTS UNDERLYING A RAIN WATER TANK MODEL

Modelling the water volume inside a rainwater tank would seem straight forward compared to, say, modelling the water balance of an open paddock of grass. A rainwater tank is usually a closed vessel so that evaporative and leakage losses can be considered negligible, and there is no direct entry of rainfall.

The water volume inside the tank could then be determined simply by the amount of roof runoff entering the tank, the amount of water drawn off (yield) to supply demand, and the amount that is lost through tank overflow. The equations needed for a rainwater tank simulation model are built upon this simple mass balance concept.

A rainwater tank system is usually comprised of a collection system (i.e., roof, gutter and downpipe) and a storage system (i.e., tank) as shown in Figure 2.1. The collection system consists of the connected roof area A (m^2), representing the area of the roof that drains to the tank, and which may be smaller than the total roof area of the household (often not all of the roof area is connected to the tank). The collection system may also include a first flush device which retains a certain volume FF (m^3). A first flush device is installed to discard the first amount of roof runoff as a contaminant mitigation measure. The storage component includes a tank with an active storage S (m^3), which excludes the dead storage volume (DS) below the supply pipe, and the area above the overflow pipe. In this closed system, water export is via the yield to household Y_t, and tank overflow O_t.

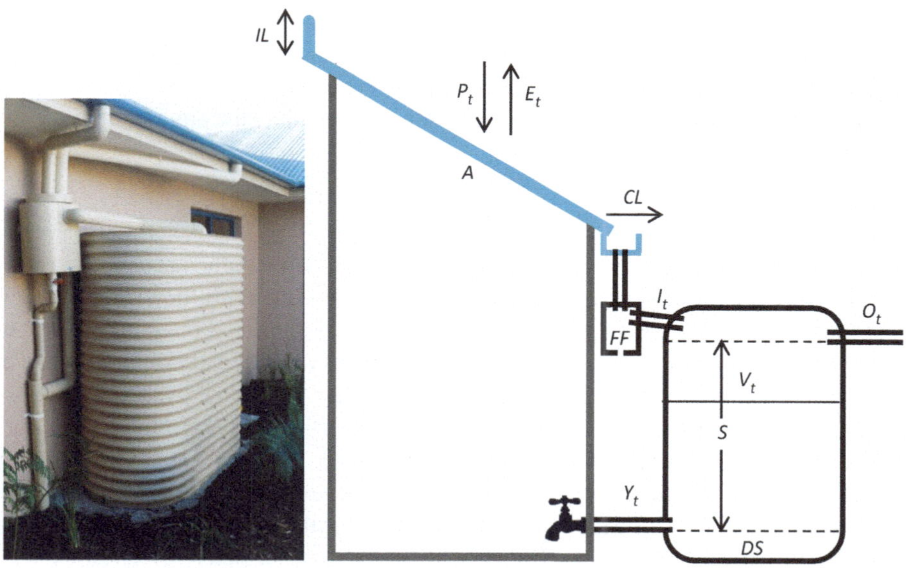

Figure 2.1 Schematic representation of a rainwater tank system.

The roof area collects rainfall P_t (mm) over a time period t. But not all rainfall that falls on the roof becomes inflow to the tank. Some of the rainfall at the beginning of a rainfall event will be lost due to wetting up of the roof surface (adhesion or detention loss), and this initial loss can be represented using a single storage of capacity IL (mm). The initial storage is also subjected to evaporation (E_t). After the initial storage is filled, the roof runoff will be further reduced by a continuing loss factor CL (%) representing a continuous percentage loss due to water being splashed or blown off the roof or gutter, and runoff capture by the first flush system. The initial and continuing losses can be represented by the following equations (Coombes & Barry, 2007; Mitchell *et al.* 2008):

$$RR_t = \max[(P_t + RST_{t-1} - IL),0] \tag{2.1}$$

$$RST_t = \max[(P_t + RST_{t-1}) - RR_t - E_t, 0] \tag{2.2}$$

$$HR_t = A * \frac{RR_t}{1000} * \left(1 - \frac{CL}{100}\right) \tag{2.3}$$

where HR_t is the harvestable roof runoff (m^3), RR_t is the roof runoff in mm, RST_t and RST_{t-1} are the roof storage level in mm in the current and previous time-step respectively and E_t (mm) is the evaporation over a time period t. If a first flush device is present, the inflow to the tank I_t is given by:

$$I_t = \max[(HR_t + FFI_{t-1} - FFV), 0] \tag{2.4}$$

$$FFI_t = \max[(HR_t + FFI_{t-1}) - I_t - LR, 0] \tag{2.5}$$

where FFI_t and FFI_{t-1} are the volume of roof runoff captured by the first flush device at time t and t–1, respectively; FFV is the capacity of the first flush device and LR is the leakage rate. The leakage rate represents the volume (m^3) at which the first flush device drips in one time-step. If the system does not have a first flush device, the inflow to the tank I_t is equal to the harvestable roof runoff HR_t.

The second part of the model uses a simple water balance to simulate the amount of water stored in the tank (Mitchell *et al.* 2008):

$$V_t = V_{t-1} + I_t - O_t - Y_t \tag{2.6}$$

where V_t and V_{t-1} are the volume of water (m^3) in the tank at the end of the current time-step and at the end of the previous time-step, respectively; O_t is the tank overflow (m^3); and Y_t is the yield or volume of water extracted from the tank during time-step t. The water balance in *Equation 2.6* assumes that the rainwater tank is a closed vessel (as mentioned previously) and hence direct rainfall and evaporation are excluded.

To solve the water balance given by *Equation 2.6* in a model simulation, overflow and yield must be calculated for each time-step. However, tank inflow, yield and overflow can occur in any order or can occur simultaneously during any time-step. If the time-steps are large, it is also possible that multiple events occur within one time-step, but the model will only be able to simulate the multiple events as one lumped event. The discrete nature of *Equation 2.6* means that an order for the inflow, yield and overflow processes must be assumed in the calculation process. If yield is assumed to occur before overflow (spill), the model type is called YBS (Yield Before Spill). If spill is assumed to occur before yield, the model type is called YAS (Yield After Spill). Therefore, for a YBS model, the yield Y_t for a time t is given by (Jenkins & Pearson, 1978; Fewkes & Butler, 2000; Liaw & Tsai, 2004):

$$Y_t = \min[D_t, V_{t-1} + I_t] \tag{2.7}$$

$$V_t = \min[S, V_{t-1} + I_t - Y_t] \tag{2.8}$$

where D_t is the demand for tank water (m^3) for a time-step t, S is the active storage (m^3) which excludes any dead storage below the water level from which yield is taken.

For a YAS model, the equations are (Jenkins & Pearson, 1978; Fewkes & Butler, 2000; Liaw & Tsai, 2004):

$$Y_t = \min[D_t, V_{t-1}] \tag{2.9}$$

$$V_t = \min[S - Y_t, V_{t-1} + I_t - Y_t] \tag{2.10}$$

As the equations for V_t constrains the tank water volume remaining at the end of the time-step to its maximum volume, the excess water is routed as overflow. Hence overflow O_t is calculated as:

For YBS model:

$$O_t = \max[(V_{t-1} + I_t - Y_t) - S, 0] \tag{2.11}$$

For YAS model:

$$O_t = \max[(V_{t-1} + I_t - Y_t) - (S - Y_t), 0] = \max[(V_{t-1} + I_t) - S, 0] \tag{2.12}$$

Comparison of the yield equations for the YBS and YAS models shows that yield will often be greater in the YBS model in each time-step, since it can include the inflow water as part of the yield to meet household demand. Consequently, the overflow will be potentially smaller in the YBS model in each time-step, since tank yield is first removed from any inflow before the remaining inflow is available for overflow. The implication of this 'order of calculation' concept on predicting rainwater tank behaviour will be discussed later.

To simulate the complete rainwater tank behaviour, these calculations of yield, overflow and water volume are simply repeated for a continuous series of time-steps, with the tank water volume from the previous time-step becoming the starting tank water volume for the next time-step. In this way, the rainwater tank model outlined above will predict the partitioning of any roof-impacting rainwater into initial roof loss, continuing roof losses and roof runoff. The runoff may then be further partitioned into loss from a first flush device and tank inflow. Any inflow is then added to the water store within the tank. Meanwhile, the tank water store may undergo further partitioning with some being extracted as yield to the household. If there is insufficient water in the tank to supply all of the household demand for a particular time-step, then the yield will be less than demand. Where the tank storage capacity is exceeded at the end of the time-step, excess tank water is partitioned to overflow.

The inputs required by this model are the connected roof area, effective tank volume, the depth of the roof storage representing initial loss, the proportion of roof runoff that is continuously lost due to splash, the capacity of the first flush device, and the first flush leakage rate. For each time-step, rainfall, evaporation and household water demand is required. The length of the time-step will determine how much variability of these inputs over time can potentially be captured by the modelling. The amount of variability that needs to be captured will depend on the purpose of the modelling, and this will be discussed later.

The outputs from this model are the predicted yield and overflow over the period simulated for the specified roof area and tank storage volume. The long term yield Y can be calculated and compared with the long term demand D to provide the tank performance measure volumetric reliability VR, the proportion of household demand that is likely to be supplied from the rainwater tank where:

$$Y = \sum_{t=1}^{T} Y_t \quad \text{and} \quad D = \sum_{t=1}^{T} D_t \tag{2.13}$$

$$VR = \frac{Y}{D} \tag{2.14}$$

where T is the total number of time-steps in the period of simulation.

Having described the general components and concepts underlying a physically rigorous rainwater tank model, we will now explore the modelling approaches that have been used to address particular objectives and issues.

2.3 ASPECTS OF SOME EXISTING RAIN WATER TANK MODELS

A selection of continuous simulation models of rainwater tank applied in various locations around the world were reviewed and were found to vary considerably in complexity, ranging from simple spreadsheet models (Roebuck & Ashley, 2007; Imteaz *et al.* 2011) to software using a spreadsheet user interface (Liu *et al.* 2006; Vieritz *et al.* 2007a,b), and to very detailed process based models (Coombes & Kuczera, 2001) that include hydraulic calculations and processes not described in our generalised model. We also mention urban water models that simulate the whole of the water cycle, including on-site detention, infiltration, wastewater, greywater and stormwater flows. Such models include rainwater tanks and are particularly common in Australia; for example, Aquacycle (Mitchell, 2005), UVQ (Mitchell & Diaper, 2006), MUSIC (CRC-CH, 2005), Urban Developer (eWater Cooperative Research Centre, 2011), and Watercress (Clark *et al.* 2002). However, these models will not be discussed in any detail in this chapter.

We will now consider the following aspects of the rainwater tank component of these models: time-step, climate data inputs, demand data inputs, roof runoff modelling, tank water balance configurations, and the use of model outputs.

2.3.1 Time-step and climate data inputs

The climate data required will depend on the time-step used in the rainwater tank model. The time-step chosen is often constrained by the availability of rainfall data, with daily data being more commonly available than sub-daily data. Hence, most of the simulation models reviewed used a daily time-step, although some used a time-step as short as 6 minutes (Coombes & Kuczera, 2001).

Climate data may be sourced from historical daily climate records provided by weather stations in the locality of interest. In Australia, interpolated climate data is available by fitting daily surfaces to climate data supplied by weather stations across the country (SILO Climate Data, http://www.longpaddock.qld.gov.au/silo/data_available.html). This climate data includes daily rainfall and evaporation values across Australia on a 0.05 degree grid, extending back to 1889.

However, Cowden (2008), when developing a model for use in a data-poor region of West Africa, found that daily rainfall records were available for only a limited number of sites, were of short-duration, and contained many missing values. This highlights a limitation in continuous simulation models. They are only useful provided reliable and continuous input data exists. One approach to addressing limited rainfall data availability is to develop rainfall models using parametric estimators of rainfall (Guo & Baetz, 2007). But a fundamental issue with parametric rainfall models is that the probabilistic relationships developed to generate a rainfall record require a certain length of historical data to exist. Secondly, the probabilistic relationships developed are generally not portable to other geographic locations, since rainfall distribution and seasonality can differ substantially from location to location. Nonetheless, improved portability has been reported when using non-parametric estimators of rainfall (Cowden *et al.* 2008; Basinger *et al.* 2010).

For models requiring sub-daily climate data, approaches include measured pluviograph data; daily rain data from which synthetic pluviograph rainfall can be generated based on nearby pluviograph rainfall records; or rainfall models such as the event rainfall model (DRIP) developed by Heneker *et al.* (2001).

All these approaches assume that the historic climate record provides an adequate prediction of the future climate, at least for the operational period of the rainwater tank system being designed. In Australia, climate projections data generated by 19 global climate models for a range of emission scenarios and climate sensitivities have been made available at http://www.longpaddock.qld.gov.au/climateprojections/about.html so that the predicted impacts of climate change can be incorporated into modelling, if desired.

2.3.2 Water demand data inputs

In rainwater tank simulations, obtaining accurate estimations of the water demand from the tank is almost as important as having access to climate data appropriate to the location. The sensitivity of tank performance predictions to demand and its temporal pattern will be demonstrated later (Section 2.4.4). Locally measured data is always the preferred source of demand data, but is not always available. Cowden *et al.* (2008), in assessing the reliability of small-capacity rainwater tank systems used in low-income households in West Africa, simply assumed 20 L/person/day, the World Health Organization's recommendation to meet basic consumption and hygiene needs (Howard & Bartram, 2003). However, assessment of rainwater tank performance in developed urbanised countries will usually require data on the non-constrained quantity of water required by the household. Household demand has been found to vary greatly due to occupancy rate, type of appliances used, behaviour and attitudes to water use, and household age structure. Such data can now be found for many locations as numerous studies have been done to quantify the amounts of water used for various internal and external end uses (e.g., Mayer *et al.* 1999; Dziegielewski *et al.* 2000; Roberts, 2004; Roberts, 2005; Willis *et al.* 2009; Sivakumaran & Aramaki, 2010; Beal & Stewart, 2011; Willis *et al.* 2011; Water Corporation, 2011; von Voigt & Mohajeri, 2013). Some models disaggregate household internal water demand into various end uses such as kitchen, laundry, bathroom, toilet, and hot or cold water (Combes & Kuczera, 2001; Vieritz *et al.* 2007a,b; Khastagir & Jayasuriya, 2010). End uses suitable for rainwater substitution are then selected and summed to estimate the demand for tank water by the household.

The household tank demand has been applied as a single value for all time-steps, either yearly or daily in a large number of rainwater tank studies (e.g., Fewkes, 2000; Fewkes & Butler, 2000; Liaw & Tsai, 2004; Ghisi *et al.* 2007; Jones & Hunt, 2010; Palla *et al.* 2011; Palla *et al.* 2012; Campisano *et al.* 2013). However, this will be adequate only where the chosen end uses do not show significant variation across time-steps, for example, if tank water is used exclusively for internal uses such as consumption (Cowden *et al.* 2008), or toilet flushing (Mitchell, 2007; KRANRRC, 2011; Neumann *et al.* 2011; Palla *et al.* 2012). External end uses (e.g., garden watering, car washing and pool top-up) usually exhibit a strong seasonal variation which must be parameterised for the model to provide realistic predictions on yield and overtopping. Coombes and Barry (2007) showed that simulations without accounting for seasonal demand underestimated rainwater yields by up to 25%, although the differences became smaller as the tank size increased.

To capture this seasonal variability in a rainwater tank model, internal and external household daily water demands are usually modelled separately. For internal household demand, Vieritz *et al.* (2007a,b) multiplied the *average* indoor water demand per person for the region by the average number of occupants per household. Burns *et al.* (2012) describes a method of using standardized measured household tank usage data to obtain daily usage distributions for each end use, from which 10-year daily usage records can be stochastically generated. Where relevant daily internal water use data is not readily available, a strategy for estimating household indoor water demand is to use an end use model to generate demands (e.g., Coombes, 2002; Duncan & Mitchell, 2008). In the empirical model developed by Coombes (2002), a regression equation was developed that related monthly daily average internal use by households in the Lower Hunter region, Australia, to the number of occupants, the average weekly income per person, monthly daily average rainfall, annual population growth, the number of days with rainfall in a month, and the mean monthly daily maximum temperature. However, these models are developed from available data for one specific region. For use in other regions, the model coefficients need to be calibrated against a set of locally measured water use data.

Where relevant external water use data is not readily available, one strategy is to use a physically based crop water use model. Jones and Hunt (2010) incorporated a plant irrigation module into their model so that outdoor demand for garden watering was determined according to the modelled soil water deficit and irrigation rules. By triggering irrigation when the soil dried to a specified moisture content, the seasonal effect of fewer irrigation events in wet periods was captured effectively in the modelling. A similar approach was adopted by Vieritz *et al.* (2007a,b) using a soil water balance based on inputs (rain and irrigation) and outputs (evapotranspiration and runoff plus deep drainage losses). Irrigation timing/volumes were then determined based on rules dependent on soil moisture and day of the week. However, outdoor water use is a behaviour moderated reaction to climate, and social factors such as garden interest and the householder's assessment of garden water needs can override soil moisture considerations (Beal *et al.* 2014). This was addressed by allowing the model user to constrain the predicted average annual external water use to an annual 'measured' value (calculated from suburban scale water meter readings and estimated internal water use). The model then automatically adjusts the irrigation amount and timing to constrain the predicted average annual external water use to the average 'measured' suburban scale value. Another strategy involves using regression equations and irrigation probability functions. Coombes (2002) developed a regression equation along similar lines to that previously described for estimating indoor water use. He then developed an improved outdoor demand generator based on the probability of irrigation taking place following various sequences of wet and dry days. Duncan and Mitchell (2008) estimated the probability of garden irrigation in a particular time-step as a function of daily maximum temperature. Combined with user-defined probability distributions (mean and standard deviation) of irrigation flow rate, garden irrigation could be estimated for each time-step. However, all these probabilistic approaches require the demand model to be calibrated against observed data. For example, Coombes and Barry (2007) used average internal and external water demands for three person households (from government survey reports) and local pluviograph rainfall records to generate daily water demands modified by climate influences for each Australian capital city using the demand generator developed by Coombes (2002).

Swimming pool top-up may be another external end use for water from a rainwater tank. In one model, the demand is based on garden watering rules, which can be adjusted to model swimming pool top-up as the water demand is related to the net evaporative demand of the local climate (Vieritz *et al.* 2007b). The pool is assumed to evaporate at a specified percentage of Class A Pan evaporation when uncovered (no pool blanket), and top-up is triggered whenever the water level in the pool falls to a specified distance below full level mark.

For other external end uses such as car washing, estimates of hosing duration for each event, hose flow rate and frequency of the event through the year can be used to estimate water use. Such data may be sourced from survey literature and field observations. For example, one U.S. study estimated that the water use for car washing was 75 L per vehicle based on field observations and simulations using a low flow nozzle (Smith & Shilley, 2009).

Once the amounts of tank water required for a range of internal and external end use options have been estimated, these are summed for each time-step to provide the demand required for the tank water balance. If the models are run at sub-daily time-steps, suitable sub-daily demand data is required. Demand generators such as the stochastic generator of Duncan and Mitchell (2008) (described further in Section 2.4) can be used to generate demand at sub-daily time-steps, or daily demand can be disaggregated into the required time-steps based on appropriate diurnal water use patterns (Coombes & Kuczera, 2001).

2.3.3 Roof runoff modelling

Some models ignore roof runoff losses and estimate roof runoff (m^3) as rainfall (m) × roof area (m^2). Others include a single roof runoff coefficient in the calculation to account for loss such that it is always a constant proportion of potential roof runoff (Cowden *et al.* 2008; Imteaz *et al.* 2011; KRANRRC, 2011). Studies have shown that roof runoff coefficients are dependent on roof materials, roof orientation relative to rain direction, as well as rainfall event patterns (e.g., Ragab *et al.* 2003). Roof runoff coefficient values typically used in modelling range from 0.7–0.95 (Fewkes, 2000; Liaw & Tsai, 2004; Ghisi *et al.* 2007), but with experimental measurements reporting values in the range 0.62–0.95 (Farreny *et al.* 2011). For some locations with semi-arid and infrequent rainfall, the coefficient of runoff can be as low as 0.5 (KRANRRC, 2011). The rainfall logger data presented in KRANRRC (2011) suggested that the low runoff coefficient may be due to the dominance of roof absorption losses for rainfall events less than about 3 mm which dominated the rainfall record (64% of events < 3mm). However runoff coefficients increased to 0.7 to 1.0 when rainfall exceeded 3 mm.

Roof runoff loss due to first flush devices has been included in some models (Clark *et al.* 2002; Roebuck & Ashley, 2007; Vieritz *et al.* 2007a,b; Basinger *et al.* 2010; Jones & Hunt, 2010; Khastagir & Jayasuriya, 2010b), and is often represented as a slowly leaking reservoir of a specified volume that must be kept filled before rain water can flow onto the storage tank. The amount of roof runoff that is discarded by the first flush device can have a significant impact on rainwater catch, especially if the rate of leakage from the device is high (Millar *et al.* 2003). A highly detailed model of a first flush device was incorporated into the PURRS model by Coombes (2002), which, together with a very short time-step (6 minutes), was able to provide accurate hydraulic analysis of the behaviour of the first flush device.

2.3.4 Tank water balance configurations

As discussed previously, both the YAS (Yield After Spill) and YBS (Yield Before Spill) order of calculations for the tank water balance (Equation 2.6) are approximations of real systems given the fact that neither considers the (plausible) scenario of yield, inflow and overflow occurring simultaneously. A number of models use the YBS assumption (Dixon *et al.* 1999; Liaw & Tsai, 2004; Basinger *et al.* 2010; Imteaz *et al.* 2011; Burns *et al.* 2012), whilst others use the YAS assumption (Roebuck & Ashley, 2007; Słyś, 2009; Palla *et al.* 2011; Palla *et al.* 2012; Campisano *et al.* 2013), or leave it to the discretion of the user to choose (Vieritz *et al.* 2007a,b).

Fewkes (2000) investigated a generic form of the YAS/YBS equations which used a parameter, θ. If $\theta = 1$, the equation reverts to the YAS form, and if $\theta = 0$, the form reverts to the YBS equation. The model using θ and a monthly time-step was then compared to daily YAS simulations and a value of θ was found for a given set of conditions as such that there was little difference between the volumetric reliability predictions (percentage of demand met by the rainwater tank) by the two models. The value θ simulating a blend of YBS and YAS rules, was considered to incorporate the effect of daily fluctuations into the monthly time-step simulation so as to improve the model's performance. However, as the value of θ is dependent on demand, tank size and rainfall, it is therefore hard to define *a priori*. The importance of this choice between YBS and YAS on a model's predictions will be discussed further in Section 2.4.1.

Some tank models include additional inputs other than roof runoff into the tank, such as: automatic mains top-up triggered when the tank water level falls below a specified threshold value (Vieritz *et al.* 2007a,b; Słyś, 2009); and trucked-in water to re-fill a tank (Vieritz *et al.* 2007a,b). In one model of a domestic water reuse system, both rainwater and grey water were included as inflows (Dixon *et al.* 1999). The water balance described by Equation 2.6 must therefore be expanded to include the new inflows. In Vieritz *et al.* (2007a,b), the components of the tank water balance are the rainwater, any top-up water (mains or trucked water), water use for internal purposes, water use for external purposes (but only if water demand for internal purposes is first satisfied), overflow, and the change in tank water volume. Top-up occurs according to rules specified by the user, with the option to specify a threshold tank water height for triggering the addition of top-up water. For the mains trickle top-up option, a trickle top-up connection allows mains water to flow into the tank when the volume of water in the tank falls below a specified level. This flow restores the volume in the tank back to this trigger level. For this daily time-step model, this level corresponds to the maximum trickle top-up volume, which is the maximum volume of mains water that can flow into the rainwater tank on any one day, and which must be at least equal to the maximum internal daily draw (L/day) on the rainwater to ensure that the household will never run out of water for internal end uses. This means that the tank volume below this trigger level is maintained with mains water, and is effectively part of the tank dead storage (*DS* in Figure 2.1) with respect to both *rainwater* collection and use when considering the active tank storage volume for tank size optimisations. In a sub-daily time-step model, trickle top-up was triggered when the tank volume reached a minimum active storage volume (Coombes *et al.* 2003). This minimum active storage volume was set to the maximum daily water use that is expected from the tank minus the maximum daily mains water top-up volume (as topping up can be simulated throughout each day). Barry and Coombes (2006) provide further details of the impact of mains trickle top-up rate and trickle top-up volume on tank performance.

2.3.5 Model outputs

In general, most rainwater tank models are used primarily to determine how much water can be supplied by the tank to meet the defined set of water demands. These estimates of yield can take various forms, such as volumetric (kL per year) based on either annual or daily yield; volumetric reliability based on the percentage of the demand that was met by rainwater; or temporal reliability based on the percentage of days the demand was met. The estimates of yield can then be used to investigate the reliability of tanks as an alternative (primary) water source, or their impact on the reduction of mains water use.

Multiple runs of the model may be used with systematic variation in roof area and storage volume (within specified ranges) to create reliability of supply contours in a graph box using roof area and storage volume as the plotting axes (Liaw & Tsai, 2004; Vieritz *et al.* 2007a,b). The contours can be used to help select optimal roof area and tank storage volume for a particular supply performance criterion (e.g., kL/year, volumetric reliability, and etc.).

The combined impact on water savings by a cluster of rainwater tanks may be modelled by performing continuous simulation of each rainwater tank system. However, simple scaling up of the results from continuous simulation of single rainwater tank system representing the 'average' system is not recommended and this will be discussed further in Section 2.5.

To avoid the need for extensive simulations, several authors (Su *et al.* 2009; Khastagir & Jayasuriya, 2010b; Palla *et al.* 2011; Palla *et al.* 2012; Campisano *et al.* 2013) proposed the use of curves relating the volumetric reliability (*aka* saving efficiency or % demand supplied) to dimensionless parameters. Palla *et al.* (2011, 2012) used two dimensionless parameters, the demand fraction (annual demand divided by annual inflow) and storage fraction (storage divided by annual inflow), to describe the mains water savings for a range of roof areas and tank storage volumes in Europe. Campisano *et al.* (2013) proposed another dimensionless parameter that takes into account intra-annual rainfall patterns, and fitted a regression equation to estimate water saving efficiency for 17 sites in Italy. Hanson *et al.* (2009) developed regression relationships to calculate required domestic storage capacity that was generally applicable for the USA. All these equations provide an estimate of yield, either in volumetric terms (kL/year) or as a proportion of demand satisfied (%), as a function of roof area, rainfall depth and storage size. Therefore, these models provide a basic means of estimating system performance, but incorporate a variety of simplifying assumptions regarding water use rates, overflow volumes, and rainfall depths. In all these cases, care must be taken as these regression models use a non-varying value to represent demand including the seasonally variable outdoor water use. This assumption can lead to substantial errors in supply predictions (Coombes & Barry, 2007). Taken overall, empirical curves and analytical formulas should be restricted to situations where the rainfall and demand patterns are similar to those patterns used in the derivation of the analytical solutions.

Once yields are calculated based on demands, rainfall, roof area and tank sizes, it is possible to extend the calculations to optimize tanks, not only in terms of yield but also on costs (Liaw & Tsai, 2004) or life cycle cost estimations which consider payback periods, expected water savings, capital and operational cost (Imteaz *et al.* 2011). Jenkins (2007) considered the cost of supplying rainwater and found that as consumption rate decreased, the unit cost of supply increased, and that this could be used to determine an economically optimum size of the rainwater tank. For further details on economic analysis of rainwater tanks, see Chapter 12.

In addition to supplying water to meet internal and external demands, rainwater tanks can also provide benefits such as reduction in stormwater runoff and pollutant export to waterways due to the reduction in volumes, as well as treatment provided by tanks (Khastagir & Jayasuriya, 2010a; Neumann *et al.* 2011). In a stormwater study, Vaes and Berlamont (2001) considered the influence of runoff reduction due to rainwater tanks and adjusted the runoff from design storms used for sizing combined sewer systems (sewage plus stormwater). Burns *et al.* (2010) assessed the impact of rainwater tanks as allotment-scale stormwater management devices for a hypothetical catchment using MUSIC for tank storage and overflow modelling, and a flood event model. The modelling indicated that tanks provided moderate, but potentially significant reductions to flood risk, with a further benefit of protecting the ecological health of urban streams degraded by the frequent stormwater runoff. (The impact of rainwater tanks on stormwater quantity and quality is discussed in detail in Chapter 13.)

2.4 INFLUENCE OF DIFFERENT VARIABLES IN THE TANK SIMULATION

Based on the rainwater tank schematic described in Figure 2.1 and the equations used to describe the hydrologic behaviour of the rainwater tank, it is clear that the model parameters that influence the yield of a rainwater tank are roof area, tank volume, demand, roof initial loss and roof continuing loss factor. In addition to these five key parameters, model results are also dependent on the total simulation length, choice of time-step and spill rule (YAS/YBS), and climate inputs (rainfall and evaporation). A brief description of the influence of each of these variables is given in this section using the rainwater tank model described by Mitchell (2007). This model is based on Equations 2.1 to 2.6 and the YAS rule.[1]

The model was run for 5000 different combinations of the model parameters (roof area, tank volume, demand, initial loss and continuing loss factor) initially using daily time-steps. Hourly and 6-minute simulations using the same parameter sets were also run to demonstrate the influence of time-step length. The rainwater tank model parameterization adopted here is identical to that used by Mitchell *et al.* (2008), in which model parameters for detached residential dwellings in Melbourne were determined by literature review and industry consultation. In that study, tank volume, roof area, roof depression storage (initial loss), and effective roof area loss factor (continuing loss factor) are modelled as truncated normal distributions

[1]The *YAS* form used in this model replaces *Equation 2.9* with $Y_t = \min[D_t, V_{t-1} + I_t]$ and replaces *Equation 2.10* with $V_t = \max[\min[S - Y_t, V_{t-1} + I_t - Y_t], 0]$. The yield for a given time step is calculated *after* the inflow to the tank is added to the existing volume, whereas in the YAS form given in *Equation 2.9*, the yield is computed *before* the addition of the inflow. This necessitates a change in the calculation of the final volume V_t to avoid negative volumes if $V_{t-1} + I_t > S$.

bounded by maximum and minimum values (to avoid physically implausible values) and are listed in Table 2.1. The initial roof loss distribution was adjusted to slightly higher values to account for the use of daily time-step as recommended by Mitchell *et al.* (2008).

Table 2.1 Assumed normal distributions of the variables used in the rainwater modelling in Melbourne using the Mitchell *et al.* (2008) model. The maximum and minimum values define the truncation values of the stochastic distributions.

Parameter	Minimum	Mean	Maximum	Standard deviation
Tank capacity (kL)	0.1	2.5	20	1.5
Roof area (m^2)	25	100	400	50
Initial Loss (mm)	0	0.5	1.7	0.5
Continuing loss (%)	0	15	30	5

The water demand data was created using a stochastic generator developed by Duncan and Mitchell (2008) and calibrated to measured Melbourne demand data reported by Roberts (2004, 2005). The demand generator uses flow rates of the various end uses, household occupancy and seasonality to generate water demands at one minute time-steps, which were aggregated to daily values. The demands considered here are toilet flushing and outdoor irrigation. The toilet end use is dependent on household occupancy, but is non-seasonal. In contrast, the garden end use has large variations in demand due to seasonality, but is relatively insensitive to the occupancy. The use of both demands therefore produces a water use pattern that has both seasonal and non-seasonal components. The toilet demand is generated using a diurnal pattern, flush volumes and number of flushes per person. The household occupancy was assumed to vary between 1 and 6 persons/household, with a distribution statistically generated to simulate household variability. The average toilet demand used was 32 litres per person per day (L/p/day), and the average outdoor use is 234 L/household/day. Based on average occupancy of 2.6 persons/household, the average yearly demand is 115 kL. The total simulation period was 50 years in order to avoid errors in estimation when using short time series (Liaw & Tsai, 2004; Mitchell, 2007). During the simulation period, the mean annual rainfall was 653 mm/year, with a mean annual potential evaporation of 1051 mm/year.

2.4.1 Time-step and spill rule

Several studies have examined the effect of time-steps on the results of rainwater tank simulation (Fewkes, 2000; Fewkes & Butler, 2000; Coombes & Barry, 2007; Mitchell, 2007) and all agreed that shorter time-steps are recommended when simulating small storage sizes. Fewkes and Butler (2000) and Mitchell (2007) compared results of year-long simulations using monthly, daily and hourly models with constant water demands, but a variety of storage size and inflows. The results from these two studies recommend a simulation time-step based on the (dimensionless) ratio of storage size to the mean annual inflow (S/I). Fewkes and Butler (2000) recommended an hourly time-step model if the S/I ratio is smaller than 0.01, with daily models recommended if $0.125 \geq S/I > 0.01$. Monthly models can be used if S/I is larger than 0.125. The study by Mitchell (2007) extended these results by considering longer simulation periods, a range of climatic zones in Australia, seasonality, and different quantum of demand. Accepting that the YAS rule with a 6 minute time-step provides the most accurate estimation of yield, Mitchell recommended using a daily time-step if the S/I ratio is equal to or higher than 0.007.

Table 2.2 shows the differences in yield estimation for different time-steps, spill rules and different *S*/*I* ratios resulting from different tank volumes (*S*) and roof sizes (i.e., different inflow, *I*), based on the parameters described in the previous section, and an annual demand of 115 kL for toilet demand and outdoor irrigation. It is clear from Table 2.2 that larger time-steps are likely to be acceptable for model simulations using the YAS rule with tank size and roof area combinations that fall within the practical ranges. Errors are small even for $S/I = 0.004$, which corresponds to a very small tank (0.5 kL) and a relatively large connected roof area (200 m^2). For a larger tank and same roof size, the error becomes negligible. However, if the model uses the YBS rule, the errors associated with different time-steps become more significant, even for the same *S*/*I* ratio and demand. The difference between a 6-minute and a daily time-step for a *S*/*I* of 0.004 is −2.9% for a YAS rule, yet increases up to 17.6% for the YBS rule. These results are in agreement with Mitchell (2007) who showed that errors due to different time-steps are larger for a model using a YBS rule than the YAS rule.

Table 2.2 The effect of simulation time-step, spill rules (YAS and YBS) and tank and roof sizes on yearly yields (kL) and its percent changes. The annual water demand was 115 kL.

Tank size (kL)	Roof area (m^2)	Inflow (kL)	*S*/*I*	Spill rule	Yield 6 min time-step (kL)	% Difference compared to 6 min	
						Hourly	Daily
0.5	200	117.5	0.004	YAS	30.6	−0.3	−2.9
0.5	100	58.8	0.008	YAS	25.9	−0.1	−2.1
5	200	117.5	0.04	YAS	64.4	0	0.2
5	100	58.8	0.08	YAS	47.8	0	0.1
0.5	200	117.5	0.004	YBS	30.7	1.0	17.6
0.5	100	58.8	0.008	YBS	26.0	0.6	13.4
5	200	117.5	0.04	YBS	64.4	0	1.1
5	100	58.8	0.08	YBS	47.8	0	0.5

The influence of time-steps in the estimation of yield is shown in Figure 2.2, with results for the estimated tank overflow for the same simulations shown in Figure 2.3. For these figures, the *S*/*I* ratio varies between 0.001 and 0.32 and as such are within the ratios for which Fewkes and Butler (2000) and Mitchell (2007) recommend a transition from 6-minute to daily and hourly time-steps. As can be seen in Figure 2.2, the degree of under or overestimation also increases with the time-step, with daily time-steps producing larger differences than hourly time-steps. In a 6-minute time-step, several demand events may occur during the day, therefore removing water from the tank and increasing the available volume to capture roof runoff, and therefore reducing overflow. As the time-step increases, individual inflow, overflow and demand events are lumped, increasing their magnitude and decreasing their frequency (Mitchell, 2007). A daily model can only have one inflow, one demand and one overflow event in a day, whereas an hourly model can have 24 occurrences of each event type in a day. Where the tank storage volume *S* is small relative to inflow (low *S*/*I* ratio), individual rainfall inflow events and demand events have a large impact on tank water volume. Hence in situations with low *S*/*I* ratio, the lumping of events required by a daily time-step model leads to a less accurate simulation of the tank water volume than shorter time-step models.

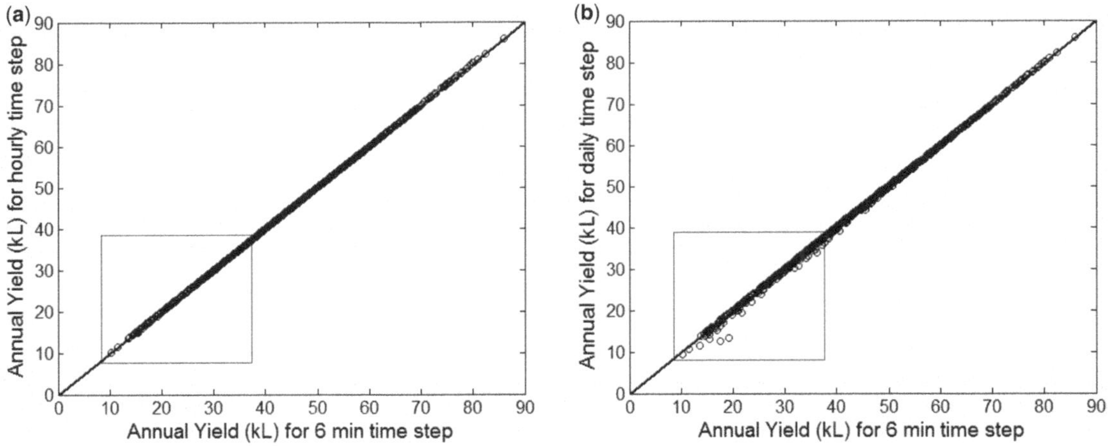

Figure 2.2 Comparison of estimated rainwater tank yield for YAS simulations using a 6-minute time-step versus simulations using hourly (a) and daily (b) time-steps. Simulations were based on the parameters described at the beginning of Section 2.4.

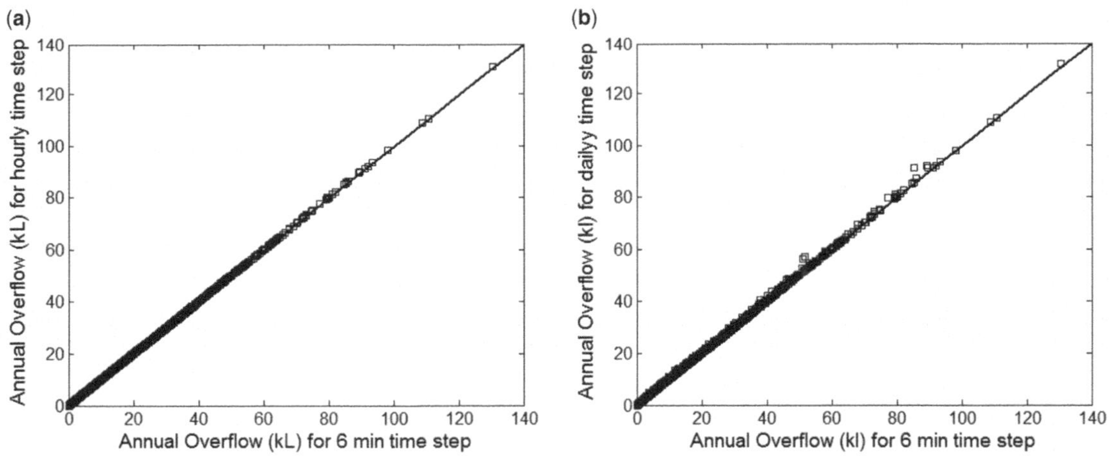

Figure 2.3 Comparison of estimated rainwater tank overflow for YAS simulations, using a 6-minute time-step versus simulations using hourly (a) and daily (b) time-steps. Simulations were based on the parameters described at the beginning of Section 2.4.

Table 2.2 shows that a YAS rule based model allows the use of hourly or daily time-steps for most practical applications as the *S/I* ratio is likely to be high if tank storage capacities of 1 kL or larger and reasonable roof sizes are used. A 5 kL tank with 100 m^2 connected roof area, which was considered as typical for planning purposes in Queensland (Department of Local Government and Planning, 2008), shows only a 0.1% difference in annual yield between daily and 6-minute time-steps. The use of smaller time-steps is recommended if detailed hydraulic analyses are needed, as flow routing and other hydraulic calculations often need short time-steps if they are to maintain accuracy. Therefore, as the required time scale of the hydraulic processes calculation is much smaller than a day, smaller time-steps have to be used for such analyses.

2.4.2 Roof area

The roof area used in rainwater tank models must reflect the area of roof actually connected to the rainwater tank. Tank location and household design may only allow for a section of the roof to be connected to the rainwater tank. Ignoring for the moment rainfall losses described by the interception and depression stores, the inflow to the tank is directly proportional to the roof area for a given rainfall. As shown in Figure 2.4, as the roof size increases, the yield from the rainwater tank initially increases linearly, but after a certain roof size, the increase falls away as the yield becomes limited either by the tank size, or the water demand from the tank. Further increases in the roof area will lead to minimal improvement in yield.

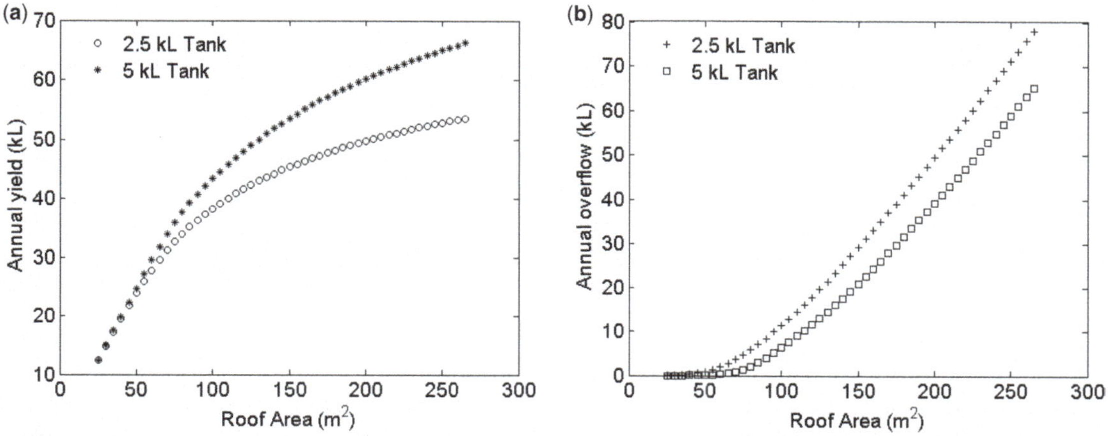

Figure 2.4 Annual yield (a) and overflow (b) for rainwater tanks of 2.5 and 5 kL for different connected roof areas (all other parameters using mean values in Table 2.1).

In contrast to the yield response, the overflow from the rainwater tank for smaller connected roof areas is nearly zero with most, if not all, inflow being used by the demands from the tank. However as the roof area, and hence roof runoff increase, the tank inflow is more than sufficient to satisfy the demand and fill the tank. At this stage, increases in the area of connected roof lead to larger increases in overflow. In such cases, larger roof area leads to increases in overflow as the system capacity is limited by storage size, not roof area.

2.4.3 Tank size

The tank capacity used in the model should be the active tank volume *S* (see Figure 2.1) and not the nominal tank volume. Any volume below the pipe from which yield (*Y*) is drawn is 'dead storage' (*DS*) as water stored in this part of the tank cannot be used. Where mains water is used to automatically top-up the tank to a specified water level, the tank volume below this level will also be effectively *DS* relative to rainwater, as discussed previously (Section 2.3.4). Furthermore, any volume above the overflow pipe will act as a temporary rain water storage buffer but that water will inevitably leave the tank as overflow (*O*) and hence cannot be used to meet demand (unless of course a demand event occurs before all water has drained away). The estimated yield from rainwater tanks of different sizes for a seasonal and non-seasonal demand (115 kL/year in total), as a function of roof area, is given in Figure 2.5 (a). The shape of the curve is very similar to the curve corresponding to the increase in roof area, with an increase in yield with tank size until a point is reached where the tank size becomes so large (about 2 to 4 kL) that it captures most of the inflow, apart from those very large events when overflow (Figure 2.5b) still occurs. At this point, increase in tank volume will lead to small increases in yield but probably large increases in cost. The overflow from the tank shows a different behaviour, with the overflow volume being inversely proportional to the tank size, as an increase in the volume leads to increased storage available, thus reducing overflow (Figure 2.5b).

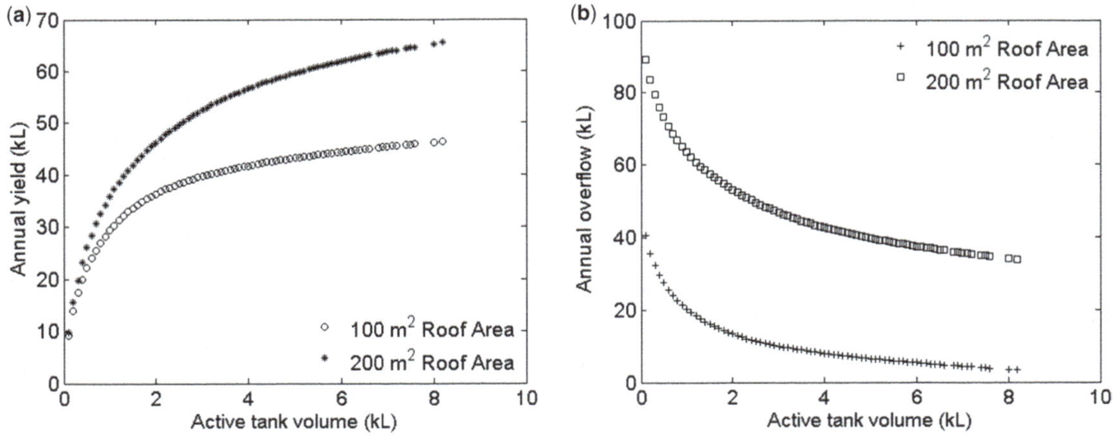

Figure 2.5 Annual yield (a) and overflow (b) for rainwater tanks of different sizes with two different connected roof areas (all other parameters using mean values in Table 2.1).

2.4.4 Demand

One of the most important variables influencing yield or overflow from rainwater tanks is the actual demand to be supplied by the rainwater tank. A large proportion of rainwater tank studies have used a constant demand, either yearly or daily (Fewkes, 2000; Fewkes & Butler, 2000; Liaw & Tsai, 2004; Ghisi *et al.* 2007; Jones & Hunt, 2010; Palla *et al.* 2011; Palla *et al.* 2012; Campisano *et al.* 2013). The impact of increasing demand for a tank with a certain volume and inflow can be seen in Figure 2.6, which shows clearly that, for a constant tank size, an increase in demand leads to an increase in the tank yield. In general, tanks with a high daily demand will result in higher yields compared to tanks with same size and inflow, but smaller daily demands.

Recent studies (Coombes & Barry, 2007; Neumann *et al.* 2011) demonstrate that if the demands have a seasonality component such as irrigation, the use of a constant demand will usually introduce errors in the estimation of yield. Coombes and Barry (2007) investigated the effect of seasonality on the supply yield of rainwater tanks supplying water to toilet (constant demand) and garden irrigation (seasonal demand) for eight Australian capital cities. For the same annual demand, the simulations ignoring the seasonal demand under estimated rainwater tank yields by up to 25%, with magnitude of under estimation becoming smaller as tank size increased.

The results of Figure 2.6 show a similar response to that of Coombes and Barry (2007) in that annual yield differences of up to 30% occur for the same value of annual demand. The same mean annual demand in Figure 2.6 has been modelled with different temporal patterns, achieved by randomly varying toilet and outdoor demand. The increase in annual yield occurs because demands for irrigation cause large draws on the tank for some days (irrigation days) and smaller demands on other days (toilet only). On days of increased demand, simulations with seasonality produce a larger drawdown in the rainwater tank, thereby increasing the capacity of the rainwater tank to capture water in the next rainfall event.

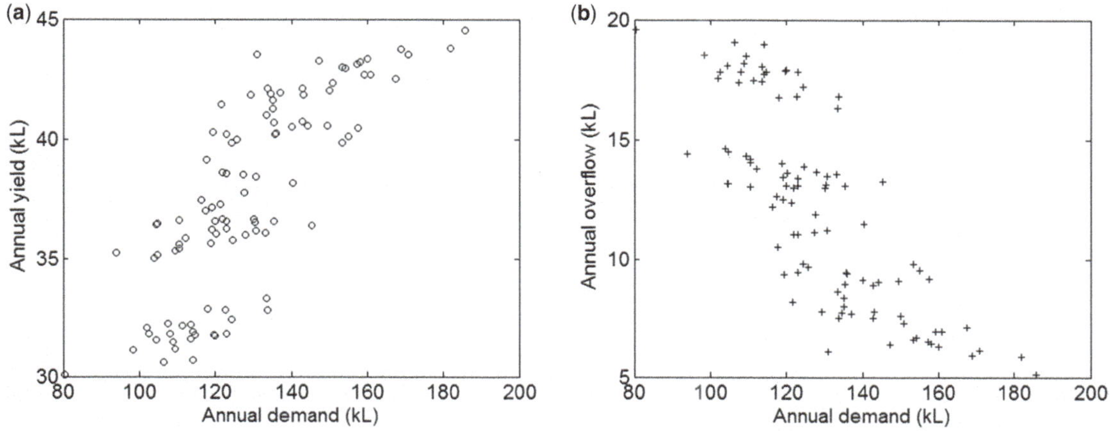

Figure 2.6 Annual yield (a) and overflow (b) for a 2.5 kL rainwater tank for different annual demands and different temporal demand patterns (all other parameters using mean values in Table 2.1).

Seasonality of demand is of particular importance for those studies which aim to determine 'optimal' tank size, or those which derive non-dimensional curves of performance based on mean annual demand and mean annual rainfall (Liaw & Tsai, 2004; Khastagir & Jayasuriya, 2010b; Palla *et al.* 2011; Palla *et al.* 2012; Campisano *et al.* 2013). Whilst the use of performance curves may be appropriate for constant or near constant demands, the variability in yield shown in Figure 2.6 for simulations with identical tank inflow and tank size suggests the approach must be used with care in cases where connected end uses show substantial seasonal variation.

2.4.5 Initial and continuing losses

Both initial loss and continuing loss will reduce the rainwater tank yield and overflow. Given the linear form described in Equations 2.1 to 2.3 for roof runoff and harvestable roof runoff, it is not surprising that for both types of losses the annual yield behaviour is closely linear, as shown in Figure 2.7 for the initial loss and in Figure 2.8 for the continuing loss. The effect of the initial loss parameter is basically to reduce the amount of rainfall for a given time-step to zero until the initial detention roof storage is full. It is important to note that the influence of the initial loss will vary depending on the distribution of rainfall. The more rainfall events there are per 100 mm of rain, the greater the effect of the initial loss.

A similar effect is observed in Figure 2.8 for the effect of continuing loss on tank yield and overflow. The continuing loss represents those losses associated with gutter overflowing, the effects of roof slope and orientation on splash losses, and evaporation during the storm event. The continuing loss coefficient is a more physically robust method to parameterise the runoff losses that are otherwise represented by a runoff coefficient in simpler models. Runoff coefficients used in daily time-step modelling typically range from 0.7–0.95 (Fewkes, 2000; Liaw & Tsai, 2004; Ghisi *et al.* 2007), with experimental measurements reporting values in the range 0.62 to 0.95 (Farreny *et al.* 2011). The equivalent runoff coefficients for the results modelled here (Figure 2.8) range between 0.63–0.89.

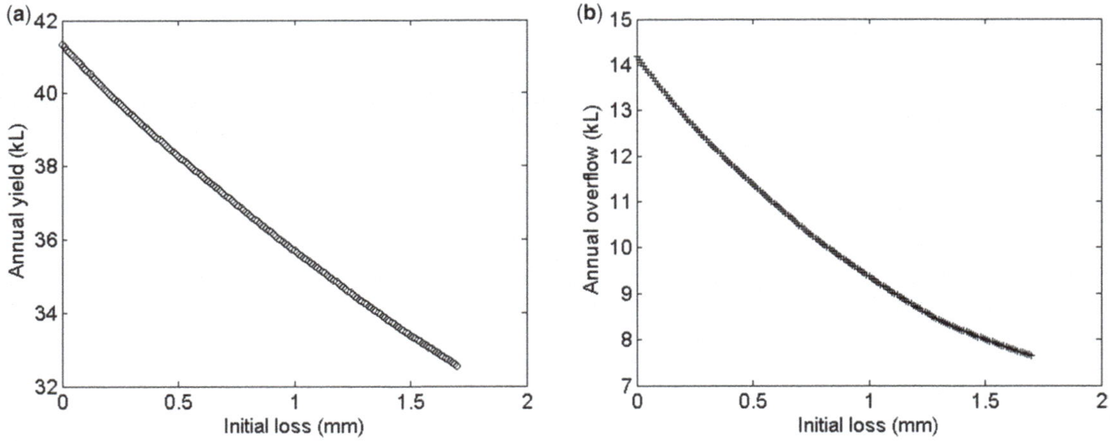

Figure 2.7 Annual yield (a) and overflow (b) for a 2.5 kL rainwater tank with different initial loss values (all other parameters using mean values in Table 2.1).

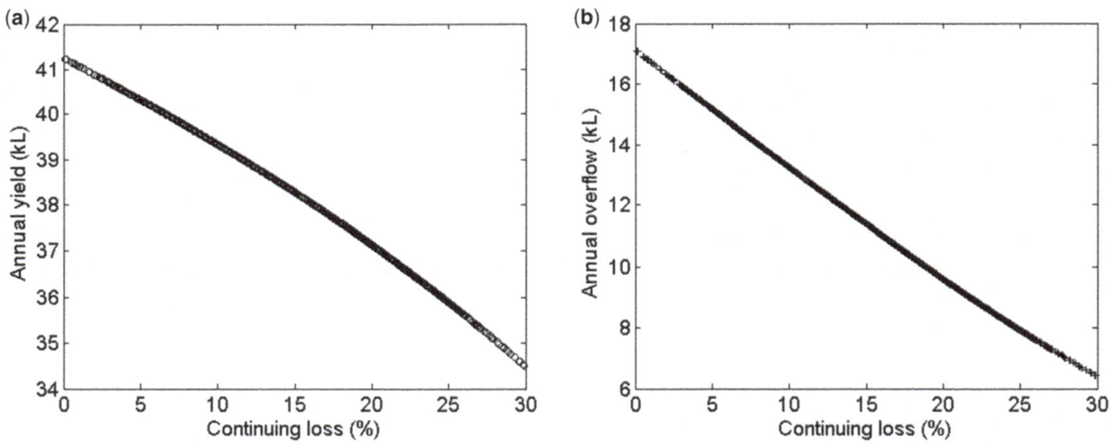

Figure 2.8 Annual yield (a) and overflow (b) for a rainwater tank for different continuing loss factors (all other parameters using mean values in Table 2.1).

2.4.6 Simulation length

To estimate volumetric reliability (percentage of demand met by the tank) of a rainwater tank system in locations with variable climates, rainfall records should be sufficiently long to capture the inter-annual and inter-decadal climate variability in the region of interest. This could mean using rainfall records of 50 years or greater. However, where computation times are an issue, use of a carefully selected portion of the rainfall record may produce acceptable results. Liaw and Tsai (2004) investigated impact of rainfall record length on the variation in the volumetric reliability for a series of rainwater tank simulations. Using one year records randomly taken from actual rainfall data, the volumetric reliability varied between 39 and 92%, with a mean of 66%. The variability depended on the mean rainfall for each rainfall year used for the simulation. As the rainfall record length (and hence simulation length) increased, the difference between the upper and lower bound of the variability estimates became smaller, reducing to less than 3% if the record length was 50 years. In a separate study, Mitchell (2007) investigated the differences in predicted yield using 1, 10 and 50-year rainfall periods in which all three periods had a similar mean annual rainfall. This study concluded that the use of 10-year periods which are representative of the long-term rainfall pattern produced small yield differences compared to those simulations using 50 years of rainfall. This difference was reduced even further if the rainwater storage was assumed to be empty at the beginning of the simulation.

However, in some cases, particularly in developing nations, the length of available historical rainfall records can be very short, and therefore, the length of simulations is limited by the length of available records rather than computational time. Approaches used to address this limitation have been briefly discussed in Section 2.3.1.

2.4.7 Validation

How accurate are the predictions from rainwater models? As described previously, the rainwater tank model is a water balance model with input parameters such as connected roof area and active tank volume that can be measured, household water demand that can be metered and rainfall and evaporation that can be measured at the site. Losses from the roof collection system need to be estimated based on results of studies measuring the runoff from various roof surfaces (see Section 2.3.3). To validate a rainwater tank model, observed (measured) rainwater use can be compared to predictions of rainwater use using measured input data. The measured data should also represent an independent dataset, that is, data that was not used in the development of the model. To provide a good test of model predictions, the data set should be long enough to capture the natural long-term variability of the rainfall pattern.

Millar *et al.* (2003) conducted an eight-month rain catch study in coastal South East Queensland on a house with three separate roof areas, each with a first flush device, connected to a rainwater tank with mains trickle top-up. Rain water inflow into the tank was firstly calculated from the monthly water balance of measured council mains inflow, measured tank water use by the household and the measured change in tank water level (and hence storage volume). Rainfall measured at roof level, combined with the sum of the projected roof areas, was used to calculate potential roof runoff so that rain catch efficiency could be calculated. These mass balance sums were complemented with a 10 minute time-step model of the loss of rainwater by retention from the leaky first flush devices, which in turn provided a prediction of rain water inflow into the tank, and hence allowed calculation of rain catch efficiency. An independently derived empirical relationship was used to describe the leakage rate from the first flush devices as a function of water volume (L) stored in their chambers. The predicted rain water inflow (catch efficiency 61%) agreed closely with the measured monthly rain water inflow (catch efficiency 58%), with a <3% error which could be reasonably attributed to initial and continuing loss not being taken into account in the water balance model. This provided confidence in the rain water inflow modelling, but the study did not extend to modelling tank yield behaviour.

Ward *et al.* (2012) compared the actual rainwater yield for a 300 person capacity building with the rainwater tanks plumbed to toilets against the yield predicted using a continuous simulation model developed by Roebuck and Ashley (2007). The model predicted that 46% of the water demand would be met by rainwater, whilst measured data indicated an average 35% of demand was met. The analysis demonstrated the sensitivity of the results to differences in building occupancy between that assumed in the design (300) and the actual occupancy (111).

South East Queensland became the subject of a number of house water use studies as a consequence of a severe drought (2006–2009) which led to rainwater tanks being mandated for new homes. Early modelling using the Coombes and Kuczera (2001) model with a pre-drought average water demand of 925 L/household/day (486 internal + 439 external) suggested that 70 kL/household/year of mains water could be saved by using rainwater from a 5 kL rainwater tank connected to 100 m^2 roof area of a house in Brisbane and plumbed to supply toilet, laundry and external demand (Le Muth, 2006). This value became established as the water savings target for South East Queensland (Department of Local Government & Planning, 2008).

The reasonableness of this target was assessed by undertaking a pairwise comparison of water utility household billing data for the year 2008 for 1100 homes with internally plumbed rainwater tanks and houses without rainwater tanks. The 1100 data pairs were distributed over three local government areas (LGA) in South East Queensland (Beal *et al.* 2012). In the absence of household occupancy data, household pairing was done on the basis of lot size and suburb information. The calculated mains water savings across the three LGA ranged from 20 to 95 kL/household/yr with an average of 50 kL/household/yr. Continuous simulation of average tank performance was undertaken using the model by Vieritz *et al.* (2007a,b) with an internal demand of 406 to 448 L/household/day assuming 30% was used for toilet and laundry (Willis *et al.* 2011) and rainfall data representative of each LGA. The predicted mains water saving ranged from 46 to 54 kL/household/yr with an average predicted value of 50 kL/household/year, which was well below the target of 70 kL/household/yr; but in good agreement with the 'estimated' average value. However, the large range in mains water savings calculated by the pairwise comparison (20 to 95 kL/hh/yr) was not reflected in the modelled water savings estimates across the LGA. This was attributed to: significant external water use in those LGA with less severe water restrictions (leading to higher tank water supply); variable penetration of water efficient appliances; and the lack of household occupancy data when matching data pairs (tank/no tank). In addition, there is some doubt that modelling an average tank to represent community water savings is a valid methodology. This issue will be addressed in the section 2.5.

A more detailed benchmark analysis of household billing data from 691 households with internally plumbed rainwater tanks was conducted by Chong *et al.* (2011) for four LGA in South East Queensland (including the three described in the Beal *et al.* 2012 study). The analysis compared average mains water usage by the known mandated rainwater users group to the respective council's average mains water consumption. The mains water billing records were supplemented with demographic survey data so that mains water usage could be normalised to household occupancy. This allowed calculation of water use per person for comparison with average mains water consumption per person for each LGA. The study found that the average annual mains water savings per household in 2009 was 59 kL/household/yr (ranging 24 to 89 across the region), and 58 kL/household/yr in 2010 (ranging from 40 to 81 across the region).

Twenty homes were then selected to represent the demographics within the South East Queensland study area (Chong *et al.* 2012). An audit of these homes established that only 45% had active tank volume ≥5 kL, and 15% had a connected roof area greater than the mandated 100 m^2. Modelling 18 of these homes using actual water use data for 2009, active tank volume and connected roof area indicated potential mains water savings ranging from 26 to 71 kL/household/yr (average 49) when using the model by Vieritz *et al.* (2007a,b), and 21 to 64 kL/household/yr (average 41) when using the UVQ model (Mitchell & Diaper, 2006). These values are 15%–30% lower than the 59 kL/household/yr calculated using 2009 billing data from 691 households (Chong *et al.* 2011). Further details regarding this auditing study are presented in Chapter 5.

The same twenty homes were intensively monitored over a 12-month period commencing in April 2011 (Umapathi *et al.* 2013). Rainwater use, main water top-up and total mains water use, as well as external water use was monitored. After infilling of missing data periods, projected mains water saving was found to be 40 kL/household/yr. However, this measured value can only be compared with the modelled averages of 41 to 49 using the earlier (2009) data from the Chong *et al.* (2012) study. Application of this 2011 data set to validate rainwater tank models was also constrained because on-site rainfall was not measured.

The above indicates that mains water savings from rainwater tanks have been shown to average 40 to 60 kL/household/yr for South East Queensland by modelling, while measurement results provided 40 kL/household/yr of mains water savings based on monitoring of only 20 homes. However, none of these studies provide data to allow a clear comparison of model predictions with their respective observed values to allow validation of the models.

Overall, it would be fair to say that there is comparatively little validation work relative to the number of models developed. Ward *et al.* (2012) noted the lack of longitudinal empirical studies that can be used to compare actual mains water savings from rainwater systems against modelled estimates. Umapathi *et al.* (2013) made a similar observation for South East Queensland.

Rainwater tank modelling is conceptually simple compared to the modelling of the water balance of paddocks. However, the cost and effort required to instrument and monitor occupied-household rainwater tanks have contributed to the lack of validation studies. One possible source of data for validating rainwater tank models may be the urban metabolism studies where water and energy flows are quantified (e.g., Beal *et al.* 2008). However, due to the logistical constraints, it is unlikely that data sets would be longer than a couple of years. Nevertheless, the expectation is that rainwater models validated with short-term data sets should be valid for the longer term, provided that important input values such as connected roof area and active tank volume do not change over that simulation period (e.g., tank and roof collection system is well maintained) and household demand is stationary. It would appear that rainwater modelling has far outpaced the studies which attempt to validate the predictions, and this represents a serious information gap.

2.5 UPSCALING OF RAINWATER TANK BEHAVIOUR TO MULTIPLE TANKS

A common method of analysing the behaviour of *multiple* rainwater tanks is a simple scaling up of the results from the continuous simulation of a *single* rainwater tank system. Therefore, for a cluster of rainwater tanks (i.e., a cluster of domestic rainwater tanks spread across a suburb or local government area), this approach uses a linear extrapolation based on the simulation of a single tank with average roof area, tank size, losses and water demand characteristics. This approach implies that the performance of all tanks is the same, or at least that the tank yield and overflow have a linear relationship to the tank parameters. Based on the yield curves versus roof area or tank volume shown in Figure 2.4 and Figure 2.5 respectively, it is clear that the assumption of a linear relationship is only valid for very narrow intervals of tank size or roof areas. It is important to note that the behaviour of the volumetric reliability (and annual overflow) is not entirely linear in relation to either tank sizes or roof area, as shown by Figure 2.9. The point where the system becomes linear depends on both the roof area and tank volume, and hence, even for a system where only those two parameters change, the system may not be linear, and hence linear up-scaling is likely to result in errors.

Rainwater tank audits in South East Queensland have demonstrated that water use demand, associated rainwater tank sizes and associated roof areas varied across the region. The variability in tank sizes and roof areas occurred despite the existence (at the time of the study) of mandatory tank sizes and connected roof areas (Biermann *et al.* 2012). For the same region, Beal *et al.* (2011) and Willis *et al.* (2009) also demonstrated that water demand varied between households due to occupancy rate, appliances used, and behaviour and attitudes of the residents to water use.

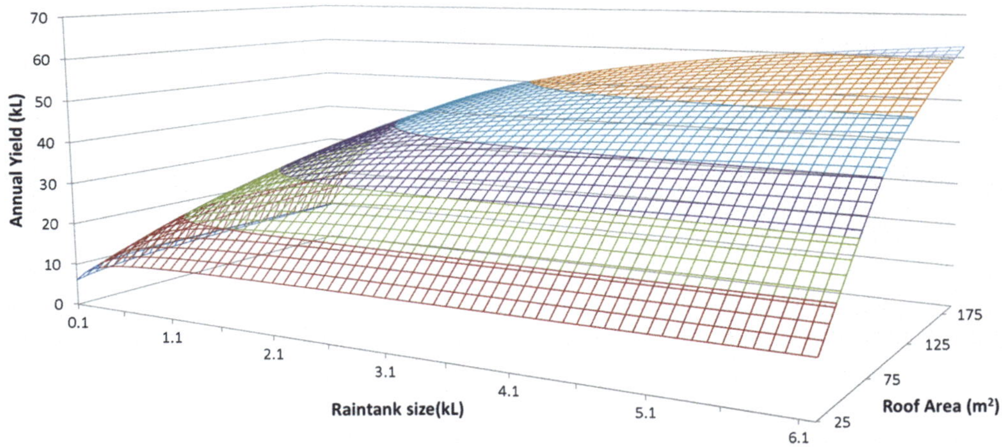

Figure 2.9 Predicted annual rainwater yield in Melbourne as a function of tank size and connected roof area for an annual demand of 115 kL/year for toilet and garden use.

The magnitude of this under and overestimation error by lumping the spatial variability under an 'average tank' was initially investigated by Mitchell *et al.* (2008). In that study, they modelled 1000 hypothetical households in Melbourne with stochastically allocated roof areas, tank sizes, roof losses (the same parameters listed in Table 2.1) to estimate the performance of a cluster of tanks. This performance was then compared to the performance of an 'average' tank system in which the roof area, tank size, demand, initial and continuing losses were the averages of the parameter values used in the 1000-house cluster. The difference in yield between the 1000 household cluster and the scaled up average tank was 14%, with the average tank method over estimating yield due to the non-linear behaviour described in Figure 2.4, Figure 2.5 and Figure 2.9. Other authors reported similar results for 5 locations in South East Queensland with different tank size, demands and climate patterns (Coultas *et al.* 2011; Maheepala *et al.* 2013). Errors varied between 10% (Gold Coast) to 22% (Ipswich), with a mean overestimation error of 15% for the five locations considered in the study. As can be seen in Figure 2.10, the simulated yields for individual household tanks show a substantial spread (from 10 to 100 kL/hh/yr), reflecting the variation in tank system characteristics and water end uses, both of which were informed by extensive inspection and survey data. For more details on input parameter values, see Chapter 5.

In addition to errors in yield estimation, Neumann *et al.* (2011) also investigated the impact of upscaling 'average tank' behaviour on overflow estimation. Given that use of average tank data *overestimates* the yield (i.e., it captures more rainfall), the authors found, as expected, that the use of average tanks *underestimated* the volume of overflow from the cluster. The difference was in the order of −37%.

Based on the non-linearity in yield with variation in roof areas and tank sizes, Neumann *et al.* (2011) suggested the use of geometric means instead of arithmetic means to represent the cluster. For the case where only roof area was varied, the use of geometric means greatly improved the estimation of yield. However when all the variables were considered, the use of geometric means resulted in even larger overestimations of yield. As such, the authors concluded that the use of 'average tank' to estimate yield and overflow is not recommended. Instead, modelling all tanks using stochastic simulation, or at least a few configurations of area/volume/demand so as to consider variability, is recommended to reduce the error in yield estimation.

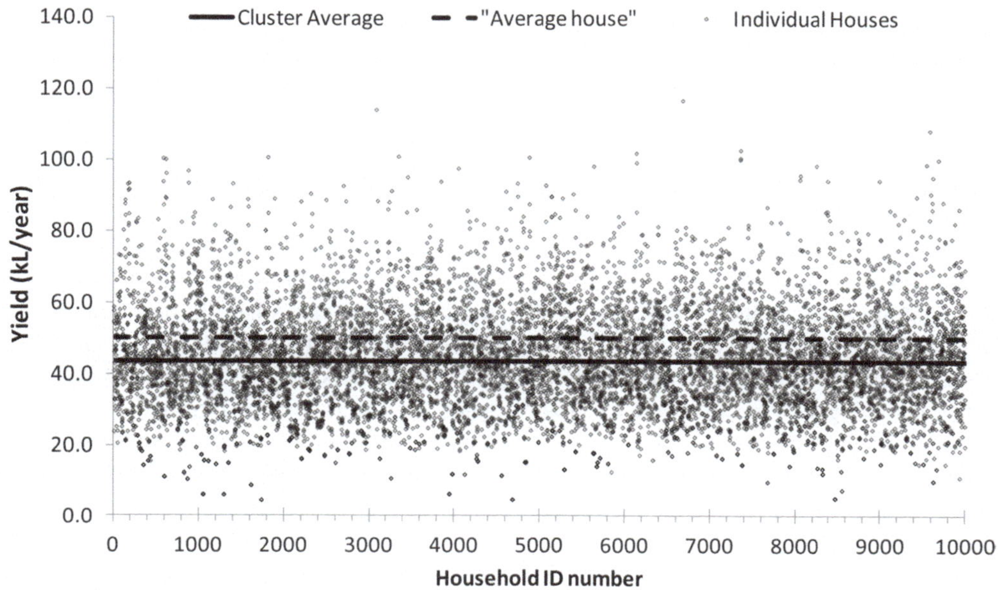

Figure 2.10 Annual tank yield for 10,000 individual South East Queensland households with various demands, connected roof areas and tank sizes. The mean yield for all tanks (cluster average) is shown as well as the average calculated from scaling up the 'average house' behavior (Maheepala *et al.* 2013).

2.6 CONCLUSIONS

The results presented in this chapter show that there are several models available for the simulation of yield and overflow from rainwater tanks. The existing models vary from simple spreadsheet models to very detailed process models which include hydraulic calculations, using time-steps from 6 minutes up to monthly. The most common modelling approach is the use of continuous simulations of the rainwater tank system over a significant period of time to minimize errors due to initial conditions, and to ensure a representation of the relevant rainfall patterns. Most, if not all, models include roof area, tank size, demand and losses as input variables that need to be considered in the model setup. In general, most models, irrespective of their level of complexity, exhibit similar patterns of yield and overflow in relation to the tank variables. Therefore, the most important consideration in modelling rainwater tanks is to ensure that the chosen model is appropriate for the scenario to be modelled so that all of the important processes of the system under study are captured, and that the assumptions behind the model are well understood. This is particularly important if the model used is a simplification based on non-dimensional parameters which are highly sensitive to demand pattern and rainfall distribution.

For the majority of practical applications in which a tank of reasonable size ($S > 1$ kL) has a connected roof area that provides a mean annual inflow (I in kL) such that the dimensionless ratio $S/I > 0.01$, daily time-step models provide sufficient accuracy, irrespective of the order of calculation spill rules, that is, YAS or YBS. If small tank sizes or very large roof areas are considered such that $S/I < 0.01$, shorter time-steps may be needed to avoid under or over estimation of yield, particularly if the model uses the YBS rule. The small time-step models, though more realistic in representing the physical processes, are more complex and more demanding of parameter values and rainfall data. However, they are the more appropriate models to use when the hydraulics of rainwater discharge (i.e., L/minute) are used in sizing the stormwater system.

Whilst input parameters for a rainwater model such as roof area and active tank volume can be directly *measured*, losses from the roof collection system, household water demand, and perhaps rainfall may need to be *estimated.* Adequate estimations of longer-term yield from rainwater tanks rely on rainfall records that are representative of the area of interest, and that ideally represent the long term average while preserving some of the long term variability. As discussed, local rainfall records are preferred as rainfall can vary significantly within short distances inside a catchment, while inherent rainfall measurement errors will also be a source of model error. Even if error-free rainfall data were available, different roof slopes and configurations influence the capture efficiency, and such losses have to be estimated. Careful consideration should also be given to the values used to represent the household water demand. The use of an annual or constant value can lead to underestimation of the tank yield if the demand has a seasonal component due to irrigation or other time varying demands. As water withdrawn from the tank effectively increases the available volume for water capture in the next rainfall event, increases in demand will lead to increases in yield (roof area permitting). Therefore, adequate representation of demand is important for the correct estimation of yield.

Since rainwater tank models are most useful to water planners when estimating potable water savings at a suburban scale or greater, quantifying the variability of the input parameters is also important, as variability in roof areas, demands and tank sizes can lead to significant variability in actual average yield. For large scale estimation involving several thousand households, the use of average parameters must be avoided as the non-linearity in yield response means the use of averaged parameters is likely to overestimate yields.

2.7 REFERENCES

Abdulla F. A. and Al-Shareef A. W. (2009). Roof rainwater harvesting systems for household water supply in Jordan. *Desalination,* **243**(1–3), 195–207.

Basinger M., Montalto F. and Lall U. (2010). A rainwater harvesting system reliability model based on nonparametric stochastic rainfall generator. *Journal of Hydrology*, **392**(3–4), 105–118. DOI: http://dx.doi.org/10.1016/j.jhydrol.2010.07.039.

Barry M. E. and Coombes P. J. (2006). Optimisation of Mains Trickle Topup Supply to Rainwater Tanks in an Urban Setting. Book of Proceedings, Grand Hyatt Melbourne, Melbourne, Australia.

Beal C. and Stewart R. A. (2011). South East Queensland Residential End Use Study: Final Report. Urban Water Security Research Alliance Technical Report No. 47. http://www.urbanwateralliance.org.au/publications/UWSRA-tr47.pdf (accessed 23 December 2014).

Beal C., Hood B., Gardner T., Lane J. and Christiansen C. (2008). Energy and water metabolism of a sustainable subdivision in South East QLD: A reality check. Proceedings of Enviro' 08. 5–7 May, Melbourne.

Beal C., Sharma A., Gardner T. and Chong M. (2012). A desktop analysis of potable water savings from internally plumbed rainwater tanks in South East Queensland. *Water Resource Management*, 26, 1577–1590.

Beal C. D., Makki A. and Stewart R. A. (2014). What does Rebounding Water Use Look Like? An Examination of Post-Drought and Post-Flood Water End-Use Demand in Queensland, Australia.

Biermann S., Sharma A., Chong M. N., Umapathi S. and Cook S. (2012). Assessment of the Physical Characteristics of Individual Household Rainwater Tank Systems in South East Queensland. Urban Water Security Research Alliance Technical Report No. 66. http://www.urbanwateralliance.org.au/publications/UWSRA-tr66.pdf (accessed 23 December 2014).

Burns M., Fletcher T., Hatt B., Ladson A. and Walsh C. (2010). Can allotment-scale rainwater harvesting manage urban flood risk and protect stream health. In: *Novatech 7th International Conference.* Lyon, France.

Burns M., Fletcher T., Duncan H., Hatt B., Ladson A. and Walsh C. (2012). The stormwater retention performance of rainwater tanks at the landparcel scale. In: *WSUD 2012: Water sensitve urban design; Building the water sensitve community; 7th international conference on water sensitive urban design, 21–23 February 2012, Melbourne Cricket Ground.* Engineers Australia.

Campbell D. (1986). WASH (Water and Sanitation for Health) Rainwater Information Center. *International Library Review*, **18**(1), 25–27. DOI: http://dx.doi.org/10.1016/S0020-7837(86)80005-X.

Campisano A., Cutore P. and Modica C. (2013). Performance of domestic rainwater harvesting systems at regional scale. World Environmental and Water Resources Congress 2013, 280–289. DOI: 10.1061/9780784412947.026.

Chong M. N., Umapathi S., Mankad A., Sharma A. and Gardner T. (2011). A Benchmark Analysis of Water Savings by Mandated Rainwater Tank Users in South East Queensland (Phase 2). Urban Water Security Research Alliance Technical Report No. 49, Queensland, Australia. http://www.urbanwateralliance.org.au/publications/UWSRA-tr49.pdf (accessed 23 December 2014).

Chong M. S., Sharma A., Umapathi S. and Cook S. (2012). Understanding the mains water saving from mandated rainwater tanks using water balance modelling and analysis with inputs from on-site audited parameters, Urban Water Security Research Alliance Technical Report No. 65. http://www.urbanwateralliance.org.au/publications/UWSRA-tr65.pdf (accessed 23 December 2014).

Clark R., Pezzaniti D. and Cresswell D. (2002). Watercress-community resource evaluation and simulation system-a tool for innovative urban water systems planning and design. In: Water Challenge: Balancing the Risks: Hydrology and Water Resources Symposium 2002, Institution of Engineers, Australia. p. 870.

Cook S., Sharma A. and Chong M. N. (2013). Performance analysis of a communal residential rainwater system for potable supply: A case study in Brisbane, Australia. *Water Resources Management*, 1–12. DOI: 10.1007/s11269-013-0443-8.

Coombes P. J. (2002). Chapter 6. An Allotment Water Balance Model. In: Rainwater Tanks Revisited: New Opportunities for Urban Water Cycle Management. PhD Thesis, School of Engineering, University of Newcastle, NSW, Australia.

Coombes P. and Barry M. (2007). The effect of selection of time-steps and average assumptions on the continuous simulation of rainwater harvesting strategies. *Water Science and Technology* **55**(4), 125–133.

Coombes P. and Kuczera G. (2001). Rainwater tank design for water supply and stormwater management. In: Stormwater Industry Association 2001 Regional Conference. Port Stephens, NSW, Australia.

Coombes P. J., Kuczera G. and Kalma J. D. (2003). Economic, water quantity and quality impacts from the use of a rainwater tank in the inner city. *Australian Journal of Water Resources*, **7**, 111–120.

Coultas E. H., Maheepala S. and Mirza F. (2011). Towards the quantification of water quantity and quality impacts of rainwater tanks in South East Queensland. 19th International Congress on Modelling and Simulation, Perth, Australia, 12–16 December 2011, http://mssanz.org.au/modsim2011/E12/coultas.pdf (accessed 23 December 2014).

Cowden, J. R., Watkins, D. W. Jr. and Mihelcic J. R. (2008). Stochastic rainfall modeling in West Africa: Parsimonious approaches for domestic rainwater harvesting assessment. *Journal of Hydrology*, **361**(1–2), 64–77. DOI: http://dx.doi.org/10.1016/j.jhydrol.2008.07.025.

CRC-CH (2005). Model for Urban Stormwater Improvement Conceptualisation (MUSIC) (Version 3). Series Editor. Series Model for Urban Stormwater Improvement Conceptualisation (MUSIC) (Version 3). Cooperative Research Centre for Catchment Hydrology, Melbourne, Australia.

Department of Local Government and Planning (2008). Queensland Development Code, MP 4.2 Water Savings Targets, Queensland Government, September 2008.

Dixon A., Butler D. and Fewkes A. (1999). Water saving potential of domestic water reuse systems using greywater and rainwater in combination. *Water Science and Technology*, **39**(5), 25–32. DOI: http://dx.doi.org/10.1016/S0273-1223(99)00083-9.

Domènech L. and Saurí D. (2011). A comparative appraisal of the use of rainwater harvesting in single and multi-family buildings of the Metropolitan Area of Barcelona (Spain): social experience, drinking water savings and economic costs. *Journal of Cleaner Production*, **19**(6–7), 598–608. DOI: http://dx.doi.org/10.1016/j.jclepro.2010.11.010.

Duncan H. P. and Mitchell V. (2008). A Stochastic Demand Generator for Domestic Water Use. Series Editor. Series A Stochastic Demand Generator for Domestic Water Use. Engineers Australia, Causal Productions, Modbury, SA, pp. 725–736.

Dziegielewski B., Kiefer J. C., Opitz E. M., Porter G. A., Lantz G. L., DeOreo W. B., Mayer, P. W. and Nelson J. O. (2000). Commercial and Institutional End Uses of Water Denver, Colorado. AWWA Research Foundation and the American Water Works Association. http://ufdc.ufl.edu/WC13511002/00001 (accessed 23 December 2014).

eWater Cooperative Research Centre (2011). Urban Developer User Guide. eWater Cooperative Research Centre, Canberra.

Farreny R., Morales-Pinzo T., Guisasola A., Taya C., Rieradevall J. and Gabarrell X. (2011). Roof selection for rainwater harvesting: Quantity and quality assessments in Spain. *Water Research (Oxford)*, **45**(10), 3245.

Fewkes A. (2000). Modelling the performance of rainwater collection systems: towards a generalised approach. *Urban Water*, **1**(4), 323–333. DOI: http://dx.doi.org/10.1016/S1462-0758(00)00026-1.

Fewkes A. and Butler D. (2000). Simulating the performance of rainwater collection and reuse systems using behavioural models. *Building Services Engineering Research and Technology*, **21**(2), 99–106. DOI: 10.1177/014362440002100204.

Ghisi E., Bressan D. L. and Martini M. (2007). Rainwater tank capacity and potential for potable water savings by using rainwater in the residential sector of Southeastern Brazil. *Building and Environment*, **42**(4), 1654–1666. DOI: http://dx.doi.org/10.1016/j.buildenv.2006.02.007.

Gould J. and Nissen-Petersen E. (1999). Rainwater Catchment Systems for Domestic Supply: Design, Construction and Implementation. Intermediate Technology Publications, London.

Guo Y. and Baetz B. (2007). Sizing of Rainwater Storage Units for Green Building Applications. *Journal of Hydrologic Engineering*, **12**(2), 197–205. DOI: 10.1061/(ASCE)1084-0699(2007)12:2(197).

Hanson L., Vogel R., Kirshen P. and Shanahan P. (2009). Generalized Storage-Reliability-Yield Equations for Rainwater Harvesting Systems. *World Environmental and Water Resources Congress 2009*, 1–10. DOI: 10.1061/41036(342)115.

Heneker T. M., Lambert M. F. and Kuczera G. (2001). A point rainfall model for risk-based design. *Journal of Hydrology*, **247**(1–2), 54–71.

Herrmann T. and Schmida U. (2000). Rainwater utilisation in Germany: efficiency, dimensioning, hydraulic and environmental aspects. *Urban Water*, **1**(4), 307–316. DOI: http://dx.doi.org/10.1016/S1462-0758(00)00024-8.

Howard G. and Bartram J. (2003). Domestic Water Quantity Service Level and Health. World Health Organization, Geneva, Switzerland.

Imteaz M. A., Shanableh A., Rahman A. and Ahsan A. (2011). Optimisation of rainwater tank design from large roofs: A case study in Melbourne, Australia. *Resources, Conservation and Recycling*, **55**(11), 1022–1029. DOI: http://dx.doi.org/10.1016/j.resconrec.2011.05.013.

Jenkins G. (2007). Use of continuous simulation for the selection of an appropriate urban rainwater tank. *Australian Journal of Water Resources*, **11**(2), 231.

Jenkins D. and Pearson F. (1978). Feasibility of rainwater collection systems in California. California Water Resources Center, University of California, Riverside, California, USA.

Jones M. P. and Hunt W. F. (2010). Performance of rainwater harvesting systems in the southeastern United States. *Resources, conservation and recycling*, **54**(10), 623.

Khastagir A. and Jayasuriya L. N. N. (2010a). Impacts of using rainwater tanks on stormwater harvesting and runoff quality. *Water Science and Technology*, **62**(2), 324–9. DOI: 10.2166/wst.2010.283.

Khastagir, A. and Jayasuriya, N. (2010b). Optimal sizing of rain water tanks for domestic water conservation. *Journal of Hydrology*, **381**(3–4), 181–188. DOI: http://dx.doi.org/10.1016/j.jhydrol.2009.11.040.

KRANRRC (2011). Harvesting For Sustainable Water Resources Development and Climate Change Adaptation in an Arid Region (N-E of Iran), Khorasan Razavi Agricultural & Natural Resources Research Center and The Institute For Global Environmental Strategies.

Le Muth N. (2006). Rainwater tank modelling investigation: South East Queensland regional water supply strategy, Document R.B15894.002.00.doc. Report prepared by WBS Oceanics Australia, Brisbane Office.

Liaw, C.-H. and Tsai, Y.-L. (2004). Optimum srorage volume of rooftop rain water harvesting systems for domestic use. *Journal of the American Water Resources Association*, **40**(4), 901–912. DOI: 10.1111/j.1752-1688.2004.tb01054.

Liu S., Makropoulos C., Butler D. and Memon F. (2006). An object based household water cycle model: concept and construction. The 7th International Conference on Urban Drainage and 4th International Conference on Water Sensitive Urban Design. Melbourne, Australia.

Maheepala S., Coultas E., Neumann L. and Sharma A. (2013). Quantification of Regional Scale Water Quantity and Quality Implications of Rainwater Tanks in South East Queensland. Urban Water Security Research Alliance Technical Report No. 104. http://www.urbanwateralliance.org.au/publications/UWSRA-tr104.pdf (accessed 23 December 2014).

Mayer P. W., DeOreo W. B., Opitz E. M, Kiefer J., Davis W., Dziegielewski B. and Nelson J. O. (1999). Residential End Uses of Water. Denver, Colorado. AWWA Research Foundation and the American Water Works Association. http://listserver.energy.ca.gov/appliances/2013rulemaking/documents/responses/Water_Appliances_12-AAER-2C/California_IOU_Response_to_CEC_Invitation_to_Participate-Toilets_and_Urinals_REFERENCES/Aquacraft_1999_Residential_End_Uses_of_Water_Study.pdf. http://listserver.energy.ca.gov/appliances/2013rulemaking/documents/responses/Water_Appliances_12-AAER-2C/California_IOU_Response_to_CEC_Invitation_to_Participate-Toilets_and_Urinals_REFERENCES/Aquacraft_1999_Residential_End_Uses_of_Water_Study.pdf (accessed 23 December 2014).

Millar G., Yu B. and Gardner T. (2003). Rainfall Catch Efficiency for Domestic Water Supply. In: *28th International Hydrology and Water Resources Symposium: About Water*, M. J. Boyd, J. E. Ball, M. K. Babister and J. Green (eds), Institution of Engineers, Australia. Barton, A.C.T., pp. 2.251–2.257.

Mitchell V. (2005). Aquacycle, a daily urban water balance model. *User Guide*.

Mitchell V., Siriwardene N., Duncan H. and Rahilly M. (2008). Investigating the Impact of Temporal and Spatial Lumping on Rainwater Tank System Modelling. Series Editor. Series Investigating the Impact of Temporal and Spatial Lumping on Rainwater Tank System Modelling. Modbury, SA, Engineers Australia; Causal Productionspp.

Mitchell V. G. (2007). How important is the selection of computational analysis method to the accuracy of rainwater tank behaviour modelling? *Hydrological processes*, **21**(21), 2850–2861. DOI: 10.1002/hyp.6499.

Mitchell V. G. and Diaper, C. (2006). Simulating the urban water and contaminant cycle. *Environmental Modelling & Software*, **21**(1), 129–134. DOI: http://dx.doi.org/10.1016/j.envsoft.2005.03.003.

Neumann L., Coultas E., Moglia M. and Mashford J. (2011). Errors in yield and overflow estimation in rainwater tank cluster modelling. In: 12th International Conference on Urban Drainage. Porto Alegre/Brazil, 11–16 September 2011.

Palla A., Gnecco I. and Lanza L. G. (2011). Non-dimensional design parameters and performance assessment of rainwater harvesting systems. *Journal of Hydrology*, **401**(1–2), 65.

Palla A., Gnecco I., Lanza L. G. and La Barbera P. (2012). Performance analysis of domestic rainwater harvesting systems under various European climate zones. Series Editor. Series Performance analysis of domestic rainwater harvesting systems under various European climate zones. 62, pp. 71–80.

Ragab R., Bromley J., Rosier P., Cooper J. D. and Gash J. H. C. (2003). Experimental study of water fluxes in a residential area: 1. Rainfall, roof runoff and evaporation: the effect of slope and aspect. *Hydrological Processes*, **17**(12), 2409–2422. DOI: 10.1002/hyp.1250.

Roberts P. (2004). Yarra Valley Water 2003. Appliance Stock and Usage Patterns Survey. Yarra Valley Water, Melbourne.

Roberts P. (2005). Yarra Valley Water 2004. Residential End Use Measurement Study. Yarra Valley Water, Melbourne.

Roebuck R. and Ashley R. (2007). Predicting the hydraulic and life-cycle cost performance of rainwater harvesting systems using a computer based modelling tool. *Water Practice & Technology*, **2**(02). http://www.iwaponline.com/wpt/002/0046/0020046.pdf (accessed 23 December 2014).

SILO Climate Data, http://www.longpaddock.qld.gov.au/silo/data_available.html (accessed 2 September 2014).

Sivakumaran S. and Aramaki T. (2010). Estimation of household water end use in Trincomalee, Sri Lanka. Water International 35 (1), 94–99. DOI:10.1080/02508060903533476. http://www.tandfonline.com/doi/full/10.1080/02508060903533476#.U2i4gfmSySo (accessed 23 December 2014).

Słyś D. (2009). Potential of rainwater utilization in residential housing in Poland. *Water and Environment Journal*, **23**(4), 318–325. DOI: 10.1111/j.1747-6593.2008.00159.

Smith D. and Shilley H. (2009). Residential Car Washing. New Data – and controversy – the state of Washington. *Stormwater*, September 2009, 1–6. http://www.stormh2o.com/SW/Articles/Residential_Car_Washing_7625.aspx (accessed 23 December 2014).

Su M.-D., Lin C.-H., Chang L.-F., Kang J.-L. and Lin M.-C. (2009). A probabilistic approach to rainwater harvesting systems design and evaluation. *Resources, conservation and recycling*, **53**(7), 393–399. http://www.stormh2o.com/SW/Articles/Residential_Car_Washing_7625.aspx (accessed 23 December 2014).

Umapathi S., Chong M. N. and Sharma A. K. (2013). Evaluation of plumbed rainwater tanks in households for sustainable water resource management: a real-time monitoring study. *Journal of Cleaner Production*, **42**, 204–214. DOI: http://dx.doi.org/10.1016/j.jclepro.2012.11.006.

Vaes G. and Berlamont J. (2001). The effect of rainwater storage tanks on design storms. *Urban Water*, **3**, 303–307.

Vaes G., Willems P. and Berlamont J. (2001). Rainfall input requirements for hydrological calculations. *Urban Water*, **3**(1–2), 107–112. DOI: http://dx.doi.org/10.1016/S1462-0758(01)00020-6.

Vaes G., Willems P. and Berlamont J. (2005). Areal rainfall correction coefficients for small urban catchments. *Atmospheric Research*, **77**(1–4), 48–59. DOI: http://dx.doi.org/10.1016/j.atmosres.2004.10.015.

Vieritz A., Baisden J. and Gardner T. (2007a). Rainwater Tank Manual, Queensland Department of Natural Resources and Mines, 2007 revision.

Vieritz A., Baisden J. and Gardner T. (2007b). Rainwater TANK model designed for use by urban planners. In: *OzWater'07*. Australian Water Association. Convention and Exhibition, Sydney, 4–8 March 2007.

Villarreal E. L. and Dixon A. (2005). Analysis of a rainwater collection system for domestic water supply in Ringdansen, Norrköping, Sweden. *Building and Environment*, **40**(9), 1174–1184. DOI: http://dx.doi.org/10.1016/j.buildenv.2004.10.018.

von Voigt T. M. and Mohajeri S. (2013). Water Demand Management Water End Use Measurements and Water Consumption Patterns in Tehran Households. Young Cities Research Briefs 09. http://opus4.kobv.de/opus4-tuberlin/files/3837/No09_Water+Demand+Management.pdf (accessed 23 December 2014)

Ward S., Memon F. A. and Butler D. (2012). Performance of a large building rainwater harvesting system. *Water Research*, **46**(16), 5127–5134. DOI: http://dx.doi.org/10.1016/j.watres.2012.06.043.

Water Corporation (2011). Perth residential water use study 2008/2009, Water Forever, Water Corporation, Western Australia.

Willis R. M., Stewart R. A., Talebpour M. R., Mousavinejad A., Jones S. and Giurco D. (2009). Revealing the impact of socio-demographic factors and efficient devices on end use water consumption: case of Gold Coast, Australia. In: *5th IWA Specialist Conference on Efficient Use and Management of Urban Water*. Australian Water Association.

Willis R. M., Stewart R. A., Williams P. R., Hacker C. H., Emmonds S. C. and Capati G. (2011). Residential potable and recycled water end uses in a dual reticulated supply system. *Desalination*, **273**(1–3), 201–211. doi:10.1016/j.desal.2011.01.022.

Chapter 3

Quantifying mains water savings from residential rainwater tanks

Cara D. Beal, Meng Nan Chong, Julian Fyfe, Andrea Turner and Ted Gardner

ABSTRACT

The premise for mandating rainwater tanks, or implementing expensive financial incentive programs such as rebates for the installation of tanks, is that meaningful savings can be achieved from the potable water supply. Whilst there is a depth of literature on modelled and theoretical savings from rainwater tanks, there are scant studies that seek to quantify the savings from multiple household 'real life' examples. The primary objective of this chapter is to present three different methods for assessing the savings in mains water use from regions of Australia that have recently installed rainwater tanks. We believe all three methods are internationally applicable. Various 'before' and 'after' comparisons are presented of mains water demand resulting from either rebated or mandated rainwater tank installations. Case Study 1 is a desktop assessment that uses water utility water billing data, lot sizes and presence or absence of an internally plumbed rainwater tank (RWT) to make pair-wise statistical inferences on the range of savings from internally plumbed tanks at a scale of local authority areas. Building on Case Study 1, Case Study 2 applies known household socio-demographic data matched with their household billing data to determine a benchmark water savings. Case Study 3 focuses on the water savings derived from a city-wide rainwater tank rebate program by comparing water consumption of each individual rebated household with a statistically-matched non-rebated household. Conclusions from all the studies focus on the need for sufficiently large sample sizes, known household occupancy, and the penetration of water-efficient appliances in households. Comparison of savings estimates highlighted the variability of rain tank yields between regions associated with climate, tank sizes and functionality, and connected end uses and roof area. Outdoor consumption is a critical end-use that will maximise savings. Thus factors such as potable water restrictions, lot size and behavioural cues (willingness to water use) are also important in determining water savings.

Keywords: modelling; harvesting; demand management; rainwater yield; tank rebates.

3.1 INTRODUCTION

3.1.1 Why quantify mains water savings?

As described in earlier chapters, the challenges in providing adequate and reliable sources of water for the urban community across the globe have been elevated due to potential climate change impacts and population growth. Various water strategies have been proposed to offset the demand from traditional potable water supplies, and one of the approaches for urban communities is the use of rainwater tanks. Rainwater tanks have enjoyed a recent resurgence in popularity due to widespread drought conditions and resultant water demand management schemes across many parts of Australia. In some Australian jurisdictions, rainwater tanks are mandated through building codes, requiring them to be installed in new developments or new buildings. The premise for mandating rainwater tanks, particularly those that supply internal household end-uses such as washing machines and toilets as well as external end-uses such as garden irrigation, is that significant volumetric savings of potable water can be achieved. These tanks will be referred as 'mandated rainwater tanks' in this chapter. The driving objective of installing a rainwater tank is essentially to reduce mains water demand through the on-site harvesting (collection, storage and use) of rainwater. In general, the underlying assumption that rainwater yield will adequately supplement household demand has yet to be convincingly demonstrated across a variety of rainwater tank configurations and regional settings. Additionally, yield (kL/household/year) will vary both spatially and temporally across a given region. Thus, the actual harvesting behaviour of rainwater tanks – internally and/or externally connected – is critical in terms of setting realistic and achievable water demand management goals.

As well as mandated rainwater tanks, many states in Australia have introduced rebate programs whereby residents are offered a financial incentive to voluntarily install a rainwater tank for internal (washing machine and or toilets) and/or external garden supply. The amount of rebate is linked to the overall internal and external end-uses. These 'rebated rainwater tanks' as they will be referred to in this chapter, have been very popular across Australia. However the success of these rebate schemes, and the justification for future similar programs, needs to be examined objectively.

Accurate data on the actual *mains water savings* from rainwater tanks is critical for assessing the range of urban water management strategies, along with improving the accuracy of demand (and revenue) forecasting models for future infrastructure planning and optimisation. This chapter addresses many of these questions by presenting three case studies which explore various approaches in determining mains water savings from residential water tanks used for either internal and/or external end-uses.

3.1.2 Previous studies on mains water savings

There are numerous studies that report predicted yields and optimal design criteria (e.g., tank size, roof catchment) for rainwater tank systems, based on water balance simulations and probabilistic methods (Campisano & Modica, 2012). Ideally however, a combination of field and desktop methods using smart metering, historical water billing records, long-term climate data, household demographics, household water end-use surveys and rainwater tank system audits, would capture of all the variables that determine rainwater yield and potable water savings. Such a holistic study would also include the energy demand associated with rainwater tanks (Siems *et al.* 2013). This level of detailed data is rarely available, thus a mixed method approach is often adopted whereby both empirical and modelled data are used to determine potable water savings. Some examples include Chong *et al.* (2012), Fyfe *et al.* (2011) and Sydney Water (2008). Others have used statistical methods with 'before and after' retrofit comparisons to identify mains water savings (e.g., Beal *et al.* 2011a; McBeth, 2011; Ghisi *et al.* 2007a; Turner *et al.* 2005). A summary of selected studies is presented in Table 3.1.

Table 3.1 Summary of some previous studies on mains water savings from rainwater tanks.

Location	Approach	Reported savings (kL/ household /year)	Comments	Reference
Various capital cities across Australia	PURRS model (Probabilistic Urban Rainwater and wastewater Reuse Simulator model)	42–90 kL	Modelling assumed rainwater was used for hot water, toilet, laundry and outdoor end-uses.	Coombes and Kuczera (2003)
Various capital cities across Australia	Water balance model	42 kL (externally plumbed only) 71 kL (internal and external)	Supply scenarios modelled included all internal end-uses (excluding cold water to kitchen and bathroom) or external only.	Marsden Jacob Associates (2007)
Sydney, Australia	Comparison between BASIX (homes that have undergone a water efficiency retrofit program) and Non-BASIX homes.	Approx. 36 kL	Volumetric savings estimated. No 'before' dataset to compare with.	Sydney Water (2008); Sullivan and Wilson (2009).
Sydney, Newcastle, Wollongong, Australia	Estimated savings from end-uses connected to rain tanks. Continuous simulation water balance model.	21–57 kL depending on tank size and location.	The end-use rates (L/p/d) were based on Sydney Water's recommended demand rates, not measured datasets.	Eroksuz and Rahman (2010)
Northern NSW, Australia	Similar to BASIX approach. Statistical analysis.	27 kL	Rebated tanks were examined.	McBeth (2011)
Sydney, Australia	Simulation modelling.	45–58 kL (irrigation use) 27–34 kL (internal and outdoor)	Assumed water-efficient appliances and fixtures. Assumed a 5 kL tank size.	Hajani and Rahman (2013)
South east Brazil	Desktop assessment using demand data and population statistics.	16–175 kL	Volumetric savings estimated from % savings, water demand and average household occupancy.	Ghisi *et al.* (2007)

3.1.3 Chapter objectives and scope

The primary objective of this chapter is to present three different, internationally-applicable methods for assessing the savings in mains water use from rainwater tanks. The case studies selected describe both theoretical modelling approaches and empirical data from in-situ measurements. They examine both mandated tanks and rebated tanks. The methodologies presented in this chapter have been chosen based on their global relevance, and thus can be applied in any part of the world where rainwater tanks are used as part of integrated urban water management to reduce reliance on mains water supply.

The different circumstances under which residential rainwater tanks are typically installed are presented in Figure 3.1. Case Studies 1 and 2 are concerned with water savings from mandated tanks, whilst Case Study 3 examines savings from tanks installed voluntarily by the householder under government rebate schemes. The methodology for each case study is discussed in detail with only a summary of results and discussion as it is the approach used, rather than the specific quantum of savings from RWTs, that is the focus of this chapter. The extended data on mains water savings and rainwater tank yields for various locations around the world can be accessed in various publications (e.g., Adeyeye, 2014; Ghisi *et al.* 2007).

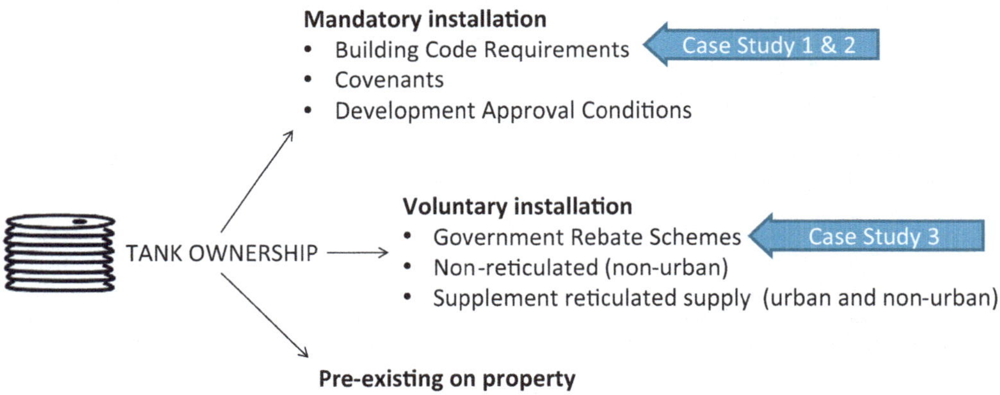

Figure 3.1 Categories of tank installation examined in the three case studies.

3.2 CASE STUDY 1 – DESKTOP ANALYSIS OF MAINS WATER SAVINGS

3.2.1 Background

In south-east Queensland (SEQ), Australia, the challenges in providing adequate and reliable sources of water for the urban community has prompted state water planning authorities to develop sustainable water planning strategies and management practices to address such important urban water issues. Different water strategies have been proposed, and one of the approaches for urban communities in SEQ is the installation of rainwater tanks. These have become an integral feature of the vast majority of detached dwellings in SEQ, either through the WaterWise Rebate Scheme commenced in 2006 (Walton & Homes, 2009) or mandated through the Queensland Development Code (QDC) MP 4.2 – *Water savings targets* (DIP, 2007). Although the requirement to achieve a mains water saving target of 70 kL per year (usually via a rainwater tank) has now been removed (as of 1st February 2013) from Queensland legislation following a change of government, there was considerable research effort directed at developing methods to assess the actual mains water savings from rainwater tanks. This knowledge provides methodologies that allow similar mandated programs on rainwater tank to be assessed by scientifically rigorous analysis of data in other regions of the world.

Mandated rainwater tanks (MRT) clearly played an important role in achieving the sustained reduction in demand, although quantifying this was not a simple task due to an absence in actual consumption data from newly constructed homes. In addition, there was difficulty in separating out the role that other water demand management strategies contributed to demand reduction. Case Study 1 presents the first of a staged methodological approach to investigate the mains water savings that can be achieved from mandated rainwater tanks. The second staged methodology to investigate mains water savings is discussed in Case Study 2.

3.2.1.1 Research objectives and hypothesis

The aim of the research was to conduct a desktop assessment using statistical analysis of the potential mains water reductions from internally plumbed rainwater tanks in new developments in the SEQ, Australia. A further objective of this desktop approach was to provide baseline data for further experimental work. The following hypotheses were used to frame the research methods:

- *Null hypothesis* (H_0)*:* Water consumption in houses with mandated rainwater tanks (MRT) is not significantly different from the water consumption for houses without rainwater tanks (No Tank).
- *Alternative hypothesis* (H_1)*:* Water consumption in houses with mandated rainwater tanks is significantly different from the water consumption for houses without rainwater tanks (No Tank).

3.2.2 Methods

3.2.2.1 Site locations and data collection

Three SEQ local government areas (LGAs) were included in this study: Pine Rivers City Council (now amalgamated into Moreton Bay Regional Council), Gold Coast City Council and Redland City Council (Figure 3.2). These local government areas were chosen as they represented a good cross-section of the socio-economic and climatic conditions in SEQ. At the last available Australia Bureau of Statistics (ABS) census in 2006, these regions collectively comprised almost 40% of the SEQ population (DIP, 2009). Further, they represented around a third of the areas marked for future greenfield development in the SEQ Regional Plan (DIP, 2009). From the council databases provided, approximately 8300 (Pine Rivers), 9100 (Gold Coast) and 1000 (Redland) new dwellings were selected, which had been approved (but not necessarily constructed) since January 1st 2007 when the QDC MP 4.2 requirements became active.

Potable water consumption data for 2008 was obtained from the water billing section of each council. Some councils had difficulties in providing complete datasets of water billings for post-2007 approved dwellings. Once the data was collected from the councils, the method described in 3.2.2.2 was applied to isolate post-2007 constructed properties with mandated rainwater tanks.

The rainfall data for Case Study 1 and 2 (both yearly and long-term) is presented in Table 3.2. For Case Study 1, the year 2008 is of interest for the regions of Pine Rivers, Gold Coast and Redlands.

*Note: Pine Rivers and Caboolture are part of the Moreton Bay Regional Council

Figure 3.2 Location of the local government areas in SEQ used for desktop study (http://en.wikipedia.org/wiki/South_East_Queensland).

Table 3.2 Rainfall data for the studies regions examined in Case Studies 1 and 2.

Region[1]	Annual rainfall in 2008 (mm)	Annual rainfall in 2009 (mm)	Annual rainfall in 2010 (mm)	Long-term annual rainfall (mm)
Pine Rivers	1201	1367	1996	1131
Gold Coast	1766	1548	2320	1372
Redland	1348	1213	1834	1192
Caboolture	1525	1971	2118	1219

[1]data taken from Bureau of Meteorology weather stations available from Climate Data Online (http://www.bom.gov.au/climate/data/).

3.2.2.2 Identification of sample cohorts

In Case Study 1, properties approved and constructed post-2007 were not able to be directly identified in the raw datasets provided. Therefore a methodology was developed to extract the relevant information from typically available household databases. The main steps and assumptions in the analysis are listed below.

(1) The raw data set was filtered for duplicate and ambiguous data (e.g., incomplete, repeated records) using MS Access™ and MS Excel™ software. This data set was then filtered for the Land Use Code representing a Class 1 building as per the Queensland Development Code mandate requirements.

Only single, detached dwellings were selected, which represent up to 60% of SEQ regional water consumption (MWH, 2007).

(2) No Tank and MRT properties were isolated by using property registration (i.e., cadastral data), meter installation and connection dates where available. In the case of Gold Coast Water, the data was supplied in predefined No Tank and MRT samples.

(3) All properties that were identified as having received a rainwater tank state government rebate were excluded. Some councils also had a field that indicated a local council rebated water tank (e.g., Gold Coast). Excluding rebated properties could only be performed where Lot and Plan data (a unique cadastral identifier for the house allotment) was supplied by council. By excluding rebated tank properties, the differences in water use between No Tank and MRT houses were likely to be maximised. Excluding rebated properties could only be performed for Pine Rivers (n = 12,342 rebated properties) and Redlands (n = 4994 rebated properties) where Lot and Plan data was supplied by council. MRT and No Tank data were divided into two lot size categories: ≤700 m^2 and >700 m^2 by filtering for lot size. The value of 700 m^2 represented the median (50th percentile) allotment size identified after developing a probability distribution curve for all councils. Water consumption between No Tank and MRT homes was analysed for the two lot size categories, where sample size allowed this. There was a trend for larger allotments to use more water, but as only limited statistically significant results occurred between regions, this data is neither presented nor discussed further in this chapter.

(4) No Tank and MRT properties were further grouped into suburbs within each lot size category. However, sample size was generally insufficient for a suburb grouping.

Only consumption data recorded in the 2008 calendar year was used for comparative analysis. This method reduced the likelihood of selecting new developments that were constructed after January 1st 2007 and were yet to be fully occupied, or developments that were approved *before* January 1 2007 but *constructed after* 2007. Billing data provided for all regions included information on the date of water meter installation and/or the date of house construction. This information was useful when differentiating between properties which were constructed pre- and post-2007. Unlike previous studies such as Turner *et al.* (2005) and the Sydney Water BASIX study (Sydney Water, 2008), a comparison of identified properties using known household occupancy data was not possible for this analysis. The final number of samples for the MRT and No Tank groups are shown in Table 3.3.

Table 3.3 Number of MRT and No Tank properties for each region of interest for pairing.

Region	MRT homes	No Tank homes
	(number of samples)	
Pine Rivers	648	32,718
Gold Coast	422	2993
Redland	112	33,117
Total	**1182**	**68,828**

3.2.2.3 Statistical analysis

Mean values were used to statistically compare water consumption for this desktop study using a two-tailed, independent Student's t Tests in MS Excel™ and SPSS© software packages. Although the distribution curves were skewed slightly to the right, the t-Test is more robust than other tests (e.g., z Test) to deviations from normality (Johnson, 1978). With the exception of comparing combined totals for water use, the *t*-Test was based on equal variance and equal samples between the No Tank and MRT properties. Further statistical descriptions can be found in Beal *et al.* (2011a).

3.2.2.4 Overcoming limitations with data availability

3.2.2.4.1 Bottom-up end use calculations

The examination of savings from mandated rainwater tanks is not an easy task, particularly given the paucity (or inaccessibility) of specific council data required for a pairwise analysis. Therefore, two other approaches have also been used to assist in evaluating and providing a 'ball park' reality check on the results of the desktop analysis. These 'cross-checks' help to set the bounds of likely potable water savings for the different end use assumptions (e.g., with and without garden irrigation).

An estimation of expected mains reductions from internally plumbed rainwater tanks was made based on internal water use data from the Gold Coast end use study in the Pimpama-Coomera region (Willis *et al.* 2010) and from a recent SEQ end use study (Beal & Stewart, 2011). These studies reported a range of consumption data for various internal fixtures including the washing machine (cold water tap) and toilet. The combined water demand from these internally connected end uses provide a baseline estimation of indoor mains water savings from a MRT (Figure 3.3). Note that whilst the statistical analysis assumes a contribution from outdoor water use, the two cross-checking approaches only consider indoor end uses. Predicting outdoor end uses with any degree of accuracy is extremely difficult due to the number of factors influencing its use (e.g., climate, lot size, garden area, turf area, soil type, personal behaviour and council water restrictions). Indoor water consumption is considered a far more homogenous dataset that has less variability and is therefore easier to predict (Makki *et al.* 2011; Fox *et al.* 2009). End use studies by Willis *et al.* (2011) and Beal and Stewart (2011) suggest external water use was atypically low during the period of our tank studies (2008–9).

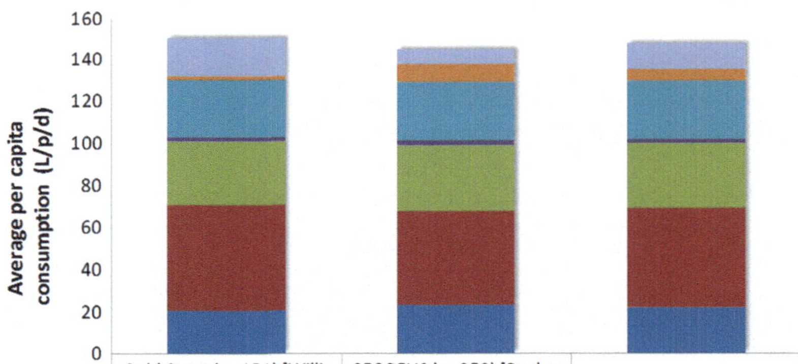

	Gold Coast (n=151) [Willis et al. 2009]	SEQREUS (n=252) [Beal et al. 2011]	Average
External	18.6	7.0	12.8
Leaks	2.1	9.0	5.6
Taps	27	27.5	27.3
Dish washer	2.2	2.5	2.4
Clothes washer	30	31.0	30.5
Shower / bath	50	44.5	47.3
Toilet	21	23.7	22.4

Figure 3.3 Summary of measured internal water end uses from a number of recent SEQ end use studies.

3.2.2.4.2 Rainwater tank modelling

The Rainwater TANK model is an Excel-based spreadsheet model linked to a FORTRAN executable file (Vieritz *et al.* 2007). Rainwater TANK simulates the capture of rain by an urban roof. The primary aim of the model is to assess the ability of the rainwater tank to meet the water demand of connected end uses. For the purposes of this study, TANK provided a first approximation of the supply performance of rainwater tanks for comparison with the statistical desktop results. Rainfall years that were used in the modelling are provided in Table 3.4.

Table 3.4 Rainfall input data used for rainwater TANK modelling.

Region	Rainfall scenario	Yearly Rain (mm)
Pine Rivers	Dry (2006–7)	850
	Av (28 yrs)	1131
	Wet (2008)	1201
Gold Coast	Dry (2006–7)	1193
	Av (28 yrs)	1372
	Wet (2008)	1766
Redland	Dry (2006–7)	956
	Av (28 yrs)	1192
	Wet (2008)	1348

3.2.3 Results

There was a significant reduction ($p < 0.05$) in mains water consumption for MRT properties in all regions (Figure 3.4). Mains water consumption for No Tank homes averaged 197.8 kL/household/year compared with an average of 148.3 kL/household/year for MRT homes. Within regions, this trend continued with Gold Coast and Redland No Tank homes consuming the most mains water at an average of 246.9 and 184.5 kL/household/year, respectively. These two council areas were operating under relaxed outdoor watering restrictions in 2008. Mains water savings varied markedly across regions, with values ranging from 20 to 95 kL/household/year, with an average of 50 kL/household/year (Figure 3.4).

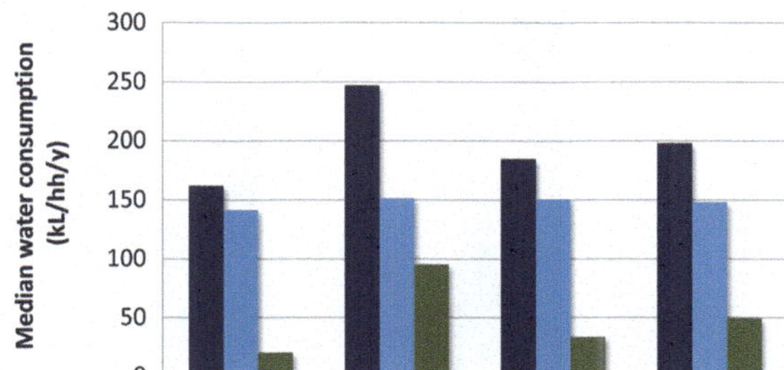

	Pine Rivers	Gold Coast	Redland	Combined
No Tank	162.0	246.9	184.5	197.8
MRT	142.0	151.7	151.4	148.3
Mains water reductions	20.0	95.2	33.2	49.5

Figure 3.4 Results from pairwise statistical analysis of water consumption from MRT and No Tank properties.

The result of the two approaches used to cross-check the statistical analyses are presented in Table 3.5. Both of these approaches only consider indoor water consumption. Assuming an average household occupancy of three people (Australian Bureau of Statistics, 2006) in new developments, tanks supplying water efficient toilets and washing machines should reduce mains water use in the range of 43 to 46 kL/household/year, regardless of outdoor uses of rainwater. Notwithstanding the high estimated savings from the Gold Coast, where there were no restrictions on external water use in 2008, the other two council areas had lower than expected mains reductions when cross-checking them with results from predicted indoor reductions, as shown in Table 3.5.

Table 3.5 Summary of mains water use reductions for 2008 compared with two independent estimates of the water savings.

Region	**Desktop Analysis of water meter records: Mean values**	**Desktop analysis of water meter records: median values**	**Water consumption based on regional end use studies (internal only)**	**TANK model predictions (internal only)**
		(kL/household/year)		
Pine Rivers	20	28		49
Gold Coast	95	52	43 to 46	54
Redland	33	41		46
Average reduction	**50**	**40**	**44.5**	**50**

A non parametric rank test was used to statistically analyse the mains water reductions between properties that were under high water restrictions compared to those under low or no water restrictions. The results show that water consumption in No Tank homes located in low or no restrictions (Gold Coast and Redland) was statistically higher ($p < 0.05$) than for No Tank homes in high water restriction areas (Pine Rivers) (Figure 3.5). This will be discussed in more detail in the section below.

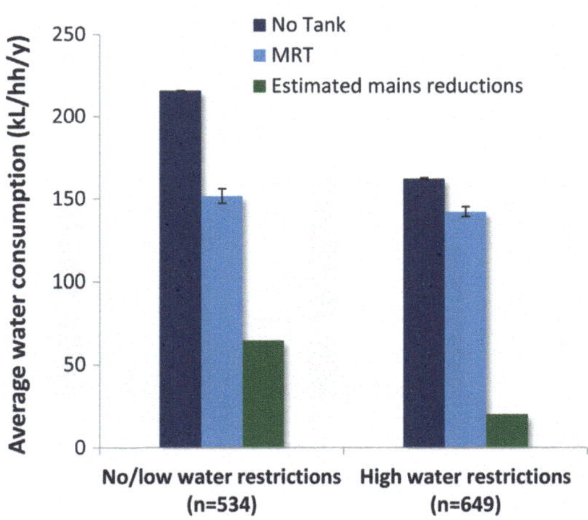

Figure 3.5 Comparison between water consumption and estimated mains reductions for regions with high and low/no water restrictions.

3.2.4 Discussion and implications

Comparative analysis of mains water consumption between No Tank and MRT properties overall clearly showed that consumption was greater for homes without MRT. There are two main factors that are likely to be influencing the lower estimated reductions calculated from the statistical analyses: the influence of water restrictions during the period of analysis (discussed below) and the limitations of the council billing data used to determine MRT from No Tank homes (Section 3.2.5).

3.2.4.1 Influence of water restrictions

The influence of water restrictions is illustrated in Figure 3.5, which showed smaller differences in water consumption between MRT and No Tank properties in those regions with high-level water restrictions (i.e., no or low outdoor watering). Conversely, the differences in mains water use (i.e., the savings) is greater for those homes located in low or no water restriction areas where rainwater had the opportunity to substitute for garden water use otherwise supplied by mains water (Figure 3.5). The strictest water restrictions in 2008 occurred in the Moreton Bay Regional Council area, which encompasses Pine Rivers. Outdoor watering using mains water was limited to hand held bucket or watering cans. This included newly established gardens or lawns. In contrast, Gold Coast City Council had no restrictions between February and November 2008 due to high rainfall events overtopping their main water supply dam (Hinze Dam). Consequently, there was no limitation to outdoor watering with mains water. Properties in Redland Shire Council were on Level 2 restrictions which allowed outdoor watering using mains water to occur with a hand held hose both for established and new gardens. Daily per capita water use for No Tank properties all exceeded the 2008 average value for areas in SEQ under water restrictions (data not shown). Conversely, average per capita water use from households with MRT was ~20% less, and was similar to the average water use for restricted SEQ regions at that time which was 128 L/p/day (equivalent to about 131 kL/household/year). When compared to homes with a MRT, high water usage from No Tank homes would maximise the main water savings able to be achieved from rainwater tanks. However, if people are frugal in their water use due to water restrictions and demand management strategies (as was the case during our 2008 study) this will compress the differences in mains water use between tank/no tank homes, and hence minimise the potential for mains water savings from rainwater substitution.

3.2.5 Limitations of Case Study 1

Although all local government regions could be confidently divided into the two groups of No Tank and MRT, and then subsequently paired for statistical testing, there still remained some important information that could not be extracted from the data provided. This absence of information for some or all of the regions created the following limitations:

- Separating the billing data into MRT and No Tank subsamples could only be done using assumptions and proxy data, as detailed in the methods section;
- Separating out the influence of MRT from other water restriction influences was not possible;
- Details on critical factors that influence residential water consumption (garden size, water efficient fixtures etc.) could not be fully taken into account; and
- Details on socio-demographic factors such as household occupancy, family makeup and income could not be controlled for in the analysis.

These limitations are likely to have had some influence on the outcomes from the analysis. Without specific knowledge of household occupancy, household water demand cannot be properly controlled for.

For example, a single person No Tank household with low total water use volumes may be matched with a six person MRT family using very high volumes of water, thereby reducing the estimated contribution from rainwater tanks. The same argument follows for controlling for outdoor water demand if garden sizes (as opposed to allotment sizes) were known. Although MRT and No Tank homes were paired based on two lot size categories, there were no strong trends in the differences in water consumption and savings between lot size categories. However, a large allotment does not necessarily translate into a large, irrigated garden area. With this knowledge, external water demand can be controlled for to some extent, although external water uses are notoriously difficult to quantify (Beal & Stewart, 2011; Wang, 2011).

Finally, the role of water-efficient household stock such as low water use (5 star rated) washing machines, low-flow shower roses and tap flow controllers have not been able to be quantified in this Case Study. Research shows that these efficient features and fixtures can be very effective in reducing domestic water consumption (Willis *et al.* 2010; Beal *et al.* 2011b). It is likely that had such data been available for the MRT properties, one would have seen a greater difference in water consumption from MRT (more savings) and No Tank properties.

3.2.6 Concluding remarks

Whilst it is clear that internally plumbed rainwater tanks will offset mains water demand, the annual volume of that offset is highly variable, and influenced by a range of factors including demand for rainwater (e.g., from external and internal water uses), rainfall, demographic factors (e.g., household size and waterwise awareness) and water efficient household appliances/fixtures. Additionally, the timing of the analysis with a drought-focussed community, where external water use was conservative due to water restrictions, made the differences due to tank supply options harder to detect. Any water saving features in new homes that are not present in pre-2007 homes will reduce mains water use and hence increase apparent rainwater contributions. Similarly, any systemic population difference between post- and pre-2007 homes will affect mains water use and hence bias the calculated tank water contributions.

Despite these acknowledged limitations, the desktop methodology presented in this Case Study has the advantages of providing a base range of savings for relatively low cost experimental inputs (e.g., no field trial costs or modelling work required). It exploits available datasets and uses a basic statistic pairwise approach to estimate the likely range of savings. At the least, it provides a 'first pass test' to estimate the range of achievable savings expected from mandating rainwater tanks in any given area. This may moderate the expectations of potable water saving from mandating rainwater tanks in other regions.

3.3 CASE STUDY 2 – BENCHMARK ANALYSIS OF MAINS WATER SAVINGS

3.3.1 Background

To improve the validity of the Case Study 1 approach, there were a number of recommended additional steps to take for a second stage assessment of mains water savings. These were focussed around improving the lack of specific knowledge on MRT and No Tank homes, socio-demographic data, knowledge of rebated tank installations and household water stock (e.g., presence or absence of water-efficient stock). This work is now presented in the following section.

3.3.2 Research aims

The aim of Case Study 2 is to provide a sound and methodical approach to validating the MRT savings target of 70 kL/hh/yr under the QDC MP 4.2 for SEQ. It will provide some contextual understanding for results discussed in Case Study 1 in achieving the mains water savings through MRT. A further aim of Case Study 2 is to document a methodological approach that can be applied globally to estimate, and subsequently justify, water supply from rainwater tanks as an alternative water supply source.

3.3.3 Methods

3.3.3.1 Data collection and participant details

The study area comprised four LGAs in the SEQ region, three of which were examined in Case Study 1: Caboolture, Pine Rivers, Redland and Gold Coast (Figure 3.2). The 2006 Australian Census described these four LGAs as containing over 40% of SEQ urban population (DIP, 2009). These regions were selected due to the availability of necessary data for this benchmark analysis. Only properties built after 2007 were included in the study to ensure only households with MRT were analysed.

A phone survey was conducted between July and August 2010 to understand the potential contribution of biophysical and social factors in achieving water saving targets as identified in Case Study 1. The results of the phone survey research are described in Chong *et al.* (2011). The participants groups, who were recruited during the phone survey study, provided their consent to access their mains water billing records from their water supply provider. Of the 15,615 targeted households, 1134 householders from the four LGAs responded to the survey satisfying the screening criterion that the household had an MRT. The water consumption data for the consenting households were obtained from the Queensland Water Commission (QWC) database. Some households were subsequently excluded from the analysis due to inconsistent or incomplete water billing data. A total of 691 households across the four council areas were ultimately found to be suitable for inclusion in the Case Study 2 analysis. Rainfall data for each of the regions is presented in Table 3.2.

3.3.3.2 Assessment procedure

A benchmark analysis approach similar to Sydney Water's BASIX approach (Sydney Water, 2008) was applied for assessing mains water savings. Figure 3.6 shows schematically the benchmark analysis approach to estimate the potential mains water savings from dwellings with mandated rainwater tanks (MRT).

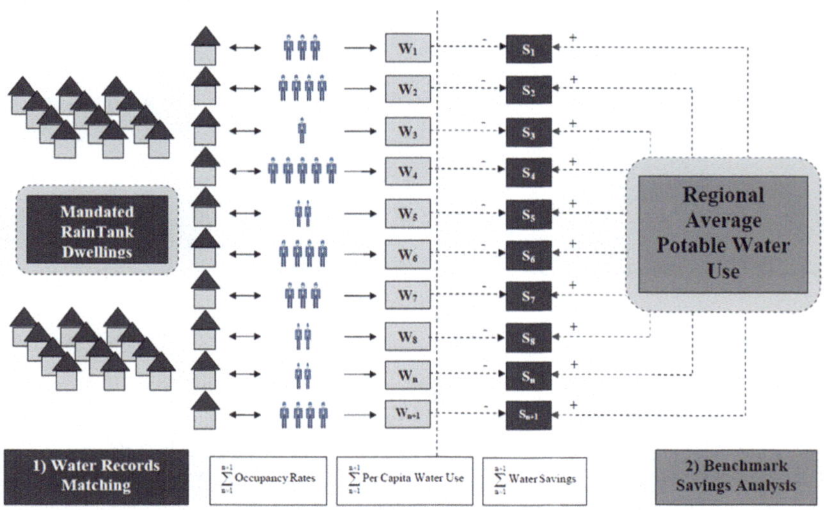

Figure 3.6 Schematic diagram for the benchmark analysis in estimating the potential mains water savings.

Mains water consumption records for each MRT dwelling were matched to their individual household occupancy number obtained from the phone survey. This was followed by the normalisation of mains water consumption to provide the per capita mains water usage (W_X) of each matched dwelling. The water usage data set (W_X) was then individually subtracted from the *average* mains water use for the respective LGA to generate the individual mains water savings (S_X). A positive sign notation (+) indicates a mains water savings from MRT dwellings. Negative values (–) indicate the mains water consumption at the particular MRT dwelling was actually higher than the regional average value. Subsequently, the average annual mains water savings from the MRT dwellings was estimated from the summation of each of the individual mains water savings (S_X) values.

3.3.4 Results and discussion

3.3.4.1 Water consumption data for MRT dwellings

The resultant mains water savings was expressed in litres/person/day (L/p/d). In order to convert the savings into kL/hh/yr, the mean occupancy rate was estimated via telephone interviews and used to provide more accurate results in determining the water consumption in these LGAs (Table 3.6). Interestingly, the known average occupancy was found to be higher than that assumed in the Case Study 1 (3 people per dwelling).

Table 3.6 Mean water consumption (L/p/d) in MRT households and average persons per household.

Region with mandated rainwater tanks	Sample size	Mean water usage 2009 (L/p/d)	Mean water usage 2010 (L/p/d)	Average person number per household assumed by Beal *et al.* (2011)	Average person number per household in this study
Pine Rivers (Moreton Bay Regional Council)	197	119.4	109.4	3.00	3.21
Caboolture (Moreton Bay Regional Council)	158	108.5	108.2	–	3.20
Gold Coast City Council	172	138.8	125.7	3.20	3.34
Redland City Council	164	129.1	121.9	2.90	3.18

3.3.4.2 Benchmark analysis of mains water savings for MRT households

The average mains water usage for the MRT cohort was compared with the SEQ average mains water consumption data for each LGA. The differences between the two data sets provide an estimate of the mains water saving for MRT households in 2009 (Table 3.7) and 2010 (Table 3.8).

The estimated per capita mean values for mains water saving from MRT dwellings were considered to be more accurate than the earlier analysis of Beal *et al.* (2011) in Case Study 1, as they are now normalised to the specific occupancy rate for every matched household. Although both the mean and median values were estimated, mean values are reported to maintain consistency with the units used in the State Government published data for the SEQ region. Since the ultimate aim of this study is to validate the 70 kL per year mains water savings target under QDC MP 4.2, the calculated annual mains water savings in L/p/d in Tables 3.7 and 3.8 were converted to kL/hh/y based on the average occupancy rates per household obtained from the phone survey (Table 3.6). Further, there is some doubt as to the relevance of savings predicted from prior Probabilistic Urban Rainwater and Wastewater Reuse Simulator (PURRS) modelling because of high water use assumed at that time, that is, 300 L/person/day (WBM Oceanics, 2006). This PURRS modelling was used to set the 70 kL/hh/year MP4.2 savings target, and was based on high external water use (around 50 L/p/d) estimated at the time (2005) in SEQ. However, this amount of outdoor use simply did not occur during and after the drought in SEQ (2007 onwards) as evidenced in Figure 3.3.

Table 3.7 Average annual water savings in MRT households in 2009 for four local government areas in SEQ (sample size in brackets).

Description	Pine rivers (197)	Caboolture (158)	Gold coast (172)	Redland (164)
Average persons per household	3.21	3.20	3.34	3.18
Average mains water consumption for all households in the LGA (L/p/d)[1]	140.4	140.4	211.4	201.5
Average water consumption in MRT households (L/p/d)	119.4	108.5	138.8	129.1
Average water savings in MRT households (L/p/d)	20.9	31.9	72.6	72.4
Average annual savings in MRT households (kL/hh/yr)	24.5	37.3	88.5	84.0
Average mains water use savings (%)	15	23	35	36
Average savings over all samples (691)		**58.8 kL/hh/yr**		

[1]*Source*: QWC data.

Table 3.8 Average annual water savings in MRT household in 2010 in four local government areas in SEQ (sample size in brackets).

Description	Pine rivers (197)	Caboolture (158)	Gold coast (172)	Redland (164)
Average persons per household	3.21	3.20	3.34	3.18
Average mains water consumption for all households in the LCA[1] (L/p/d)	143.3	143.3	192.0	183.1
Average water consumption in MRT households (L/p/d)	109.4	108.2	125.7	121.9
Average water savings in MRT households (L/p/d)	33.6	34.8	66.3	61.2
Average annual savings in MRT households (kL/hh/yr)	39.7	40.9	81.0	71.0
Average water use savings (%)	24	25	35	33
Average savings over all samples (691)		**58.2 kL/hh/yr**		

[1]*Source*: QWC data.

Results from 2009 modelling (Table 3.7) demonstrate that households with MRT substantially reduced mains water use in all the studied LGAs. Variation between LGAs could be driven by factors such as rainwater tank yield including factors related to rainfall, socio-demographic factors (water wise awareness and household water conservation behaviour) and water efficient household appliances and fixtures. The average mains water savings for Pine Rivers and Caboolture in 2009 were 20.9 L/p/d and 31.9 L/p/d respectively, which were significantly lower than the water savings for Gold Coast and Redland (72.6 and 72.4 L/p/d respectively). These data reflect the continued low water consumption in Pine Rivers and Caboolture in 2009 in the aftermath of the severe water restrictions placed on those regions in 2008.

Table 3.8 presents the average mains water consumption for IPT dwellings in 2010. The average annual mains water savings per household per year across the four council areas were found to range from 39.7 kL/hh/yr (Pine Rivers) to 81.0 kL/hh/yr (Gold Coast). Per capita reduction in mains water consumption per day ranged from approximately 24 to 35% (Table 3.8). The overall average water savings across the four regions in 2010 (for 691 households) were 58.2 kL/hh/yr. Interestingly, it was found that the mains water use pattern for the quarters in 2010 are quite different from quarters in 2009 where the inverse of higher water consumption rate towards late 2010 was observed. As discussed for Case Study 1, higher potable water savings for Gold Coast and Redland, which approximate the PURRS predictions, are probably due to much higher external water use, as these areas had minimal water restrictions compared with the other 2 LGAs.

3.3.5 Challenges and limitations

Although some challenges faced in Case Study 1 have been addressed in this analysis, there remained difficulties in obtaining complete data sets for some households. This limitation is likely to be a globally common problem. Typical difficulties associated with data gathering include:

(1) Many local authorities often had partially complete or missing billing information for households;
(2) Some datasets had been merged or removed for various reasons;
(3) The period of time for which water consumption was billed was not consistent, for example, quarterly versus six monthly; and
(4) Privacy issues can severely delay or prevent obtaining identified data.

As for Case Study 1, some inconsistencies in datasets made matching of data pairs more challenging and resulted in a reduced sample size.

3.3.6 Concluding remarks

Case Study 2 demonstrated that MRT households could reduce their reliance on mains water supplies in all the studied LGAs, albeit with substantial variation among LGAs. Case Study 2 (benchmark with empirical data) was designed to build on the results from Case Study 1 (desktop study with billing data), and to identify the advantages of this approach in more accurately quantifying mains water savings from rainwater tanks. The key difference between the two approaches is that known household occupancy rates (from the phone survey) were matched to the individual water billing records in Case Study 2. Conversely, Case Study 1 did not have access to this data, thus relied on using the average household occupancy rate from the 2006 Australian Bureau of Statistics (ABS) Census District Data for cross-checking the pairwise statistical analyses. It is anticipated that the Case Study 2 methodology can be used for most urban areas of the world, although the exact uses of the rainwater should be known. For example, if there is only internal uses (toilet and clothes washing) and little or no outdoor use, the savings from mains supply will be reduced. End uses studies reported by Beal *et al.* (2011b, 2013) are very valuable in understanding the quantum of potable water savings expected from rainwater tanks. Additionally, consumption and end-use also should be matched with socio-demographics and socio-economic status as this strongly influences water use per person per day (wealthy people typically use more water!).

3.4 CASE STUDY 3 – WATER SAVINGS FROM REBATED RAINWATER TANKS

3.4.1 Background

This case study presents the estimated mains water savings from installation of rebated rainwater tanks in Canberra and the broader Australian Capital Territory (ACT) based on analysis of water billing data. As part of its *Think Water, Act Water* strategy, the ACT Government subsidised the cost of purchasing and installing tanks. Initially run by the local water utility, the rebate program commenced in 1997, offering subsidies for installing medium to large tanks (>4 kL), but with no requirement for plumbing tanks to indoor connections (Fyfe *et al.* 2011). In 2004, the ACT Government took over administration of the program, adding rebates for indoor connections to new and existing tanks, and reducing eligible tank size threshold to 2–4 kL. Rebate incentives were adjusted four times between 2004 and 2007, and from July 2006 indoor connections were made a requirement for all rebates.

Throughout the majority of the program, the ACT experienced drought conditions and residents were subject to mandatory water restrictions. From 2005 to 2007, when restrictions were at their tightest, peak summer demand in the ACT dropped from 250–300 ML/d (unrestricted) to 150–170 ML/d (Fyfe *et al.* 2011). This demand reduction was in part due to customer response to water restrictions and associated public campaigns. Additional factors were the national Water Efficiency Labelling Scheme (WELS) (Australian Government, 2014), local water sensitive urban design projects, and a number of efficiency programs such as home retrofits of water-efficient devices under the *Think Water, Act Water* strategy (ACT Government, 2004).

3.4.1.1 Research aims

The central aim of the evaluation study conducted for the ACT Government (Fyfe *et al.* 2011) was to produce robust estimates of water and energy savings, and associated reductions in greenhouse gas emissions from the various efficiency programs under the *Think Water, Act Water* strategy. A key component of the research was to validate the methodologies used to generate the estimates. The research presented in this section focuses on the water savings derived from the rainwater tank rebate program.

3.4.2 Methods

3.4.2.1 Data sources and pre-processing

Data identifying rebate participants (all voluntary), their address, rebated tank size and connection details were provided by the ACT Government. The data were filtered to remove duplicates, incomplete records and participants that had participated in *other Think Water, Act Water* efficiency programs. The data for the remaining participant households were linked to quarterly water billing (metered consumption) data provided by the ACT water (and electricity) utility using lot, block, section and suburb identifiers. Only individually metered dwellings were analysed, causing most multi-residential dwellings to be excluded. The utility also supplied dates of changes to dwelling occupants (identified by changes to electricity account holders[1]), allowing the analysis to focus on households that occupied

[1] In Australia, electricity is typically billed to actual household occupants whilst water is billed to property owners who can choose to pass on the charges to their tenants.

a property both before and after receiving a rebate. The connected roof area was not known for the households examined, but as the ACT study examined retro-fitted tanks, it was assumed to be lower than for the MRT homes in SEQ. Water billing data for all non-participant households in the ACT was also supplied by the utility to provide a pool of 'controls' information. All billing data were screened for negatives, missing records and statistical outliers before being converted into monthly values using the 'binning' algorithm explained in Fyfe *et al.* (2010). Binning is used to overcome the problem of households having differing billing cycles where for example group X's household quarterly bills might end on 5th April, whilst group Y's household bills ends on 20th May. The process regularises the consumption data on a pro-rata basis so that it conforms to calendar months, allowing direct time-based comparisons between households.

3.4.2.2 Analysis procedure

The methodology used to estimate savings is based on a pair-matching approach similar to that used in Case Study 1 (Section 3.2), except matching was performed using historical consumption patterns rather than lot size and location. The matched pairs means comparison (MPMC) method compares the consumption of *each rebated household* with *every non-rebated household* in the entire utility based on data generated within the period between 3 and 14 months *prior to tank installation.*[2] The strongest match is determined by the lowest root square error (RSE) result calculated as:

$$\sqrt{(N_{-14} - R_{-14})^2 + (N_{-13} - R_{-13})^2 + \ldots + (N_{-3} - R_{-3})^2} \tag{3.1}$$

where R = monthly average day consumption of the (future) rebated household (kL/d), N = monthly average day consumption of corresponding non-rebated household (kL/d) and subscripts indicate the month relative to the participant's rebated tank installation. A perfect match will produce an RSE of zero.

The matched non-rebated household is assumed to have similar characteristics and responses to external demand drivers as the rebated household, and is adopted as a control. Matching is performed for each participant household in a random sequence until every rebated participant has its own control household. Matches are then subjected to several statistical tests to check the veracity of the match, which are described in detail in Fyfe *et al.* (2010).

Savings in month m of year y were then calculated as:

$$(N_i - R_i)_{m,y} - (N_i - R_i)_{m,Y} \tag{3.2}$$

where R_i is consumption of rebated household i in month m of post-installation year y or the pre-installation year Y, and N_i is consumption of the matched non-rebated control in the same month.

Repeated measures t-Tests were applied to the paired household differences for each month to test against the null hypothesis that the population of monthly savings had a mean of zero (i.e., no discernable water savings). Household savings typically showed a non-normal distribution, thus, Wilcoxon signed rank tests were also applied as a non-parametric (non-normally distributed) alternative.

[2]Since ACT water bills span three months, monthly consumption data produced by the binning process is influenced by consumption that occurred up to two months before or after any given month. Thus a distinct intervention month could not be isolated in the consumption data and the two months data before and after installation had to be excluded from the analysis.

3.4.3 Mains water savings results

Over the life of the program, 2744 rainwater tank rebates were paid to residents. Linking and pre-processing the data reduced the sample to 1913 households, and filtering through the MPMC method further reduced the sample size to 1410 households, with consumption data that ranged from October 2001 through to March 2011. The global mean savings estimate for rebated tanks between April 2003 and March 2011 (inclusive) was 40 ± 25 L/household/day, equivalent to 15 ± 6 kL/household/year. This is equivalent to 5% of average participant household water consumption in the pre installation period. Savings estimates for particular tank configurations based on analyses of subsets of the full data set are given in Table 3.9.

Table 3.9 Sample sizes, median tank sizes and savings estimates for different configurations of rebated rainwater tanks (April 2003–March 2011).

Rebated tank configuration	Sample size		Median tank size	Mean annual savings		Saving as a % of average participant consumption before installation	
	Households	Monthly consumption data points	kL	(kL/hh/year)			
All tanks	1410	66,116	5	15	±6	5	±2
Indoor plumbed	176	4837	5.3	9	±15*	3	±5*
Outdoor plumbed	845	45,176	5	10	±7	3	±2
Indoor and outdoor plumbed	182	4516	5.6	21	±16	7	±5
Tanks <4kL capacity	242	10,998	2.25	7	±11*	2	±3*
Tanks ≥4kL and <9kL capacity	660	30,794	5	13	±8	4	±3
Tanks ≥9kL capacity	478	23,016	10	20	±11	6	±3

**Notes*: Error bounds are 95% confidence intervals; not statistically significant at the 5% level.

The plot of savings for all tanks in Figure 3.7 shows that there is no clear long-term decay or growth, or seasonal pattern. Note that Figure 3.7 does not include data for the year 2003 as savings estimates were grossly exaggerated by outliers in small samples and were not statistically significant. The same absence of seasonality is evident in savings from tanks with exclusively outdoor connections (data not shown). Savings were not consistently statistically significant ($p > 0.05$), exhibiting a large dip from October 2006 to January 2007 (summer) following an extended period of low rainfall that also led to the introduction of stricter stage 3 water restrictions (after a year of relaxed restrictions) and low overall water consumption.

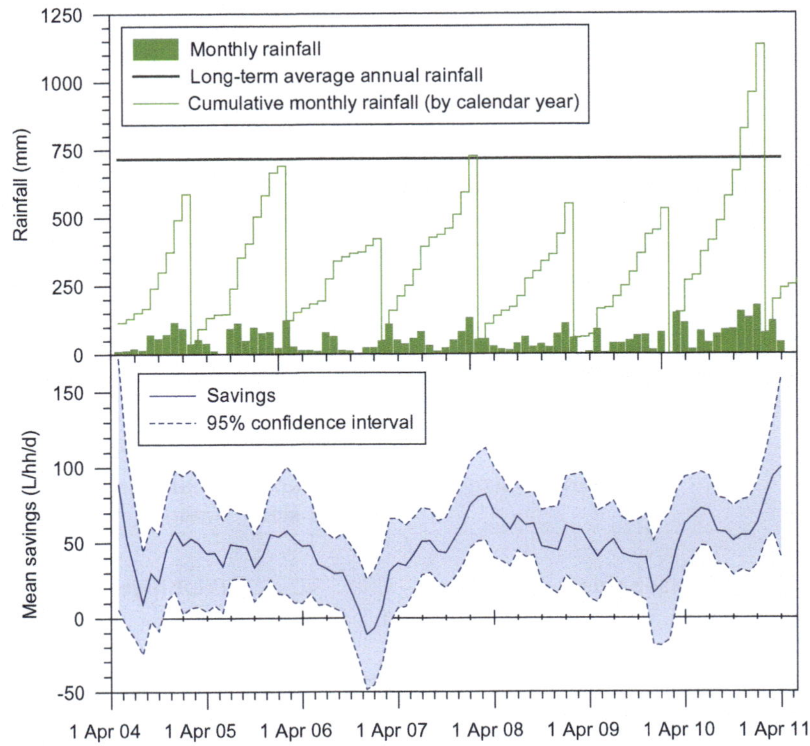

Figure 3.7 Mean monthly savings for all rebated rainwater tanks (bottom) and observed and long-term average monthly rainfall (top) over time.

3.4.4 Interpretation and implications

Savings were generally lower than anticipated and were not statistically significant for tanks with only an indoor connection, and for tanks of <4 kL capacity. Combined indoor and outdoor connections produced the greatest potable water savings (21 kL/hh/yr), and despite the small sample, the savings estimate was statistically significant ($p < 0.05$) and close to the sum of the separate indoor and outdoor connection savings. The climate of the ACT region may be characterised as Mediterranean, with relatively low rainfall throughout the year and hot, dry summers. Accordingly, rainwater tank yields cannot be expected to be as high as in the sub-tropical region of SEQ (Case Studies 1 and 2). However, savings for outdoor connected tanks (10 kL/hh/yr) were notably lower than the theoretical yield of 19 kL/year for a median-sized (5 kL) tank in the ACT region assuming a small roof catchment (50 m^2) and a relatively small 100 m^2 irrigated garden/lawn.[3] As shown in the upper plot of Figure 3.7, annual rainfall was below average (716 mm) in 5 of the 7 years s, which would have reduced yield from all tank installations. With water restrictions in force, access to mains water for irrigation was heavily constrained, thereby reducing apparent mains water savings. That is, the substitution of potable water with rainwater for outdoor end uses would not have been reflected in mains water savings for tank-owning houses.

[3]Derived using the water balance model described by McKibbin and Fane (2011).

Savings associated with indoor connections (9 kL/hh/yr) were also considerably lower than theoretical yield for connections to toilets and/or laundry (between 15–31 kL/year for 5 kL tanks connected to a 50 m^2 roof catchment). This supports the findings of Mukheibir *et al.* (2013), who recommend applying a 'functionality factor' of between 0.5 and 0.7 to theoretical rainwater tank yields to account for reduced catch efficiency associated with compromised installation quality, operational failures and behavioural issues. The scale of the yield impairment would appear high in this case, particularly when contrasted with the good agreement between measured and predicted savings in the SEQ case studies. However, this could in part be due to the fact that the tanks were retro-fitted to existing dwellings such that many would have had sub-optimal roof catchments, guttering and plumbing. In contrast, the SEQ houses with MRT as per the 2007 building code, were likely to have greater area of connected roof to the tanks, thus further contributing to the higher mains water savings observed in Case Study 1 and 2.

3.4.5 Challenges and limitations

By using prior consumption patterns to match tank participants to controls, the MPMC method circumvents the need for collecting data on household characteristics such as number of occupants, lot size, income and plumbing fixtures and appliances. It also implicitly controls for external factors such as restrictions and price changes. However, it cannot be applied to new homes with no water use history. Also it is not immune to the vagaries of internal household dynamics such as changes in appliances, new or departing occupants and voluntary behaviour change. Thus, it relies on a sample size of several hundred or more households to obtain robust savings estimates. Four of the six tank configuration subsets reported in Table 3.9 comprised less than 500 households, which meant that monthly average savings estimates (within the time series) were at times not statistically significant (see Figure 3.7). The extensive longitudinal component of the data set helped those sub-samples produce statistically significant global savings estimates, but these have large confidence bounds, making inferences more indicative than definitive. The analysis did benefit, however, from having household occupancy details verified using electricity accounts, thereby ensuring savings estimates were not biased by changes in ownership or tenancy.

3.4.6 Concluding remarks

The analysis of household billing data confirms that the installation of rebated rainwater tanks in existing homes of the ACT has achieved measurable potable water savings, but that those savings are significantly less than theoretical yield estimates. Recent research undertaken by Mukheibir *et al.* (2013) found that rainwater tank installation, maintenance and usage is often sub-optimal, resulting in impaired yields. In the case of rebate programs such as this one, poor tank functionality would be exacerbated by the difficulties associated with retrofitting tanks to existing dwellings such as limited accessible roof catchments and deteriorating guttering. Small yields caused by low rainfall and substandard functionality would produce low apparent mains water savings, which would have also been suppressed by reduced water usage amongst the broader community stemming from water restrictions and acute awareness of water scarcity. Functionality issues and low rainfall are also likely to be behind the notable differences between savings observed in this case study and those reported in Case Studies 1 and 2.

Nonetheless, the MPMC method is a robust method, and can be considered the 'Gold Standard' of treatment comparisons in estimating actual savings achieved by implemented programs, including rainwater tank rebate programs, provided prior water use behaviour of the house cohorts is available. It has been used on a variety of efficiency programs across Australia (Turner *et al.* 2013). It is recommended that a minimum 28-month billing dataset comprising 14 months either side of implementation is available before a robust analysis can be conducted using a 3-month billing cycle. Longer billing cycles require proportionally larger datasets. Based on the observed magnitude (~5%) and variation of the savings signal, the analysis precision will benefit from a sample size of more than 500 households, particularly when yield is likely to be lower due to small tank sizes or singular (indoor or outdoor) plumbing connections.

3.5 KEY CONSIDERATIONS IN QUANTIFYING MAINS SAVINGS

Having presented and critiqued three approaches to quantifying mains water savings from installing both internally and externally supplied rainwater tanks, there are a number of key points to consider when designing an approach to quantifying mains water savings. Ultimately, the goal is to have a large sample size based on desktop and field data of high quality. This is not always possible due to resource and time constraints. A method evaluation chart is presented in Figure 3.8 which assess costs against sample size, method approach and data quality. The larger the circle, the greater the costs, but usually, the higher the accuracy of outcomes. Figure 3.8 suggests that if only one approach is used for determining mains water savings, then the accuracy can be improved with a dataset of large sample size and high quality. Similarly, if the quality of the data is not detailed, but two or more approaches are being used based on a large sample size, a reasonably accurate outcome can be achieved.

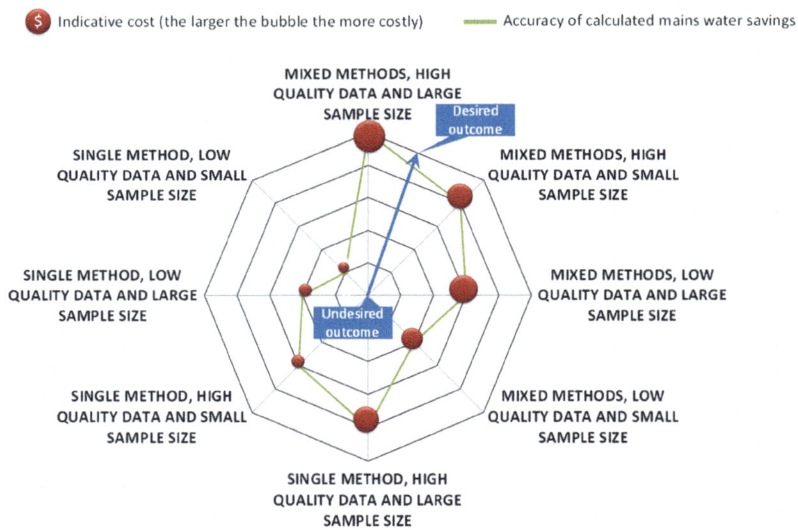

Figure 3.8 Method evaluation chart to assist in study design.

Each of the three main variables considered in Figure 3.8, data quality, methodological approach and sample size, are described below.

3.5.1 Quality of the datasets

3.5.1.1 Desktop approach only

Having access to quality data is obviously paramount in any field of research. A desktop approach using modelling and statistical analysis can certainly be valid, and even more so if it is coupled with at least one other method that allows for some empirical data to underpin the analysis (i.e., measured water consumption data). However, in the (often likely) absence of this possibility, a desktop approach alone may provide a reasonably accurate range of mains water savings, provided that it uses good quality data that can be applied with confidence to subsequent analysis. 'Good quality' data for estimating mains savings can be defined as having at least some or most of the following:

- *High resolution information on residential property* for example: dwelling configuration (detached, multi-unit, townhouse), people per dwelling, lot size, size of rainwater tank, rainwater end uses (internally or external only), date of tank installation, other water supply options on the property (dual reticulation/greywater system), connected roof area and degree of water-efficient appliances and fixtures.
- *Estimating external water use* is an important component as this end use has a high impact on the volume of rainwater used per year. As a first estimate, it is the difference between meter billing data and estimated internal water use for non-tank homes (as described in Case Study 1).
- *Large sample size* ($n = \geq 500$) of homes with *and* without rainwater tanks (and the configurations of the tanks).
- *A complete dataset of billing information* of water consumption (ideally at a three-month interval or less).
- *A spatially variable* dataset containing all of the above to allow some control for climate and biogeographical factors during analysis.
- *A longitudinal dataset* to also consider different seasons, water restriction regimes and water use activities (e.g., irrigation, school holidays, Christmas). Long-term data (≥3 years) for homes with and without rainwater tanks is critical to ensure representative water consumption patterns that encompasses both water restriction and non-water restriction regimes for example.

3.5.1.2 Field measurement approach

Where it is feasible to design a field measurement methodology, it is desirable that as many relevant parameters are measured for subsequent modelling and/or statistical analysis, particularly if an objective is to validate a desktop/modelling study. Chapter 4 presents a detailed section on the instrumentation for actual measurement for validation of the water savings. Below are some suggested ways to improve the quality of the data and accuracy of the method used for assessing mains potable savings from MRT (as per Figure 3.7):

- *Water consumption* – ideally both total and end-uses from the mains supply. End-use data will confirm the proportion of demand that can potentially be offset by rainwater. Smart metering equipment on both the mains meter and rainwater offtake should allow, at minimum, the total volumes of water supplied from each source. External water use, as emphasised throughout, is critical to estimate or measure as accurately as possible.
- *Socio-demographic data* – household occupancy has been emphasised as a very important parameter throughout this chapter. If there is no prior water consumption data, it is recommended to identify, as accurately as possible, the actual number of people in a household.
- *Household water-efficient stock* – if possible, the key water-related fixtures and appliances in the sample households should be identified as best as possible, and can be done simultaneously with smart meter installation if this is a feasible design option.
- *Water use behaviour* – the field methodology could also include a short survey on some water use behaviours around irrigation and outdoor use in general. For example, water behaviour information such as how often irrigation occurs and method of application (e.g., hand hose vs dripper system) will provide further opportunity to correctly match 'like with like'.

3.5.2 Mixed method and analyses

The types of data-gathering methods such as desktop (accessing council data on water consumption and other relevant information), modelling (using known or assumed input parameters), field instrumentation (direct measurement of water consumption) and stock audit and survey (water-efficient fixtures and outdoor irrigation) will strongly influence the accuracy and representativeness of the results. Ideally, it is recommended a mixed method approach be adopted, whereby a desktop/modelling exercise (based on council billing data), is followed by a field study validation (instrumentation and survey).

In terms of statistical data analysis, a pair-matching approach can facilitate before-after control-intervention analysis design, which supports robust savings estimation. Ideally, well controlled, household pair-matching for both pre-intervention (No Tank dwellings) and post-intervention (MRT dwellings) is the ideal scenario for statistical comparisons of likely water use savings. Naturally, the higher the resolution of pair-matching the more informative the outcomes can be.

3.5.3 Sample size v quality of datasets

As shown in all three case studies, determining mains water savings often relies on a third party dataset (e.g., water utility billing data) of potentially doubtful quality for pair matching. Therefore, larger sample sizes are desirable as there can be considerable noise in both billing data and rainwater consumption rates from this third party dataset. For example in Case Study 1, a starting sample of council billing data for nearly 29,000 homes was reduced to 2800 possible matched pairs for MRT and No Tank, and down further to <790 if lot size category was being matched.

If there is a field study component, where good quality metering data is available for each home, and end-uses of rainwater and household occupancy is available, then a lower sample size is likely to be sufficient. However, this may be at a higher project cost. This is also true for modelling the savings from rainwater tanks, where high quality input parameters can improve the accuracy and reliability of the model outputs.

3.6 SUMMARY AND CONCLUSIONS

The primary objective of this chapter is to present and critique alternative statistical methods that can be widely used for assessing the savings in mains water use from rainwater tanks. Three case studies were selected which incorporated both theoretical modelling approaches and empirical field data for both mandated rainwater tanks and rebated voluntary rainwater tanks. Some key conclusions from this chapter are:

- Outdoor consumption is the critical end-use that will maximise savings. Thus, factors such as water use restrictions, lot size and behavioural cues (willingness to use water outdoors) are very important in determining savings.
- The methods employed to assess savings will depend on desired outcomes, availability of good quality data and resources. A large sample size ($n > 500$) can be partially substituted for by good quality data where household occupancy, type of RWT end uses, potable water use restrictions and lot size are known.
- A desktop approach using statistical analysis should be coupled with at least one other independent approach to underpin the confidence of the empirical data analysis.
- All three case studies focussed on the need for large sample sizes, known household occupancy and the level of water-efficient stock in households.

- Actual rainwater yields can vary significantly between regions due not only to climatic factors, but also tank sizes, connected end uses, connected roof area, and the level of functionality (related to the quality of the installation). The functionality component is an important factor to be considered in *ex-ante* assessments of yield from rebate programs. For example, internal mains water savings in Case Study 3 were significantly lower than theoretical yields, indicating compromised tank system functionality.
- Models can provide a valid range of potential savings data provided they use realistic end use consumption figures and household population estimates. Nonetheless, validation by some level of field work (e.g., phone survey, instrumentation) is ideal.
- The pre-intervention pair-matching approach is the most statistically robust method to estimate savings as the same households are used in the post-intervention analysis.
- Statistical analysis will benefit from a sample size of more than 500 households (matched pairs), particularly when yield is likely to be lower due to small tank sizes, or singular (indoor or outdoor) plumbing connections.

3.7 REFERENCES

ACT Government (2004). Think Water, Act Water Strategy: Strategy for sustainable water resource management in the ACT. Australian Capital Territory Government, 2004, http://www.thinkwater.act.gov.au/.

Adeyeye K. (ed.) (2014). Water Efficiency in Buildings: Theory and Practice. John Wiley & Sons Ltd, UK.

Agudelo-Vera C. M., Keesman K. J., Mels A. R. and Rijnaarts H. H. M. (2013). Evaluating the potential of improving residential water balance at building scale. *Water Res*, http://dx.doi.org/10.1016/j.watres.2013.10.040.

Australian Government (2014). http://www.waterrating.gov.au/. Website (accessed 15 July 2014).

Beal C. D. and Stewart R. A. (2011). South East Queensland Residential End Use Study: Final Report. Urban Water Security Research Alliance Technical Report No. 47, November 2011.

Beal C. D., Gardner T., Sharma A., Barton R. and Chong M. (2011a). A Desktop Analysis of Potable Water Savings from Internally Plumbed Rainwater Tanks in South East Qld. Urban Water Security Research Alliance Technical Report No. 26, Queensland, Aust.

Beal C. D., Stewart R. A., Huang T.-T. and Rey E. C. (2011b). SEQ residential end use study. *Australian Water Association Journal*, **38**(1), 80–84.

Beal C. D., Sharma A., Gardner T. and Chong M. (2012). A Desktop Analysis of Potable Water Savings from Internally Plumbed Rainwater Tanks in South-East Queensland, Australia. *Wat Res Man*, **26**(6), 1577–1590.

Chong M. N., Umapathi S., Mankad A., Sharma A. and Gardner T. (2011). A Benchmark Analysis of Water Savings by Mandated Rainwater Tank Users in South East Queensland (Phase 2). Urban Water Security Research Alliance Technical Report No. 49, Queensland, Aust.

DIP (2007). Queensland Development Code. Queensland Government. January 2007.

DIP (2009). South East Queensland Regional Plan 2009–2031. Department of Infrastructure Planning, Queensland Government, 2009.

Eroksuz E. and Raham A. (2010). Rainwater tanks in multi-unit buildings: A case study for three Australian cities. *Resources, Conservation and Recycling*, **54**(12), 1449–1452.

Fyfe J., Mohr S., May D. and Rickwood P. (2010). Statistical Evaluation ofWater, Electricity and Greenhouse Gas Savings from the Think Water, Act Water Residential Efficiency Programs, report prepared for the ACT Environment and Sustainable Development Directorate, Institute for Sustainable Futures, University of Technology, Sydney.

Fyfe J., May D., Tuner A. and White S. (2011). Techniques for Estimating Water Saved Through Demand Management and Restrictions. In: Integrated Resource Planning for Urban Water — Resource Papers, Waterlines report, National Water Commission, Canberra.

Gardner T. and Vieritz A. (2010). The role of rainwater tanks in Australia in the twenty first century. *Architectural Science Review*, **53**(1), 107–125.

Ghisi E. and Ferreira D. F. (2007). Potential for potable water savings by using rainwater and greywater in a multi-storey residential building in southern Brazil. *Building & Environment*, **42**, 2512–2522.

Ghisi E., Bressan D. L. and Martini M. (2007). Rainwater tank capacity and potential for potable water savings by using rainwater in the residential sector of southeastern Brazil. *Building & Environment*, **42**, 1654–1666.

Hajani E. and Rahman A. (2013). Effectiveness of Rainwater Harvesting System in Peri-Urban Region: A Case Study for the Greater Sydney. *Journal of Hydrology and Environment Research*, **1**(1), 21.

Johnson N. J. (1978). Modified *t* Tests and confidence intervals for asymmetrical populations. *J Amer Stat Ass* **73**(363), 536–544.

Marsden Jacobs Associates (2007). The Cost-Effectiveness of Rainwater Tanks in Urban Australia. A report prepared for the National Water Commission – Waterlines, March 2007.

Makki A., Stewart R. A., Panuwatwanich K. and Beal C. (2011). Revealing the determinants of shower water end use consumption: enabling better targeted urban water conservation strategies. *Journal of Cleaner Production*, (doi:10.1016/j.jclepro.2011.08.007).

McKibbin J. and Fane S. (2011). Water Balance: a new integrated water management tool for households, businesses, and precincts. In: International Conference on Integrated Water Management, Perth, 2–5 February.

Mukheibir P., Boyle T., Moy C. and White S. (2013). Estimating the reliable residential water substitution from household rainwater tanks. In: Proceedings of the 7th IWA International Conference on Efficient Use and Management of Water, Paris, 22–25 October.

Siems R., Sahin O., Talebpour R., Stewart R. and Hopewell M. (2013). Energy intensity of decentralised water supply systems utilised in addressing water shortages, European Water Resources Association 8th International Conference. 26–29 June, European Water Resources Association, Porto, Portugal.

Sullivan J. and Dawson R. (2008). BASIX: Building Water Efficiency in NSW: Results from the First 5 years of BASIX: Sydney Water, NSW Department of Planning.

Sullivan J. and Wilson S. (2009). 'BASIX: The proof is in the water savings', AWA Ozwater 2009 Conference and Exhibition, Melbourne Convention and Exhibition Centre, Melbourne, Victoria, 16–18 March 2009.

Sydney Water (2008). BASIX Monitoring Report: Water Savings for 2007–8, Final Report November 2008, Sydney, New South Wales, Australia.

Talebpour M. R., Stewart R. A., Beal C. D., Dowling B., Sharma A. and Fane S. (2011). Rainwater tank end usage and energy demand: a pilot study. *Journal of the Australian Water Association*, **38**(1), 85–89.

Turner A., Fyfe J., Rickwood P. and Mohr S. (2013). Are we there yet? The importance of evaluating efficiency programs. *Water*, **40**(3), 60–65.

Vieritz A., Gardner T. and Baisden J. (2007). Rainwater TANK Manual. Department of Natural Resources and Water, 2007, Queensland.

Walton C. and Holmes K. (2011). How much water efficiency does $321 million buy? In: Proceedings of the 5th IWA Specialist Conference 'Efficient 2009', eds. International Water Association (IWA) and Australian Water Association, Sydney, Australia.

Wang Y. C. (2011). Separating indoor and outdoor water consumption. *Journal of the Australian Water Association*, **38**(3), 63–67.

WBM Oceanics (2006). Rainwater Tank Modelling Investigation: South East Queensland Regional Water Supply Strategy. WBM Oceanics Australia. Document ID: R.B15894.002.00.doc.

Willis R., Stewart R. and Emmonds S. (2010). Pimpama-Coomera dual reticulation end use study: pre-commission baseline, context and post-commission end use prediction. *Water Science and Technology: Water Supply*, **10**(3), 302–314.

Chapter 4

Monitoring of household rainwater tank systems for rainwater usage

Shivanita Umapathi, Reid Butler, Matthew Ferguson, David Pezzaniti and Ashok K. Sharma

ABSTRACT

The purpose of this chapter is to describe monitoring methods for household rainwater tank systems to provide evidence-based data on their effectiveness. Rainwater tanks are incorporated in urban environments around the world under integrated urban water management (IUWM) and water sensitive urban design (WSUD) approaches. Individual homeowners have been increasingly embracing rainwater tanks as one of the most easily adoptable alternate water supplies for 'fit-for-purpose' uses. Various modelling tools are available to estimate the rainwater usage based on tank size, connected catchment roof area, rainwater demand based on occupancy rate and connected appliances, and local climate data. However, the actual rainwater usage depends upon a number of factors such as occupants' behaviour towards using the various appliances connected with rainwater supply, installation of rainwater tank pumping system, system reliability and losses. Savings in mains water achieved by using rainwater are plausible; however, analyses on quantifying the actual reduction in mains water demand are still at an early stage and lack confidence for their incorporation into strategic water plans. The most important challenge to monitor household rainwater tanks is to properly instrument and accurately collect data for analysis and performance assessment of household rainwater tank systems. The main focus of this chapter is to describe approaches that address these challenges. The rainwater tank system monitoring examples discussed in this chapter come from Australian case studies. However, the methodology described to measure and monitor water savings in households with internally plumbed rainwater tanks should have worldwide applicability, and hence will be of value to water professionals.

Keywords: data collection; logging; monitoring; rainwater tanks; main water saving.

4.1 INTRODUCTION

Although rainwater is a significant source of fresh water in both natural and human-managed ecosystems, the resource is greatly underutilised (UNEP, 2009). Rainwater harvesting involves standard components that are universally applicable, due to similar underlying concepts in rainwater capture, supply and usage (Umapathi *et al.* 2013). One of the most widely accepted and commonly used rainwater harvesting techniques in urban environments is the rainwater tank collecting rainfall from household roof area. Depending on the end uses and the configuration of the roof, tanks can be set up in a variety of ways. In urban environments, end uses for rainwater are often chosen to allow minimal treatment requirements under an integrated water management system. For example, supplying the non-potable portion of the total water demand on a 'fit-for-purpose' basis (Cook *et al.* 2013; Jones & Hunt, 2010; Sharma *et al.* 2013a; Sharma *et al.* 2013b; Umapathi *et al.* 2013; Zhang *et al.* 2009). This approach complements a centralised mains water supply, and uses the collected rainwater to meet daily demands from toilets, laundry, garden taps, and in some cases hot water supply.

Urban dwellers and service providers alike are increasingly embracing a variety of demand management strategies by including decentralised systems such as rainwater tanks. Although the water saving potential of rainwater tank systems through modelling approaches is well known, the actual supply volumes and limitations of operating a rainwater tank system are still open for debate. In order to manage ever increasing water demand, measures such as metering, water accounting and loss control, pricing and education are being implemented (Boyle *et al.* 2013). If adequate and reliable data can be garnered using these measures, there exists a great potential to interpret, analyse and understand the water saving implications of using rainwater tanks that are integrated into modern-day urban water supply strategies. Detailed consumption data analysis can be utilised to not only help water utilities improve customer services, reduce water losses and manage demand, but also to provide reliable information at the consumer level in order to help the utilities make informed decisions on water management (Aravinthan *et al.* 2012; Boyle *et al.* 2013; Davison, 2008).

Monitoring of rain water tanks provides an opportunity to acquire rich details on the collection efficiency, the effects of water end use on yield, seasonal influence on rainwater use, and associated energy use that influence rainwater collection and usage. The following sections will discuss the major components of rainwater tank systems that form the basis of monitoring studies, followed by discussions on experimental methods and monitored data assessments. The chapter draws examples from studies undertaken in some of Australia's most urbanised areas.

4.2 MONITORING OF WATER USAGE IN RAINWATER TANK SYSTEMS

One of the earliest attempts to optimise the design of rainwater collection systems using water demand patterns was by Fewkes (1999) who analysed household rainwater consumption on a 'fit-for-purpose' basis. This was followed by various research methodologies to study and estimate the performance characteristics of domestic rainwater harvesting systems (Beal *et al.* 2012; Coombes *et al.* 2000; Coombes & Kuczera, 2003; Ghisi & Mengotti de Oliveira, 2007; Villareal & Dixon, 2005; Willis *et al.* 2011a), whilst other studies have analysed the social, economic and energy implications (Mankad *et al.* 2012; Moglia *et al.* 2012). Techniques to disaggregate water flows using smart water meters within rainwater tank pumping systems have emerged over the last few years (Heinrich, 2008; Talebpour *et al.* 2011; Willis *et al.* 2011b). Disaggregation of end-uses as a means of identifying the water use characteristics within urban residences was first studied extensively in the United States (Mayer *et al.* 1996; Mayer *et al.* 1999) and used to monitor single-source centralised urban water supply systems. Smart metering or automatic meter reading (AMR) (Fane *et al.* 2011) involves key elements such as real time monitoring, high-resolution interval metering (≥10 seconds), automated data transfer (drive by, GPRS, 3G) and remote access to data from the internet (Aravinthan *et al.* 2012; Giurco *et al.* 2008). Although high resolution water flow meters are discussed in this study, the use of high resolution smart meters is beyond the scope of this study. Energy meters are used to estimate electricity consumption in supplying rainwater through rain tank pumping systems.

Using similar monitoring methods, more detailed studies have been conducted within individual urban settings in recent years through the disaggregation of multiple water sources that are an intrinsic part of an integrated approach to manage water demands. Ferguson (2012) and Umapathi *et al.* (2013) have highlighted the use of water flow metering/automatic meter reading (AMR) methods to conduct end-use analyses of rainwater collected in domestic rainwater harvesting systems using small time-steps ($t \leq 1$ minute). In addition, the water-energy relationship of the rainwater system can also be studied as there is some concern high energy use could be a perverse outcome of such decentralised water supply systems.

Due to ongoing changes in water use patterns resulting from increasing urban densification, changing methods of water supply coupled with introduction of new technologies for water supply and end use, water efficient devices, analyses of end-use 'water behaviour' using high resolution monitoring methods is the key to demand forecasting (Turner *et al.* 2010). Water flow monitoring methodologies are clearly superior to theoretical assessments of, say, peak day water demand, and analysis of actual use patterns should provide a deeper understanding of the constantly evolving water supply milieu. Water efficiency and energy performance of fixtures and appliances can also be extensively investigated through these advanced metering methods.

4.2.1 Drivers for monitoring

Desktop studies such as estimation of urban residential water usage using modelling approaches have often been used as a basis to explore the feasibility of rainwater tank systems. However, a practical approach to understand the role of alternative water supply technologies on existing water supply infrastructure has been largely unexplored until the previous decade. Wide-scale urban water supply issues such as regulatory framework, life cycle costing and community acceptance using actual real-time water demand were not explored in-depth until early 2000s (Lloyd, 2001).

Water meter readings undertaken for billing purposes for residential customers are usually recorded on a quarterly basis. Additionally, meter read timings differ between different households to complete any given billing cycle (Fane *et al.* 2011). Disaggregated information on rainwater use for toilets or garden irrigation could have a good influence on changing consumer water consumption attitudes and behaviours by improving water literacy (Boyle *et al.* 2013; Giurco *et al.* 2010). Utilities can also benefit from smart metering technologies as it enables acquisition of data on rainwater consumption in households (Neenan & Hemphill, 2008) and also more accurate usage and billing data from mains supply.

Our knowledge of the complex interactions of key drivers such as urban form, demographics, life-style and end-use water demand, technology adoption, climate and water supply can be greatly improved and add to the argument for increasing use of alternate water supplies (Kenway *et al.* 2008). The most important driver for monitoring actual rainwater consumption from household rainwater tank supply system is the use of such data in developing more accurate strategic water plans for a region.

4.2.2 Case studies in Australia

Flow metering technologies have been used to conduct water efficiency audits since 1996 by consultants across Australia, including Sydney Water (Aravinthan *et al.* 2012). Extensive end-use studies on urban residential water consumption using flow monitoring methods were conducted in South East Queensland (SEQ), Australia (Beal *et al.* 2011; Beal *et al.* 2013; Umapathi *et al.* 2013; Willis *et al.* 2011a). A brief summary of some of these recent studies conducted in Australia for household rainwater supply monitoring are outlined in the following sections.

4.2.2.1 Sydney Water, Sydney

Sydney Water recognized a considerable knowledge gap in existing research on the impact of using rainwater tanks on the mains water consumption in Sydney households that were designed under the State Government's Building Sustainability Index (BASIX) Scheme (Ferguson, 2011). The main driver for the study was to determine the performance of household rainwater tanks in saving mains water. Sydney Water conducted an 18-month study of 52 households around Sydney using rainwater. Rainwater usage and mains water top-up for non-potable demands and the corresponding energy consumption of rainwater pumps were monitored remotely, all at one-minute intervals.

4.2.2.2 UWSRA, South East Queensland

The SEQ study involved monitoring, over a cumulative 12-month period, 20 households with rainwater tanks (Umapathi *et al.* 2012; Umapathi *et al.* 2013) installed to meet fit-for-purpose household demands. Similar to Sydney Water's investigation, the SEQ households were monitored for mains water use (including rainwater tank top-up), rainwater use and rainwater pumping energy consumption. The main drivers for this study were to determine the effectiveness of the rainwater tanks; to validate the savings expected from the rainwater tank systems based on local building codes; and to assess the corresponding energy consumption. Earlier desktop studies, conducted in 2011 by Beal *et al.* (2012) and Chong *et al.* (2011), reported that the rainwater tanks installed in new households under the Queensland Development Code (QDC) MP 4.2 failed to meet their predicted water savings target of 70 kilolitres per household per year (kL/hh/year), prompting a detailed investigation into water and rainwater demand in selected rainwater tank households distributed across the region. Further detail on these studies can be found in Chapter 3.

4.3 RAINWATER SYSTEM COMPONENTS, ACCESSORIES AND CONFIGURATIONS

The majority of rainwater tank systems consist of an above ground tank connected to a rainwater catchment area (the roof) together with a combination of system components comprised of filtering, pumping and backup equipment to supply water to specific end-uses (Australian Government, 2013). Some household rainwater tanks may be built underground to save space. However, this is usually a more expensive option as underground tanks cost more and can hinder maintenance and ongoing monitoring efforts in case of system faults. It can also be difficult to measure tank volume if the tank specifications are unknown prior to monitoring (Australian Government, 2013).

In Australia, different configurations for household rainwater tank systems have been identified which include gravity fed systems with pumps, pumping systems, dry systems, wet systems and gutter storage systems (Australian Government, 2013), of which, uncharged conveyance systems (dry systems) and charged conveyance systems (wet systems) are more commonly found. These systems are described in detail in Chapter 5. An above-ground household rainwater tank system setup in South East Queensland is shown in Figure 4.1.

A typical household rainwater collection system (Figure 4.1) consists of roof catchment areas that have attached gutters, which are in turn connected to downpipes that transport water through gravity flow into the rainwater tank reservoir. The size and slope of roof catchment areas can be measured to determine the efficiency and catchment losses of collection areas. This data is used in conjunction with monitored flow data to determine the relationships between weather patterns and rainfall and the reliability of rainwater tank supply. Methods to estimate catchment areas in the absence of building plans were described by Chong *et al.* (2014) and Chong *et al.* (2012) as well as in Chapter 5. Aerial photographs were used during on-site household inspections to match the roof catchment slopes with those downpipes that were plumbed to the rainwater tank. This method can also determine the approximate roof area connected to rainwater tank when there is more than one downpipe connected along a run of gutter. Collection system components such as first flush devices and mesh guards may be present to improve water quality during rainfall event (Figure 4.1). These components are discussed in further detail in Chapter 5.

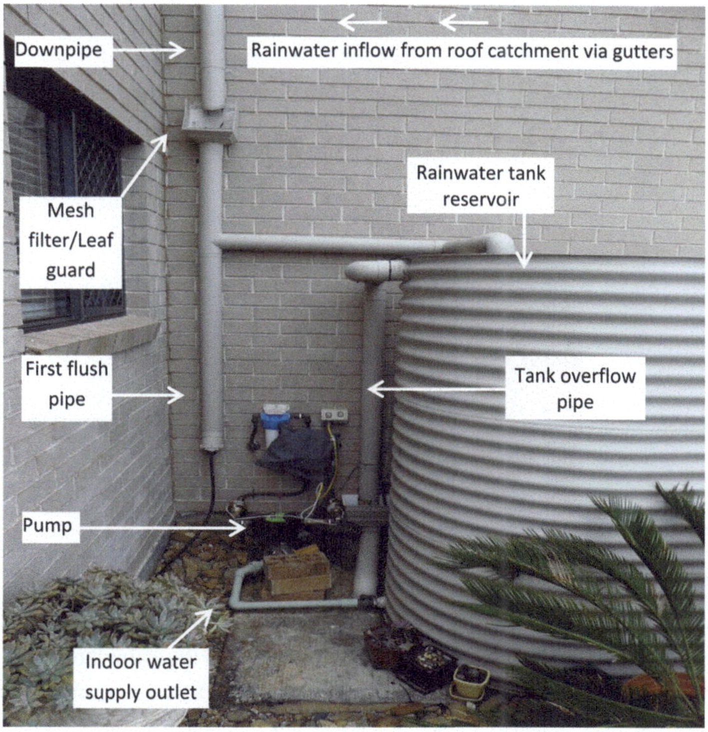

Figure 4.1 An aboveground household rainwater tank system setup in South East Queensland, Australia.

The storage system is obviously a critically part of a rainwater tank system. The volume of rainwater inflow into the storage tank (Figure 4.2) is dependent on weather conditions including rainfall, wind speed and direction and characteristics of the roof catchment area, such as the slope angle. Outflow of rainwater from the tank is dependent on supply factors such as the pumping capacity, supply pipe characteristics and most importantly, connected end-uses such as toilets or washing machines.

Pumps used in rainwater tanks systems come in different sizes (kilowatt ratings) and specifications including pressure cut off settings; hence, the water supply rate and energy used by the pumps vary. The performance of pumps also depends on characteristics external to the pumping system such as the distribution plumbing system. Monitoring of pumping systems can help in optimising pump sizing in rainwater tank systems which are generally known to be highly energy intensive. From an experimental monitoring perspective, it is preferable for the pump to be plugged into a general power outlet (GPO), rather than hard wired into the household electricity system.

In most cases, rainwater tank pumping systems also include tank level sensors (generally float based) and top-up devices including trickle top-up systems or switching device systems that provide mains water backup. Backup systems are an important component of rainwater tank systems; they deliver uninterrupted water supply to end uses when the water level in the rainwater tank is too low for pumping. The type of backup system employed will impact the water demand profiles from the rainwater tank and mains water back up supply. Detailed discussions on pumping systems, backup systems and the energy consumption aspects are discussed in Chapters 5 and 6. Pumping systems may also have other ancillaries such as pressure vessels, header tanks and so on (Retamal *et al.* 2009; SEWL, 2009) which are also discussed in Chapter 6.

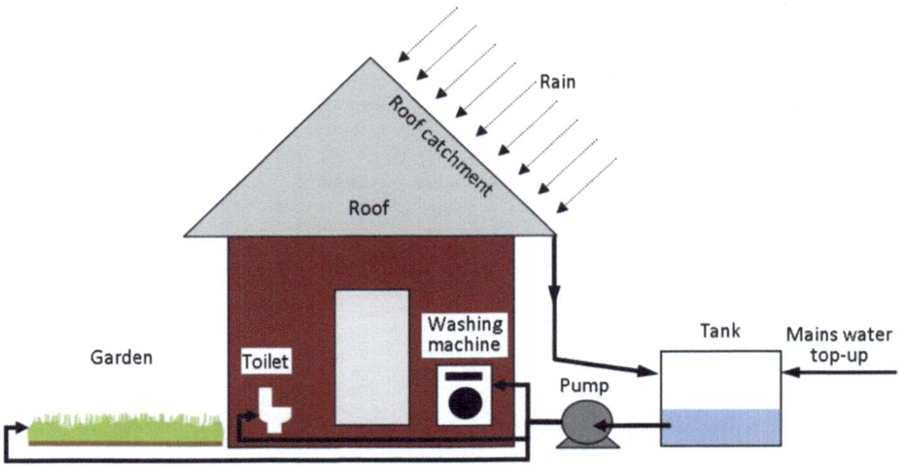

Figure 4.2 Typical end use connections for household rainwater use.

Household end use connections that receive water supply from the rainwater tank form the demand-end of the system. From a monitoring perspective, information on the water and energy demands for the various end uses can yield many useful insights for designing supply systems suited to unique end use configurations. Characteristics unique to any rainwater supply system, such as energy intensity of the pumps, reliability of the rainwater tank and sizing of collection, storage and pumping accessories, can all be estimated based on end use demand. Toilets, washing machines and garden taps are the most common end uses of rainwater (Figure 4.2). Hot water systems have also been connected in some cases. As there is often minimal or no water treatment requirement for these non-potable end uses, water quality concerns are addressed by limiting rainwater use on a fit-for-purpose basis.

4.4 EXPERIMENTAL APPROACHES

4.4.1 Monitoring methods

In developing an experimental methodology, it is necessary to identify the different water sources (e.g., rainwater and mains water) in the system and the corresponding end uses associated with the water sources. Figure 4.3 depicts a general schematic outline of a simple water and energy monitoring setup in a household setting. As rainwater is collected in the storage tank, the change in volume of water in the tank can be measured using water level sensors. Water flow meters monitor supply from the rainwater tank to indoor and outdoor end uses, total mains water supply to the household, and mains water backup to the rainwater tank. An energy meter is installed to measure the energy consumption by the rainwater pump.

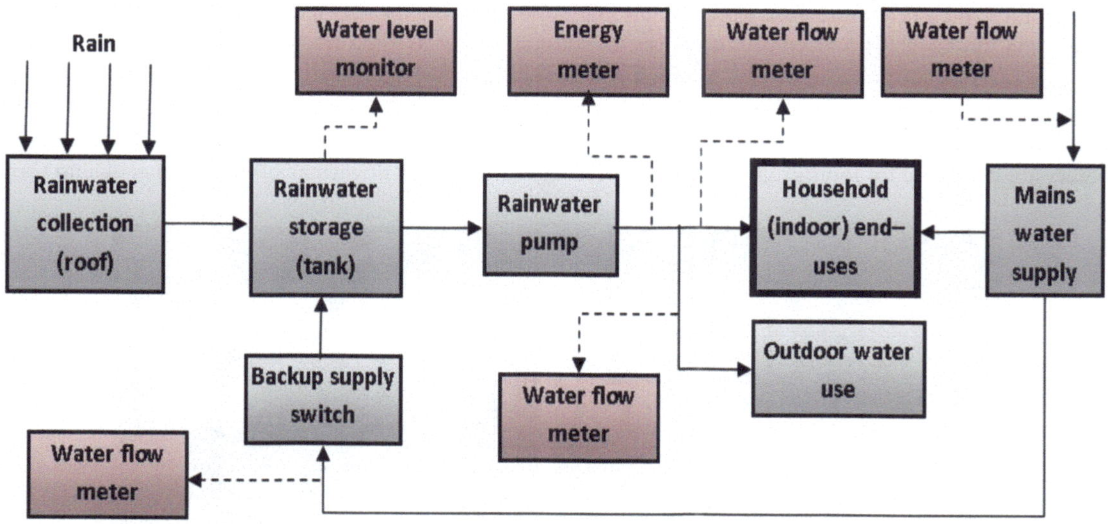

Figure 4.3 Process flow diagram of the water and energy monitoring setup in a typical household rainwater tank.

A data logger continuously records water flow through the meters and this data can be downloaded remotely using wireless connections. Water flow and energy consumption data are generated when water uses such as washing machine, showers or toilet flushing create electrical pulses in the water meter (via a reed switch) that can be logged against time stamps of a pre-determined frequency, for example, every 60 seconds The data collected using meters can vary in resolution (i.e., pulses per litre of flow) based on the specifications of the meters and data loggers. Metering of residential water supply can be tailored to meet the information requirements of researchers and other stakeholders. Case studies based on standard monitoring methodologies are further discussed in later sections.

An alternate method to installing multiple meters is to employ a high resolution meter and high frequency logger that can be used to record water supply from a primary water source (such as mains water supply and/or the rainwater tank supply) wherein the data in the form of high resolution flow traces can be disaggregated for various end uses using available commercial software. More details on this method of analysis can be found in Mayer *et al.* (1999) and Talebpour *et al.* (2011).

4.4.2 Instrumentation

4.4.2.1 Water meter

For a basic monitoring setup, two water meters, one for the mains water inflow (into the tank) and another for total water out of the tank pumping system, are minimum requirements to enable comparisons between rainwater and mains water consumption. The simple setup would also help analyse the impact of rainwater on diurnal water consumption, including volumetric reliability of rainwater supply and percentage contribution to total water usage. For intensive monitoring, additional meters to measure individual end uses (such as different household appliances and garden water taps) can also be installed.

Residential main meters are typically sized at 15, 20 or 25 mm. Rainwater tank connections are either copper pipes or HDPE and these are generally 20 or 25 mm. The best meters to use are positive displacement meters (Figure 4.4), which function by water flowing into compartments of a known volume, which continuously fill and empty, turning the counter wheel as they move. As the counter rotates it causes a reed switch to open or close, which generates a low voltage pulse. The pulse ratio varies with the make and size of the meter ranging from 1 pulse per 0.5 litre to 1 pulse per 5 litres depending on the model of meter. For most rainwater tank and domestic use studies, 20 mm pipe diameter, 0.5 L/pulse meters are preferred for their greater accuracy in measuring the low flow rates of most domestic end uses.

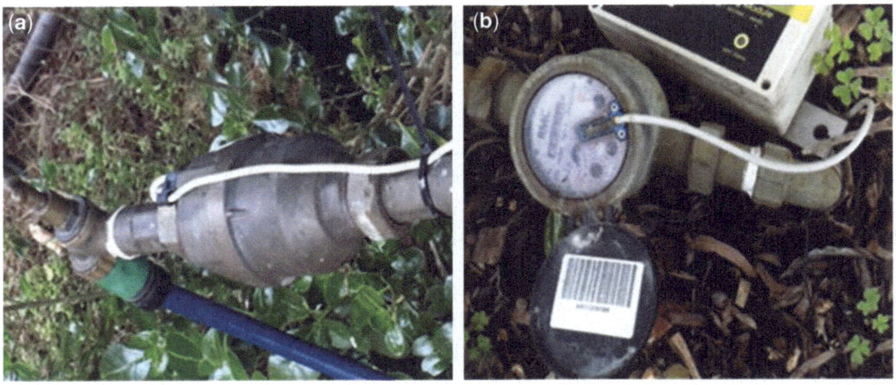

Meter type	Positive displacement (A)	Rotary piston measurement (B)
Description	Water flows into compartments of a known volume, which continuously fills and empties	Rotary piston measurement
Size range	15 mm- 40 mm	15 mm-25 mm
Pulse	0.5 to 5 L/pulse	0.014 L/pulse
Minimum flow detection	0.048 L/min - 0.6 L/min	0.25 L/min - 0.6 L/min

Figure 4.4 Flow meters used to measure residential water flow (for demonstration purposes only).

Higher resolution meters (Figure 4.4), such as rotary piston measurement type meters, can be used to achieve greater resolution of data. This is required where a single flow meter is used to measure and estimate (by signal deconvolution) consumption of different downstream end uses. Pulse ratios in these high resolution meters can be as low as 0.014 L/pulse and are usually used in studies in conjunction with the flow trace software such as Trace Wizard (Mayer *et al.* 1996) to determine water end use characteristics (see Talebpour *et al.* (2011).

4.4.2.2 Electricity meter

In addition to water flow monitoring, rainwater pumping energy can be monitored at a small cost. Desktop studies have typically shown between 0.9 and 2.3 kWh/kL of energy intensity for rainwater use, a much lower range compared to in-situ studies in Australia which showed anywhere between 0.4 kWh/kL to 11 kWh/kL (Tjandraatmadja *et al.* 2012). Electricity meters used for domestic metering by some of the notable electricity generating companies in Australia such as Energy Australia, measure root mean square (RMS) power to accurately gauge true power consumption. They emit a pulse for each Watt-hour used and are able to capture quiescent loads. They are rated at +/–1% accuracy, which is standard industry practice for single phase meters.

Figure 4.5 shows an electric meter, which is a direct connect, single phase, static Watt-hour meter used to measure electrical energy. It delivers a pulse for each Watt-hour that passes through, which is recorded by a data logger (refer Figure 4.6). Measuring instruments of various makes/types are available in the market. Users are advised to thoroughly investigate these instruments based on their needs. For a tank monitoring setup, standard residential electricity meters (Figure 4.5) with pulse output capability have been modified by adding an inlet and outlet power cord, and a cable for the pulse outputs. The meters are usually modified to ensure the units are waterproof, and are made electrically safe for residential use by sealing the rear terminal connectors, and checking and tagging before deployment in a residency. It is important that the rainwater pumps being monitored are 'plug-ins' into power outlets and not hard-wired into the domestic power circuit, thereby avoiding the need for an electrician to visit each site. This power connection set-up needs to be confirmed during the household recruitment process.

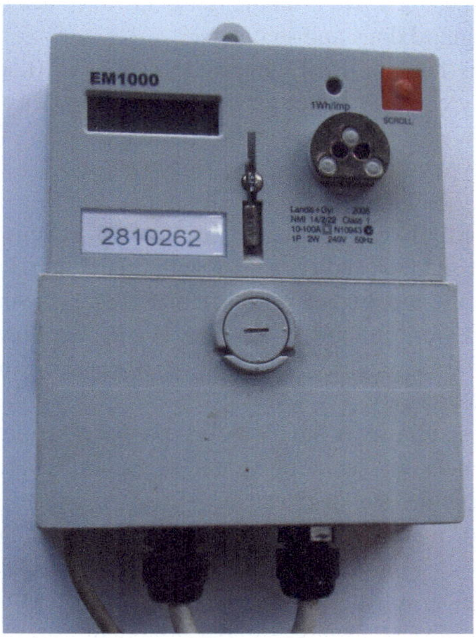

Figure 4.5 Electricity meter attached to data logger setup (for demonstration purposes only).

Figure 4.6 Remote terminal data logger (for demonstration purposes only).

4.4.2.3 Data logger

Data loggers are essential to record the measured electrical pulses over a given time period or frequency. A basic monitoring setup consists of a four-channel data logger, which is set up with a subscriber identification module (SIM) card and batteries with easy connectivity to a power outlet. The reed switches generating pulse outputs from the water and electricity meters are wired to the switch closure input channel of the data logger. Additional loggers may be required if the flow meters are located too far apart (due to signal interference and /or decay) to be wired to one logger. Raw data can be retrieved directly from the loggers or converted to engineering units of litres (for flow meters) and Watt-hours (for energy meters) before or after download from the logger. The units of data will be cumulative litres, or Watt-hours, in each monitoring interval.

On-site data loggers can be accessed either on-site or remotely downloaded. Remote logging is necessary for the studies where access to the study site is challenging due to issues of privacy, remote site location, household access difficulties or guard dogs and so on.

Remote logging equipment is favoured for intensive (e.g., every 30 second interval) data collection studies to ensure logger memory capacity is not exceeded and the ability to remotely detect any failure in the data logging system. Two types of remote metering data loggers have commonly been used for these studies in Australia. One type of data logger transmits the collected water meter pulse data by radio frequency to a hub, which can then be accessed via the phone line for data downloads. These smaller units need to have a clear line of radio communication to the hub, meaning that they need to be in fairly close proximity (less than 1 km) and have a clear line of sight to the hub. Loggers can be either individual or multichannel units based on the number of input channels (i.e., sensors) they need to monitor. Data loggers have finite internal memories (quantities of bytes), hence data flow rates and data duration of the samples being measured must be taken into consideration when choosing the combination of internal storage capacity and download frequency.

Other types of data logging systems use an inbuilt mobile phone transmitter which requires a unique SIM card supplied by a mobile phone network carrier/provider (Figure 4.6). These units can be located anywhere in the surrounding area that has a strong mobile phone signal, and are able to download data to remote servers on a regular basis, either daily or more frequently depending on the battery life and study requirements. They use a low voltage circuitry, so a unit transmitting twice a day is capable of remaining in the field for five years or more without a change in batteries. There are several brands of data loggers available which are capable of monitoring 1, 2, 4, 8 or 16 meters. Some can be mains powered while others can be battery charged by a solar panel for longer deployment periods. Users should investigate available loggers in the market based on their needs. The data logger shown in Figure 4.6 connects to sensors in the field and collects pulse outputs from water meters and electricity meters and transmits the data via a GPRS/CDMA/3G mobile telecommunications network to a central server at pre-set intervals. The data is then viewed on the internet and can be manipulated and analysed. Water and energy data are collected in 1 minute intervals.

4.4.2.4 Rainfall measuring devices

Rainfall data is an essential part of the monitoring system. It can be obtained from a local weather station but more usually is measured on site using a tipping bucket rain gauge or pluviometer (Figure 4.7). This type of rain gauge consists of a collection funnel of standard diameter (e.g., 200 mm) which directs collected rainwater to a small 'tipping bucket' that tips to alternating sides with every pre-set amount of precipitation, which in turn generates a pulse output signal from a reed switch similar to that of the water meters. These pulses are countered and recorded by the data logger, which can be inbuilt as part of the rain gauge (Figure 4.7).

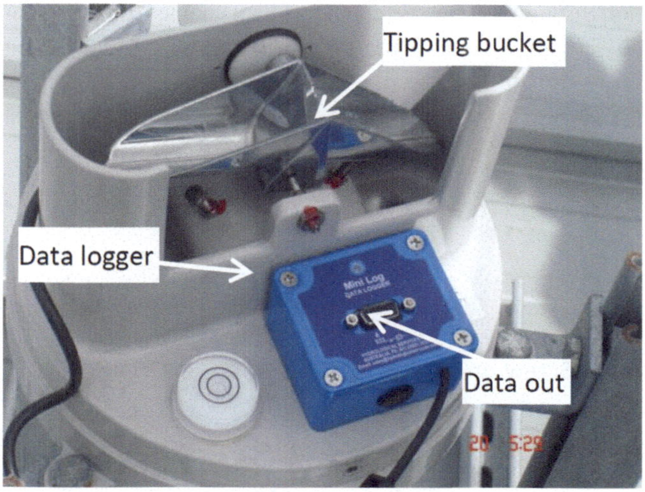

Figure 4.7 The internal tipping bucket and data logger of a typical pluviometer (for demonstration purposes only).

Monitoring studies can also comprise rain gauge stations in order to obtain localised rainfall for individual buildings; however, it is recommended that the gauges are located away from areas that are likely to collect dust and debris (such as rooftops), which may cause disturbances to the data being logged. A typical rain gauge station used in monitoring studies is programmed to record data for every 2 mm of rainfall, which is equivalent to ten pulses by the tipping bucket. A cylinder of standard diameter (200 mm) goes around this *chassis* to funnel rainfall into the tipping buckets. Sub-hourly data from rain gauges can be used to supplement daily metrological data that is usually available at a detailed spatial scale for most cities in Australia. Temperature measurements may also be useful to study evaporative effects of roof materials on the volume of rainfall collected.

4.4.2.5 Other monitoring equipment

Water level monitoring in tanks is very useful to compare rainfall (in millimetres) with tank catch (in litres). Continuous monitoring of water levels within rainwater tanks often use a hydrostatic pressure monitor that have submersible, differential pressure transducers. These are available with inbuilt data loggers and generally require manual downloads.

Differential pressure sensors measure the pressure difference between the reference location of the sensor submersed in the tank and the outside atmospheric pressure. One side of the differential pressure unit is exposed to the air through a vent tube exiting the top of the tank. The other side is in contact with the water in the tank. Differential pressure monitoring is a convenient method to monitor rainwater tank levels and requires little to no processing before data display and analysis.

Absolute pressure sensors, on the other hand, are generally housed in a submersible casing and measure the pressure (and in some cases the temperature) at the sensor suspension reference point in the tank. However, due to the absence of venting tubes, they do not measure the changes in atmospheric pressure. Hence, an atmospheric reference sensor of the same configuration needs to be suspended in the air above the water surface to allow correction for day-to-day variation in atmospheric pressure. Figure 4.8 shows a typical absolute pressure sensor setup configured to download data to a computer. The pressure sensor is connected to a base station through a *magnetic coupler* which enables the downloading of data through an infra-red sensor. The software used is a propriety product of the logger manufacturer.

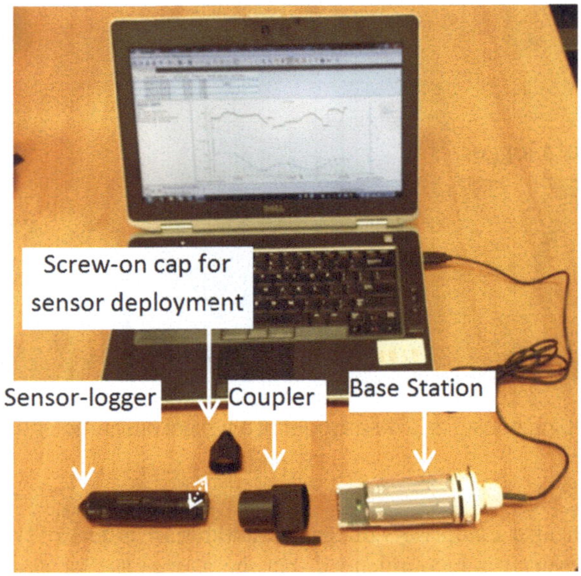

Figure 4.8 An absolute pressure sensor setup to download data to a laptop computer. The sensor has been retrieved from the rainwater tank (for demonstration purposes only).

A capacitance water level sensor and data logger (Figure 4.9) uses the varying potential between two concentric plates within the logger to determine the depth of water in the tank. The levels are recorded every 15 minutes and stored in the data logger until the data can be downloaded to a computer. Capacitance water level sensors are applicable to both absolute and differential type pressure sensors. Users should investigate the suitability of the commercially available sensors based on their needs. The water level sensors were not used in the case studies described in this chapter, however Moglia *et al.* (2014) used these sensors for rainwater tank water level monitoring in Melbourne.

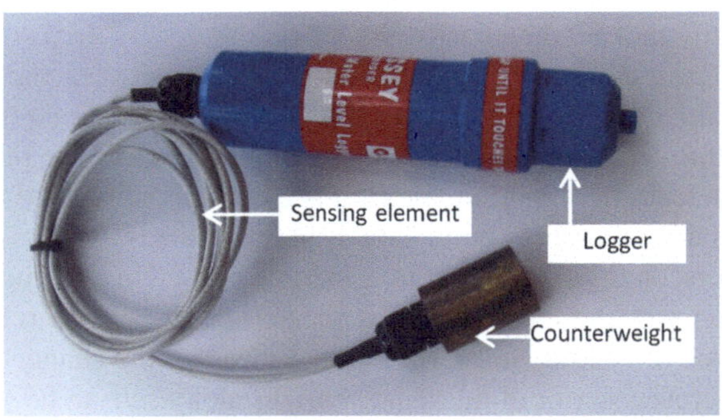

Figure 4.9 A capacitance water level sensor (for demonstration purposes only).

4.4.2.6 Protective casing

Outdoor monitoring equipment must be well protected from possible environmental damage. Equipment can be encased in a protective enclosures or casing to shield from weather damage caused due to rainfall, humidity, dust, extreme temperatures and from direct sunlight. Ideally, all equipment that is susceptible to weather damage should have an Ingress Protection (IP) rating (IEC, 2004). Equipment specifications should be consulted when deciding the level of protection required ensuring safe operating conditions are provided.

Disruptions in data logging have been observed in the past as a result of accidental damage to the wiring, hence data monitoring equipment should preferably be placed away from common reach. Metering and logging equipment malfunction due to accidental or environmental damage could also be a contributing factor towards discrepancies in logged data. Enclosing equipment casing in waterproof bags can also provide added protection.

4.4.2.7 Recent developments in monitoring instrumentation

An advanced real-time water and energy monitoring system has been set up to study the water usage in an urban development (Lochiel Park) in South Australia (Whaley *et al.* 2010). The houses and apartments (total of 106) in the development are designed to receive their water supply from three sources (mains, rainwater and recycled stormwater). All connected water sources in the development are being monitored at medium (1-minute) to high (5-second time intervals) resolution, with a pulse ratio of 1 litre per pulse. The digital outputs from each sensor are connected to an in-home touch screen display (EcoVision) (Ecovision Systems, 2014).

This is a part of a detailed monitoring system consisting of various analogue (rain tank level, temperature, relative humidity) and digital (water use, electricity use, gas and photovoltaic) sensors (Whaley *et al.* 2010). The EcoVision display unit (Figure 4.10) allows homeowners to monitor their water consumption in real time, together with electricity and gas usage. This feedback information is highly beneficial in aiding consumer self-management of their energy and water use, including detection of leaks within their water supply plumbing.

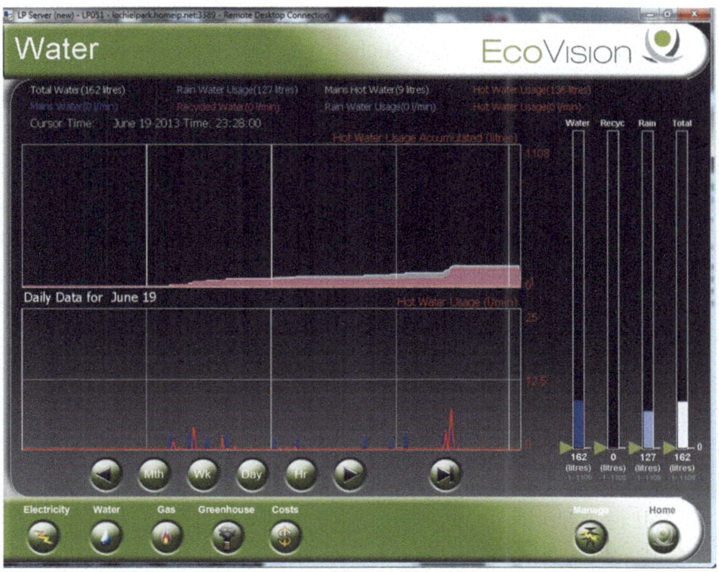

Figure 4.10 Screenshot of the in-home 'EcoVision' display screen (for demonstration purposes only).

A system such as the Ecovision system is very sophisticated and requires installation by qualified plumbers and electricians.

4.4.3 Sample number and ethics approval

For research purposes, a statistically representative sample sizes is required to ensure that population error is within acceptable limits of precision. The number of participants required to represent the behaviour of an area's entire population of rainwater tanks owners can be estimated using Equation 4.1.

$$n \geq \frac{\left(\frac{1}{\varepsilon^2}\right)N}{\frac{1}{\varepsilon^2}+N} \tag{4.1}$$

where, n is the sample size, N is the population size, and ε is the sample error expressed as a fraction of the mean value and can be chosen as anywhere between 0.01 and 0.2 (Chowdhury, 2013; Ghisi & Ferreira, 2007). In this case, the population size 'N' is the number of households with rainwater tanks in a local council area and 'n' is number of households required to achieve representative monitoring.

An important aspect of participant recruitment for research activities involves adhering to the appropriate research conduct guidelines, including obtaining necessary ethics approvals from the concerned research ethics committees or similar bodies. Methods adopted for participant recruitment may differ between organisations and institutions and from country to country, as will ethics code requirements for recruiting the participants. A letter of offer detailing the monitoring procedures, together with a concise description of research outcomes that are expected from the study, and any related ethics code consideration should be provided to prospective participants. In order to be accepted, participants are required to acknowledge to the letter of offer granting approval for the proposed monitoring/audit activities.

Initial involvement of the participants is necessary during the planning of equipment installation. However, participation cooperation may also be required during later stages in case of the need of replacing failed sensors and so on. Participation incentives such as gift vouchers can be considered if required.

4.4.4 Site inspections and audits

On-site activities involving monitoring equipment are not confined to setting up meters and loggers. Necessary inspections of the rainwater tank system and other attributes that are external, but related to the system, such as the rainwater pumping system, building setup, tank connections, rainwater catchment areas, and associated end uses (appliances and fixtures) need to be audited to gain a better understanding of the system. This in turn helps tailoring the monitoring setup for each specific household. Detail on household rainwater tank audits is presented in Chapter 5.

4.4.5 Correlation with weather and seasonal patterns

Monitoring studies of household rainwater use and yield can be correlated to the local rainfall in 'case study' areas. Previous research on household water consumption (including those with a focus on rainwater tanks) have treated rainfall as a homogenous entity over the study area (Khastagir & Jayasuriya, 2010). Monitored households with rainwater tanks are generally distributed over a suburb or a township, and may often have different rainfall patterns. Weather and rainfall data from the nearest weather station can be acquired from organisations such as the Bureau of Metrology in Australia (www.bom.gov.au).

Studies in Australia can also access rainfall data from an enhanced climate database, SILO, hosted by the Science Delivery Division of the Department of Science, Information Technology, Innovation and the Arts (DSITIA, 2013) which contains Australian climate data from 1889 in various formats that are suitable for research.

4.4.6 Monitoring duration and intervals

Determining the best time-step for continuous monitoring is directly associated with the objectives of the monitoring study. Monitoring at very short time-steps or intervals (i.e., 1 second) will provide the most detailed outcome but data management and analysis of this fine grained data can become very difficult over long monitoring periods.

For studies that focus mainly on the supply aspect of rainwater usage in households, time-steps up to 1-minute data intervals is satisfactory. Time-steps of longer than 1 minute will limit the user's efficiency to cleanse the data.

Smaller time-steps may improve end use analysis for high flow rate water use events, but the resulting large data files may be very difficult to manipulate over longer monitoring periods. Table 4.1 summarises the limitations associated with the duration of monitoring-based analyses.

Table 4.1 Data monitoring frequency and Limitations of data interpretation.

Monitoring period	Limitations
>1 yearly basis	Change in ownership or household population confound results
Yearly	As above + Annual rainfall difference
Half-yearly (6 months)	All the above + Seasonal differences in water use
Monthly	All the above + Effect of rainfall/dryness + Public holidays, school holidays, Christmas effects
Weekly	All the above + Weather effect (e.g., hot week) + Data loss
Daily	All the above + Weekend vs weekday effects
Sub-daily	All the above + End user effects + Impact of fixtures and appliances

Demand distribution at annual intervals may be influenced by water saving schemes or legislation affecting water consumption in households. Demand distribution measured at monthly intervals can provide information on seasonal effects such as rainfall, temperature and household usage patterns, including use of gardening equipment, swimming pools, frequency of showers and so on. Intensive monitoring (small intervals) for short periods (such as 2 weeks) during each season can provide insights into water use behaviour and rainfall yield if costs of continuous long-term monitoring are not affordable. This approach is often used in detailed end use studies that analyse water use signal data (using software tools such as Trace Wizard) to disaggregate the data from one water meter into the multiple end uses (Nguyen *et al.* 2013).

4.5 DATA MANAGEMENT

Data loggers are installed on-site to store information collected by the water flow meters. Recorded data files can be retrieved by wireless telemetry at a time frequency determined by study requirements. The instrumentation can include mains water, rainwater, mains top-up supply, rainwater end uses and energy consumption data. Initially collected data should be checked at short intervals (i.e., weekly) to ensure correct functioning of the meters and loggers. In order to avoid long-term system malfunctions, data loggers can be equipped with alarms that report when no water flow is recorded for a specified time period. This enables early intervention of monitoring equipment malfunction before any significant loss in data occurs.

There are possibilities for loggers to be accidentally disconnected post installation. During case studies conducted in SEQ and Sydney by Umapathi *et al.* (2012) and Ferguson (2011) respectively, loggers were found disconnected due to cut wires and other types of circuit breakage. In some cases, loggers malfunctioned or were damaged by water penetration, and required replacement. It is highly recommended that regular on-going review of data should be undertaken to ensure the proper functioning of monitoring instruments.

Data management and processing is critical if monitoring data are to be used effectively. The type of monitoring undertaken, whether periodic or continuous, will affect the protocol for data management. Loggers can aggregate recorded data over a specified period for recording purposes based on the study needs and format.

4.5.1 Data cleansing and validation

Data 'cleansing' is the most time consuming part of data analysis. As problems inevitably occur over time, collection of high quality data requires constant tracking of the loggers' performance. Data from continuous monitoring will require a large amount of cleaning before any analysis. This is due to a number of factors discussed below.

Some perceived discrepancies in data may be more apparent than real. For example, changes can occur in water consumption patterns within individual premises due to the absence of occupants over some part of the monitoring period. Change in ownership or occupant numbers may also change the pattern of water consumption due to varying water use habits/demand.

Although remote monitoring is a convenient and reliable method for assessing the performance of rainwater tanks, regular visual and logical cleansing of data is essential for reliable results. In some previous case studies (Ferguson, 2011; Umapathi *et al.* 2012), one-minute water flows and energy consumption data logged for each home were reviewed for discrepancies and validated. Collected records were checked for accurate data, consistency, range and format by both manual screening or program codes of which, one of the most accessible and easily programmable is the 'macros'/VBA function in Microsoft Excel®.

A reliable method to check the validity of monitoring outcomes is by collating and comparing all recorded data into simple graphical outputs using a daily time-step, a short period (i.e., 15 minute) time-step, and a high resolution time-step (i.e., 1 minute). Smoothing of data may eliminate noise in the data series. Removal of a recurring peak water use identified as a water leak is an example of data smoothing. Elimination of selective recorded data to achieve a smoothing effect for effective data comparison is termed as 'binning' (Fane *et al.* 2011). Binning helps address minor discrepancies in the data, sometimes through replacement with representative values without causing significant changes to the overall outcome of future data analyses.

Comparing water use across the graphical outputs with different time bases can help detect ambiguities in the data. In addition, the ability to compare mains water and rainwater uses is critical in identifying underlying changes to water use patterns, such as identifying and differentiating between homeowners on holiday vs. pump faults. A data preparation process needs to be carried out using reliable data analysis software as large and comprehensive data sets need to be organised and assessed following the data validation process.

4.5.2 Data analysis

A clear understanding and visualisation of monitored data prior to data analysis is important. There are various methods that can be used for analysis. A commonly used method is water balance analysis.

Water balance analyses are conducted on a basic mass balance principle where water flows within a residential development are measured at the entry and exit points to determine consistency of total water use. The energy consumption by rainwater pumps can also be recorded and assessed to determine the corresponding energy efficiency (i.e., kWh/kL) of the system. Monitoring studies can be used to evaluate the accuracy of residential efficiency programs by matching the detailed monitored data with consumption data from water billing agencies to check for data reliability (Umapathi *et al.* 2012) using cross-sectional analysis.

For more detailed analyses, household appliances such as washing machines can be assessed for their unique water consumption patterns. In such cases, their distinct usage cycles can be obtained from the manufacturer (Aravinthan *et al.* 2012). Knowing the typical water flow rates of toilet flushes and taps is also helpful in assigning water flow patterns to the related end uses. The following sections discuss the data analysis methodologies and outcomes of two Australian case studies that investigated water supply and use from rainwater tanks in South East Queensland and Sydney.

4.6 CASE STUDY 1: RAINWATER TANK MONITORING, SOUTH EAST QUEENSLAND, AUSTRALIA

The study focussed on validating the expected reduction of 70 kilolitre/household/year (kL/hh/yr) in potable water consumption in newly build households with internally plumbed rainwater tanks in South East Queensland. They were constructed under the requirements of the Queensland Development Code MP 4.2, which is detailed in Chapter 5.

The water flow data from 20 such households were recorded over a period of 12 months to assess the rainwater systems for their volumetric reliabilities and to determine their water use profiles. Monitoring a sample size sufficiently large to represent the South East Queensland region would have been too expensive because of the high cost in installing monitoring equipment and remote data transfer. Hence, this study can be considered as a pilot project to provide finer grained detail on the water savings estimated from billing data methods (see Chapter 3).

Monitoring data was initially collected at four instalments over a period of two months on a fortnightly basis, followed by monthly intervals (Umapathi *et al.* 2012) to ensure the system was working effectively before the collected data was analysed. The instrumentation setup (Figure 4.11) recorded water flows associated with the rainwater system in individual households including: (1) mains water supply; (2) tank top-up from mains water supply; (3) total water flow out of the rainwater tank; and (4) water supply to the external garden tap. All monitored water flows were measured using water flow meters that had a pulse ratio of 0.5 L/pulse. The mains water meters, installed by the water utility, were configured at 5 L/pulse. Each monitored household also had energy meters installed to record the energy consumption by the rainwater pumps. The system generated 1 pulse per watt-hour. The water flow and energy consumption data were recorded in 1-minute time intervals. Recorded data was retrieved on a monthly basis using wirelessly enabled data loggers.

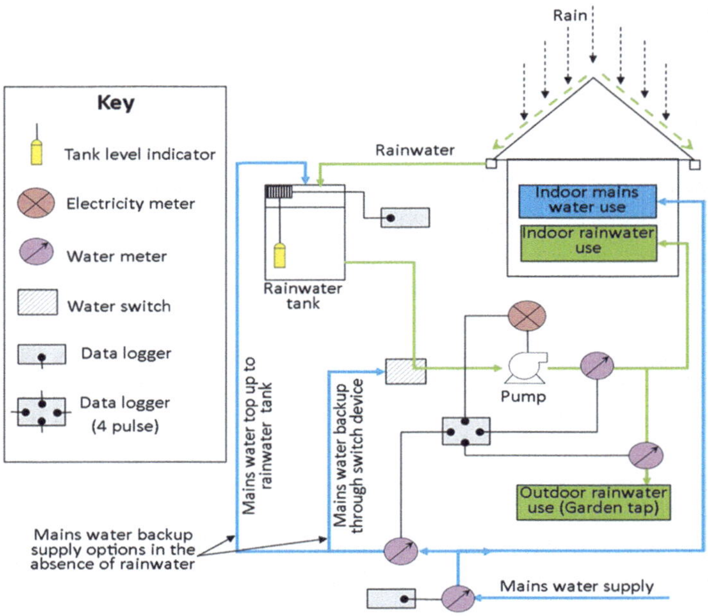

Figure 4.11 Schematic of the metering and logging system setup at 20 households with rainwater tanks in South East Queensland.

The monitored data for each household was assessed for mains water usage, mains water savings as a result of rainwater consumption, rainwater backup required in the absence of rainwater, household rainwater consumption (internal and external), and energy consumption by the rainwater pumps (Umapathi *et al.* 2012, Umapathi *et al.* 2013). The study also examined diurnal pattern and peak water demand analysis wherein the water use data for the initial 4-month period was converted to visual diurnal demand patterns. Hourly demand patterns were also generated to obtain average household water use behaviour for the 20-home cluster. The energy use monitoring also allowed measurement of the two types of top-up systems used in the 20 households, that is, the trickle top-up type (requiring pumping water from the rainwater tank) and the switching device type (which continuously consumed electricity, but result in mains water by-passing the tank). These systems are described in Chapter 5.

The study also examined correlations between measured rainfall patterns and the rainwater collected in individual tanks (Umapathi *et al.* 2012).

4.6.1 Data analysis and results

The average rainwater consumption per household (which equates to mains water savings) was 36.1 kL/hh for a total period of the 11 months during which a complete data set was available over the 12-month monitoring period (Umapathi *et al.* 2012). The total water use per household (combination of rainwater and mains water) was 136 kL/hh. Scaling up the data to a 12-month period gave an average rainwater use of 40 kL/hh/yr and a total water demand of 151 kL/hh/yr (Figure 4.12). As shown in Figure 4.12, the rainwater consumption in the households varied between 0.6 to 69 kL/hh/yr and the total water consumption from 50 to 398 kL/hh/yr. The validity of the monitored potable water use data was verified through cross-correlation with the water billing records obtained from the local water utility. The correlation coefficient was 0.98.

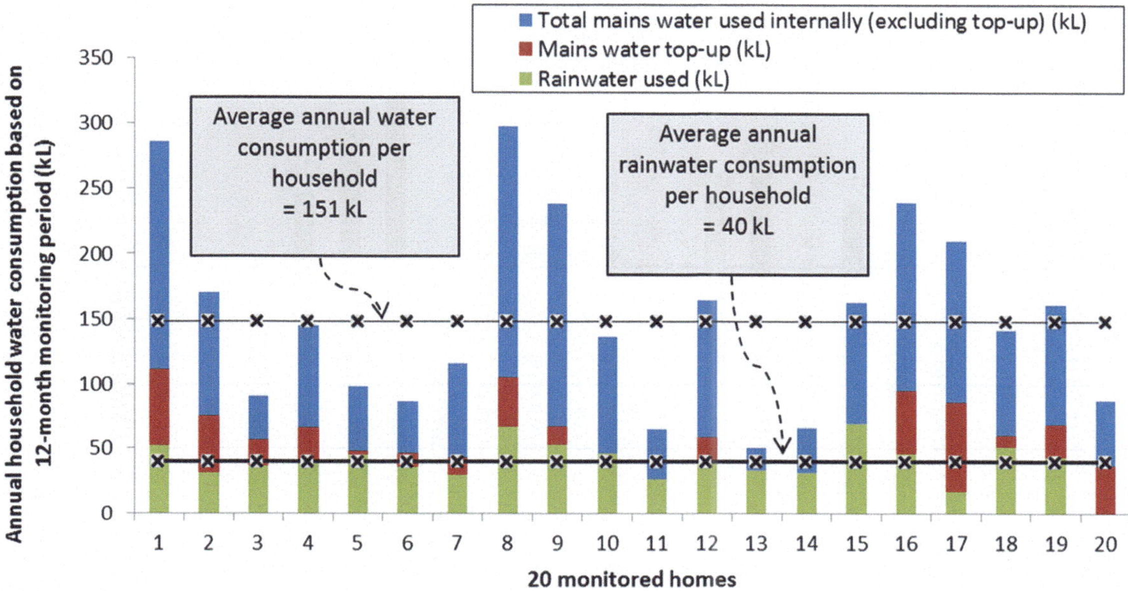

Figure 4.12 Partitioning of the 12 month consumption for 20 monitored homes in SEQ. Data includes total potable, total rainwater, and mains top-up to the tanks. Source: (Umapathi *et al.* 2012).

The results obtained are in contrast with findings from desktop studies (Chong *et al.* 2011) of 691 households using rainwater tanks under very similar circumstances in South East Queensland, which identified an average annual mains water saving of 58 kL/hh/yr and 59 kL/hh/yr for 2009 and 2010 respectively. It may be due to the small sample size of 20 homes. Although the study represented a smaller sample, the results obtained gave a better indication of the various factors that can be involved in influencing the water consumption patterns in everyday households. These factors were: differences in per capita water usage; effects of locally imposed water use restrictions; differences in external water use (irrigation) habits; and variations in local rainfall which varied substantially between households that were located only a few kilometres apart.

The rainwater supply as a percentage of total household water demand for 20 homes is shown in Figure 4.13. The average overall rainwater supply based on the 20 households was 31% and the contribution of rainwater top-up (backup supply) was 14% of total household water usage (Umapathi *et al.* 2012). The results suggests that physical factors such as rooftop collection areas and rainwater tank sizes have a direct impact on the mains water offset, and hence the rainwater reliability of the rainwater in the system. The rainwater tank system could be better designed to reduce backup supply (from 14% of total demand) by either increasing the tank sizes, or the rainwater catchment areas, or a combination of both.

Figure 4.14 depicts the household rainwater demand on the tank for intended use and the corresponding supply of rainwater from the tank. Household rainwater tank systems on, or close to, the optimum demand-supply line demonstrate high volumetric reliability. The rainwater demand was almost completely satisfied by rainwater supply in 6 households as shown by their proximity to the water demand-supply line in Figure 4.14 also corresponding to very low mains water top-up for these households, as shown in Figure 4.12. The homes farthest from the optimum demand-supply line were the least rainwater sufficient.

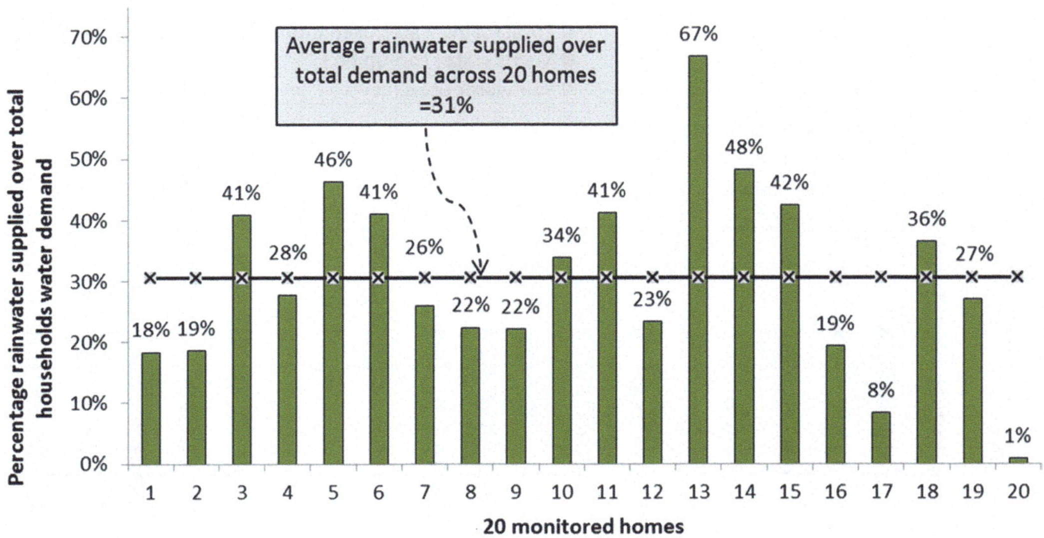

Figure 4.13 Rainwater reliability of rainwater tank systems at 20 monitored homes.

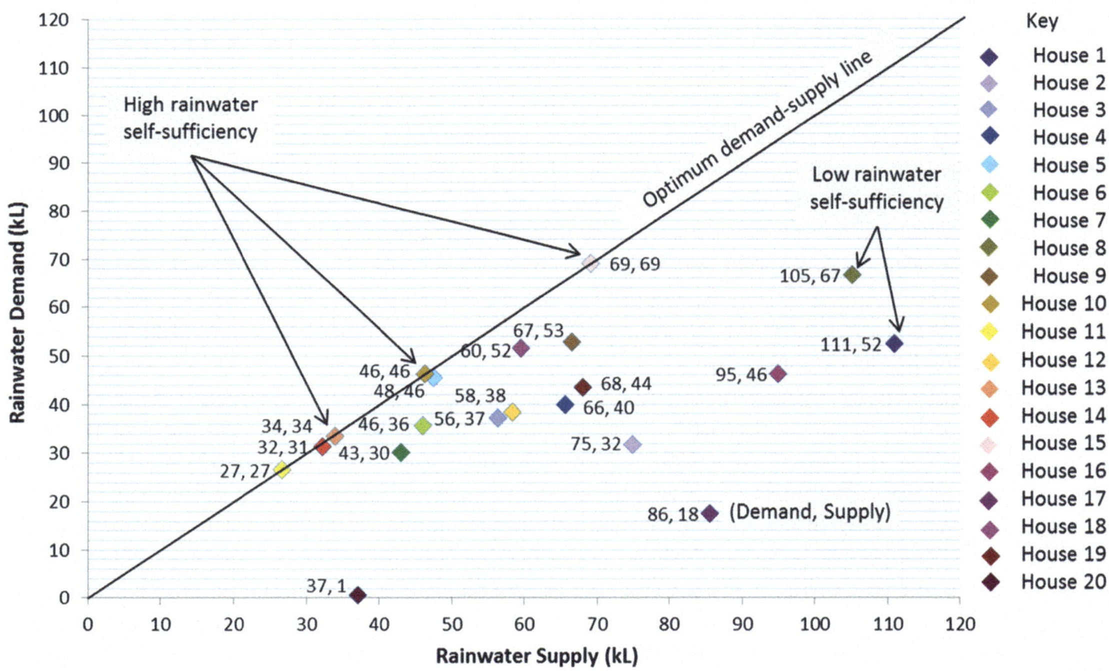

Figure 4.14 Rainwater demand and rainwater supply of 20 households.

Figure 4.15 shows the hourly diurnal water demand pattern for one of the 20 studied households using data recorded over a 4-month period. The graph shows the average hourly water demand of 29.1 L/day, with average peak water usage of 61.4 L/day. The peaking water use factor (ratio of average peak demand divided by average hourly demand) is 2.1, which is low compared to the peak factor of 5 suggested for the design of water supply network systems for developments with a population below 2000 (Swamee & Sharma, 2008). This may be due to a very low external water demand for garden irrigation recorded during the study period. Figure 4.15 also shows the diurnal pattern water supplied from the mains for top-up to the rain tank (MIRW), and the water supply from rainwater tank (TRW).

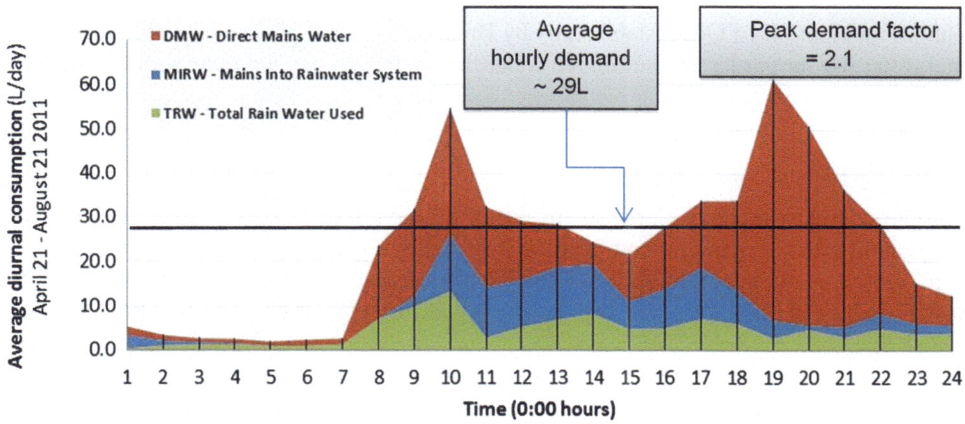

Figure 4.15 Diurnal water use pattern averaged over 4 months for one of the 20 households in the South East Queensland monitoring study.

On the energy front, the average specific energy (SE) for the pumping systems at 19 homes was 1.52 kWh/kL (one home was omitted due to a faulty pump). The average SE for homes with trickle top-up system was slightly higher than those with automatic switching devices, at 1.59 kWh/kL and 1.46 kWh/kL respectively. In the case of the trickle top-up systems, the total water (rainwater + mains water top-up) supplied from the tank would require pumping. In comparison, other alternative potable water supplies such as indirect potable reuse require more than 2.8 kWh/kL and energy required to make freshwater from seawater by desalination is around 3.5 kWh/kL. Thus, rainwater is the least energy intensive alternative water source compared to desalination and indirect potable reuse. Traditional potable water supply from catchment reservoirs, which is often gravity assisted, is less energy intensive than rainwater supply, requiring less than 0.9 kWh/kL (Tjandraatmadja, 2012).

Further information on the performance of pumping systems and the energy use associated with rainwater tank systems is detailed in Chapter 6.

4.7 CASE STUDY 2: SYDNEY WATER, SYDNEY, NEW SOUTH WALES, AUSTRALIA

The New South Wales Government introduced BASIX (the Building Sustainability Index) policy in 2004, which required all single and multi-unit residential buildings to be designed to use less potable water and emit fewer greenhouse gases (NSW Department of Planning, 2008). BASIX set the reduction targets at

40% less for potable water use and 40% fewer greenhouse gas emissions than the average NSW dwelling (Ferguson, 2011). The main objective of the 12-month study conducted by Sydney Water in 2011 was to confirm the water savings being achieved for 52 newly built BASIX compliant households spread broadly across the Sydney basin to capture the variation in climatic range (Ferguson, 2011). The study also aimed at identifying opportunities for improving the water saving capacity and reducing the pumping energy use at the rainwater tank.

The metering arrangement was setup to monitor: 1) the water supply from the rainwater system for connected non-potable end uses; 2) mains water demand for the top-up system; and 3) total mains water use in the household (including top-up), and energy demand from the rainwater tank pumping system. The instrumentation setup was similar to that shown in Figure 4.11. The monitoring set-up used in this study was adopted in the South East Queensland study by Umapathi *et al.* (2013), which additionally monitored the external water use for garden irrigation. All water flow data were logged in one-minute intervals using meters that generated 0.5 L/pulse.

4.7.1 Data analysis and results

Using a sample size of 40 detached households out of 52 monitored homes, Ferguson (2011) found that the total household water demand ranged from 84 to 556 kL/yr, with a mean demand of 197 kL/yr. These results were shown to coincide closely with the potable demands for new homes built under BASIX regulations. The water demand from the rain tanks ranged from 5 to 161 kL/year with a mean demand of 59 kL/yr.

The water savings achieved from rainwater use in 46 households with complete data ranged from 0 to 96 kL/yr (Figure 4.16), with a mean savings of 38 kL/yr (median of 39 kL/yr). Hence, although the demand for water from the rainwater tank was on average of 30% of the total household water demand, only 19% was met by rainwater from the tanks. The difference was supplied by mains water top-up.

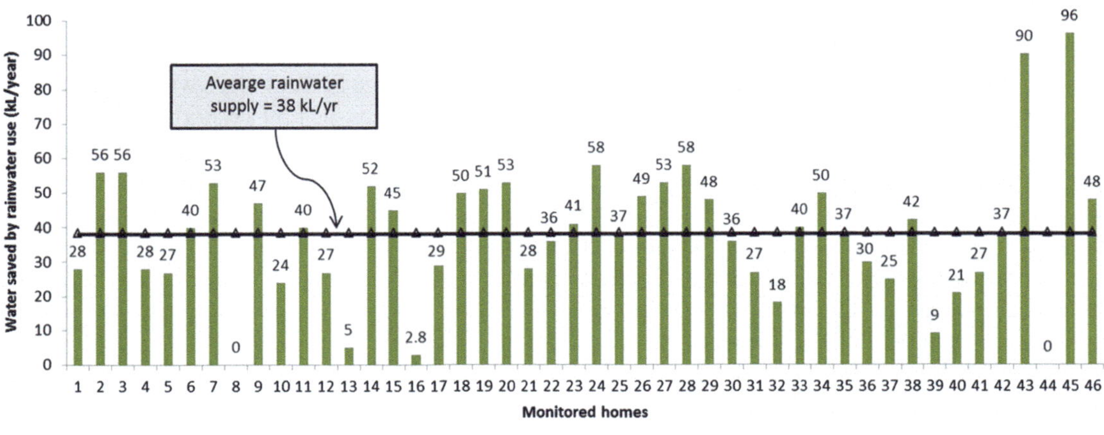

Figure 4.16 Water savings from rainwater tanks in 46 monitored households in the Sydney study.

The study found that the connected roof area (rain catchment area) has a major impact on the volumes of rainwater collected in the tanks. A case study on a single home with a 5 kL rainwater tank and 140 m^2 of connected roof area found that the demand for rainwater was 161 kL/yr out of a total household demand of 556 kL/yr (Ferguson, 2011). Modelling using household specific data with local rainfall showed that the water savings (i.e., rainwater yield) could be increased by 12 kL to 81 kL/year by increasing the roof collection area to 210 m^2.

Modelling assessment of another single household indicated the underperformance of the rainwater tank system that achieved a savings of 28 kL per year, compared with a potential of 45 kL per year. This could be attributed to an unusually high cut-in level in the tank for activation of the backup system, thus leaving less capacity in tank for captured rainwater. Hence, notwithstanding the system having adequate roof catchment area and storage capacity, there is a need to have a clear understanding of settings for the other system components which are determined at the time of tank installation.

The 12-month study period also allowed researchers to assess the seasonality of water use in the households. Water demand for indoor end uses such as toilets and washing machines was found to be constant throughout the year in contrast to outdoor end uses that showed seasonal fluctuations (water demand was higher in summer season than the winter season). The study also found that the water demand between November 2009 and January 2010 (summer months in Australia) appeared to be met by both potable (mains water) and non-potable (rainwater) water sources.

A major part of the study focussed on linking the water use from the rainwater tank systems with the energy consumed in supplying the assigned end uses. Energy use by rainwater pumps was found to be an average of 78 kWh/hh/yr per household per year, with the median energy intensity of 1.48 kWh/kL. The median active pumping energy (energy used for pumping water only) intensity was 1.42 kWh/kL. The dormant (stand-by) energy use was a negligible 4 kWh/year. Due to the detailed nature of the data obtained, significant uses of dormant energy (up to 8 Wh/minute) were identified in some households which warranted further investigation as they suggest faulty pumps or leaks in the rainwater system.

Results also showed that most household water uses had low flow rates, with over 75% of the household uses being 10 L/minute or less and 50% were 6 L/minute or less. The maximum water flow rates measured ranged from 10 to 22 L/minute. However, some of the installed pumps (more specifically, submersible pumps) were designed for much higher flow rates, which suggest they were oversized for the task and hence used more energy than needed to provide the acceptable level of service.

The majority of water use events were between 4.5 to 9 L per event, which were most likely full-flush toilet events. Flow events between 50 and 90 L accounted for 15% of the demand and were most likely washing machine end uses, whereas large water use events (>99 L) that contributed about 20% of the total demand may have been associated with some (top loading) washing machines and garden uses. The study highlighted the significance of choosing the correct pump capacity (kW) matched to the flow rates of the connected end uses. The study also suggested the positive benefits of connecting large volume, high flow rate end uses, such as garden taps, to rainwater tanks to ensure effective use of tank storage (adequate storage space preceding rainfall events) and efficient pumping operation. These aspects in details are also covered in more detail in Chapter 6.

4.8 OTHER COST CONSIDERATIONS ASSOCIATED WITH MONITORING

Apart from the obvious costs involved in the procurement of monitoring equipment such as loggers, meters, cables and so on, there are some other factors that affect the overall cost of the monitoring, and these should also be considered in the contingencies for planning the study.

Participant recruitment – This can be done by a contractor or the water utility with a mail out to selected properties known to have rainwater tanks. The records of rainwater tanks are varied, depending on the location. In the Australian state of New South Wales (NSW), the Department of Planning holds records on BASIX certification and details on size and end use for rainwater systems. Councils and utilities may also have records of tank installations as a result of rebates.

Installation scheduling – The installation process requires careful coordination with participant homeowners. Maps are used to determine the location of households, the time required to install equipment, and then the travel time to the next location.

Other factors – Other issues can arise during the installation process which adds to the time and cost of the project, these include:

- *Rescheduling:* Occasional delays in installation of equipment may be expected due to unforeseen circumstances which may prolong installation time periods.
- *Damage or failure of equipment:* Equipment, particularly for mains meter monitoring, can be damaged by vandals, lawn mowers and even cars throughout the project. These faulty units need to be replaced as soon as practicable to maintain dataflow.
- *Mobile phone reception:* Some areas have insufficient mobile phone reception to regularly upload the data, requiring site visits for manual downloads.
- *Inaccessible or inappropriate systems:* There have been cases where the plumbing required to install the meters on the inlets and outlet lines cannot be accessed because the pipes are located within walls or covered by other materials, or the pipe fittings were of a particular material unavailable to the installer. There are about 15 different pipe types (in Australia) with different fittings, which need different crimping tools. In many cases, it was easier to move on to another house than purchase a new set of tools and fittings for just the one installation.

4.9 CONCLUSION

The increasing integration of rainwater tanks and any alternative water supply technologies is expected to have an impact on the future demands of centralised systems. Therefore, the need to validate and assess the influence of these alternative systems on existing centralised infrastructure is important. Monitoring the rainwater tanks will help determine the reliability of these systems and enable better planning by policy makers. Studying the rainwater use patterns within households enables governments and utility planners to develop suitable guidelines for installing rainwater tanks to reduce the reliance on potable water.

Nonetheless, monitoring is a comprehensive process. Monitoring programmes may face a range of setbacks, from exorbitant costs to the inability to obtain statistical significance in analytical outcomes due to variations in the samples, or simply the failure of all or part of the monitoring equipment. The chapter has also briefly discussed an emerging trend in instrumentation available for water consumption assessments. Experimental and analytical methodologies, and also issues associated with monitoring based assessments have been discussed using studies conducted in a few major cities across Australia as examples. The monitoring studies have identified a gap between modelled and monitored data on household rainwater usage, providing valuable information for water planners for developing regional strategic water plans. The main conclusions are summarised below:

- Technology allows sub-hourly monitoring of water use, tank water supply and energy consumption for investigating the rainwater usage, diurnal patters of water supplied by various sources and specific energy consumption in rainwater supply.
- High cost of installation (approximately $5000 per house) limits the number of monitoring sites. Hence, these studies are generally conducted at pilot scale and the data collected may not be representative of suburb-scale tanks behaviour.
- Technology supports rich data collection of rainwater supply, which can be done remotely. However, this can be labour intensive and more time consuming if the study sites are wide

spread. Daily or sub-daily data collection will also be impossible, thus limiting the scope of the study.

- Rainwater tank monitoring allows checking on compliance with government policies and building codes, for example, the Queensland Development Code and BASIX. The South East Queensland and the Sydney Water studies both showed significant rainwater supply undershoot (40 kL/hh/yr and 38 kL/hh/yr respectively).
- Water flow data collected at hourly, monthly or even daily time-steps (intervals) are sufficient if the objective of the research is to quantify the gross water consumption in households. However, smaller sub-hourly intervals, such as one-minute intervals (or lower), help in understanding the finer details associated with end use consumption characteristics and can provide insights into leakage occurrence, malfunctioning of monitoring equipment or of the end-use appliances, as well in the design of reticulation pipe sizing in new suburbs with mandated tanks. Nevertheless, analysis of fine resolution data can be cumbersome and thus the selection of the time interval should be based on the study objectives.
- One minute monitoring intervals were chosen in this case study to avoid problems attributed to larger time intervals, however study indicated that 6-minute monitoring intervals will also be suitable for understanding diurnal flow patterns.
- Tank water level monitoring provides rain catch figures on an event basis, which can be compared with modelled rain capture information data. Very few studies of this type are reported in the literature.

4.10 REFERENCES

Aravinthan V., Ho G., Gray J., Butler R. and Connell D. (2012). What are smart meters? In: St Leonards, Australia, Australian Water Association, pp. 1–4, https://www.awa.asn.au/Fact_Sheets/ (accessed January 2015).

Australian Government (2013). Your Home: Water Use 7.3 Rainwater. www.yourhome.gov.au (accessed October 2013).

Beal C., Sharma A., Gardner T. and Chong M. (2012). A desktop analysis of potable water savings from internally plumbed rainwater tanks in South-East Queensland, Australia. *Water Resources Management*, **26**(6), 1577–1590.

Beal C., Stewart R A., Huang T. and Rey E. (2011). SEQ residential end use study. Smart Water Systems and Metering, *Journal of the Australian Water Association*, **38**(1), 80–84.

Beal C., Stewart R. A. and Fielding K. (2013). A novel mixed method smart metering approach to reconciling differences between perceived and actual residential end use water consumption. *Journal of Cleaner Production*, **60**, 116–128.

Boyle T., Giurco D., Mukheibir P., Liu A., Moy C., White S. and Stewart R. (2013). Intelligent metering for Urban water: A Review. *Water*, **5**(3), 1052–1081.

Chong M. N., Umapathi S., Mankad A., Gardner E. A., Sharma A. and Biermann S. (2011). Estimating water savings from mandated rainwater tanks in South East Queensland. In: Begbie D. K. and Wakem S. L. (eds) (2011) Science Forum and Stakeholder Engagement Building linkages, Collaboration and Science Quality, Urban Water Security Research Alliance 14–15 September, Brisbane, Queensland, pp. 31–34, http://www.urbanwateralliance.org.au/publications/forum-2nd-2011/science-forum-2011-program-abstracts.pdf (accessed July 2014).

Chong M. N., Sharma A., Umapathi S. and Cook S. (2012). Understanding the Mains Water Saving from Mandated Rainwater Tanks using Water Balance Modelling and Analysis with Inputs from On-Site Audited Parameters. Urban Water Security Research Alliance Technical Report No. 65, Urban Water Security Research Alliance, Available: http://www.urbanwateralliance.org.au/publications/technicalreports/ (accessed August 2014).

Chong M. N., Sharma A. K., Umapathi S. and Cook S. (2014). Performance of Plumbed Rainwater Tanks in Urban Australia: Understanding the potential mains water saving using integrated water balance modelling, regression and response surface analysis. Manuscript submitted.

Chowdhury R. (2013). A stochastic model of domestic water consumption and greywater generation in the Al Ain city. In: 20th International Congress on Modelling and Simulation, Adelaide, Australia, 1–6 December 2013.

Cook S., Sharma A. and Chong M. (2013). Performance analysis of a communal residential rainwater system for potable supply: A case study in Brisbane, Australia. *Water Resources Management*, **27**(14), 4865–4876.

Coombes P. J. and Kuczera G. (2003). Analysis of the Performance of Rainwater Tanks in Australian Capital Cities. In: Paper Presented to the 28th International Hydrology and Water Resources Symposium, 10–14 November 2003, Wollongong, New South Wales, Engineers Australia.

Coombes P. J., Argue J. R. and Kuczera G. (2000). Figtree place: a case study in water sensitive urban development (WSUD). *Urban Water*, **1**(4), 335–343.

Davison G. (2008). Down the gurgler: historical influences on Australian domestic water consumption. In: Troubled Waters: Confronting the Water Crisis in Australia's Cities, P. Troy (ed.). Australian National University Press, Canberra, Australia, pp. 37–65.

DSITIA (2013). SILO Climate Data. The State of Queensland Department of Science, Information Technology, Innovation and the Arts.

Ecovision Systems (2014). Ecovision 2000 Series. http://www.ecovisionsolutions.com.au/products/ (accessed 13 January 2014).

Fane S., Turner A., McKibbin J., May D., Fyfe J., Chong J., Blackburn N. and Patterson J. W. S. (2011). Integrated Resource Planning for Urban Water - Resource Papers, Waterlines Report, National Water Commission. Canberra, University of Technology Sydney.

Ferguson M. (2011). Rainwater tank monitoring report – A 12-month one-minute interval data study of rainwater tank water savings and energy use for 52 real life installations, Sydney Water, https://www.sydneywater.com.au/web/groups/publicwebcontent/documents/document/zgrf/mdu2/~edisp/dd_056073.pdf (accessed July 2014).

Ferguson M. (2012). A 12-Month Rainwater Tank Water Savings and Energy Use Study For 52 Real Life Installations. In: Paper Presented to the Oz Water '12 Sydney, Sydney, Australia, 8–10 May 2012.

Fewkes A. (1999). The use of rainwater for WC flushing: the field testing of a collection system. *Building and Environment*, **34**(6), 765–772.

Ghisi E. and Ferreira D. F. (2007). Potential for potable water savings by using rainwater and greywater in a multi-storey residential building in southern Brazil. *Building and Environment*, **42**(7), 2512–2522.

Ghisi E. and Mengotti de Oliveira S. (2007). Potential for potable water savings by combining the use of rainwater and greywater in houses in southern Brazil. *Building and Environment*, **42**(4), 1731–1742.

Giurco D., Carrard N., McFallan S., Nalbantoglu M., Inman M., Thornton N. and White S. (2008). Residential end-use measurement Guidebook: a guide to study design, sampling and technology. Prepared by the Institute for Sustainable Futures, UTS and CSIRO for the Smart Water Fund, Victoria.

Giurco D. P., White S. B. and Stewart R. A. (2010). Smart metering and water end-use data: conservation benefits and privacy risks. *Water*, **2**(3), 461–467.

Heinrich M. (2008). Water use in Aukland Households, Aukland Water Use Syudy (AWUS) Final Report. BRANZ Ltd, Judgeford, New Zealand.

IEC (2004). IEC 60529. In: Degrees of Protection by Enclosures (IP Code) (Identical National Adoption), International Electrotechnical Comission, Geneva, Switzerland.

Jones M. P. and Hunt W. F. (2010). Performance of rainwater harvesting systems in the southeastern United States. *Resources, Conservation and Recycling*, **54**(10), 623–629.

Kenway S., Turner G., Cook S. and Baynes T. (2008). Water-energy Futures for Melbourne: the Effect of Water Strategies, Water Use and Urban form, CSIRO.

Khastagir A. and Jayasuriya N. (2010). Optimal sizing of rain water tanks for domestic water conservation. *Journal of Hydrology*, **381**(3–4), 181–188.

Lloyd S. D. (2001). Water Sensitive Urban Design in the Australian Context – Synthesis of a conference held 30–30 August 2000, Cooperative Research Centre for Catchment Hydrology, Melbourne Australia.

Mankad A., Chong M. N., Gardner T. and Sharma A. (2012). Examining biophysical and socio-demographic factors across mandated tank users in urban Australia: A linking step towards achieving best practices. *Water Resources Management*, **26**(7), 1983–1998.

Mayer P. W., DeOreo W. B. and Alexander M. (1996). Trace Wizard Water Uses Analysis Software. Aquacraft Inc., Boulder, Colorado.

Mayer P. W., DeOreo W. B. and Foundation A. R. (1999). Residential End Uses of Water. American Water Works Association Research Foundation(AWAA), Denver, Colorado.

Moglia M., Tjandraatmadja G., Sharma A., Walton A., Umapathi S. and Gardner J. (2012). Proposed Strategy Portfolio for the Management of Rainwater Tanks: the South East Queensland case. Urban Water Security Research Alliance Technical Report No. 75, Urban Water Security Research Alliance. http://www.urbanwateralliance.org.au/publications/technicalreports/ (accessed August 2014).

Moglia M., Tjandraatmadja G., Delbridge N., Gulizia E., Sharma A. K., Butler R. and Gan K. (2014). Survey of Savings and Conditions of Rainwater Tanks. Smart Water Fund and CSIRO, Australia, Melbourne.

Neenan B. and Hemphill R. C. (2008). Societal benefits of smart metering investments. *The Electricity Journal*, **21**(8), 32–45.

Nguyen K. A., Zhang H. and Stewart R. A. (2013). Development of an intelligent model to categorise residential water end use events. *Journal of Hydro-environment Research*, **7**(3), 182–201.

NSW Department of Planning (2008). BASIX Monitoring Report Water Savings for 2007–08: Final report prepared by Sydney Water for the NSW Department of Planning.

Sharma A. K., Cook S. and Chong M. N. (2013a). Monitoring and validation of decentralised water and wastewater systems for increased uptake. *Water Science & Technology*, **67**(11), 2576–2581.

Sharma A. K., Tjandraatmadja G., Cook S. and Gardner T. (2013b). Decentralised systems-definition and drivers in the current context. *Water Science & Technology*, **67**(9), 2091–2101.

Swamee P. K. and Sharma A. K. (2008) Design of Water supply Pipe Networks, John Wiley & Sons, New Jersey, USA. http://onlinelibrary.wiley.com/doi/10.1002/9780470225059.fmatter/pdf (accessed July 2014).

Talebpour M. R., Stewart R. A., Beal C., Dowling B., Sharma A. and Fane S. (2011). Rainwater tank end usage and energy demand: a pilot study. *Water: Journal of the Australian Water Association*, **38**, 97–101.

Tjandraatmadja G., Pollard C., Sharma A. K. and Gardner E. A. (2012) Optimisation of Energy Use in Household Rainwater Supply Systems. Urban Water Security Research Alliance Technical Report No. 89, Urban Water Security Research Alliance. http://www.urbanwateralliance.org.au/publications/technicalreports/ (accessed August 2014).

Turner A., Willetts J., Fane S., Giurco D., Chong J., Kazaglis A. and White S. (2010). Guide to Demand Management and Integrated Resource Planning. Water Services Association of Australia (WSAA): Sydney, Australia, pp. 1–174.

Umapathi S., Chong M. N. and Sharma A. (2012). Investigation and Monitoring of Twenty Homes to Understand Mains Water Savings from Mandated Rainwater Tanks in South East Queensland. Urban Water Security Research Alliance Technical Report No. 63, Urban Water Security Research Alliance. http://www.urbanwateralliance.org.au/publications/technicalreports/ (accessed August 2014).

Umapathi S., Chong M. N. and Sharma A. K. (2013). Evaluation of plumbed rainwater tanks in households for sustainable water resource management: a real-time monitoring study. *Journal of Cleaner Production*, **42**, 204–214.

UNEP (2009). Rainwater Harvesting: a Lifeline for HumanWell-being. Chapter 6, Rainwater Harvesting for Water Security in Rural and Urban Areas 2009: United Nations Environment Programme and Stockholm Environment Institute.

Villareal E. L. and Dixon A. (2005). Analysis of a rainwater collection system for domestic water supply in Ringdansen, Norrköping, Sweden. *Building and Environment*, **40**, 1174–1184.

Whaley D. M., Saman W., Halawa E. and Mudge L. T. (2010). Lessons Learnt from Implementing Intelligent Metering and Energy Monitoring Devices in a new Housing Development. Canberra.

Willis R. M., Stewart R. A., Giurco D. P., Talebpour M. R. and Mousavinejad A. (2011a). End use water consumption in households: impact of socio-demographic factors and efficient devices. *Journal of Cleaner Production*, **60**, 107–115.

Willis R. M., Stewart R. A., Williams P. R., Hacker C. H., Emmonds S. C. and Capati G. (2011b). Residential potable and recycled water end uses in a dual reticulated supply system. *Desalination*, **272**(1–3), 201–211.

Zhang Y., Grant A., Sharma A., Chen D. and Chen L. (2009). Assessment of rainwater use and greywater reuse in high-rise buildings in a brownfield site. *Water Science & Technology*, **60**(3), 575–581.

Chapter 5

Physical verification of household rainwater tank systems

Sharon Biermann and Reid Butler

ABSTRACT

When rainwater harvesting is promoted or even mandated by governments in order to achieve potable water savings, it is critical that harvesting systems are installed correctly to achieve the planning objectives and targets. Whilst there are numerous design guidelines available for the installation of systems, there is little evidence of post installation monitoring of rainwater tank installation to assess if the *as built* systems follow the *as designed* specifications, on which all modelled gains of water supply are based. Responding to this knowledge gap in installation monitoring, and focussing on the physical installation factors of collection, storage and connections to designed end uses, this chapter describes a generic rainwater tank installation compliance audit protocol, and describes its application in South East Queensland (SEQ), Australia. The on-site physical assessment established that installed rainwater tank storage capacity, in kilolitres (kL), is mostly above requirements (5 kL), with only 16% of sites below requirements; connected roof catchment area (100 m^2 or 50% of roof area) does not meet requirements in 40% of cases; whilst connection to toilets, washing machine and external tap meets requirements in most cases. Except for the installation of backflow devices that prevent contamination of reticulated town water supply, requirements for water quality protection are adequately met. A high level of compliance exists for the requirement for a continuous supply of water to internal fixtures supplied from a rainwater tank by installing a mains water supply backup system.

Keywords: Rainwater tank installation; installation monitoring framework; verification.

5.1 INTRODUCTION

Capturing and using rainwater for the purposes of reducing the demand for potable water, boosting water supply and reducing negative environmental impacts and costs is well documented, particularly in the contexts of increasing population, dry and variable climatic conditions, limited or unreliable access to potable water and fiscal constraints. Governments and water authorities have promoted rainwater capture and use through public campaigns, financial incentives such as rebates, and in some cases, regulatory enforcement. Some form of rainwater harvesting is mandatory for buildings and homes in various cities and states in India, Catalonia in Spain, Flanders in Belgium (UN-HABITAT, 2005; Environmental Agency UK, 2008) and in Jordan (Abdulla & Al-Shareef, 2009). In North America, rainwater harvesting is mandatory in new buildings in Tucson, Arizona, Santa Fe County in New Mexico and several Caribbean islands. In Australia, the states of South Australia, New South Wales (NSW) and Queensland, require all new buildings to install a rainwater harvesting system or some other alternative source of water supply (New South Wales Government, 2004; DLGP, 2008; Government of South Australia, 2010), although this regulation has recently become optional in Queensland (Department of Housing and Public Works, 2013).

There has been considerable attention given to the planning and design aspects of rainwater harvesting. Determining the potential feasibility and cost-effectiveness of roof rainwater harvesting, mainly at the scale of cities, has been a primary focus of investigation involving modelling and projections of potential potable water savings under contextual climatic, household demographic and tank design regimes (Hermann & Schmida, 1999; Coombes & Kuczera, 2003; Coombes *et al.* 2003; Handia *et al.* 2003; Villareal & Dixon, 2005; Ghisi, 2006; Ghisi *et al.* 2007; Abdulla & Al-Shareef, 2009; Aladenola & Adeboye, 2010; Tam *et al.* 2010). Studies to improve tank performance have been undertaken largely at the household or building scale, focussing on optimising the collection, storage and use of rainwater under specific contextual conditions, usually involving simulation modelling (Fewkes, 1999a, 1999b; Vaes & Berlamont, 2001; Ghisi, 2010; Khastagir & Jayasuriya, 2010; Imteaz *et al.* 2011). Numerous best practice guidelines have been prepared in support of policy implementation (Sehgal, 2005; Texas Water Development Board, 2005; Chapman *et al.* 2008; Commonwealth of Australia, 2010; Environment Agency UK, 2010; Canada Housing and Mortgage Corporation, 2012).

However there is much less evidence relating to measuring the actual on-the-ground implementation of policy objectives, for example, potable water savings. Evidence of actual water savings through performance monitoring has been provided at the building and individual settlement levels (Coombes *et al.* 1999; Zaizen *et al.* 1999; Coombes *et al.* 2000; Coombes *et al.* 2003; Coombes *et al.* 2004; Coombes *et al.* 2006; Domènech & Saurí, 2011; Umapathi *et al.* 2012). At the wider city level, the most comprehensive evidence of actual potable water savings achieved is found in research undertaken in South East Queensland (SEQ), Australia (Beal *et al.* 2012; Chong *et al.* 2011b). Other than the research in SEQ, on which this chapter is based (Biermann *et al.* 2012), and another 'on-ground truthing' water savings monitoring study in Sydney (Sydney Water, 2012), no other published evidence has been found of implementation or installation monitoring. In situations where rainwater tanks are mandatory, and authorities are depending on the savings achieved in order to defer the construction of additional potable water infrastructure, it seems to us essential that post-installation monitoring studies be conducted to ensure success of planning expectations and optimisation studies. In addition to monitoring the achievement of planning goals, post-installation monitoring is important for the following crucial quality reasons:

- If a tank is not installed or connected properly, it will not function as designed and may result in the tank being turned off or removed by the owner.
- A tank may be assumed to be functioning correctly while actually not, and the benefits will either be reduced or lost.
- Non performing tanks will damage the credibility of rainwater tanks in general and impact on future uptake.
- A badly installed tank may pose a health risk through poor water quality or mosquito breeding allowing the transmission of arboviruses.

A rainwater tank system normally includes a rainwater tank connected to a roof area as the rainwater catchment for rainwater harvesting through roof gutters and downpipes, a pump for supplying rainwater at the required flow and pressure to connected end uses, and a backup mechanism to supply mains water when the tank water level falls below a certain level. Any aspect of this system, if not installed properly, will result in reduced performance benefits. There is agreement in the literature that the critical factors influencing the water supply performance of rainwater tanks include:

- *Tank size* – 'The most important factor relating to the efficiency of the rainwater system is the correct sizing of the rainwater tank' (Ghisi, 2010, p 2381). 'Water saving efficiency is more sensitive to changes in storage capacity [than collection area]' (Villarreal & Dixon, 2005, p. 1180). '... increases in water supply benefits from rainwater tanks diminish with larger tank volumes while stormwater management benefits increase with tank volume' (Coombes & Kuczera, 2003, p. 242).
- *Connected roof area* – Ghisi (2010a) reported a good correlation between potable water savings and rainwater tank capacity for cities with high rainfall or where roofs are greater than 200 m^2. In Western Sydney, Coombes and Kuczera (2003, p. 241) reported that 'mains water savings provided by the rainwater tanks increase with larger roof areas'.
- *Demand for non-potable water* – 'We argue that the volume of rainwater consumed and the economic feasibility are determined by the end-uses given to rainwater, which are frequently limited to a few purposes.' (Domènech & Saurí, 2011, p. 598). The economic feasibility increases with the increase in rainwater consumption.
- *Rainfall supply factors* – Imteaz *et al.* (2011) found that in the cases of two underground rainwater tanks at Swinburne University in Melbourne, the tanks were effective in wet and average years but less effective in dry years. Vaes *et al.* (2001) concluded that 'rainfall is the most important input for many hydrological and hydraulic design calculations'. 'Brisbane and Sydney, with larger annual rainfall depths, provided greater yields from rainwater tanks' (Coombes & Kuczera, 2003, p. 242).

This chapter relates to the physical factors of tank size, connected roof area and the physical aspects of increasing demand through the connection of appliances to rainwater supply. Responding to the identified knowledge gap in auditing the installation of rainwater tank systems, the purpose of this chapter is to present a generic rainwater tank installation audit protocol, and describe its application in SEQ, Australia. The objectives are to identify any installation issues which could adversely impact on achieving potable mains water savings, and also to highlight potential improvements to the audit protocol based on the learnings gained from its application.

5.2 RAINWATER TANK INSTALLATION COMPLIANCE MONITORING FRAMEWORK

The audit protocol used to evaluate physical installation compliance in SEQ was based on the initial assessment framework developed as part of a pilot study to test the compliance of new housing units in Sydney, NSW with the Building Sustainability Index BASIX certification requirements (New South Wales Government, 2004). This initial audit protocol was further developed for application in SEQ by incorporating the requirements of the 2007 Queensland Development Code Mandatory Part 4.2 (QDC MP 4.2) (DLGP, 2008).

5.2.1 Sydney BASIX assessment

BASIX, the Building Sustainability Index, was introduced in 2004 by the NSW Government and required all new houses and units to achieve a 40% reduction in potable water and energy consumption, compared to the average 'pre-BASIX' home. Each complying development needs to complete an on-line application for a BASIX certificate. This certificate outlines the requirements the household needs to achieve to comply with the regulation. Examples of requirements include installing a toilet of a certain minimum water efficiency (i.e., minimum water star rating), installing a rainwater tank of a certain size, and connecting certain internal end uses (e.g., toilets and washing machines) to the tank.

BASIX requires that all tanks are installed according to Australian Standard 3500 of the National Plumbing and Drainage Code (Standards Australia, 2003) and the installing plumber can certify whether this has been done. Inspection of the tank installation, however, is often overlooked during other building inspections as tanks are not necessarily fully installed until all other building work is complete. When local council or private certifiers assess whether all BASIX requirements are met, which includes elements such as insulation, glazing, window shading and water efficient and energy efficient electrical fixtures, these can be accepted as complete based on receipt of specification details, and no site check is carried out. This means that several critical design elements including the connected roof area, number of connected fixtures and filter systems are not checked. Also, the simple task of turning on the rainwater pump can be missed when a home owner moves into their new home, with the assumption that it is already running. In NSW, the Department of Fair Trading inspects plumbing installations; however, this often does not include rainwater tanks, which can be installed after the inspection.

An assessment of the potable water savings attributable to rainwater tanks was undertaken in 2009 at a pilot scale of 52 households to test whether this sample of properties was in compliance with BASIX requirements, and to quantify whether the expected 40% water saving was being achieved (Sydney Water, 2012). This included monitoring of water use from the rainwater tanks and is detailed in Chapter 4. Another on-ground truthing study was undertaken in 2010 to check the installations and water using fixtures in a larger sample of 475 houses in Sydney. The information collected will help clarify whether the BASIX savings are likely to be maintained in the long-term. The analysis results are not yet (2014) publically available from Sydney Water.

5.2.2 Queensland Development Code Performance criteria

The QDC MP4.2 required all detached residential dwellings built after 1 January 2007 to save mains water through the use of supplementary water sources (DLGP, 2008). In SEQ, the water savings target was 70 kilolitres/household/year (kL/hh/yr). Installing a rainwater tank internally plumbed to the washing machine cold water tap, toilet cisterns and at least one outdoor tap for external use was the typical compliance method chosen. Other acceptable solutions included communal rainwater tanks, greywater treatment plants, dual reticulation of recycled water and stormwater reuse, or a combination of these sources.

However, buildings in Queensland no longer have to meet compulsory water savings targets, following the repeal of laws mandating the installation of water supply systems on 1 February 2013 (Department of Housing and Public Works, 2013). Provisions have been made for local governments to opt-in to water savings requirements in recognition of Queensland's varying climatic conditions and regional circumstances. Builders in these local government areas still need to comply with water savings requirements.

Under the (now superseded) QDC MP 4.2, Performance Criteria P2-P6 related to the requirements for rainwater tank installation, capacity and water quality protection for Class 1 buildings (detached dwellings) supplied directly from the reticulated town water supply. Acceptable solutions (A2-5) listed in the QDC for achieving the performance criteria for a detached Class 1 dwelling, where rainwater tanks are selected as the alternate source for achieving water savings targets, were as follows:

- Water savings targets:
  - *Capacity* – minimum storage capacity of 5000 litres (L).

    - *Catchment* – installed to receive rainfall from a minimum of 50% of total roof area or 100 m^2 (whichever is the lesser).
    - *Connection to* – (1) toilet cistern/s and washing machine cold water taps (other than those connected to a greywater treatment plant or alternative water substitution measure); *and* (2) an external use.
- Water quality protection:
    - *Prevention of contaminants from entering rainwater tank* – (1) screened downpipe rainhead having a screen mesh 4–6 mm to prevent leaves from entering downpipe; *and* (2) discarding a minimum of 20 litres of first flush runoff where rainwater is connected to showers, wash basins, kitchen or hot water services.
    - *Prevention of mosquito breeding and vermin entering the rainwater tank* – (1) either mosquito-proof screens (suitable material and not coarser than 1 mm aperture mesh); or flap valves at every opening of tank; *and* (2) a vermin trap; or (3) in case of wet systems, mosquito-proofing in accordance with Standard HB230 for rainwater tank installation and design.
    - *Continuous supply of supplementary water to internal fixtures from the reticulated town water supply* – by either (1) an automatic switching device; or (2) a trickle top-up system.
    - *Prevention of contamination of the reticulated town water supply from the rainwater tank* – a backflow prevention device is installed.

5.2.3 Rainwater tank installation audit protocol

This chapter focuses on a study undertaken as part of the Urban Water Security Research Alliance (UWSRA, www.urbanwateralliance.org.au) project on decentralised systems, which, *inter alia*, aimed at understanding the effectiveness of achieving the mandated 70 kL/hh/yr water saving target specified under the SEQ Water Strategy (2010) (Queensland Water Commission, 2010). Whilst other parts of the project considered demographic and behavioural factors influencing water savings, this chapter relates to that part of the study which considered physical installation factors.

The audit protocol initially developed for the Sydney BASIX compliance assessment (Sydney Water, 2009) was adapted to accommodate the requirements of the SEQ-specific QDC MP 4.2 water savings target. This protocol formed the basis of the household inspection protocol undertaken in SEQ in 2011. The final framework was the result of a number of iterations based on inputs obtained from a number of key stakeholders, including the water service provider, after piloting the survey for ten households. The final protocol covers the following aspects (Figure 5.1):

- characteristics of individual dwellings (e.g., dwelling type, total roof area, property dimensions);
- installation of water efficient appliances and fixtures (e.g., showerheads, washing machine);
- information on the rainwater tank systems (e.g., tank volume, roof area connected, type of rainwater conveyance system – wet or dry, pump size);
- internal connections for rainwater supply (e.g., plumbing connections to/from the tank); and
- other water related features on the property (e.g., swimming pool, spa).

For each of these components, a number of detailed parameters was identified and a measurement method determined (Table 5.1). The protocol has been designed to be inclusive of water quality aspects as well as quantity. However, in the application of the protocol to the case study area, the focus of the assessment was on water quantity aspects because the specific research question related to potable water savings.

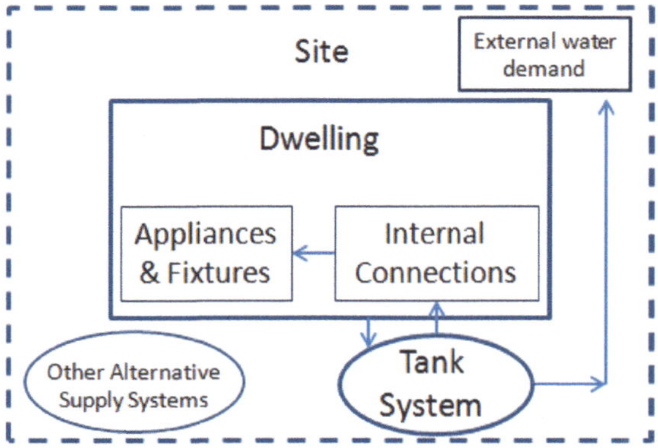

Figure 5.1 Physical components comprising the rainwater tank installation compliance audit protocol.

Table 5.1 Tank installation compliance audit protocol for SEQ.

Parameter	Measurement method
Characteristics of Individual Dwelling	
Dwelling type	Visual inspection. The dwelling types are: detached house; semi-detached house; townhouse; studio/ granny flat; or other (described).
Total roof and garden area	The total roof area for each site to be determined using aerial photography, where available, to mark out and measure the garden and roof catchment areas.
Number of bedrooms	Visual inspection and count of all of the bedrooms in the house.
Water Efficient Appliances and Fixtures	
Number of showers and showerheads – check/ calibrate the flow rate in litres per minute for each showerhead	The number of showers and shower heads determined through inspection of each bathroom and asking the owner. Occasionally there are showers located in the garden. For each shower head, determine the flow rate using a stopwatch and measuring jug. Record the time required to fill a 3 L measuring jug.
Number of washing machines, including type, brand, model and load size (kg) for each machine	The manufacturer's plate has the information required.
Are any of the washing machines connected to the rainwater tank? If yes, Is it connected to the cold tap? Specify other washing machine cold tap Configuration (if any)	Two effective methods of assessment. The first is turning on the washing machine and checking whether the rainwater pump is activated. The second is to turn off the rainwater supply while leaving the mains water supply on and check whether water flows into the washing machine. Also, some washing machines are plumbed with a separate rainwater cold tap.

Table 5.1 Tank installation compliance audit protocol for SEQ (*Continued*).

Parameter	Measurement method
Number of toilets – check /calibrate the flush volumes for half and full flushes	The number of toilets determined through inspection and asking the owner. To determine the flush volumes of the toilet, turn off the water leading into the cistern, noting the water level and then flushing the toilet. A measuring jug is then used to fill the cistern back up to the previous level and the water required to do this is noted.
Is/are the toilet(s) connected to the rainwater tank?	The same method used as for testing the connection of the washing machine. Either the toilet will be flushed and the system examined for rainwater pumping, or the rainwater supply will be switched off and the toilet checked to see if mains water fills the cistern.
Information on Rainwater Tanks	
Number of rainwater tanks	Inspection and ask the owner, check for underground/hidden tanks.
For each tank, record the type and shape of tank	Visual inspection – either underground, above ground or bladder style.
Tank dimensions (height × length × width)	Measure in millimetres using a tape measure. Underground tanks cannot be measured in this way and the assessor must rely on the manufacturer's certificate.
Tank diameter	Measured using a tape measure.
Tank volume (claimed)	If the tank has a manufacture's volume stamp on it, record this.
Tank materials (e.g., Polyethylene, galvanised steel, stainless steel, concrete)	Determine by visual inspection. Unless convincing evidence can be given by the owner (such as a receipt), the material of underground tanks will not be able to be determined.
Actual tank volume in storage	Determine by noting the shape, measuring the height of the off-take, cut-off switch, overflow and total height of the tank. It may not be possible to measure the float level where there is a trickle top-up system or the water level threshold used to activate the automatic switching to the reticulated water supply. These are located in the tank and accessing them is a safety issue. The measurements that are able to be obtained will give valuable information on the effective tank volume and any dead spaces in the tank.
Size of roof catchment area connected to the tank	If the system is uncharged (that is, the downpipe drains completely into the tank, with no water held in the pipe – Figure 5.2a), it is relatively clear which

(*Continued*)

Table 5.1 Tank installation compliance audit protocol for SEQ (*Continued*).

Parameter	Measurement method
	downpipes feed rainwater to the tank. Once this has been determined, examination of the roof shows which area drains to the downpipe. Where a section of roof is drained by two downpipes, one draining to the tank and one not, the slope of the gutter will be used to determine which part of the roof drains to which downpipe and from there the tank.
	Charged systems, (where the downpipes feeding the tanks are full of water – Figure 5.2b), can be checked by tapping the pipes to hear if they are full. After that, the same method outlined above can be used to determine the roof catchment area connected to tank.
	Mark this information on a satellite image or aerial photographs of the house to determine the area.
Rainwater pump brand and model number	Determine from the manufacturer's plate on the pump. Where plates are in an awkward position to access, damaged or lack required data, pump parameters can be determined from manufacturer's specifications.
Minimum level setting in rainwater tank (in litres)	This will be either the level of the cut-off switch or the off-take height (see Figure 5.3).
Is there an automatic switching device or a trickle top-up system providing supplementary water from the reticulated town water supply?	Note type of potable top-up system (description and model). The difference between trickle top-up and automatic switching devices can be determined by a simple examination of the presence of a switching system on the pump *or* a mains top-up valve on the top of the tank. The switching device has an in-tank water level sensor that causes the switching valve to change over to mains water based on the water level in tank. That is, no mains water enters the tank. The trickle top-up system operates through a float valve that allows mains water to recharge the tank between two defined water levels.
Is there any backflow prevention system?	Backflow prevention systems are located either immediately upstream of the mains input into a switching device, or on the mains meter. They are quite distinctive fittings.
Is there any screened downpipe rainhead, having screen mesh 4–6 mm and designed to prevent leaves from entering each downpipe?	Rainheads are easy to identify visually. Use a measuring tape to measure the screen mesh.

Table 5.1 Tank installation compliance audit protocol for SEQ (*Continued*).

Parameter	Measurement method
Is there any first flush device installed. If yes, what is the volume capacity of the device?	First flush devices are easily recognisable visually. Volume of first flush devices can be estimated either by measuring the length and diameter of the structure and calculating the volume, or by taking a photo, writing down the product name and comparing with manufacturers specifications.
Is there a mosquito-proof screen of brass, copper, aluminium or stainless steel gauze not coarser than 1 mm aperture mesh or flap valves at every opening of the rainwater tank?	This can be assessed through visual inspection.
Is there a vermin trap, wet system used to harvest rainwater, mosquito proofing in accordance with Standard HB230?	Visual assessment.
Are the rainwater tank's openings constructed to prevent ingress of surface stormwater and groundwater?	This can only be assessed to a limited degree for underground tanks. Where possible, assess all water entry points into the tank to ensure there is no potential contamination from groundwater or surface water. This may include assessment of invert levels, quality of fittings and quality of lid design and construction.
If multiple tanks exist, are they connected to each other?	Visual inspection of connections, usually at the base of tanks.
Number of external taps connected to each rainwater tank	Assessed by turning on the tap and checking for flow from the rainwater pump.
Is the rainwater tank connected to the hot water system?	This is done by turning off the rainwater tank system and operating a hot water tap. No flow indicates a rainwater source.
Is the rainwater used for drinking water and other household supply?	Turn off the rainwater tank system and operate a kitchen tap, laundry tap, bathroom tap, and so on for flow. No flow indicates a rainwater source.
Is the overflow connected to the stormwater drain? If yes, is there a physical air break or non-return valve on the rainwater tank overflow outlet before connecting to the stormwater drainage system?	Visual assessment. From NSW experience, this has been almost impossible to determine without either digging up the pipes or running enough water through the rainwater overflow to discharge into the stormwater drain.
Condition of Tanks, Gutters and Pumps	
What is the rainwater tank condition (good, satisfactory or poor)	Visual inspection including: leaks from the tank; rust or corrosion; and quality of foundation
Rainwater tank installation year	Assessed by questioning the owner. There is rarely an installation date placed on the tank.

(*Continued*)

Table 5.1 Tank installation compliance audit protocol for SEQ (*Continued*).

Parameter	Measurement method
Condition of the gutters (good, satisfactory, poor)	Assess gutters with respect to the quality of joins, sufficient fall to tank, and amount of sediment. Sediment and litter can be assessed using a mirror on the end of a pole.
Condition of the pump (good, satisfactory, poor)	Assess the pump for: whether the pump is working; presence of rust; leaking of pipes; previous downtime; and improper running.

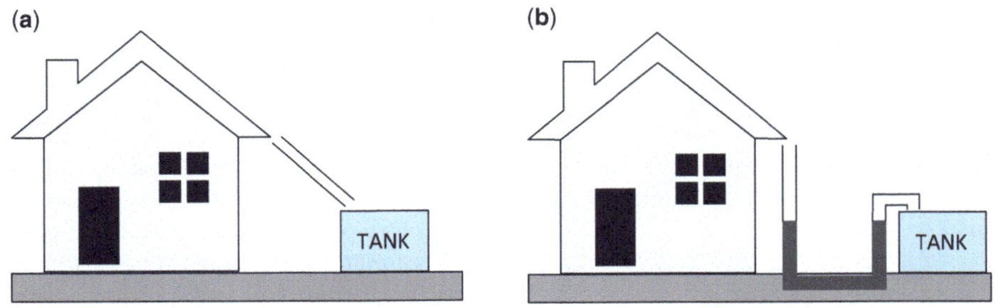

Figure 5.2 Schematic of (a) uncharged and (b) charged rainwater conveyance systems from roof to tank.

5.3 APPLICATION OF THE METHOD TO THE SEQ CASE STUDY AREA

5.3.1 Context

Under the (now repealed) QDC MP 4.2 for detached dwellings in SEQ, the water savings target was 70 kL/hh/yr (DLGP, 2008). The combination of a 5 kL rainwater tank connected to at least 100 m^2 of roof area was deemed to be an acceptable method to achieve the water saving target. According to the QDC, the tank must also be internally plumbed to toilet cisterns, the cold water washing machine tap in the laundry, and one external tap, and have either a trickle top-up or an automatic switching device installed to provide back-up supply of potable water to internal fixtures supplied from the tank.

Research to validate the efficacy of individual rainwater tanks in augmenting the mains water supply, and delivering the water saving as per QDC MP 4.2, is important. This information can build confidence in, and help inform and guide the SEQ Water Strategy (Queensland Water Commission, 2010). It can also help guide the development of policy and regulations addressing issues such as an inspection and maintenance protocols. Knowledge and understanding of how rainwater is perceived among the general public and how it is used in urban SEQ is important in developing and encouraging future sustainable water practices in SEQ.

Desktop studies of mains water savings by Beal *et al.* (2012) and Chong *et al.* (2011b), using statistical analysis of potable water billing data, both revealed that a combination of 5 kL tanks and 100 m^2 connected roof area may not achieve the expected mains water saving target. Both studies reported savings of less than 60 kL/hh/yr. Possible reasons for these lower than expected savings are: (1) water restrictions which may reduce external water use; (2) the possible use of other water resources (e.g., recycled water systems) to achieve the QDC MP 4.2 targets; (3) different household demographics including household occupancy rates (Chong *et al.* 2011a); (4) different attitudes, behaviours and perceptions toward the utilisation of rainwater (Mankad *et al.* 2012); and (5) non-compliant installation of mandated rainwater tanks at individual residential dwellings (e.g., inappropriate tank size and connected roof area, tanks not internally plumbed, etc.).

The aim of the SEQ study was to assess the degree to which the 223 selected Class 1 (detached) residential dwellings in four Local Government Areas (LGA) of SEQ, met the physical tank installation requirements deemed necessary to achieve the SEQ Water Strategy's water savings target of 70 kL/hh/yr.

5.3.2 Data collection

An on-site inspection was conducted of 223 consenting individual households with mandated rainwater tank systems in the four LGA of Caboolture, Gold Coast, Pine Rivers and Redland. These particular LGA were selected for the study as part of a wider suite of research projects conducted in SEQ at the time, investigating a range of aspects relating to water savings from individual rainwater tanks. These areas were originally selected on the basis that they are high growth residential areas and hence were expected to have a high proportion of dwellings constructed after January 2007 when mandated rainwater tanks came into effect. Previous research indicated discrepancies in water savings results, particularly in Caboolture and Pine Rivers where lower than expected water savings were identified, and so, larger samples were drawn from these areas (Table 5.2).

Table 5.2 Sample and population sizes.

LGA	Number of sites inspected	Number of dwellings with tanks
Caboolture	59	4000
Gold Coast	45	3300
Pine Rivers	78	5000
Redland	41	3300
Total	**223**	**15,600**

Households who had previously indicated a willingness to allow an on-site inspection as part of a separate, broader study into water savings from individual rainwater tanks (443 households), were sent a follow-up letter providing details of the intended research and requesting that a consent form be completed, signed and returned. Written consent was provided by 223 households and these were subsequently contacted by email and post, provided with details of the intended site inspection, and asked to provide telephone contact details for the purpose of appointment scheduling. The site inspections were conducted during the third quarter of 2011.

It was found during the site visits that connected roof area could not be obtained for those systems with charged (wet) downpipes due to low rainfall. This was particularly the case for charged systems connected to underground tanks, 10% of the sample population. The survey contractor revisited these sites during late January 2012 following plentiful rain and was able to measure connected roof area.

Drawing on the known population of households with mandated tanks, it is possible to calculate a statistical measure of the accuracy provided by the sample used in this study. This calculation is based on the size of the sample, the size of the population from which the sample is drawn, the level of statistical confidence desired (typically 95%), and the variability in the sample (typically estimated from a worst-case assumption of a sample proportion of 50%). In the results reported below, all the proportions reported for the full sample (e.g., the proportion of all the inspected installations that are fully compliant, etc.) are 95% certain to be an accurate reflection of the true proportion in the entire population of installations, with a margin of error of +/–6.5%. Proportions reported for specific LGA are less accurate, given their smaller sample sizes. The margins for error for individual LGA range from +/–11.1% for Pine Rivers to +/–15.4% for Redland. Clearly, larger samples would provide greater precision, and a smaller margin for error. For example, a sample size of 1000 would provide a margin of error of +/–3%, whilst a sample of around 6000 households is required for a +/–1% error. Due to funding constraints, it was not possible to inspect such large sample sizes.

5.4 RESULTS

5.4.1 General rainwater tank characteristics

For the purposes of this study, a rainwater tank system is defined as a single tank or group of tanks on a site that share a common catchment area or pump. If multiple tanks on one property do not share a common pump or catchment area, they are classified as belonging to two different tanks systems. In all cases of tank systems with multiple tanks in this study, the tanks were found to have similar shape and size.

Of the 223 sites inspected, most had only one rainwater tank system, with 14 sites having two rainwater tank systems and two sites having three rainwater tank systems to collect rainwater. Whereas most (93%) single tank systems at households comprised only one tank, 6% had two tanks and two sites in Caboolture had three tanks. Redland had the greatest number of two-tank systems. All second and third rainwater tank systems comprised only a single tank.

Most tank systems comprised cylindrical shaped tanks (47%), while 37% of tank systems comprised rectangular shaped tanks. Irregular shaped tank, including one bladder tank, made up 5% of the tank systems audited. It was not possible to determine the shape of the 24 underground systems (10%).

5.4.2 Rainwater tank storage volume

Although QDC MP 4.2 is not specific about which measure of volume it is referring to when it specifies a minimum storage capacity of 5000 L, it is generally accepted that this refers to the net volume *below* the overflow pipe, and this is referred to as storage volume in this paper. Compliant storage volume is calculated from the cross sectional dimensions of the tank, whether rectangular or cylindrical, and the height of the overflow level.

It was possible to calculate storage volumes for 180 sites, with 43 sites having insufficient dimension data available due to tanks being underground, inaccessible for measurement, or irregular in shape. Storage volume, calculated from the basic geometric dimensions of the tank, was on average 6659 L with a median of 5716 L (Table 5.3). There are 15 sites which have a capacity greater than 10,000 L. It was considered likely that these households were aiming for self-sufficiency in water supply, rather than compliance with the 5000 L specified in QDC MP 4.2, and were accordingly excluded from the average compliance analysis following discussions with main stakeholder (Queensland Water Commission). If the tanks with a volume larger than 10 kL are excluded from the analysis, storage volume reduces to an average of 5521 L with a median of 5385 L (Table 5.3), which is still greater than the requirement specified in QDC MP 4.2.

Manufacturer's claimed volume is based on the same calculation used in this study to calculate storage volume. It is expected that manufacturer's claimed volumes would be more accurately calculated, accounting for exact shaping of tanks (such as rounded edges of rectangular tanks) and resulting in slightly

Table 5.3 Calculated on-site rainwater tank storage volumes.

LGA	Number of sites per tank volume range (kL)								Compliance		Tank volume (L)	
	No vol.	<4	>4–<5	>5–<6	>6–<7	>7–<10	>10	Total	% Comply	20% Variance[1]	Average	Median
Caboolture	12	4	7	26	4	1	5	47	77%	91%	6818	5664
Gold coast	12	0	4	17	7	3	2	33	88%	100%	7575	5766
Pine rivers	10	4	3	43	11	3	4	68	90%	94%	6174	5733
Redland	9	2	5	19	1	1	4	32	78%	94%	6509	5445
Total (all tanks)	43	**10**	**19**	**105**	**23**	**8**	**15**	**180**	**84%**	**94%**	**6659**	**5716**
Excl. tanks >10KL											**5521**	**5385**
Calculated volume[2]											**7126**	
Claimed volume											**6372**	

[1] The % compliance increases if tank volumes within 20% of the target volume of 5000 L are included.

[2] Average calculated volume for 86 sites where both claimed and calculated storage volumes were available, including large tank sites, showing a 12% overestimation compared to the manufacturer's claimed volume.

lower net claimed volumes than the storage volumes calculated in this study, which do not account for exact shapes. A direct comparison between sites where both claimed and calculated storage volumes were available (86 sites), including large tank sites, confirms that, on average, calculated storage volume (7126 L) is 12% higher than claimed volume (6372 L) and this should be factored in when considering compliance (Table 5.3).

Over all four LGA, 84% of all sites with calculated storage volume were at or above the QDC MP 4.2 minimum compliance volume of 5000 L, with most sites (58%) falling into the 5000 L to 6000 L class and 13% into the 6000 L to 7000 L class (Table 5.3). Only 16% of sites had storage volumes less than the required 5000 L. Including those sites which are within a 20% variance of the minimum volume, 94% of sites are at or near the minimum compliance level, but there are still 10 sites (0.5%) which fall outside the 20% variance level (Table 5.3). Caboolture has the lowest level of compliance at 91%, compared with 100% compliance at the Gold Coast after allowing volume criteria relaxation of 20% variance (Table 5.3). Allowing for the 12% overestimation in calculated storage compared to the manufacturer's notified volume, average compliance in all LGA is well above the required 5000 L as specified in QDC MP 4.2. There are, however, 16% of tanks with volumes below the required 5000 L, most of which occur in Caboolture (Table 5.3).

Effective or active volume, which excludes the 'dead' volumes both above the tank overflow level and below the outlet or pump cut-off switch levels, was also considered to illustrate the real volume available for use by the household. Two methods are available for calculating effective volume. In Case 1, a reasonable estimate of the effective volume is obtained by simply deducting the dead volume below the outlet pipe to the pump *and* the air space above the height of tank overflow. However, the pump cut-off water level (measured with a pressure transducer) used to activate the switching valve to the reticulated water supply (Figure 5.3a), *or* the float level that triggers the trickle top-up system (Figure 5.3b) are set higher than the tank outlet to prevent pump cavitation and ensure adequate water supply to the house during trickle top-up events. The threshold water level is almost certainly higher than the installed level of the pressure transducer. Therefore for Case 2, a more accurate estimate of effective volume is to exclude the volume below the pressure transducer or float valve level rather than just considering the dead volume below the pump outlet pipe (Figure 5.3). It is suggested that readers should also look at Figure 6.5 in Chapter 6 for additional information.

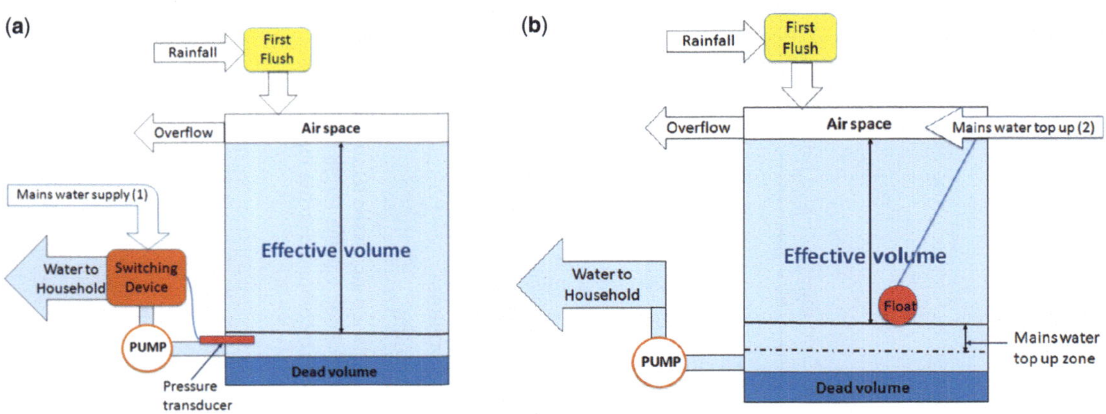

Figure 5.3 Diagram showing effective volume of a tank using Case 2 for (a) an automatic switching device and (b) for mains water top-up.

Average effective volume for Case 1 is 6342 L per site, with a median of 5451 L (Table 5.4), which is still above the QDC MP 4.2 requirement of 5000 L. However, there are 48 sites which have a lower effective volume than the minimum required volume of 5000 L, although 34 of these are within a 20% variance (Table 5.4). More than half the sites (89 sites, or 52%) fall into the >5000–<6000 L category (Table 5.4).

The second estimate of effective volume (Case 2) is less than for Case 1 due to the location of the pump cut-off (pressure transducer) or the float level that triggers the automatic switching device or trickle top-up system, respectively. Only 60 sites had sufficient measurements for calculating site level effective volume for Case 2, so these results should be considered with caution. The reason for this low data return is that, in addition to underground and inaccessible tanks, the pressure sensor and/or the float are located inside the tank. Accessing them is a safety issue, so it is not always possible to measure the level the pump cut-off or trickle top-up system is activated. Similarly, the geometry of submerged pumps cannot be measured.

Nonetheless, the effective volume for Case 2 is greater than the QDC MP 4.2 minimum capacity requirement of 5000 L, but this time only marginally so at an average of 5019 L, with the median now just below the minimum volume at 4947 L (Table 5.5). Using this method to calculate effective volume reduces the average volume by approximately 20% compared to Case 1, with only 43% of sites having volumes of at least 5000 L (Table 5.5). Including height to pump cut-off level reduces effective volume to the extent that 8 of the 60 sites (13%) are now below the 4000 L volume mark (Table 5.5).

The results indicate that all of the different gross and net measures of volume return, on average, volumes mostly above 5000 L. However, when pump protection cut-off effects are included, indications are that volumes start to reduce to below the 5,000 L minimum capacity. Measuring the level at which pump cut-off takes place is difficult. Manufacturer's claimed volumes (which already account for the 'dead' volume above overflow level) may be adequate for future modelling, possibly allowing for a reduction of around 10% to account for the dead volume below the outlet level (difference between claimed volume and effective volume Case 1 averages), or around 20% if pump cut-off factors are also included (as described for Case 2). Further investigation of the effect of pump cut-off levels on effective volume are required before this proposed 20% reduction factor could be accepted with certainty. The sample size of tanks with cut-off levels provided ($n = 60$) was too small in this survey to provide an adequate level of confidence.

In the absence of manufacturer's claimed volumes, if storage volumes are calculated using in situ physical tank dimensions, calculated volume should be reduced by a further 12% (approximately) to account for volume lost due to tank shapes which generally round off edges and are not perfectly square or round.

5.4.3 Connected roof area

To achieve the water savings target, QDC MP 4.2 requires that rainwater tanks be connected to either a minimum of 50% of total roof area or 100 m^2, whichever is the lesser. A simple approach was developed to measure connected roof area as rainwater catchment for the tank. The connected roof area was estimated using the position and direction of the roof drainage system, the direction of the gutter slope, and location/number of downpipes connected to rainwater tanks. The gutter slopes were determined either visually in case of double storey dwellings where access was an issue, or with a spirit level for single storey dwellings. The inflexion point of the change in slope direction of the gutter between downpipes was marked on the working aerial photograph sheets for households (Figure 5.4). All these working sheets were processed in the office with either Google Earth or NearMap, to mark the location and inflexion points of gutters, as well as determining the tank catchment area, and the total roof area. An example of a marked up aerial photograph used as the working sheet is shown in Figure 5.5 (from Chong *et al.* 2012).

Table 5.4 Effective site rainwater tank volume using overflow and outlet heights (Case 1).

LGA	No volume	Number of sites per tank volume range (kL)						Total	% Comply	Compliance after a 20% variance allowance	Average (L)	Median (L)
		<4	>4–<5	>5–<6	>6–<7	>7–<10	>10					
Caboolture	12	4	10	17	3	0	4	38	63%	89%	6393	5380
Gold coast	11	1	7	17	3	3	3	34	76%	97%	7276	5623
Pine rivers	10	5	9	40	8	2	4	68	79%	93%	5939	5570
Redland	9	4	8	15	0	1	4	32	63%	88%	6130	5202
Total	**42**	**14**	**34**	**89**	**14**	**6**	**15**	**172**	**72%**	**92%**	**6342**	**5451**

Table 5.5 Effective site rainwater tank volume using overflow and pump cut-off heights (Case 2).

LGA	Number of sites per tank volume range (kL)								% Comply	Compliance after a 20% variance allowance	Average (L)	Median (L)
	No cut-off	<4	>4–< 5	>5–< 6	>6–< 7	>7–< 10	>10	Total				
Caboolture	50	0	6	2	0	0	1	9	33%	100%	5562	4947
Gold coast	30	0	7	5	2	1	0	15	53%	100%	5256	5258
Pine rivers	52	3	12	9	1	1	0	26	42%	88%	4834	4937
Redland	31	5	1	3	1	0	0	10	40%	50%	4656	4377
Total	**163**	**8**	**26**	**19**	**4**	**2**	**1**	**60**	**43%**	**87%**	**5019**	**4947**

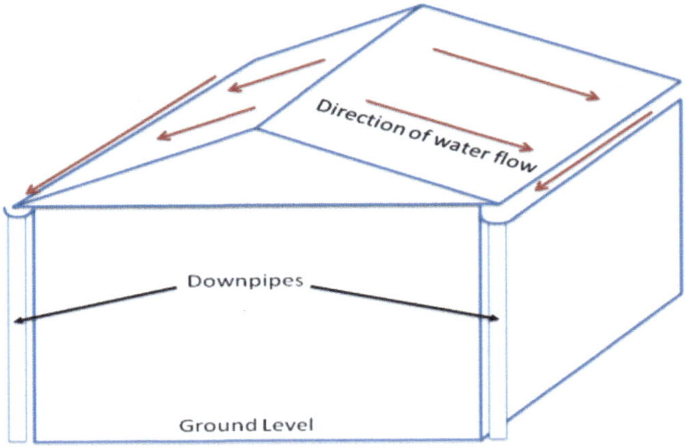

Figure 5.4 Identifying slope and direction of flow in gutters and location of downpipes.

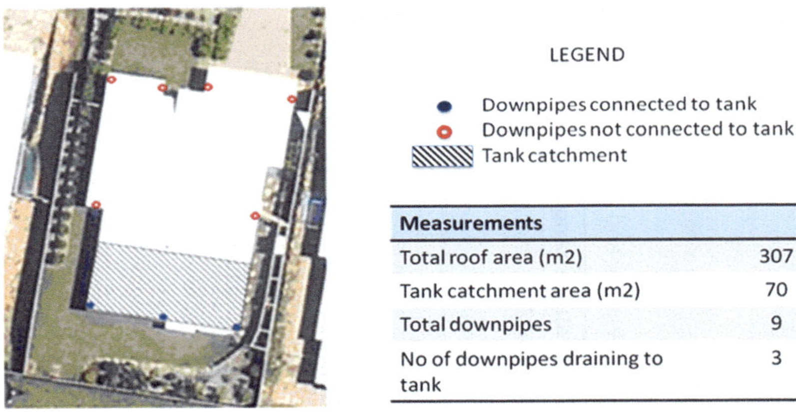

Measurements	
Total roof area (m2)	307
Tank catchment area (m2)	70
Total downpipes	9
No of downpipes draining to tank	3

Figure 5.5 Marked up aerial photograph to estimate tank catchment area.

The average roof area across the four LGA is 300 m^2 (Table 5.6), which means that, on average, houses would have 150 m^2 of roof area connected to the rainwater tank system if the 50% minimum option is applied. However, this is very difficult to achieve in practice due to the location of downpipes and their connection to the rainwater tank whist achieving an aesthetically pleasing outcome. A charged pipe system is one option for connecting the larger roof area, but it is often expensive. With an average measured connected roof area of 118 m^2, it is not surprising, therefore, that the most common option to achieve compliance across the four LGA was the lower 100 m^2 requirement rather than the 50% of total roof area option (Table 5.7).

Of all sites, 58% have connected roof area of >100 m^2, with most sites belonging to the 100–200 m^2 category (Table 5.7). Only 24% of sites have more than 50% of their roof area connected, with 55% of sites falling into the <40% roof area connected category (Table 5.7).

To assess overall compliance with the QDC MP 4.2 'either/or' rule, the following method has been adopted. For the house population with roof area >200m^2, houses that have at least 100 m^2 connected to the tank will be compliant. For the population of houses with roof area <200m^2, houses that have 50% or greater of their roof area connected to the tank will also be compliant. The weighted sum of these numbers is the percent compliance for the population of houses. Applying the 'either/or' rule for the SEQ study, there is an overall compliance of only 60% across all four LGA, with Gold Coast being the least compliant at 54%, and Caboolture being the most complaint at 64% (Table 5.7).

Table 5.6 Average detached house roof area (m^2) across four LGA.

LGA	Average roof area (m^2)
Caboolture	310
Gold coast	326
Pine rivers	281
Redland	294
Total	**300**

Unlike the situation with tank volumes which include sites within a 20% variance of compliance criteria, a similar relaxation of connected area requirements did not significantly improve overall compliance. A large percentage of sites still remain in the lowest category of compliance (either <80 m^2, or <40% connected area), after allowing for a 20% variance factor.

5.4.4 Connection to toilets, washing machines and external use

The more household appliances that are plumbed or connected to the rainwater tank, the more rapidly the water level in the tank will be drawn down, resulting in a higher probability there will be a larger air space to capture a larger fraction of the subsequent rainfall event. Hence, both rainwater utilisation and mains water savings increase. QDC MP 4.2 (A2(c)) requires that rainwater tanks be connected to: (1) toilet cistern and washing machine cold water taps; and (2) an external use.

Most of the sites inspected were fully compliant with the required connection to household appliances and to an external tap. All homes inspected had one washing machine present and only 2% of households (five sites in Caboolture) did not have their washing machine connected to the rainwater tank system. 93% of sites had two toilets connected to the rainwater tank, with only five sites not having any toilet connected at all. Except for one site, all homes inspected had at least one external tap connected to the rainwater tank system.

5.4.5 Continuous supply

As part of the section relating to water savings targets, QDC MP 4.2 (P5) requires a continuous back-up supply of mains water to internal fixtures supplied from a rainwater tank through either a trickle top-up or an automatic switching device (see Figure 5.3). The most commonly used device is an automatic switching valve (70% of audited sites), with 23% employing a trickle top-up system and 6% having no device (Table 5.8).

5.4.6 Water quality protection

Although water savings – related issues were the main focus of this study, information regarding the presence of the QDC water quality protection measures (Performance Criteria 3 (P3), P4 and P6) was also collected as part of the site inspection.

Table 5.7 Number of sites for each class of connected roof area (m^2), or percentage of total roof area connected to rainwater tanks, and overall compliance level.

LGA	No. sites per connected roof area class (m^2)						No. sites per % roof area connected class					
	<80	80–<100	100–200	>200	Av.	% Comply >100	<40%	40–<50%	≥50%	Av.	% Comply	% Overall comply*
Caboolture	14	5	28	6	119	64%	26	16	11	44%	21%	64%
Gold coast	12	7	15	5	136	51%	23	6	10	41%	26%	54%
Pine rivers	23	10	37	6	110	57%	42	17	17	39%	22%	58%
Redland	9	7	18	4	113	58%	22	4	12	41%	32%	63%
Total	**58**	**29**	**98**	**21**	**118**	**58%**	**113**	**43**	**50**	**40%**	**24%**	**60%**

*% overall compliance is based on the lesser of 50% of total roof area or 100 m^2 (as per QDC MP 4.2).

Table 5.8 Type of continuous back up water supply device installed.

	No.	%
Trickle top-up	51	23%
Automatic switch	155	70%
None	14	6%
Total	220	

QDC MP 4.2 (A3) requires screened downpipe rainheads and first flush devices as suitable measures to prevent contaminants from entering the rainwater tank. 95% of rainwater tanks inspected in the study had a screened downpipe rainhead installed. Only 37% of rainwater tanks inspected had a first flush device installed, but this requirement is only necessary when the rainwater tank is connected to showers, wash basins or hot water services, which are not the end uses for most of the rainwater tanks inspected.

QDC MP 4.2 (A4) requires mosquito-proof screens, flap valves, or, in cases of wet systems, mosquito-proofing in accordance with Standard HB230 (MPMSAA, 2008) and vermin traps to prevent mosquito breeding and vermin entering the rainwater tank system. Of the sites inspected, 96% had some form of mosquito breeding prevention device installed, and 95% had a vermin trap installed.

Similarly, QDC MP 4.2 (P6) requires a backflow prevention device to prevent contamination of the reticulated town water supply with tank water. All tanks with a trickle top-up system (23% of households) had an air gap present between the overflow and the trickle top-up device and this operates as a backflow prevention device. Only 45% of rainwater tanks inspected had an additional backflow prevention device present. Accounting for both forms of backflow prevention, 68% of tanks were compliant.

5.5 SUMMARY AND CONCLUSION

In response to an identified knowledge gap in post tank installation monitoring, this chapter has presented a generic rainwater tank installation compliance audit protocol, and described its application in SEQ, Australia, to assess compliance of rainwater tank systems and validation of water savings and quality objectives required by QDC MP 4.2.

This chapter has summarised the results of applying the audit protocol in the assessment of household rainwater systems of 223 Class 1 detached dwelling sites with mandated rainwater tanks in SEQ. The audit identified installation factors which may impact adversely on achieving the 70 kL/hh/yr water savings target for detached dwellings in SEQ.

The three main installation requirements of QDC MP 4.2 relate to: rainwater tank storage capacity, roof catchment area and connection to appliances. The results of the on-site assessment have shown that:

- Installed storage capacity is mostly greater than the required 5000 L, although 16% of sites inspected had storage volumes less than 5000 L.
- Roof catchment area does not meet requirements in 40% of cases, either by having <100 m^2 or <50% of total roof area (whichever is the lesser), connected to the tank.
- Connection to toilets, washing machine and external tap meets requirements in most cases.

The study has concluded that in estimating effective tank volume from manufacture's claimed volumes, claimed volume should be reduced by about 10% to account for the dead volume below the outlet level. In addition, the active storage volume should be further reduced due to the water level setting on the pressure switch and the operating levels of the float valve, with an overall total reduction in effective tank volume of about 20% if pump water level cut-off factors are also included. In the absence of information on claimed volumes, storage volumes can be calculated using in situ physical tank dimensions (after allowing for dead volumes etc.), but should first be reduced by about 12% to account for volume lost due to tank shapes which generally round off edges and are not perfectly square or round.

Of greater concern is the higher level of non-compliance in roof area catchment connected to the rainwater tank system, which we report with a reasonable degree of confidence. Even relaxing the level of compliance to within a 20% variance does not significantly improve the compliance. Certainly, in any policy and management intervention to improve the yield from rainwater tank systems in SEQ, this aspect of increasing the connected roof catchment area is the single most important physical installation factor to be considered. This level of non-compliance in connected roof area has implications for potable water savings across the region, most likely resulting in lower actual savings than those predicted, and thereby reducing the time period of deferred infrastructure investment. Specification of the area of the roof to be connected, rather than the percentage of roof area, is a more easily achievable target in situations when average roof area is greater than $200m^2$.

The practical reasons for this low compliance probably stem from construction phase issues such as cost, difficulty in getting the downpipes to connect to the tanks past windows, doors and other obstructions, and the aesthetics of having horizontal pipes running along external walls. Clearly, it is important to consider the overall design of the house when planning rainwater tank installation so that the anticipated benefits can be realised.

Charged systems, where the downpipes are usually full of water and run along the ground or beneath the house before rising up to discharge into the tank, are often used to achieve the maximum roof area connection. A caution is given here in that several sites with charged systems had the potential of cracking and delivering water to the house foundations. Silting up with leaf debris over time is also a problem if adequate inspection is not carried out and cleaning points are not installed.

The lack of detailed inspections by a dedicated agency of all aspects of rainwater tank installation has the potential to both reduce the anticipated savings and to cause health and safety problems in the future. Inspection of connected roof area is probably the single most important aspect as this seems to be most problematic in terms of compliance.

QDC MP 4.2 also includes requirements for water quality protection through prevention of contaminants and vermin from entering the rainwater tank, prevention of mosquito breeding and prevention of contamination of reticulated town water supply. Except for the installation of backflow devices that prevent contamination of reticulated town water supply, requirements are adequately met. The results of the on-site inspection have shown that:

- Screened downpipe rainhead installation largely meets requirements.
- First flush device installation were infrequent but were not required as rainwater was generally not connected to showers, wash basins or hot water services.
- Mosquito breeding prevention device installation meets requirements.
- Vermin trap installation mostly meets requirements.

Backflow prevention device installation does not meet requirements, with only 45% of those rainwater tanks with switching valves having a backflow device installed. However, all tanks with a trickle top-up system (23%) had an air gap present between the overflow and the trickle top-up device which serves the purpose of a backflow prevention device.

The results indicate a high level of compliance with QDC MP 4.2 (P5), which states the requirement for a continuous back-up supply of potable water to internal fixtures supplied from a rainwater tank using either a trickle top-up or an automatic switching device.

In applying the rainwater tank installation compliance audit protocol in SEQ, a number of shortcomings were identified. The protocol needs further enhancement in relation to the measurement of underground tank systems, and the timing of the assessment in relation to rainfall events is important so that measurement occurs when conveyance pipes are sufficiently charged with water.

5.6 REFERENCES

Abdulla F. A. and Al-Shareef A. W. (2009). Roof rainwater harvesting systems for household water supply in Jordan. *Desalination*, **243**, 195–207.

Aladenola O. O. and Adeboye O. B. (2010). Assessing the potential for rainwater harvesting. *Water Resource Management*, **24**, 2129–2137.

Beal C., Sharma A., Gardner T. and Chong M. (2012). A desktop analysis of potable water savings from internally plumbed rainwater tanks in South East Queensland. *Water Resouce Management*, **26**, 1577–1590.

Biermann S., Sharma A., Chong M. N., Umapathi S. and Cook S. (2012). Assessment of the Physical Characteristics of Individual Household Rainwater Tank Systems in South East Queensland. Urban Water Security Research Alliance Technical Report No. 66.

Canada Housing and Mortgage Corporation (2012). Guidelines for Residential Rainwater Harvesting Systems Handbook. http://www.homebuildercanada.com/_images_Pub/20130216_CMHC_Rainwater.pdf (accessed September 2013).

Chapman H., Cartwright T. and Tripodi N. (2008). Guidance Manual for the Design and Installation of Urban Roofwater Harvesting Systems in Australia. 1st edn, Cooperative Research Centre for Water Quality and Treatment, Research Report 39.

Chong M. N., Mankad A., Gardner T. and Sharma A. (2011a). Baseline Demographic Analysis of Mandated Rainwater Tank Users in South East Queensland, Urban Water Security Research Alliance Technical Report No. 48.

Chong M. N., Umapathi S., Mankad A., Sharma A. and Gardner T. (2011b). A Benchmark Analysis of Water Savings by Mandated Rainwater Tank Users in South East Queensland (Phase 2). Urban Water Security Research Alliance Technical Report No. 49.

Chong M. N., Sharma A., Umapathi S. and Cook S., (2012). Understanding the Mains Water Saving from Mandated Rainwater Tanks using Water Balance Modelling and Analysis with Inputs from On-Site Audited Parameters. Urban Water Security Research Alliance Technical Report No. 65.

Coombes P. J. and Kuczera G. (2003). Analysis of the Performance of Rainwater Tanks in Australian Capital Cities [online]. In: M. J. Boyd, J. E. Ball, M. K. Babister and J. (E)Green (eds). 28th International Hydrology and Water Resources Symposium: About Water; Symposium Proceedings, Barton, A.C.T., Institution of Engineers, Australia.

Coombes P. J., Kuczera G., Argue J. R., Cosgrove D., Arthur H. A., Bridgman H. A. and Enright K. (1999). Design, monitoring and performance of the water sensitive urban development at Figtree Place in Newcastle. Proceedings of the 8th International Conference on Urban Storm Drainage, 1319–1326. Sydney, Australia.

Coombes P. J., Argue J. R. and Kuczera G. (2000). Figtree place: a case study in water sensitive urban development. *Urban Water*, **4**(4), 335–343.

Coombes P. J., Kuczera G. and Kalma J. D. (2003). Economic, water quantity and quality impacts from the use of a rainwater tank in the inner city. *Australian Journal of Water Resources*, **7**(2), 101–110.

Coombes P. J., Spinks A., Evans C. and Dunstan R. H. (2004). Performance of Rainwater Tanks in the Inner City During Drought. WSUD2004 Conference. Adelaide, Australia.

Coombes P. J., Dunstan R. H., Spinks A., Evans C. and Harrison T. (2006). Key Messages from a Decade of Water Quality Research Into Roof Collected Rainwater Supplies. 1st National HYDROPOLIS Conference. Perth, Australia.

Commonwealth of Australia, Department of Health and Ageing (2010). Guidance on Use of Rainwater Tanks. http://www.health.gov.au/internet/main/publishing.nsf/Content/DD676FA1241CDD0DCA25787000076BCD/$File/enhealth-raintank.pdf.

Department of Housing and Public Works (2013). Web site accessed February 2013. http://www.hpw.qld.gov.au/construction/BuildingPlumbing/Building/WaterSupplySystems/Pages/default.aspx.

DLGP (2008). Queensland Development Code MP 4.2 – Water Savings Targets. Department of Local Government and Planning, Queensland Government.

Domènech L. and Saurí, D. (2011). A comparative appraisal of the use of rainwater harvesting in single and multifamily buildings of the Metropolitan Area of Barcelona (Spain): social experience, drinking water savings and economic costs. *Journal of Cleaner Production*, **19**, 598–608.

Environment Agency, UK Government (2010). Harvesting Rainwater for Domestic Uses: An Information Guide.

Fewkes A. (1999a). The use of rainwater for WC flushing: the field testing of a collection system. *Building and Environment*, **34**(6), 765–772.

Fewkes A. (1999b). Modelling the performance of rainwater tank collection systems: towards a generalised approach *Urban Water*, **1**(4), 323–333.

Ghisi E. (2006). Potential for potable water savings by using rainwater in the residential sector of Brazil. *Building and Environment*, **41**, 1544–1550.

Ghisi E. (2010). Parameters influencing the sizing of rainwater tanks for use in houses. *Water Resource Management*, **24**, 2381–2403.

Ghisi E., Montibeller A. and Schmidt R. W. (2006). Potential for potable water savings by using rainwater: An analysis over 62 cities in southern Brazil. *Building and Environment*, **41**, 204–210.

Ghisi E., Bressan D. and Martini M. (2007). Rainwater tank capacity and potential for potable water savings by using rainwater in the residential sector of southeastern Brazil. *Building and Environment*, **42**, 1654–1666.

Government of South Australia, Department of Planning and Local Government (2010). Water Sensitive Urban Design Technical Manual for the Greater Adelaide Region. Government of South Australia, Adelaide.

Handia L., Tembo J. M., Mwiindwa C. (2003). Potential of rainwater harvesting in urban Zambia. *Physics and Chemistry of the Earth*, **28**, 893–896.

Hermann T. and Schmida U. (1999). Rainwater utilisation in Germany: efficiency, dimensioning, hydraulic and environmental aspects. *Urban Water*, **1**, 307–316.

Imteaz M. A., Shanableh A., Rahman A. and Ahsan A. (2011). Optimisation of rainwater tank design from large roofs: a case study in Melbourne, Australia. *Resources, Conservation and Recycling*, **55**, 1022–1029.

Khastagir A. and Jayasuriya N. (2010). Optimal sizing of rain water tanks for domestic water conservation. *Journal of Hydrology*, **381**, 181–188.

Mankad A., Chong M. N., Umapathi S., and Sharma A. (2012). The Role of Biophysical and Psychological Factors in Understanding Rainwater Tank Maintenance. Urban Water Security Research Alliance Technical Report No. 64.

Master Plumbers and Mechanical Services Association of Australia (MPMSAA) (2008). Rainwater Tank Design and Installation Handbook. National Water Commission, Australian Government.

Master Plumber and Mechanical Services Association of Australia (2008). Requirements for Installation of Rainwater and Greywater Systems in Australia, Waterlines Report, National Water Commission, Canberra, Australia.

New South Wales Government (2004). State Environmental Planning Policy (Building Sustainability Index: BASIX) (the BASIX SEPP).

Queensland Water Commission (2010). South East Queensland Water Strategy. Queensland Water Commission, the State of Queensland, Brisbane.

Sehgal J. D. (2005). A Guide To Rainwater Harvesting in Malaysia Rotary Club of Johor Bahru, District 3310 of Rotary International.

Standards Australia (2003). National Plumbing and Drainage Code Australian Standard AS 3500.

Standards Australia, MPMSAA, ARID (2006). Rainwater Tank Design and Installation Handbook, First Edition January 2006, Standards Australia, Master Plumbers and Mechanical Services Association of Australia, Australian Rainwater Industry Development Group. HB 230 – 2006, Sydney, NSW, Australia.

Sydney Water (2012). BASIX Water Savings Monitoring for 2010–11. https://www.basix.nsw.gov.au/basixcms/images/water_monitoring_report-2010-11.pdf.

Tam V., Tam L. and Zeng S. (2010). Cost effectiveness and trade-off on the use of rainwater tank: an empirical study in Australian residential decision-making. *Resources, Conservation and Recycling*, **54**, 178–186.

Texas Water Development Board (2005). The Texas Manual on Rainwater Harvesting Third Edition. Austin, Texas.

Umapathi S., Chong M. N. and Sharma A. (2012). Investigation and Monitoring of Twenty Homes to Understand Mains Water Savings from Mandated Rainwater Tanks in South East Queensland. Urban Water Security Research Alliance Technical Report No. 63.

UN-HABITAT (2005). Blue Drop Series on Rainwater Harvesting and Utilisation. Meeting Development Goals in Small Urban Centres—Water and Sanitation UN-HABITAT, Nairobi, Kenya.

Vaes G., Willems P. and Berlamont J. (2001). Rainfall input requirements for hydrological calculations. *Urban Water*, **3**, 107–112.

Villarreal E. L. and Dixon A. (2005). Analysis of a rainwater collection system for domestic water supply in Ringdansen, Norrkoping, Sweden. *Building and Environment*, **40**, 1174–1184.

Zaizen M., Urakawa T., Matsumoto Y. and Takai H. (1999). The collection of rainwater from dome stadiums in Japan. *Urban Water*, **1**(4), 335–359.

Chapter 6

Understanding energy usage in rainwater tank systems through laboratory and household monitoring

Grace Tjandraatmadja, Monique Retamal, Shivanita Umapathi and Guenter Hauber-Davidson

ABSTRACT

There is limited understanding by rainwater tank owners and water planners about the energy requirements associated with rainwater tank supply. The design and set-up of rainwater systems can be improved to reduce the energy requirements for rainwater supply in urban areas whilst maintaining service requirements. This chapter examines the factors that influence the energy use in commonly used rainwater pump systems (pump characteristics, end uses, system and infrastructure design and, indirectly water policy). It provides recommendations on how system configuration could be improved to reduce pumping energy requirements, and how it can be influenced by regulatory policies. The typical set-up adopted for domestic rainwater systems has pumps operating well outside their optimal energy efficiency due to a mismatch between pump design and the low flow rates required for most of the end uses in urban dwellings. Increased focus on the energy footprint and awareness of the energy consumption should lead to improved system design and optimise the energy requirements for residential rainwater pumping, making rainwater supply more sustainable and the least energy intensive of all the alternative urban water supplies.

Keywords: Energy footprint; pumps; rainwater tanks; specific energy; rainwater harvesting.

6.1 ENERGY USAGE IN RAINWATER SYSTEMS

Energy usage in water supply contributes to the generation of greenhouse gases, and to the overall cost efficiency and sustainability of water services.

The energy and costs associated with traditional centralised water supply and distribution services in urban areas have been thoroughly characterised (Kenway *et al.* 2008; Apostolidis, 2010), however limited data is available on the energy associated with residential rainwater supply systems. This is partly because, historically, rural dwellings that depended on rainwater as their sole water supply (i.e., no mains water sources available) have not been concerned with rainwater pump energy usage, but rather with the level of service from water delivery. However, as rainwater systems are increasingly implemented in urban environments where they co-exist and/or complement other water supplies (e.g., mains water, recycled water, desalinated water, stormwater) there is greater need to understand their energy contribution to water supply and their environmental impact.

A few studies have estimated energy savings to the water industry or costs for pumping rainwater to residential dwellings using average energy values reported in the literature, 0.9–2.0 kWh/kL (Marsden Jacob Associates, 2007; Hall *et al.* 2009; Tam *et al.* 2010; Apostolidis, 2010). Apostolidis (2010) examined the energy-water nexus for water supply across Australia. Others compared rainwater energy use of centralised and alternative water services in Australia (Lane *et al.* 2010) and overseas (Proença *et al.* 2011; Chiu *et al.* 2009). Proença *et al.* (2011) and Chiu *et al.* (2009) considered rainwater pumping energy a relatively minor component compared to other energy needs in the water industry and extolled the potential savings that rainwater harvesting could bring to water providers. However, Hall *et al.* (2009, 2011) have shown that in areas of high rainwater tank uptake, such as South East Queensland, Australia, the combined energy consumption by rainwater tanks could equate to the amount of energy used in the centralised water supply system. Furthermore, the uncertainty associated with energy estimates for residential rainwater supply is high, +/–50%, whilst energy estimates for centralised water supply have a much lower uncertainty margin of +/–15% (Hall *et al.* 2011). In the United Kingdom, the operational energy for rainwater supply has been reported to range from 0.6 to 5 kWh/kL (Parkes *et al.* 2010).

Likewise, across Australia, in-situ studies on the energy consumption for rainwater tank pumping systems have been conducted for 164 residential dwellings in a range of climatic zones and jurisdictions (Gardner *et al.* 2006; Beal *et al.* 2008; Retamal *et al.* 2009; Cunio & Sproul, 2009; Hauber-Davidson *et al.* 2010; Hood *et al.* 2010; Talebpour *et al.* 2011; Umapathi *et al.* 2012; Ferguson, 2011). These studies have covered dwellings located from tropical to temperate climates, on flat to sloping topography, with a wide range of characteristics (single and double storey and various sizes) and various system set-ups (pumps, storage and control types), occupancy (from single to multiple occupants), from standard dwellings to ecovillages, and various rainwater end uses (from single uses to all household uses).

The specific energy for rainwater pumping reported in those studies ranged from 0.4 to 11.6 kWh/kL of rainwater used, indicating a large variability in the energy for supply of rainwater to urban dwellings. However, the median specific energy was around 1.4 to 1.8 kWh/kL, which compares favourably with other alternative supplies such as desalination or recycled water (>2 kWh/kL) as shown in Figure 6.1. Dwellings that used rainwater for all end uses in the house, recorded a median energy consumption of 1.4 kWh/kL, which is within the same average range verified for selected end uses (Hood *et al.* 2010).

Figure 6.1 compares the energy requirements for water supply using various sources. The energy associated with conventional centralised water supply in major Australian capital cities in 2006/07 ranged from 0.09 to 1.85 kWh/kL, with the minimum and maximum applicable to the cities of Melbourne and Adelaide respectively and the capital cities average of 1.03 kWh/kL(Retamal *et al.* 2009; Kenway *et al.* 2008). Recycled water and desalinated seawater supply tend to have energy requirements in excess of 2 kWh/kL. Hence, by reducing the spread of energy for rainwater supply, the energy requirements for rainwater supply becomes comparable to the requirements of other alternative water supply sources (Figure 6.1).

By understanding the actual operation of rainwater tank systems and the factors that influence energy consumption in rainwater pumping, a range of opportunities to optimise the energy associated with rainwater supply can be identified (Cunio & Sproul, 2009; Retamal *et al.* 2009; Hauber-Davidson & Shortt,

2011; Tjandraatmadja *et al.* 2013). This, in turn, is important for increasing the operating efficiency of rainwater harvesting in comparison to other alternative water sources. The following sections summarise the key factors in the design and set-up of rainwater harvesting systems that impact energy consumption and examine how it can be reduced.

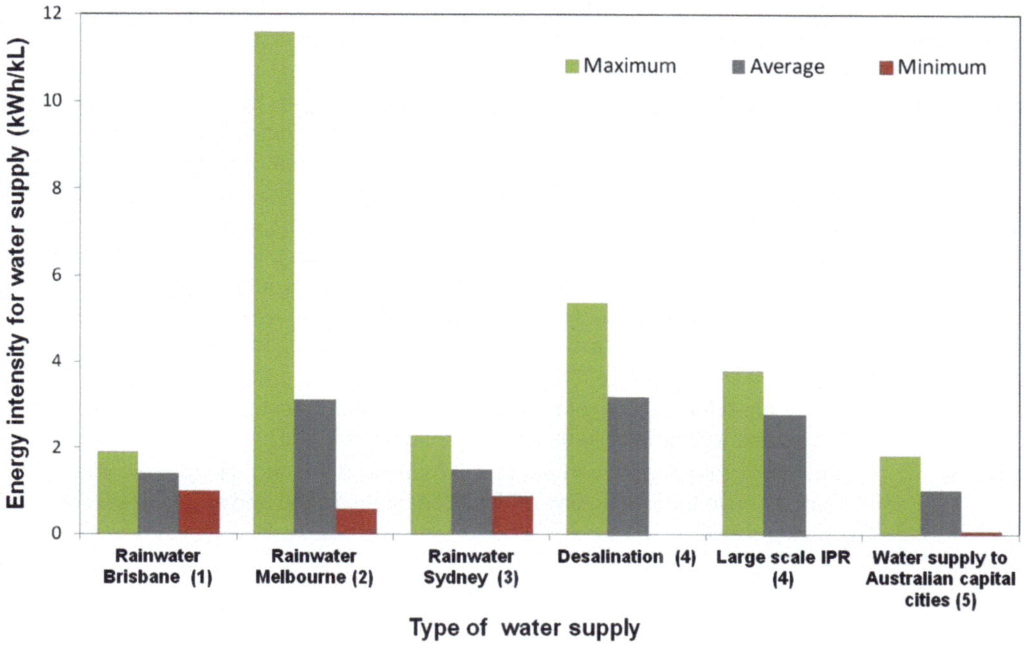

Figure 6.1 Examples of energy requirements for water delivery in major Australian capital cities. (*Sources*: (1) Gardner *et al.* (2006); (2) Hauber Davidson *et al.* (2010); (3) Retamal *et al.* (2009); (4) NSW Government (2006); (5) Kenway *et al.* (2008)).

6.2 RAINWATER SYSTEM SET-UP

Rainwater supply systems are typically designed for residential, commercial and industrial applications. A rainwater system comprises a series of components including tanks, pumps, flow control devices, switching valves, back-up systems, and ancillary components such as pressure vessels and filters, as listed in Table 6.1 (Retamal *et al.* 2009; Hauber Davidson *et al.* 2010; Umapathi *et al.* 2012). A more comprehensive review of key system components and configurations adopted in rainwater supply can be found in Retamal *et al.* (2009).

In urban settings, the most common system adopted consists of an above ground rainwater tank with a fixed-speed pump (either external or submersible) and a trickle top-up or an automatic switching valve to allow mains water back-up. Figure 6.2 shows a typical set-up, with rainwater flowing from the roof into the tank after passing through a leaf-guard to remove leaves. The bottom of the tank is connected to an external fixed-speed pump equipped with an automatic switching device for mains water back-up. The pump provides rainwater for non-potable, fit-for-purpose water use applications such as toilet flushing, laundry cold water taps and garden irrigation. When the tank is full, rainwater overflow is discharged to the stormwater drain.

Table 6.1 Components that can be adopted in a rainwater system.

Component	Description
Pump	Pumps water from the rainwater tank to the dwelling under pressure. There are various types: Fixed-speed external or submersible, variable-speed, jet pump, and so on (Retamal *et al.* 2009).
Rainwater tank	Tanks are supplied in a range of shapes and volumes including above ground, below ground and bladder tanks.
Back-up system	Ensures continuous water supply. There are two types: (a) trickle top-up which fills a tank with mains water via a float activated mechanical valve; or (b) automatic mains switch valve that sources water directly from mains water if the rainwater tank falls below a threshold water level measured with a pressure transducer.
Pressure vessel	Device that releases water from a vessel whilst maintaining the pressure in the line from the pump to the vessel above the threshold value that would otherwise activate the pump. They are particularly useful to prevent pump activation for small water demand events such as hand washing or water leakage. They come in a range of sizes from 3 L to >100 L. Pressure vessels used in residential dwellings typically have small volumes (<10 L).
Filter	Device used for removal of particulate matter before supply to the dwelling. Resistance to flow increases as they become clogged.
Header tank	Balancing storage between a rainwater tank and the end use of a dwelling. It is located above the ceiling line and is filled by the pump and supplies water to end uses by gravity.

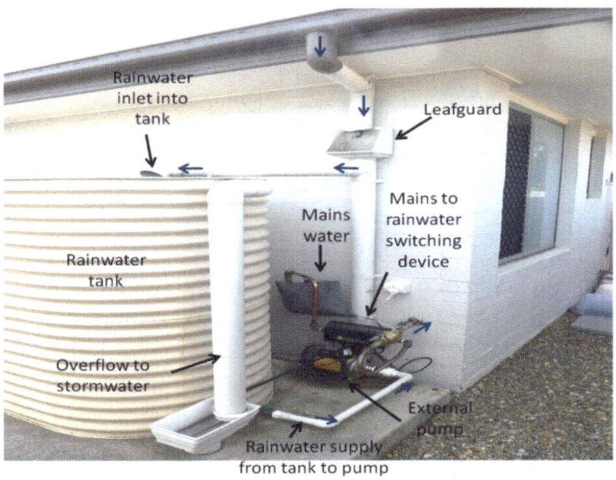

Figure 6.2 Example of a typical residential rainwater system set-up with an external fixed-speed pump equipped with an automatic rainwater to mains water switch. The rainwater flow path into the dwelling is shown by the blue arrows. (*Source*: Umapathi *et al.* 2012).

Fixed-speed pumps are much more common than variable-speed pumps as they are significantly lower in cost (Retamal *et al.* 2009; Tjandraatmadja *et al.* 2012). Most importantly, a fixed-speed pump draws almost the same amount of electrical energy irrespective of the water flow rate. The typical energy associated with pump operation is characterised by three stages, as shown in Figure 6.3: (a) *Start-up* as the inertia of the pump motor is overcome; (b) *Operation* at steady flow and pressure; and (c) *Over-Run* as the flow is stopped and the system is re-pressurised to the set pressure. The start-up stage is characterised by a transient peak in power lasting less than 1 second, followed by constant power draw during the steady water supply regime (Retamal *et al.* 2009; Tjandraatmadja *et al.* 2012). The energy required for the start-up, operation and over-run stages are a function of the design of each pump. This is illustrated in Figure 6.4 which shows the energy consumed during each of those stages for three pumps during the filling of a toilet cistern after a half and a full flush. In Figure 6.4, the start-up energy contributed to a larger share of pump A's total energy use than for pumps B and C, whilst for pump C, the over-run energy was higher than for the other pumps. Analysis of 11 pumps sold in Australia showed that the over-run energy can contribute from 5% to 25% of the total energy required to fill a toilet cistern after a full flush depending on the pump model and its pressure settings (Hauber-Davidson & Shortt, 2011).

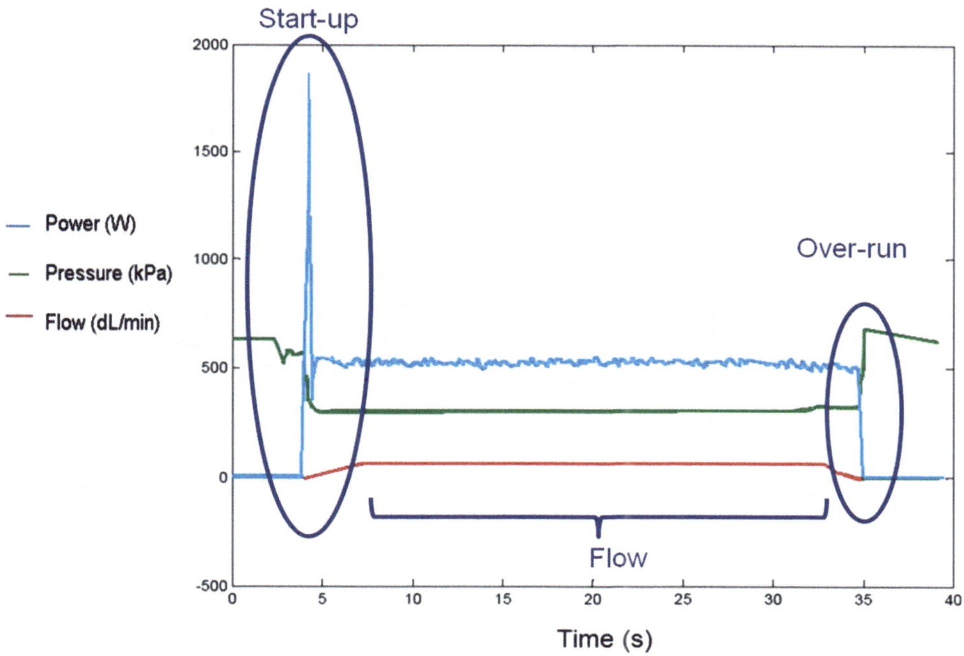

Figure 6.3 Typical operation profile for a fixed-speed pump showing the power consumption, pressure and flow delivered by the pump at the start-up, operation (flow) and over-run of pump operation. (*Source*: Tjandraatmadja *et al.* 2011).

Rainwater tanks adopted as a supplementary water supply system are usually fitted with a back-up system, that is, a mechanism to switch to the mains water when the tank water level falls below a threshold level set by a float or pressure transducer. There are two types of back-up systems commonly used: trickle top-up and switching device. A trickle top-up system (Figure 6.5a) uses a mechanical float device within the storage tank, so that when the water level reach a threshold, a mechanical valve opens and mains water is trickled into the tank to maintain water supply to the household. A trickle top-up system has adverse energy implications as mains water is depressurised when it enters the tank, after which it is re-pressurised by the household pump. A switching device (Figure 6.5b) uses a pressure sensor to measure the water level in the tank, and once a threshold value is reached, it signals a solenoid valve to open which connects directly to the mains water supply. The switching device is placed on-line after the pump and does not depressurise the mains water supply. However, energy is used to operate the sensor circuitry and the stand-by energy use of the devices in the market vary substantially. In some cases, the non-pumping related energy (stand-by and over-run time) can contribute up to 20 to 40% of the total energy consumption depending on pump type and operation (Water Conservation Group, 2010). However Umapathi *et al.* (2013) examined 20 dwellings with rainwater tanks with the two different mechanisms and found a slightly higher energy consumption for the trickle top-up system (1.59 kWh/kL) compared to the switching device (1.46 kWh/kL).

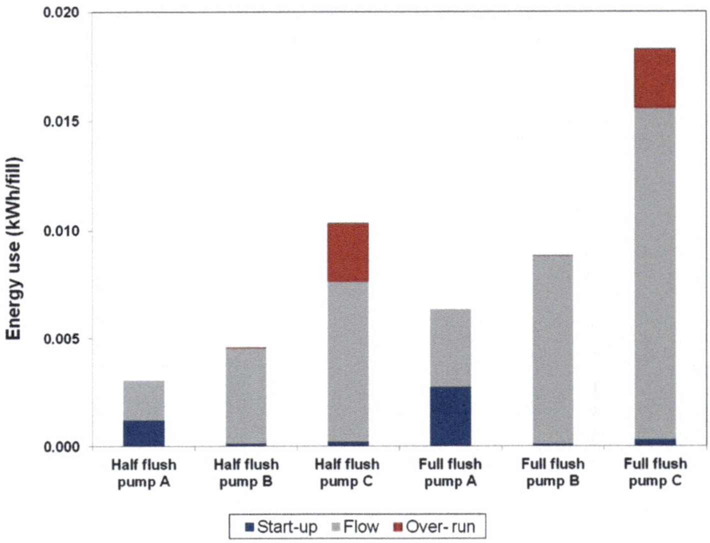

Figure 6.4 Energy use during supply of rainwater to a toilet cistern by three different pumps A, B and C with motor capacities of 0.2 kW, 0.55 kW and 0.75 kW respectively. (*Source*: Tjandraatmadja *et al.* 2011).

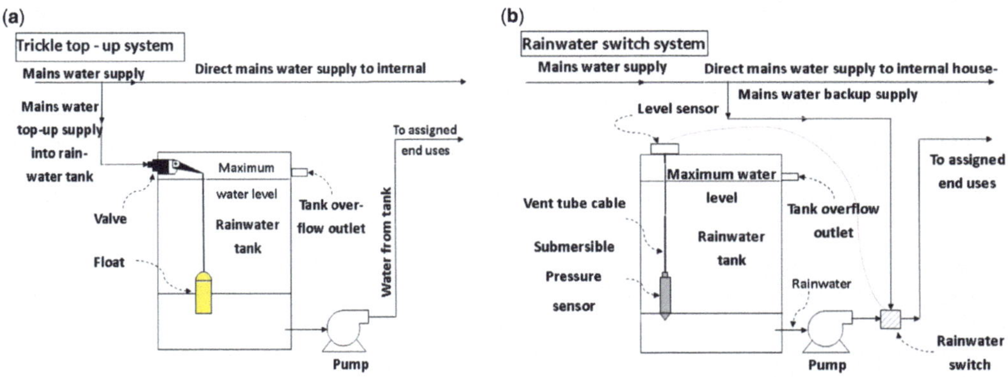

Figure 6.5 Illustrative diagram for a rainwater tank system working on: (a) a 'Trickle top-up' mechanism; and (b) a 'Rainwater Switch' mechanism.

The energy usage for rainwater supply is mainly dependent on the pumping of water from a rainwater tank to the end uses in a building (Retamal *et al.* 2009). Hence, thoughtful system design and the proper selection of ancillary components can help to reduce the energy usage for rainwater pumping.

6.3 ENERGY USE ESTIMATION

6.3.1 Factors impacting energy consumption

Energy needs vary due to seasonal change, storage capacity and intensity of rainwater use. Studies conducted in urban areas across Australia showed that rainwater end uses vary across the country due to system configuration, household characteristics and home owner motivation and preferences. However, legislation, standards and codes promoting the use of rainwater tanks can also play an important role in determining rainwater system configurations adopted (Gardiner *et al.* 2008; Gardiner, 2009).

Energy usage for rainwater supply is influenced by a range of factors: pump selection, system configuration and infrastructure, end use requirements, appliances and fittings in the dwelling, and dwelling occupancy. To understand energy performance of a rainwater system and to evaluate the system in its entirety it is necessary to understand how rainwater is used in a dwelling and the role of the various system components (Talebpour *et al.* 2011; Beal *et al.* 2011). So, in order to theoretically estimate pump energy use for a household rainwater system, two key pieces of data are required:

- A relevant pump power curve to understand the relationship between the electrical power draw of the pump and water flow rates; and
- A profile of household water consumption with information on the frequency, duration and flow rate of water end-uses supplied by the tank.

The need for such information is explained in the following sections.

6.3.2 Pump performance

The performance of a pump is summarised in its power curves. Figure 6.6a provides an example of a pump power curve. The flow rate and power curve data can be used to estimate the total energy consumed by a rainwater system as well as the rainwater volume delivered over a given period of time. Each pump has an optimal range of water flow rates where the pump operates most efficiently, the Best Efficiency Point, shown in Figure 6.5a.

In addition, to compare the energy efficiency of a system to others with different water use and pump characteristics, the *Specific energy* or *Energy intensity* is calculated. This is an expression of the total energy consumed by a pump (e.g., in kilowatt-hours (kWh)) divided by the total volume of rainwater delivered (e.g., in kilolitres (kL)) at a set flow rate as shown in Figure 6.6b. The specific energy (in kilowatt-hours per kilolitre (kWh/kL)) decreases as pump flow rate increases.

The other key parameter adopted for the evaluation of the energy use for water delivery in a dwelling is the *Total energy consumption* over a set period of time in KWh per time unit (e.g., year). This measures the energy footprint for rainwater supply over a set period of time irrespective of differences in pump operating systems.

6.3.3 Pump design and characteristics

The individual design and characteristics of each pump influence the energy use. Retamal *et al.* (2009) demonstrated that pump type and application impact its specific energy use, as shown in Figure 6.7 for a selection of fixed-speed, submersible and variable-speed pumps and high and low flow uses in urban dwellings. Figure 6.8 compares the specific energy for running a range of pumps at two set flow rates, 8 L/min and 17 L/min, and shows that, as expected, pumps of higher motor power capacity tend to have higher specific energy requirements, for example, Davey HP45-05 and Davey HM60-10. However, there are examples where pumps of the same type with the same nominal power capacity differ in specific energy requirements depending on the manufacturer, such as Storm and Onga SMH55 (Hauber-Davidson & Shortt, 2011).

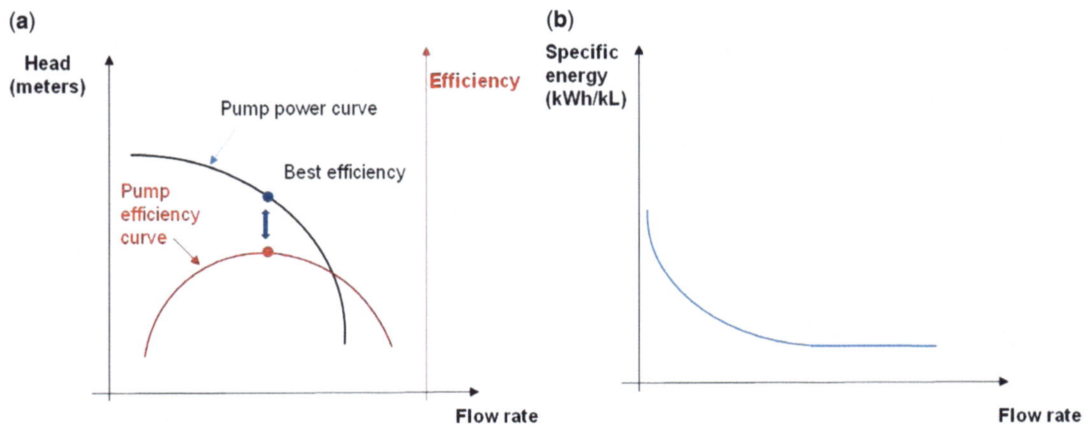

Figure 6.6 (a) Pump power curve for a fixed-speed rainwater tank showing the relationship between water head versus flow rate (black curve) and pump efficiency (red curve) and the Best Efficiency Point; (b) Specific energy (kwh/kL) used by the pump as a function of flow rate.

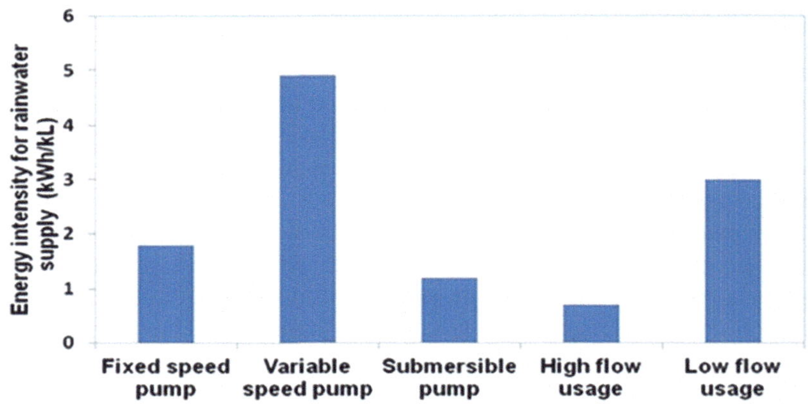

Figure 6.7 Specific energy use for rainwater supply using a fixed-speed, submersible and variable-speed pumps in Sydney homes. (*Source*: Retamal *et al.* 2009).

Clearly individual pump design impacts energy use, and each pump has its own specific energy curve as shown in Figure 6.9 for a range of pumps sold in Australia (Hauber-Davidson & Shortt, 2011). The figure also shows that the specific energy for pump operation decreases as the flow rate increases, particularly for flow rates >10 L/min.

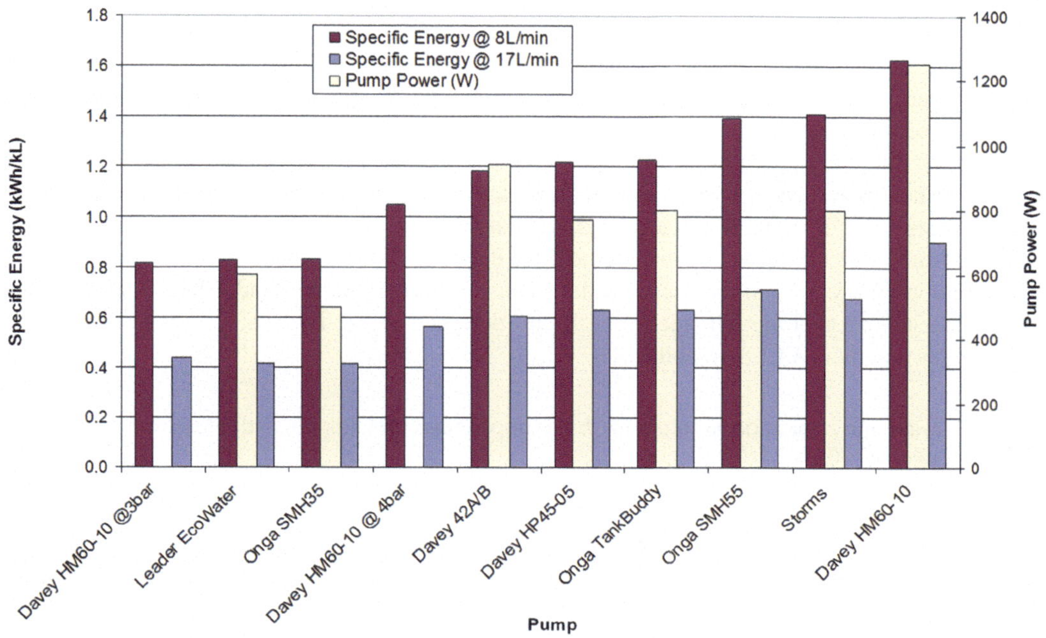

Figure 6.8 Comparison of the specific energy for operation of various pumps at flow rates of 8 L/min and 17 L/min. (*Source*: Hauber-Davidson & Shortt, 2011).

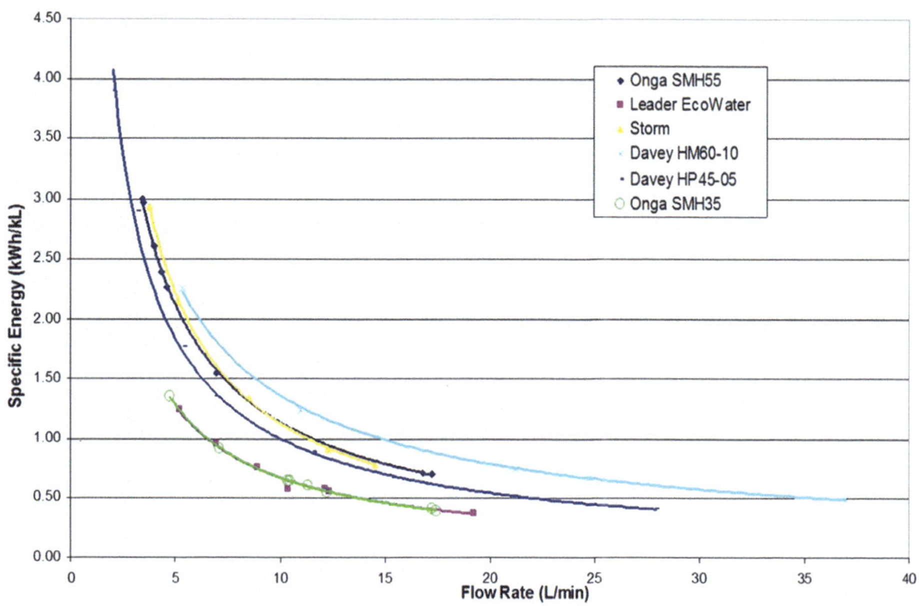

Figure 6.9 Comparison of specific energy curves for a selection of rainwater supply pumps sold in Australia. (*Source*: Hauber-Davidson & Shortt, 2011).

6.3.4 End use water requirements

The number of times a pump starts, the duration of its operation, and the flow and volume of water supplied will vary with the water use pattern for each individual appliance, and with the pump settings.

Water end uses in a household have specific service requirements for volume, pressure, flow and pattern of water supply. Table 6.2 provides a summary of typical operating requirements for appliances such as washing machines, dishwashers, showers and taps. Water fittings such as taps, showerheads and irrigation devices are designed to operate within a pre-set flow rate range. Likewise, household appliances, such as washing machines, toilet cisterns and dishwashers are designed to operate within a specific range of flow and pressure conditions. In addition, in Australia water efficient appliances and fittings that comply with the Water Efficiency Labelling Scheme (WELS) are designed to use less water and/or supply water at lower flow rates than non-WELS appliances.

Table 6.2 Service and design parameters for water appliances and fittings. (Adapted from Tjandraatmadja *et al.* 2013).

Appliance or fitting	Specifications set by manufacturer		Service conditions	
	Minimum operating pressure (kPa)	Maximum operating pressure (kPa)	Flow rate (L/min)	Volume of water (L/use)
Washing machine	40–100[1]	800–1000[1]	<13[4]	WELS[6]: 5–30L/kg load Non-WELS: >30L/kg load
Dishwasher	30–150[2]	800–1000[1]	<4.1[4]	16[4]
Toilet cistern	35–50[2] (conventional)	400[1]	<6.3[4]	Full/half flush[6]: 9/4.5L, 6/3 L and 4.5/3 L
Tap	n.a.	n.a	Indoor tap: Non-WELS: 15–18 WELS[6]: 2–7	Standard: 15–18[3] Hand basin: 3.3[5] Kitchen sink: 19.4[5] Laundry through: 26[5]
			Outdoor tap: Non-WELS: 16–30 WELS: Automatic irrigation device: max. 9[6] Hand held hose trigger: 7–12[6]	Spray type[7]: Full: 6 L/min Shower: 3.8 L/min Jet: 4 L/min Soaker: 6 L/min Mist: 2.2 L/min
Shower head	n.a.	n.a	Non-WELS: 15–25 WELS: 6–12	

Note: [1]Manufacturer specifications, [2]Standards Australia (1999), [3]Australian Government (2011), [4]Measured (adapted from Tjandraatmadja *et al.* 2011), [5]Measured by Roberts (2005), [6]Water Efficiency Labelling Scheme (WELS) is a water efficient rating scheme for all major water using fittings and appliances in Australia (Australian Government, 2011 (http://www.waterrating.gov.au/index.html)), [7]Measured by Water Conservation Group (2010), n.a. Not applicable.

The most common end use applications for rainwater in dwellings are toilet flushing, washing machine supply and outdoor irrigation. Toilet flushing and washing machines typically require low flows, less than 6 L/min and 13 L/min respectively, and limited volumes of delivery, up to 4 to 6 L for a toilet cistern, 40 to 60 L for a front loader washing machine, and 100 to 150 L for a top loader washing machine (Tjandraatmadja *et al.* 2012).

The pattern of water supply also varies with end use (Tjandraatmadja *et al.* 2012). For instance, washing machines require a pump to stop and start multiple times during a wash. Tjandraatmadja *et al.* (2012) showed that the wash cycle in a top loader washing machine comprised two to three major water supply episodes of 40 ± 18 L each, and four to five short water top-up episodes in the spray cycle (Figure 6.10a). The rainwater pump operated on average for 21 ± 4 min. In comparison the wash cycle of a front loader washing machine was characterised by three to four water supply episodes of 11 ± 0.5 L each, and up to five small injections of water (Figure 6.10b). The rainwater pump operated on average for 5 +/−2 min during the cycle. In contrast filling a toilet cistern is a short duration single event where the pump operated for 0.5 min to 1 min depending on a half or a full flush, (Figure 6.10c).

Laboratory and in-situ studies show that the flow rate associated with end uses is the key influence on the energy intensity of pump operation, with low flow end uses typically using more energy per kL than high flow uses. In addition, Hauber-Davidson has also shown that pumping energy is barely affected by the pumping distance to the various end uses supplied, in a double storey house (Water Conservation Group, 2010). Retamal *et al.* (2009) estimated that the energy intensity associated with individual end uses varied from 0.4 to 2.9 kWh/KL for pumps with motor sizes between 0.5 to 0.9 kW. Talebpour *et al.* (2011) examined the energy associated with individual end uses in five dwellings equipped with a popular pump and switching valve system and measured energy intensities from 1.04 to 1.67 kWh/kL for garden irrigation and toilet flushing respectively. Hauber-Davidson and Shortt (2011) determined the energy for the operation of a washing machine, hose spray for outdoor irrigation, toilet cistern and a hot-cold water mixing tap and found specific energies ranging from 1.13 to 4.73 kWh/kL. Tjandraatmadja *et al.* (2012) verified that the specific energy for rainwater supply to various appliances (dishwasher, toilet cistern, top and front loading washing machines and header tank) depended on the end use and on the pump motor capacity, ranging from 0.39 kWh/kL to fill a header tank with a 0.2 kW pump, to 5.3 kWh/kL for a dishwasher with a 0.75 kW pump, as shown in Figure 6.11.

Hauber-Davidson and Shortt (2010) and Tjandraatmadja *et al.* (2012) have shown that the typical operating requirements for rainwater end-uses in urban dwellings (toilet flushing, washing machine and irrigation) are below the best efficiency point for the majority of pumps adopted in rainwater systems. This is shown in Figure 6.12, which reports on the service flow requirements for typical end uses, and the associated specific energy requirements for pumps of 3 different sizes (0.2 to 0.75 kW). Figure 6.12 shows more generally that low flow water supply events of short duration, such as toilet flushing and opening a tap briefly for hand washing, result in higher specific energy use than longer duration events with higher flows, such as filling a washing machine or garden irrigation. In comparison the best efficiency point for most pumps was achieved at flow rates greater than 25 L/min.

Ferguson (2011) verified, in a one-year study of 52 dwellings constructed after 2009 in Sydney, that 40% of rainwater pumping events are short duration events supplying volumes of 2.5 to 9.5 L, such as toilet flushing, whilst less than 20% required over 100 L (for irrigation and washing machine use), which are characterised by higher flow rates. Similarly, toilet flushing was also identified as the predominant end use for rainwater in Melbourne studies (Hauber-Davidson & Shortt, 2011).

Thus, a fixed-speed pump will often operate inefficiently and outside its optimal range for many of the end uses. Consequently, it is almost impossible to select a fixed – speed pump that is suitable for all water uses using the pump curve alone (as shown in Figure 6.6a).

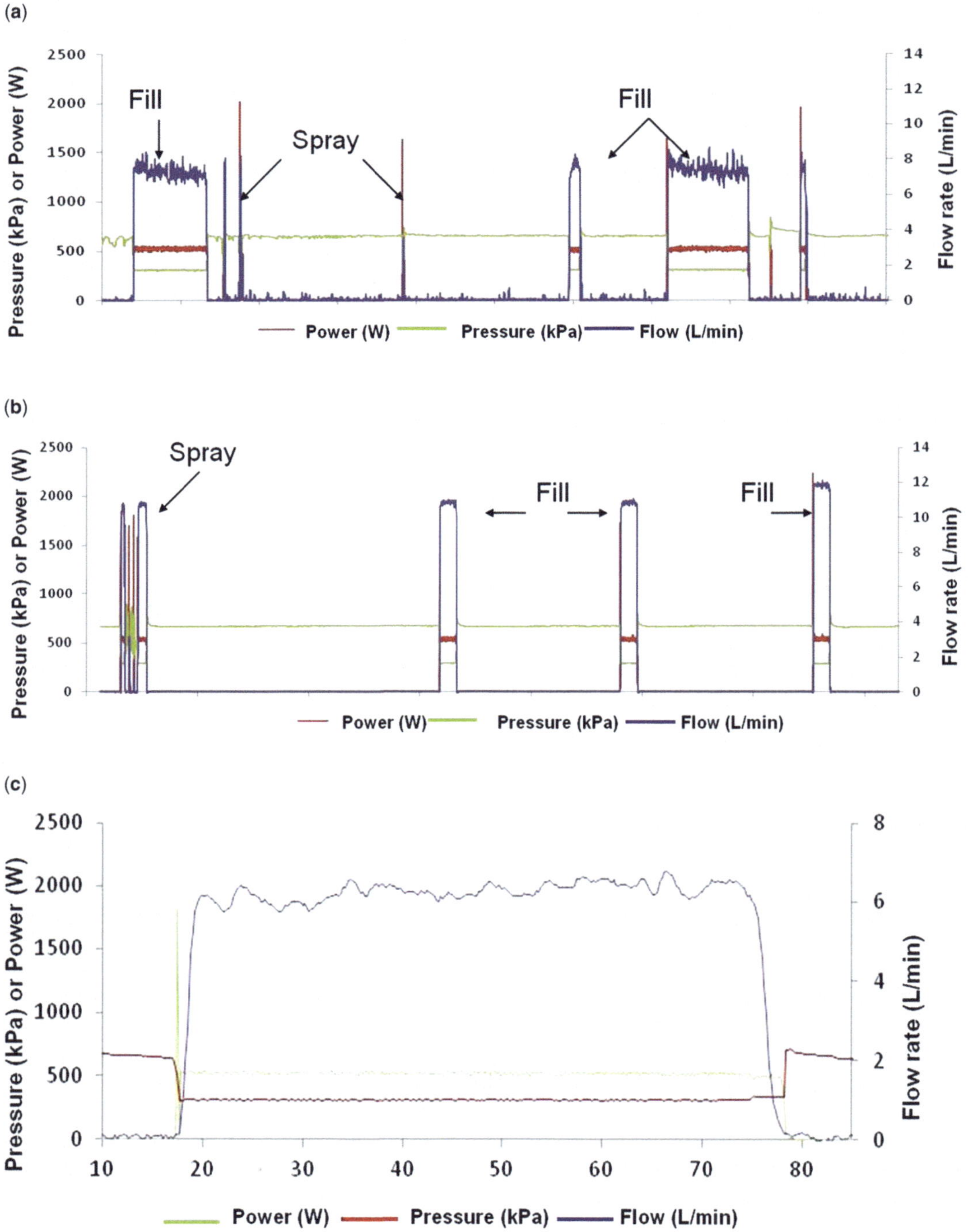

Figure 6.10 Pump energy use and water supply flow and pressure patterns during operation of (a) Top loader washing machine, (b) Front loader washing machine and (c) Filling of a toilet cistern. Note that the x-axis (time) does not have the same scale for all three events. (*Source*: Tjandraatmadja *et al.* 2012).

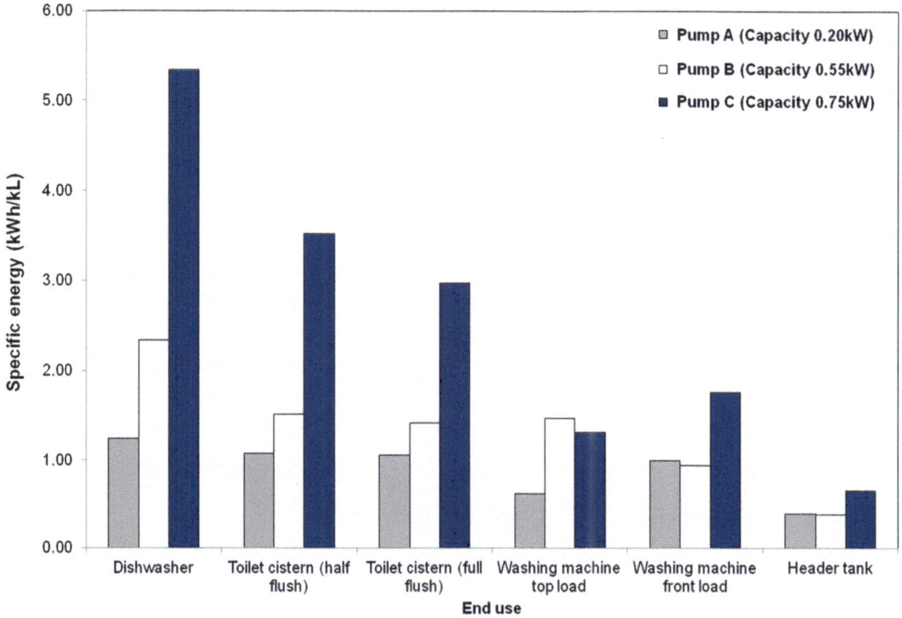

Figure 6.11 Specific energy requirements for rainwater supply to a dishwasher, washing machine, toilet cistern and a header tank using pumps of motor capacities ranging from 0.2 to 0.75 kW. (*Source*: Tjandraatmadja *et al.* 2012).

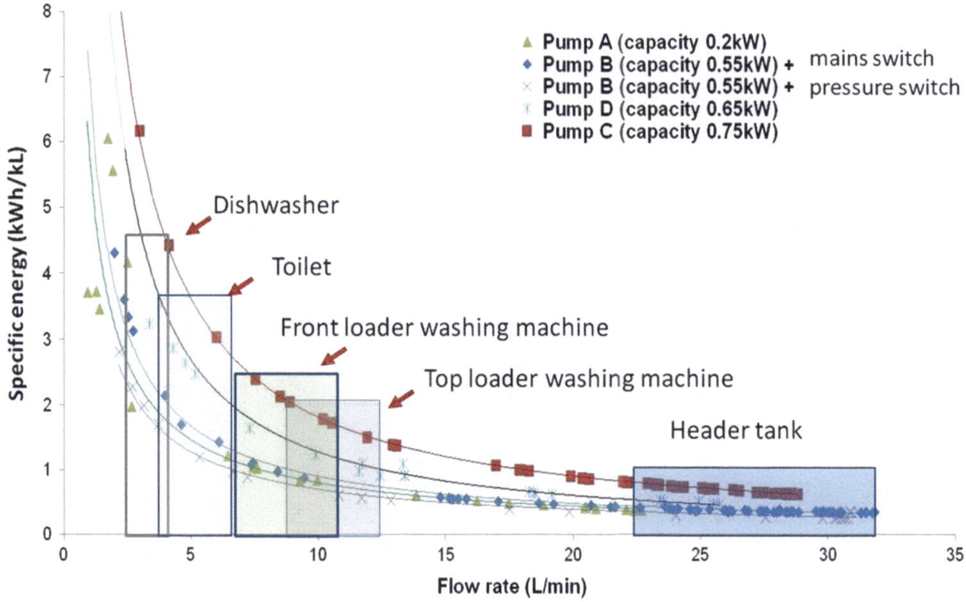

Figure 6.12 Relationship between specific energy and flow required for rainwater uses in a dwelling. (*Source*: Tjandraatmadja *et al.* 2012).

Yet, pump motor size also matters as demonstrated in Figure 6.11. In Figure 6.12, pumps of various motor sizes fulfilled the individual service requirements of the common household end uses. In addition Pump B was tested also with an automatic water mains switch (i.e., when the tank is empty mains water is provided to the dwelling). In particular, the smaller size pumps were able to deliver a satisfactory level of service at a lower specific energy (Hauber-Davidson & Shortt, 2010; Ferguson, 2011; Tjandraatmadja *et al.* 2012, 2013).

Cunio and Sproul (2009) showed that under-sizing a pump runs the risk of limiting the flow rate and pressure of the rainwater supply. However, the common perception that a 'one size bigger is better' for pumps has resulted in many pumps being oversized and not optimised for their operating range, particularly in urban areas where rainwater supply is connected to limited end uses instead of to the entire household demand. However the quality of service (flow rate & pressure) does not vary markedly between a 0.2 kW and a 0.75 kW pump for supply to end uses such as toilet cisterns and washing machines (Tjandraatmadja *et al.* 2012). Most household rainwater users typically only notice the performance of their rainwater pump when they water the garden and expect a high flow rate and pressure similar to their mains water supply.

6.3.5 Dwelling occupancy and characteristics

Household occupancy, that is, the number of inhabitants in a dwelling, and the water use habits of the household impact the pattern and the frequency of rainwater use, and hence the *total energy* use in a dwelling. Dwellings with more occupants generally use larger volumes of rainwater and consequently consume more energy as shown in Figure 6.13. On the other hand, the pattern of rainwater use and the end uses will also differ among dwellings. For instance, a pump that supplies the toilet cistern alone tends to be less energy efficient than the same pump supplying both a toilet and a washing machine. This will also be influenced by how often the washing machine is used. In addition, indirect factors, such as climate, season and lifestyle, can influence water use patterns also impact energy use indirectly (Beal *et al.* 2010; Umapathi *et al.* 2012).

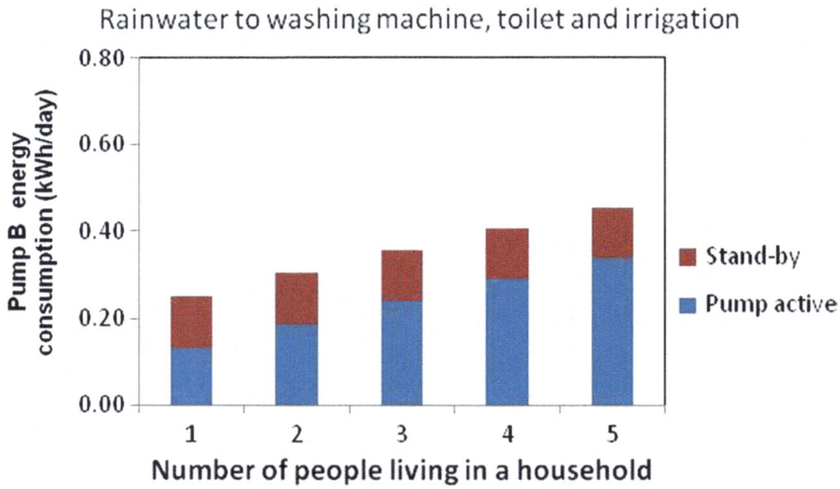

Figure 6.13 Estimated daily energy consumption for stand-by use and pump active use to supply rainwater to households of various sizes using pump B (0.55 kW) assuming a same end use pattern. (*Source*: Tjandraatmadja *et al.* 2012).

Examination of individual tank set-ups confirmed that the high specific energy consumption in a number of dwellings was caused by incorrect system configuration and oversizing of pumps for the end use needs (Retamal *et al.* 2009; Hauber-Davidson *et al.* 2010; Ferguson, 2011).

6.3.6 Friction losses

The friction loss expected in an actual dwelling is determined by the plumbing characteristics and the flow rate of water. Details on how to estimate friction losses are provided in Cook *et al.* (2012), Swamee and Sharma (2008) and Tjandraatmadja *et al.* (2012). In domestic dwellings, rainwater tanks tend to be placed at a variety of locations and hence the distance between the rainwater tank and end-use appliances will vary with the design of the house, its land size and property layout. However, energy losses due to pipe friction as water travels through a rainwater system into a dwelling are considered to be minimal. Instead, the majority of the energy is used for running a pump for rainwater supply in the typical dwelling (Retamal *et al.* 2009; Tjandraatmadja *et al.* 2012). Retamal *et al.* (2009) estimated that losses in pumping energy in a typical domestic dwelling attributed to friction losses during delivery of water amounted to only 2%.

However, pump motor size can also impact on friction losses and care is required to prevent the oversizing of pumps for a required end use. For example, Cunio and Sproul (2009) reported an experiment where use of an oversized (high pressure) pump for the delivery of water to a toilet cistern resulted in 90% pressure losses in friction alone. However, on the other hand, Tjandraatmadja *et al.* (2012) verified in a laboratory model house that the friction losses within the pipes for a given system set-up did increase with pump capacity, but rather ranged from 1% to less than 6% of the total pressure supplied for pumps of motor sizes 0.2, 0.45 and 0.75 kW.

Overall, whilst friction losses are expected to constitute only a minor energy loss component, they also reinforce the benefit of matching pump size to end use and system requirements.

6.3.7 Other components

6.3.7.1 Pressure vessels

Pressure vessels can be used to mitigate the problem of inefficient pump operation, particularly where a pump is switching on and off frequently for short duration low flow rate water use events. A pressure vessel is a tank that contains an internal diaphragm which is partially filled with air as shown in Figure 6.14a. Pressure vessels are placed in-line after the pump and essentially store pressurised water. When the diaphragm in the pressure vessel is empty and a tap is turned on, the pump will start operating initially at maximum flow and will continue to pump until the pressure vessel is full as shown in Figure 6.14b. The next time a tap is turned on, water is drawn from the pressure vessel until the internal pressure reaches a threshold value, thereby delaying the start-up of the pump. The volume of water delivered by a pressure vessel is a function of the nominal volume of the vessel and the pump trigger pressure settings. However it is always less than the nominal volume as shown in Figure 6.15a.

Pressure vessels come in a range of sizes (nominal volumes of 5 L to over 1000 L). Small pressure vessels of 5 to 8 litres volume are sometimes installed with pumps to reduce the number of pump start-ups caused by small leaks in the water reticulation of a dwelling. The larger sizes are usually adopted for industrial applications.

A properly sized pressure vessel can reduce pumping energy by two mechanisms:

- It reduces the number of pump starts provided the volume required for a given end use is less than the water volume contained in the vessel.

- The flow rate at which the pressure vessel is filled is often greater than that at which an appliance is filled. This is particularly advantageous for low flow end uses such as filling a half-flush toilet cistern or brief operation of a tap for hand washing a glass of water (Tjandraatmadja *et al.* 2012).

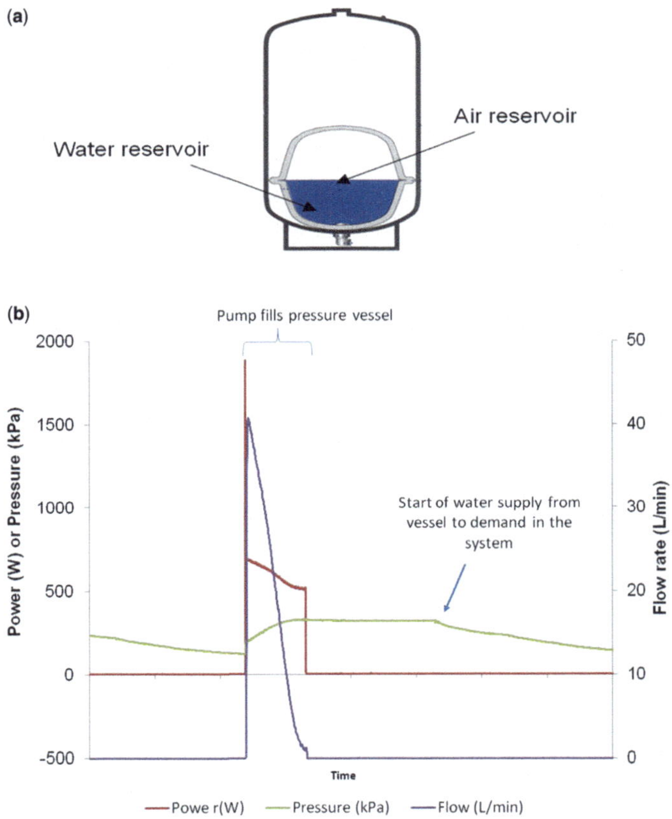

Figure 6.14 Pressure vessel details: (a) Schematic diagram of pressure vessel components and (b) Pump operation with pressure vessel operation (*Source*: Tjandraatmadja *et al.* 2012).

Energy savings are generated if the volume of water that the pressure vessel provides is equal to or greater than the total volume of water required for an end use. For instance, a 5 L pressure vessel provides low volumes of water (1 to 2 L) before the pump starts, reducing stop-starts associated with small leaks in the system. However it offers negligible energy savings (Water Conservation Group, 2010). In comparison a larger pressure vessel, such as an 18 L pressure vessel with a 0.75 kW pump, can provide up to 6.3 L of water prior to the start-up of a pump, thereby reducing the energy used for rainwater pumping, particularly for low duration and high energy intensity water uses. This is shown in Figure 6.15b where the addition of an 18 L pressure vessel halves the specific energy required for low volume end uses such as the supply to a tap for a short duration or dishwasher operation (Tjandraatmadja *et al.* 2011).

Some caution is required regarding the set-up for pressure vessels. Installation of a pressure vessel in a system with a mains switching device can cause the pressure vessel to malfunction as the switching device controls the pump on-off cycling via a flow sensor (Retamal *et al.* 2008).

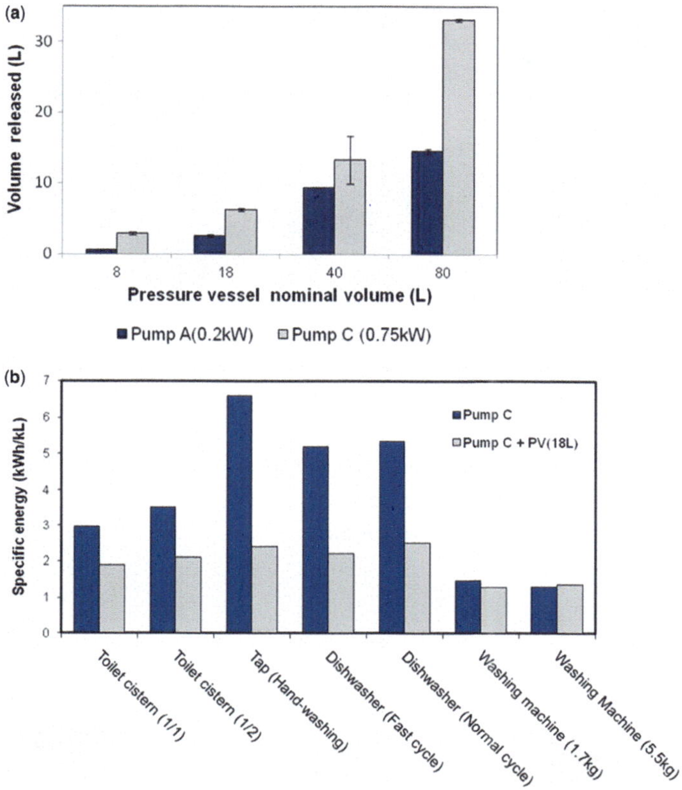

Figure 6.15 Impact of a pressure vessel on pump operation: (a) Volume of water released by pressure vessels of nominal volumes of 8 to 80 L coupled with pumps of 0.2 kW and 0.75 kW capacity; (b) Specific energy for rainwater supply to common end uses by a 0.75 kW pump with and without an 18 litre pressure vessel coupled. (*Source*: Tjandraatmadja *et al.* 2012).

6.3.7.2 Header tanks

A header tank is a localised storage vessel on the roof (or at similar height in a building). Water is pumped to the header tank and that supplies water by gravity to various end uses. This set-up is adopted in many parts of the world, for example, Asia, UK and South America. In Australia, header tanks can be found in commercial or industrial settings, but are not common in domestic residences.

Header tanks can provide large storage, (e.g., 100 to 300 L), reducing the number of pump start-ups compared to direct pump supply, and the pump operates at high flow rates to fill the header tank, increasing the energy efficiency. A properly sized header tank could provide the daily water needs of a dwelling with a single pump start-up per day, provided a proper switch is used. Thus in principle, header tanks could offer high energy savings for rainwater supply.

Laboratory studies in a simulated dwelling, shown in Figure 6.16, estimated that energy savings of 58% to 79% could be achieved by a 300 L header tank when compared to direct supply to individual appliances by pumps ranging from 0.2 to 0.75 kWh motor capacity (Figure 6.17). This is equivalent to a reduction in the specific energy for rainwater supply ranging from 0.39 to 0.66 kWh/kL depending on pump size (Tjandraatmadja *et al.* 2012).

Figure 6.16 Simulated dwelling for evaluation of energy use for rainwater supply (a) Overview of major end uses (washing machine, toilet, taps and dishwasher), (b) Set-up for rainwater supply with a 300 litre header tank, (c) Rainwater tank supply and monitoring instrumentation.

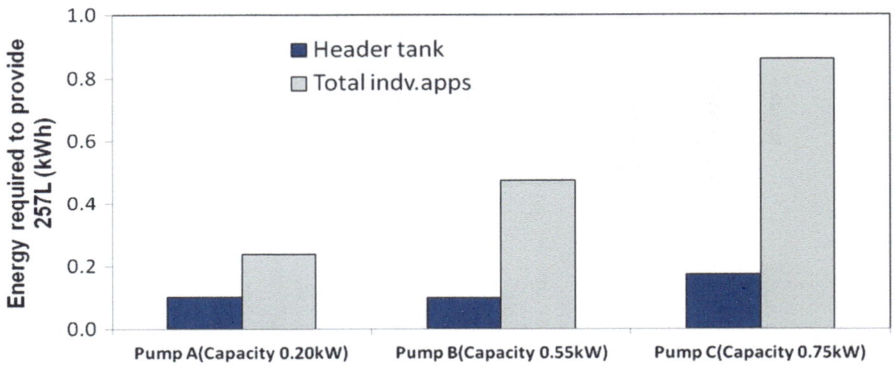

Figure 6.17 Energy use for rainwater supply using a 300 litre header tank compared to direct supply of individual appliances (*Source*: Tjandraatmadja *et al.* 2012).

The caveat, however, is that a minimum height is needed for gravity supply from the header tank to provide the minimum pressure to open the solenoid valves that control water ingress into the appliances. The installation adopted in Figure 6.14b placed the header tank at a height of 2.7 m above floor level, corresponding to the ceiling height, and provided a service pressure of <25 kPa, which was well below the minimum pressure requirements for operation of household appliances in Australia (31–100 kPa). However, by increasing the header tank to 5m, (higher than the ceiling height for a single storey dwelling) an adequate minimum service pressure of 50 kPa for toilet cistern filling would have been achieved. The appliances can also be manufactured to work on low pressures if there is a significant market for the industry to consider.

Therefore, in designing a header tank set-up, building design needs to allow adequate height for the header tank installation to generate sufficient hydrostatic pressure for solenoid operation (Tjandraatmadja *et al.* 2013). Alternatively, changes to the solenoid valve design in common appliances could be considered to allow for low pressure operation.

6.3.7.3 Different types of storages (under-floor bladders, gutter storage)

In urban areas, a variety of different rainwater storage methods have been adopted to overcome space constraints. For example, large plastic 'bladders' are sometimes used to store rainwater beneath the floor of older style houses and have the advantage of a flexible shape in addition to making use of 'dead' space. In one Sydney home with an under-floor bladder that was monitored, the rainwater system used a more powerful Venturi-style pump (a fixed-speed ejector pump that operates by creating a vacuum), to draw up the rainwater from beneath the house (Retamal *et al.* 2009). A combination of very efficient water use and a powerful pump caused the energy intensity of this particular system to be high at approximately 5 kWh/kL.

Some other innovative storage types also have the potential to reduce energy intensity, for example, gutter storages which provide some gravitational pressure, reducing pumping requirements (Retamal *et al.* 2009). However, other operational factors such as the potential mosquito breeding hazard may limit the uptake of such options. Overall, the type and location of storage also needs to be considered for optimising pumping energy.

6.4 REDUCING ENERGY USE FOR RAINWATER SYSTEMS – LESSONS FROM AUSTRALIA

Since 2007, Australia experienced a high uptake of rainwater tanks in urban areas. Rainwater tanks are found in 23% of suitable dwellings across major capital cities increasing to 43% in selected capital cities. Rainwater tank installation is particularly strong in new dwellings, with approximately 57% of all dwellings less than 1 year old in south east Queensland connected to rainwater supply (Government of Queensland, 2009; ABS, 2010). Australia has also had the largest number of studies which examined the energy associated with rainwater pumping in urban settings.

Marsden Jacob Associates (2011), Stewart (2011), Gurung *et al.* (2012) and Gurung and Sharma (2014) have examined the life cycle costing of individual tanks systems. Electricity costs for operation of a rainwater tank are considered minor and estimated to represent only 2% of total operating costs over the life of the tank in Gurung *et al.* (2012). Ferguson (2011) recorded low energy consumption for rainwater supply in 52 dwellings in the Sydney area, with a median energy consumption of 62 kWh per dwelling per year. In monetary terms, this is equivalent to AUS$15 per year assuming an energy intensity of 1.48 kWh/kL and current electricity prices of A$0.20 /kWh.

Notwithstanding the low costs associated with rainwater harvesting, the end use requirements for rainwater in urban dwellings in Australia cause pumps to operate well below their best energy efficiency point. Thus, there are significant benefits still to be gained by improving the energy efficiency in rainwater pumping through better matching end use service requirements and pump operation, and the use of ancillary devices such as pressure vessels and header tanks.

By adopting a smaller 0.2 kW pump instead of a 0.75 kW pump, electricity usage for rainwater supply in a typical dwelling can be reduced from 213 kWh per year to 66 kW per year, that is, a 73% reduction in energy consumption (Tjandraatmadja *et al.* 2012).

The use of devices such as pressure vessels and header tanks that could further reduce the energy consumption is uncommon, however their potential for energy savings has been demonstrated in laboratory studies (Tjandraatmadja *et al.* 2012) and in examples of dwellings where the energy intensity for pumping was 30–36% lower with the use of pressure vessels in-situ (see Section 6.3.7) (Retamal *et al.* 2009). The larger the pressure vessel with respect to the end use volume requirements, the least often the pump needs to start. However, almost no information is available to the public on the performance of pump and pressure vessels combinations.

Furthermore, header tanks coupled with an adequate level switch as previously discussed in section 6.3.7 have the potential to generate the largest energy savings of all ancillary devices and should be further examined.

In addition, significant improvements could be achieved by further investment in product design in areas such as the development of pumps customised for low flow urban end uses; improving the design of mains switching valve systems to reduce energy consumption, and facilitate their integration with pressure vessels; and the redesign of solenoid valves in household appliances for low pressure operation.

Investing in education, greater emphasis in current design and set-up of rainwater pumping systems followed by the development of benchmarks and guidelines that can inform and assist consumers to better select pumps could further reduce the energy requirements and lead to more sustainable practices.

6.4.1 Policy considerations for rainwater system energy use

Policy and legislation can play a significant role in the uptake and set-up of rainwater systems and subsequently on associated impacts such as energy needs. In Australia, severe drought and increase in demand on water resources due to population growth resulted in the introduction of demand management, water use efficiency and the diversification of water sources through policy and legislation (Tjandraatmadja *et al.* 2012). Across Australia, education campaigns, financial incentives and/or mandatory water demand reduction targets have driven the uptake of water efficient appliances and rainwater tanks (Tjandraatmadja *et al.* 2013).

Because adoption of rainwater tanks has focused on reducing mains water consumption, little consideration has been given to energy by industry, government or rainwater users. In addition, the awareness and expectations of tank owners regarding the operation of rainwater systems varies markedly in urban settings compared to rural settings where tanks are the sole source of water supply (Gardiner *et al.* 2008; Gardiner, 2009).

Overall, there is a lack of benchmarks and policy in Australia and around the world on energy efficiency for rainwater pumps and there is limited awareness and guidance to tank owners on how to minimise their energy usage or how to compare pump performance based on energy efficiency (Tjandraatmadja *et al.* 2012; Hauber-Davidson & Shortt, 2011). Typically, home dwellers rely on the knowledge of rainwater system and pump distributors and installers for system selection (Tjandraatmadja *et al.* 2012).

At the same time, the design of household pumps has not evolved at the same pace as water efficient appliances and demand management approaches. Whilst water requirements for end uses have decreased over time, the pumps sold for rainwater pumping are designed to achieve optimal performance at high flow rates, which is at odds with the low flow rates needed for most internal end uses as discussed in section 6.3.4.

The uptake of rainwater tanks in urban areas as a supplementary water source resulted in a large variety of pumps and system set-ups with large variance in the energy performance and many of the systems currently installed in dwellings may be oversized for their service requirements.

6.5 CONCLUSIONS

There is a need for a holistic perspective to assess the overall performance of rain supply systems, including cost and environmental implications such as energy consumption. Traditionally, rain tanks have been introduced as a water saving measure, but with little concern about optimising energy requirements until tens of thousands of systems had been installed (in Australia). Reducing the energy associated with rainwater supply will lead society closer to sustainable practices.

The current experience shows that:

- In assessing the energy efficiency of a rainwater supply system, it is important to consider not only the pump but the whole rainwater supply system including the intended end uses in a dwelling;
- Pumps are designed so that high flow rates correspond with lower specific energies. However the typical residential end uses often require low flow rates, causing pumps to operate well below their best efficiency point;

- Pump size, (i.e., kW), has a dominant effect on specific energy. Undersizing a pump can risk providing a poor level of service. However, more often, pumps in rainwater systems are oversized, leading to unnecessary energy consumption;
- Total energy use per day for a dwelling can be calculated for the various appliances connected to rainwater. Whilst the literature shows large variability in the energy associated with rainwater pumping, the median specific energy across Australian studies ranges from 1.4 to 1.8 kWh per kL. Whilst this is in the upper range for mains potable water supply in Australian capital cities (0.06–1.84 kWh/kL), it is much lower than the energy required for recycled water using reverse osmosis (2.8 kWh/kL) and desalinated sea water (3.5 kWh/kL);
- The energy use from internally plumbed rainwater systems represents only a small fraction of total household energy use (about 2% of average household use) (Gurung *et al.* 2012). However the overall energy and its variability could be reduced further. A number of simple measures can assist in reducing the energy requirements;
- Header tanks have the potential to significantly reduce specific energy. Laboratory studies in a simulated dwelling identified energy savings of 58% to 79% could be achieved by a 300 L header tank. However, their installation height in a dwelling needs to ensure that the hydrostatic water pressure will be sufficient to operate the solenoids on most domestic appliances. Alternatively the design of appliances could be modified to work at low water pressures. Header tanks do not seem to be viable in single story dwellings under current conditions;
- Pressure vessels reduce specific energy values by reducing the number of start-ups but more importantly because of the high flow rates on their refill cycle. In-situ tests undertaken at two households with pressure vessels reduced the overall energy intensity of rainwater supplied by 30–36% and lab studies have shown that energy savings will be proportional to the pump, vessel size and end uses in a dwelling, for example, the specific energy for filling a top loading washing machine using a 0.55 kWh pump is 1.47 kWh/kL, however this can be reduced by 7% and 50% respectively by the addition of vessels of 40 and 80 L capacity. This will further require life cycle cost analysis for comparison.

Overall, better education can reduce the installation of oversized pumps. It is important to advance the awareness and education among manufacturers, the plumbing industry, consumers and government on the energy associated with rainwater pumping systems. Such measures could lead to significant improvements in the design, and the overall efficiency and sustainability of rainwater systems and their components.

6.6 REFERENCES

Apostolidis N. (2010). Australian experience in water and energy footprints. *Water Practice and Technology*, **5**(4), 76–87.

Australian Bureau of Statistics (2010). Environmental issues: Water use and Conservation March 2010, 4602.0.55.33, http://www.abs.gov.au/ausstats/abs@.nsf/Lookup/96D8ED5A95E96BEDCA2577660017E318.

Australian Government (2011) WELS Products, 2011. Available from: http://www.waterrating.gov.au/products/index.html, updated May 2011 (accessed May 2011).

Beal C. D., Hood B., Gardner T., Lane J. and Christiansen C. (2008). Energy and water metabolism of a sustainable subdivision in South East Queensland: a reality check. Enviro 2008, Melbourne.

Beal C. D., Stewart R. A. and Huang T. (2010) South East Queensland Residential End Use Study: Baseline Results – Winter 2010. Urban Water Security Research Alliance Technical Report No. 31, http://www.urbanwateralliance.org.au/publications/UWSRA-tr31.pdf (accessed July 2014).

Beal C., Gardner T., Sharma A. Barton R. and Chong M. (2011) A Desktop Analysis of Potable Water Savings from Internally Plumbed Rainwater Tanks in South East Queensland. Urban Water Security Research Alliance

Technical Report No. 26. http://www.urbanwateralliance.org.au/publications/UWSRA-tr26.pdf (accessed July 2014).

Chiu Y. R., Liaw C. H and Chen L. C. (2009). Optimizing rainwater harvesting systems as an innovative approach to saving energy in hilly communities. *Renewable Energy*, **34**, 492–498.

Cook S., Sharma A., Gurung T. R., Chong M. C., Umapathi S., Gardner T., Carlin G. and Palmer A. (2012). Performance of Cluster Scale Rainwater Harvesting Systems: Analysis of Residential and Commercial Development Case Studies. Urban Water Security Research Alliance Technical Report No. 68, http://www.urbanwateralliance.org.au/publications/UWSRA-tr68.pdf (accessed July 2014).

Cunio L. N. and Sproul A. B. (2009). Low energy pumping systems for rainwater tanks, Solar09-47th Annual Conference of Australian and New Zealand Solar Energy Society (ANZSES), Townsville, 29 September – 22 October.

Ferguson M. (2011). Rainwater Tank Monitoring Report – A 12 Month One-Minute Interval Data Study of Rainwater Tank Water Savings and Energy Use for 52 Real Life Installations, State of New South Wales through Sydney Water Corporation.

Gardiner A. (2009). Domestic rainwater tanks: usage and maintenance patterns in South East Queensland. *Water*, Feb. 2009, 73–77.

Gardiner A., Skoien P. and Gardner T. (2008). Decentralised water supplies: South-East Queensland householders' experience and attitudes. *Journal of the Australian Water Association*, February 2008, 53–58.

Gardner E. A., Millar G. E., Christiansen C., Vieritz A. M. and Chapman H. (2006). Energy and water use at a WSUD subdivision in Brisbane, Australia. *Australian Journal of Water Resources*, **10**(3).

Gurung T. R. and Sharma A. (2014). Communal rainwater tank systems design and economies of scale. *Journal of Cleaner Production*, **67**, 26–36.

Gurung T. R., Sharma A and Umapathi S. (2012). Economics of Scale Analysis of Communal Rainwater Tanks. Urban Water Security Research Alliance Technical Report No. 67. http://www.urbanwateralliance.org.au/publications/UWSRA-tr67.pdf (accessed July 2014).

Hall M., West J., Lane J., de Haas D. and Sherman B. (2009). Energy and Greenhouse Gas Emissions for the SEQ Water Strategy. Urban Water Security Research Alliance Technical Report No. 14. http://www.urbanwateralliance.org.au/publications/UWSRA-tr14.pdf (accessed July 2014).

Hall M. R., West J., Sherman B., Lane, J. and de Haas D. (2011). Long-term trends and opportunities for managing regional water supply and wastewater greenhouse gas emissions. *Environmental Science and Technology*, **45**(12), 5434–5440.

Hauber D., Shortt G., Jl Wetherall B. (2010). Energy Consumption in Domestic Rainwater Harvesting, Water Conservation Group, Melbourne, Australia. Report available on request from South East Water.

Hauber-Davidson G. and Shortt J. (2011). Energy consumption of domestic rainwater tanks – why supplying rainwater uses more energy than it should. *Water*, **38**(3), 72–76.

Hood B., Gardner T., Barton R., Gardiner R., Beal C., Hyde R. and Walton C. (2010). The Ecovillage at Currumbin – a model for decentralised development. Ozwater 2010, Brisbane, Australia, 8–10 March 2010.

Kenway S. J., Priestley A., Cook S., Seo S., Inman M., Gregory A. and Hall M. (2008). Energy use in the provision and consumption of urban water in Australia and New Zealand, CSIRO: Water for a Healthy Country National Research Flagship. http://www.clw.csiro.au/publications/waterforahealthycountry/2008/wfhc-urban-water-energy.pdf (accessed July 2014).

Marsden Jacob Associates (2007). The Economics of Rainwater Tanks and Alternative Water Supply Options. Australian Conservation Foundation, Nature Conservation Council and Environment Victoria.

New South Wales Government (2006). A Sustainable Water Supply for Sydney. New South Wales Legislative Council, Sydney.

Parkes C., Kershaw H., Hart J., Sibile R. and Grant Z. (2010). Energy and Carbon Implications of Rainwater Harvesting and Greywater Recycling, Report SC090018, Environment Agency, August 2010, UK.

Proença L. C, Ghisi E., Tavares D.daF. and Coelho G. M. (2011). Potential for electricity savings by reducing water consumption in a city scale. *Resources, Conservation and Recycling*, **55**, 960–965.

Retamal M., Glassmire J., Abeysuriya K., Turner A. and White S. (2009). The Water-Energy Nexus: Investigation into the Energy Implications of Household Rainwater Systems, (prepared for CSIRO) Institute for Sustainable

Futures, University of Technology, Sydney. http://epress.lib.uts.edu.au/research/bitstream/handle/10453/20456/retamaletal2009waterenergynexus.pdf?sequence=4 (accessed July 2014).

Standards Australia (2008). HB230-2008, Rainwater tank design and installation handbook.

Stewart R. A. (2011). Verifying Potable Water Savings by End Use for Contemporary Residential Water Supply Schemes. Waterlines Report Series No. 61, National Water Commission, October 2011.

Swamee P. K. and Sharma A. K. (2008). Design of Water Supply Pipe Networks. John Wiley and Sons Inc., NJ, USA.

Talebpour M. R, Stewart R. A, Beal C., Dowling B., Sharma A. and Fane S. (2011). Rainwater energy tank end usage and energy demand: a pilot study. Water, March 2011, 97–101.

Tam V. W. Y., Tam L. and Zeng S. X. (2010). Cost effectiveness and trade-off on the use of rainwater tank: an empirical study in Australian residential decision-making. *Resources, Conservation and Recycling*, **54**, 178–186.

Tjandraatmadja G., Pollard C., Sharma A. and Gardner T. (2011). Dissecting Rainwater Pump Energy Use in Urban Households. In: Science Forum and Stakeholder Engagement Building linkages, Collaboration and Science Quality, D. K. Begbie and S. L. Wakem (eds) Urban Water Security Research Alliance 14–15 September, Brisbane, Queensland, 35-40. http://www.urbanwateralliance.org.au/publications/forum-2nd-2011/science-forum-2011-program-abstracts.pdf (accessed July 2014).

Tjandraatmadja G., Pollard C., Sharma A. and Gardner T. (2012) Optimisation of Energy Use in Household Rainwater Supply Systems. Urban Water Security Research Alliance Technical Report No. 89, http://www.urbanwateralliance.org.au/publications/UWSRA-tr89.pdf (accessed July 2014).

Tjandraatmadja G., Pollard C., Sharma A. and Gardner T. (2013). How supply system design can reduce the energy footprint of rainwater supply in urban areas in Australia. *Water Science and Technology: Water Supply*, **13**(3), 753–760.

Umapathi S., Chong M. N. and Sharma A. K. (2012) Monitoring of residential rainwater tanks in South East Queensland to investigate mains water savings and volumetric reliability. In: Science Forum and Stakeholder Engagement: Building Linkages, Collaboration and Science Quality, D. K. Begbie, S. J. Kenway, S. M. Biermann and S. L. Wakem (eds), Urban Water Security Research Alliance, June 2012, Brisbane, Queensland, p. 88. http://www.urbanwateralliance.org.au/publications/forum2012/UWSRA%20Science%20Forum%20-%2019-20%20June%202012.pdf (accessed July 2014).

Umapathi S., Chong M. N. and Sharma A. K. (2013). Evaluation of plumbed rainwater tanks in households for sustainable water resource management: a real time monitoring study. *Journal of Cleaner Production*, **42**, 204–214.

Water Conservation Group (2010). Energy Consumption of Domestic Rainwater Harvesting, (prepared for South East Water Limited), Water Conservation Group, Melbourne, Australia. Report available on request from South East Water, Melbourne, Australia.

Chapter 7

Management and operational needs for urban rainwater tanks

Magnus Moglia, Andrea Walton, John Gardner and Grace Tjandraatmadja

ABSTRACT

Urban rainwater tanks are infrastructure assets that provide important public and private benefits, but which also have operational and maintenance requirements to avoid key risks. Adequate asset management should be in place to address any potential risks. The risks of inadequate management of urban rainwater tanks relate primarily to non-achievement of water savings targets as well as public health risks. Public health risks arise from water quality issues, but also from the risk of mosquitoes breeding, thus potentially spreading arbovirus diseases such as dengue fever to humans. Failure of rainwater tanks due to poor maintenance or installation practices can lead to the risk of collapse of walls or fences due to rainwater tanks leaking or leaning onto them. The management requirements are not onerous for a single rainwater tank; however ensuring that the entire stock of rainwater tanks is maintained is a complex task, given their distribution in private backyards as well as ownership. The main difficulty for urban water planners and health officials is that private owners of rainwater tanks are, in many cases, not motivated to undertake maintenance. Therefore there is a need to find strategies, preferably in cooperation with owners, to ensure that private rainwater tanks remain both functional and safe.

Keywords: Asset Management; Risk Assessment; Socio-technical analysis; Water Supply.

7.1 THE NEED FOR MANAGING RAINWATER TANKS

This chapter focuses on the management and maintenance aspects of rainwater tanks; a topic that is scarcely covered in academic literature although some basic guidelines exist authored by the WHO (2008). A handful of household surveys have partially considered this topic (Rodrigo *et al.* 2010; Karim, 2010; Gardiner, 2009). These surveys have explored self-reported maintenance behaviours, but only one report has covered the physical inspection of rainwater tanks (Biermann *et al.* 2012). In the context of rural Uganda, Baguma and colleagues (2010) have shown a positive correlation between experience in managing rainwater tanks, self-reported maintenance behaviours and public health outcomes; but it is noted that these communities use rainwater for potable purposes. In Botswana, Gould (1996) has argued that maintenance requirements and associated problems is one reason for the poor level of uptake and diffusion of this technology at the time. With the rapid and widespread uptake of rainwater tanks in Australian urban settings, research has been undertaken to understand maintenance and management needs for rainwater tanks (Moglia *et al.* 2013a; Moglia *et al.* 2012b; Moglia *et al.* 2013b) and this research is reported in this chapter. Whilst the chapter content is influenced by the situation in Australia, many of the conclusions and findings should be applicable to other locations globally as well.

Rainwater tanks are water management infrastructure assets, contributing to household water supply in both urban and rural settings. An advantage of rainwater tanks when compared to other water supply technologies is the relatively small scale of the systems; they do not require any economy of scale to become viable except optimal rainwater tank sizing, based on demand and supply modelling. This makes them particularly popular in rural areas where population densities are low. They have also become increasingly popular in urban areas, even when rainwater tanks are not always the cheapest option, in the belief that rainwater tanks are 'more environmentally friendly'. In several states in Australia, the mandated requirements for rainwater tanks in new dwellings have led to a very large increase in the number of rainwater tanks in Australian cities.

Unlike rural settings, the risks to an urban household's water supply of not maintaining a rainwater tank are relatively small. In rural settings, with no alternative water supply, failure of the rainwater tank will lead to a loss of supply, the need for immediate attention, and potentially having to truck water onsite in the interim. In such settings, rainwater is often used for drinking water, and thus poor water quality potentially poses significant and immediate health risks to the consumers of the water. In contrast, in urban settings, there is almost always an alternative water supply, and failure of the rainwater tank may go un-noticed. Furthermore, it is relatively unusual that urban dwellers, at least in places like Australia, would use rainwater for drinking purposes. This means that in urban settings the risks of rainwater tank failure are relatively low because the consequences of failure are low. This situation is referred to as the 'low stakes' setting of urban rainwater tanks. Due to the low stakes setting of urban rainwater tanks, it is thought that the motivation for maintaining rainwater tanks in urban settings is low. There is a growing body of evidence confirming this premise. However, although the risk of failure of an urban rainwater tank is less than that for rural rainwater tanks, a number of important and related risks persist in rainwater tank systems, and these are summarised in Table 7.1.

7.2 ISSUES TO CONSIDER IN THE MANAGEMENT OF URBAN RAINWATER TANKS

The management of rainwater tanks across a city requires the understanding of a number of issues:

- Irrespective of whether there is a serious problem with rainwater tanks or not, there is a need to develop management strategies if it is reasonable to expect that many rainwater tanks are not kept in a good condition. This can have serious consequences for public health risks and/or for water planning.
- The maintenance needs, which vary depending on how people use the collected rainwater, the risk of pollution and the public health risk of mosquitoes breeding in the rainwater tank.
- The behaviour of rainwater tank owners, which is influenced by:
    - The reasons for installation of rainwater tanks across the urban landscape, as these influence maintenance behaviours.
    - The historical context of rainwater tanks: the past experiences with tanks in the local area.

Table 7.1 Risks of urban rainwater tank related failure events. The consequences, likelihoods and risks are rated as L (Low), M (Medium), or H (High).

Failure events	Consequence (L, M, H)	Likelihood (L, M, H)	Risk (L, M, H)
Foundation becoming unstable or leaning	H: With a rainwater tank's weight, an unstable or leaning foundation may lead to collapse of walls or fences, or the rainwater tank itself may fall off, with potentially fatal consequences.	L-M: If overflow has not been correctly installed, the water may erode the foundation or make it unstable. The foundation may also be poorly installed to start with.	M-H: The risk of this type of event may be catastrophic, but the likelihood can be managed through good design and installation and checks after installation.
Faulty switching or top up valves	M: Switching valves are often installed to allow mains top up into rainwater tanks or a direct connection to mains supplies. When these devices fail, the water savings potential is significantly reduced.	H: Indications are that this is a very common fault.	H: With a severe consequence of considerably limiting any water savings, and being a common fault, this is a high risk issue.
Broken pump	M: When a pump fails, it makes it difficult to use the rainwater in the rainwater tank. But the pump can be repaired or replaced (at a cost), thus restoring the capacity of the tank.	M-H: Data indicates that problems with pumps are common, and that it may be necessary to replace pumps every 3–5 years.	M-H: The remaining risk is due to the fact that broken pumps are often not noticed (particularly in systems equipped with automatic switching between rainwater and mains water), and thus making the tank ineffective as a capture/storage device. This risk can be managed through inspections and repairs.
Blocked gutters	L: Blocked gutters and screens limit the flow of water into the tank and can also impact on the water quality in the tank.	H: The likelihood of this type of event is very high; especially if there are over-hanging trees.	M: This is a low-consequence event, but due its high frequency, if it is not managed, only a small amount of water will be collected.
Removal or damage of mosquito meshing	H: Mosquitoes in tanks in urban settings are generally considered by health departments as a serious issue, potentially leading to the spread of diseases such as dengue fever.	L: Good design and rigorous installation should minimize the likelihood of this type of event. A regular check of meshing can easily be carried out and replacement of the meshing is inexpensive.	M: Whilst the likelihood of this event is low and manageable, the consequence can be very high, particularly in places where serious mosquito born disease may spread.

(*Continued*)

Table 7.1 Risks of urban rainwater tank related failure events. The consequences, likelihoods and risks are rated as L (Low), M (Medium), or H (High) (*Continued*).

Failure events	Consequence (L, M, H)	Likelihood (L, M, H)	Risk (L, M, H)
Poor water quality	L-M: People can get ill if they ingest poor quality water. Poor water quality may also lead to discolouration of toilets, other fixtures or washing.	M: The likelihood of poor water quality is situation dependent. Discolouration can be managed by regular cleaning of gutters although leaves may contaminate water in wet downpipe systems. Where small mammals such as possums and birds can access the roof, a number of pathogens are likely to be found in the water.	M: Water quality in tanks is difficult to manage. If alternative potable water sources are available, rainwater should not be used for drinking purposes. Alternatively, proper disinfection treatment is recommended. If so, this risk is limited.
Incorrect plumbing	L-M: The consequences depend on the type of error. Failure to properly install backflow prevention can lead to contamination of drinking water.	L: This appears to be a relatively common problem but should be identified in post-installation inspections and therefore the likelihood is Low, subject to such checks.	L: The consequences depend on what error has been made, but the risk is manageable.
Sedimentation in tank	L: Unless water is used for drinking purposes this only really becomes a problem if severe sedimentation exists and water quality is affected or pipes become blocked.	M: Sedimentation is inevitable but it is usually a very slow process and can be managed with regular desludging.	L: This is not a serious issue, but desludging is necessary at regular intervals (typically at intervals of minimum 3 years).

- The practice of installation and design of rainwater tanks, which may or may not be influenced by guidelines and regulations.

Furthermore, if plans for rainwater tank management are to be established, it is also important that strategies are developed which are:

- Accepted by stakeholders;
- Effective in achieving outcomes; and
- Aligned with the current legislation and regulation.

The stakeholders in this instance are those individuals or organisations with a strong interest or concern relating to rainwater tanks. Rainwater tank owners, plumbers and tank industry professionals, government agencies, local government and urban water planners, property developers and house builders, and health officials all have an interest in rainwater tank installation and maintenance.

Strategy outcomes would usually be defined by the stakeholders, in what they hope to achieve or avoid with the rainwater tanks. These issues are described in detail in this Chapter for the benefit of the wider water professional community across the globe. A methodology for exploring the relevant issues and the associated management strategies is described. The application of this methodology will be demonstrated using a case study in South East Queensland (SEQ), Australia. SEQ is a primarily urban area. It is acknowledged that extrapolation of study results to other countries is influenced by the context of the case study, but it is also noted that there is little written on the topic for other locations. This is an emerging topic with very little prior research to report on.

7.3 PRIVATE OWNERSHIP VS PUBLIC BENEFIT

Why is it a public concern if private rainwater tanks are not functional and safe? Why is it not just the problem of the individual owners? We argue the following premises make it a public concern:

- Rainwater tanks are sometimes a legal requirement for home owners under local policies. For example, in some circumstances local building regulations require rainwater tanks to be installed, or rainwater tanks have been a key feature in getting building approvals.
- Rainwater tanks that are not maintained may pose public health risks, via poor water quality (if used for drinking), mosquitoes breeding, or structural collapse of foundations, walls or fences. This is not acceptable, as health risks may impact on others and not just the rainwater tank owners.

There are compelling reasons why urban water planners have a strong interest in keeping rainwater tanks functional. Rainwater tanks can play an important role in ensuring that water supply is adequate for the water demand of the city; the potential for water savings is further described in Chapter 3. By installing rainwater tanks, homeowners are reducing the load on mains water supply. This means that if rainwater tanks are not maintained, increased demand is placed on mains water supplies. To deal with increased demand, new supply infrastructure may be required or fast-tracked with a consequent increase in the price of water. Water planners also see rainwater tanks as helping to reduce urban stormwater flows and the pollution of the city's waterways; this is considered in more detail in Chapter 13.

Furthermore, if rainwater tank owners have received public sector rebates for the installation of their rainwater tank, ensuring the ongoing optimal functioning of the rainwater tank will maximise the return of public investment in rainwater tank assets. It could be argued that non-subsidised tank owners have invested in public infrastructure providing urban water supply at their own expense. This can be perceived, although disagreement about this exist, as an implicit social contract between citizens and the government, given it is the government's responsibility to provide basic services to the urban community. Viewed from this perspective, the government is morally and perhaps even financially obligated to rainwater tank owners.

An implication of private ownership of rainwater tanks is that it is likely to be perceived by some as over-stepping the mark if government and/or water utilities penalise rainwater tank owners that do not look after their rainwater tank adequately. Similar community attitudes may arise if such organisations were to mandate owners to maintain their rainwater tank.

In situations when water tariffs are based on consumed water alone, with no separate volumetric tariff for wastewater disposal, water service providers may also argue that rainwater tank owners should be charged for rainwater used for indoor purposes such as toilets or laundry as they pose a burden on wastewater collection, treatment and disposal systems. Moreover collection, treatment and disposal of wastewater is often more expensive than providing water in the first place.

These issues combine to create a situation that is ripe for controversy and discussion. However, it can also be argued that the motivations of urban planners, health professionals and rainwater tank owners are aligned. Also, it is in the best interest for rainwater tank owners to ensure their tanks provide value rather than become a liability, especially as rainwater tanks are often perceived to increase the value of a property. This means that stakeholder motivations for keeping rainwater tanks maintained and well-managed are aligned, and that collaborative solutions are recommended rather than 'command and control' type strategies, which are likely to alienate rainwater tank owners and engender non-cooperation. Some sort of regulation of certain aspects of rainwater tank management may still be required in addition to collaborative solutions; for example, if there is a high risk of mosquito born disease. As an example, for onsite sewage systems, the public health risk from treatment failure is considered so high that maintenance is mandated by most state regulators or local authorities at the owner's expense. Such moves were initially resisted by the community, but this obligation is now widely accepted.

7.4 FACTORS THAT INFLUENCE CONDITION

Rainwater tanks are relatively straightforward infrastructure systems, when considered on an individual basis. In fact, a well-designed rainwater tank that has been adequately installed and maintained is likely to operate in a safe and effective manner, although individual components of the system will break down at the end of their asset life. Through our research, it has been found that water experts agree that maintenance, design and installation are the fundamental factors that impact on the likely condition of rainwater tanks.

7.4.1 Maintenance

Rainwater tanks need to be maintained to ensure their efficient water capture and to minimise public health risks. The maintenance needs are relatively well-established, as shown in Table 7.2. The World Health Organisation notes similar maintenance activities, but are not specific regarding frequencies (WHO, 2008).

7.4.2 Design

In Australia, the Rainwater Tank Design and Installation Handbook (Standards Australia, 2008) is the key reference document providing guidance on design and product issues. There is a vast array of designs, materials and types for rainwater tanks and those differences can impact on performance, ease of access, as well as the need for maintenance. A number of plumbing professionals and industry representatives have indicated that, among the various rainwater tank systems, there have been a number of components with common faults, such as poor design restricting access for cleaning and maintenance, poor durability or inadequate quality (Moglia *et al.* 2012b). There are, however, quality products available at competitive prices that fulfil their intended purpose and are easy to maintain. In each local context, different products and services are on the market – and it is important that there is guidance to customers on the performance of different products, but this is currently scarce, limited to specialist magazines or rainwater tank enthusiasts' forums. Future innovation in products and designs also has the potential to remove much of the maintenance needs of rainwater tanks, for example by developing switching valves that require minimal maintenance (inspection programs are yet to confirm what types of switching valves that require minimal maintenance) and development of monitoring systems that will enable notifications to the owner of any faults.

Table 7.2 Recommended inspection and maintenance activities for rainwater tanks.

Frequency	Activity	Maintenance required
3 months	Inspect and clean gutters.	Remove leaves and debris.
	Inspect and clean first flush devices and leaf guards on rainheads.	Clean, repair or replace if necessary.
	Check screens on overflow outlet.	Repair or replace if necessary.
	Check switching valves	Repair or replace if necessary.
6 months	Check roof and flashings for defects and remove overhanging branches.	Repair if necessary and remove overhanging branches.
	Checks tank for defects, ensure screens and lids are in place and functional.	Repair if necessary.
	Check water quality.	Identify cause for quality change, if any.
	Check rainwater taps have correct signage.	Repair or replace if necessary.
	Check pump for operation, noise, pressure, leaks and acoustic enclosure if applicable.	Repair or replace if necessary.
Annual	Check tank support for structural integrity.	Repair or replace if necessary.
2–3 years	Check sediment level in tank, and desludge if necessary[1].	Organise removal with a qualified contractor if sediments pose a risk to block tank outlet.

Source: Adapted from guidelines authored by Queensland Health (2007) and Standards Australia (2008).
[1]*Note:* HB230 (Standards Australia, 2008) recommends placing of rainwater outlets to dwellings at a minimum height to prevent uptake of sludge upon water extraction. In addition, it recommends desludging at a frequency of 2–3 years, however sludge build-up varies with tank set-up and gutter maintenance. Note also that, in urban areas, mandated uses are restricted to non-potable uses (toilet flushing, washing machine and outdoor irrigation), hence the risk of ingestion by users is considered low.

7.4.3 Installation

Installation of domestic rainwater tanks can be of variable standard and can present problems to the rainwater tank owner if installed incorrectly (Moglia *et al.* 2011c; Moglia *et al.* 2012a). Rainwater tanks are usually installed by plumbers or licensed tradespersons. However, many are installed by the householder. Problems with installation may lead to issues such as faulty drainage and plumbing connections, inadequate pump sizing, and incorrectly installed mosquito meshing or first flush devices. Different industry codes govern rainwater tank installation and vary between different jurisdictions. For example in Queensland, Australia, the location of our case study, certification of rainwater tank installations is two-part. An accredited building certifier certifies the installation because rainwater tanks are defined as structures under the Building Code; and a plumber inspects the connections to water fixtures and piped water supply because this falls under the local Plumbing Code. Despite such differences, post-installation inspection of rainwater tanks is a good way to ensure that rainwater tanks are appropriately installed, although this requires the inspection protocols to be sound, and for inspectors to be knowledgeable to ensure standards are maintained.

7.5 LOCAL CONTEXT

The local context is an important factor in rainwater tank management. Specifically, it needs to be acknowledged that rainwater tanks exist in a regulatory and legal environment; and any strategy will need to consider the socio-psychological aspects of the domestic rainwater tank owners. Regardless of situation, without the broad cooperation of rainwater tank owners, any strategy to manage rainwater tanks is likely to fail. Even if maintenance were mandated as per onsite sewerage systems, it is likely that upholding such regulations would still be largely dependent on community cooperation.

7.5.1 Local regulatory environment

For the purpose of our topic, the regulatory environment relating to rainwater tanks in a given location is important. In some locations, regulatory or legal frameworks relating to rainwater tanks may not exist, or if such frameworks do exist, may not be widely adhered to. In other locations, such regulations may be stringent and people tend to follow them. The key components of the regulatory environment include:

- Whether domestic rainwater tanks are mandated (i.e., a legal requirement for new dwellings) or rebated (incentives for the purchase and installation of rainwater tanks), and so on. The reasons why people install rainwater tanks influence the behaviour of tank owners (Mankad, 2012; Mankad *et al.* 2012; Mankad *et al.* 2013) (see Chapter 8).
- The available technical guidelines to support rainwater tank owners, developers and plumbers in installing and maintaining rainwater tanks (Standards Australia, 2008; Standards Australia, 2004; WSAA, 2002; Queensland Government, 2011; Queensland Health, 2007).
- The approval process to certify that rainwater tanks have been designed and installed appropriately. These may follow the process outlined in Table 7.3. For example in the case of South East Queensland, a number of agents are involved throughout the process (Moglia *et al.* 2012a).
- Guidelines and legal requirements regarding public health issues. What is the legal requirement to ensure that mosquitoes are not breeding in the rainwater tank? What are the guidelines and legal responsibilities relating to using the rainwater for drinking and other purposes? What organisation is responsible for making sure that domestic rainwater tank owners comply with public health related legal requirements? In the case of South East Queensland, local councils are responsible for this.
- Stakeholders. What agencies have vested interests in making sure that rainwater tanks are adequately maintained? Urban water planners and health officials, but also environmental managers, should have an obvious interest in this topic. It is likely that it is these stakeholders that will have to drive a management strategy – assuming they are given adequate resources.

The roles of various agencies, departments and stakeholders vary between jurisdictions. There are also relevant guidelines regarding design and installation, as discussed in Sections 7.4.2–7.4.3.

Table 7.3 Process for installation of rainwater systems.

Step	Relevant documentation	Agent responsible for step
Government/Regulatory requirements	Building codes of Australia • Setback distances • Height restrictions • Source of rainwater/stormwater catchment • Plumbing code of Australia and AS/NZD series • Rebates eligibility	Government
Site conditions assessment and product selection	System design including • Tank type • Pump • Filtration	Developer, builder or homeowner
Install tank and connections	• HB230 for technical guidelines • Local guidelines • Pump • Filtration, and so on	Developer, builder, third party or homeowner
Provide documentation	• Regulatory reporting • Rebates (must be certified by plumber) • Certificate of installation by certified building inspector • Informing client of on-going maintenance program	Developer, builder or homeowner
Ongoing maintenance	• Local guidelines • Manufacturer's instructions	Homeowner, third party

Source: Adapted from HB230 (Standards Australia, 2008).

7.5.2 Understanding behaviour

Tank maintenance is a voluntary behaviour, unless (or until) some form of regulation is introduced. Tank maintenance actually reflects a suite of different activities, performed at regular but infrequent intervals (see Section 7.4.1). Psychological research suggests that the drivers of this sort of voluntary behaviour are twofold: a *motivation* for keeping the tank maintained; and a sense of one's own capability (or *self-efficacy*) in being able to undertake maintenance activities (Walton & Gardner, 2014; Walton *et al.* 2012).

7.5.2.1 Motivation

The motivations that drive tank maintenance can be grouped into three broad sources. Firstly, individuals are motivated to take action to the extent that their perceived benefits outweigh the costs of keeping the tank maintained. The 'costs' in this situation reflect concerns for the effort, inconvenience, and time required to conduct maintenance, as well as financial cost. The benefits of rainwater tank system maintenance also reflect a range of factors, for example, it may provide access to a personal source of water where the rainwater tank owner has autonomy regarding its use for garden watering, swimming pool top-up, or an indoor connection to the washing machine and toilets. Many rainwater tank owners also regard tank water as cheaper than mains water – 'it comes for free as rain from the sky'. This perceived cost advantage, although probably incorrect (see Chapter 12), is seen as a benefit to the home owner, especially in a situation of increasing mains water prices and other cost of living pressures. Some rainwater tank owners also strongly support water conservation, and see rainwater tank ownership as a way of drawing maximum benefit from water that would otherwise be 'wasted', thereby helping to mitigate future drought impact.

As well as these personal benefits that are associated with keeping a rainwater tank maintained, some tank owners feel what is best described as a sense of 'moral obligation' towards rainwater tank system maintenance. Such feelings relate to the notions of preventing public health risk associated with mosquito breeding in unmaintained rainwater tanks, and of reducing the overall demand for mains water. This type of motivation is especially strong in people who have experienced severe drought and/or grew up in rural areas and were fully dependent on rainwater tank water. Further, the idea of maximising the return on money invested in the rainwater tank can also appeal to a person's sense of obligation and act as a motivator to keep the rainwater tank maintained. Many people received their rainwater tank through a government subsidy or rebate and are aware of the considerable public money invested in providing households with roof-water harvesting facilities. People who paid for their own rainwater tank system, often as part of building their own home, may also feel a sense of obligation in keeping it maintained and not wasting their own investment.

The third source of motivation for keeping a tank maintained is the rainwater tank owner's self-image: a perception of themselves as someone who keeps things well maintained and in sound working order. For some people, this self-image helps to shape their behaviour, prompting water tank maintenance along with other household maintenance tasks, like pest control, swimming pool and garden maintenance. Self-reported confidence to maintain things may also reflect a capability to undertake practical tasks.

These three sources of motivation obviously can vary among tank owners, and can vary depending on the proposed use of the tank water. Research has indicated that motivations for keeping a tank maintained were lower in those tank owners that only used their water for outdoor use, and higher amongst owners that had their tanks connected to indoor devices. This finding supports other research, which suggests that the level of personal contact someone has with a water source is an important driver of how they behave towards that water (Hurlimann, 2011; Mankad & Tapsuwan, 2011). For example, when recycled water is used for outdoor use, people find it more acceptable than if it is for indoor use. As personal contact with the water increases, for example using the water for showering or drinking, the more people become concerned about aspects of water quality such as colour, odour and safety (Nancarrow *et al.* 2010).

7.5.2.2 Self-efficacy

This is the second broad driver of tank maintenance: this construct revolves around issues of knowledge, skills, and capacity for performing the maintenance tasks. The starting point for a sense of efficacy is a degree of awareness of the problem, in this case awareness that water tanks need to be maintained. The majority of tank owners are not aware that a rainwater tank needs to be regularly maintained (Walton & Gardner, 2014). Also, self-efficacy requires knowledge of the actual maintenance tasks: how and when they are to be done. The tank owner then needs to feel a sense of confidence in being able to perform these tasks. This confidence may involve the physical capability to actually perform the tasks, for example depending on age and fitness levels, such as clearing gutters and checking first flush devices. Another factor is the financial capability and willingness to outsource the tasks, such as desludging the tank. Successful outsourcing will also depend on access to suitable services. Self-efficacy can be viewed as a 'necessary condition' for tank maintenance: even if motivation levels are high, if a person does not feel capable of performing the required tasks, the tank may go unmaintained.

As well as the internal drivers of tank maintenance described above, there are other external factors that could influence tank maintenance behaviour.

7.5.2.3 External influences

Regulations, penalties, incentives, education programs and awareness campaigns are all policy instruments that may potentially influence tank maintenance behaviour. However, these approaches may or may not result in increased tank maintenance because they interact with a person's views and opinions of the approach, as well as with their own levels of motivation and self-efficacy. For example, a person who feels very motivated and capable of keeping a tank maintained may view any form of intervention by an authority as inappropriate interference. Indeed, the intervention may actually decrease motivation for keeping the tank maintained. Furthermore, research suggests that people who were 'forced' to get a tank as part of adhering to building codes have different levels of motivation towards maintaining their tank compared to those who chose to retrofit a tank, especially if they also received a subsidy from the government to install the tank (Gardiner, 2009; Mankad & Greenhill, 2014). Those householders who were mandated to install a tank were less motivated to maintain a tank than those who had retrofitted their tank. Increased government intervention to maintain the tank could again act as a further de-motivator in the mandated tank owner case. Thus, the public's acceptance of a policy instrument designed to change behaviour is important if the intervention is to succeed.

Research indicates that people's judgments about the features of a policy will influence acceptance, particularly judgments about how fair and how effective they feel the intervention will be. The most widely studied policy feature in the environmental psychology literature is the type of behaviour change mechanism embodied in the policy, which can be more or less coercive. The literature uses a variety of names to describe these mechanisms (Garling & Schuitema, 2007; House of Lords Science and Technical Committee, 2011; Osbaldiston & Schott, 2012). At the non-coercive end, there are 'soft' and 'pull' approaches. *Soft approaches* include education and awareness campaigns, and activities to encourage and facilitate voluntary undertaking of the target behaviour. *Pull policies* are those that provide incentives and rewards as ways to encourage a change towards the target behaviour.

In contrast, the more coercive policy options include 'hard' and 'push' approaches, and use increasing levels of regulation to bring about change. *Hard approaches* can include regulatory mechanisms designed to enforce behaviour, and *push policies* use disincentives, such as penalties and increased taxes or prices, to bring about change. In general, less coercive mechanisms are perceived to be more acceptable, fairer, and more effective. Perceptions of fairness involve the concepts of equality, equity, and personal freedoms. The notions of personal freedom and fairness are important to tank owners who participated in surveys and focus groups, where the freedom to 'do what I want with my water' is seen as an important benefit of having a rainwater tank (Walton *et al.* 2012).

7.6 RESEARCH FRAMEWORK

Strategies for managing rainwater tanks need to address the broad topics of design, installation and maintenance of tanks; whilst also considering the local context, legislation and regulation (if present and followed) as well as behavioural considerations relating to tank owners. Within these broad areas, there are some choices to be made before a strategy can be formulated. As such, research needs to be undertaken to collect information about community attitudes, preferences of stakeholders, possible creative and effective local solutions, appropriate success criteria, and the resourcing of solutions. Strategies need to be evaluated to assess if they are likely to achieve success criteria. This research needs to be carried out in acknowledgment that management of rainwater tanks can be a contentious issue. There is public interest in making sure that tanks are well maintained, but householders and the community may not be motivated to cooperate. The need to find collaborative solutions and to address a potentially contentious issue suggests that a participative approach to research is warranted.

A methodological framework for this type of approach to research is outlined in Table 7.4 and address three important research functions fundamental to achieving the following (Jones *et al.* 1999):

- Increasing the legitimacy of the process of knowledge generation.
- Allowing greater integration of more sources of knowledge and information, hence improving the capacity for problem solving.
- Trying to help build collaborative relationships to assist with implementation of strategies and for reducing conflict.

Table 7.4 Methodological implications of developing collaborative solutions around a contentious issue.

Consideration	Methodological implications – there is a need to...
Stakeholders	Identify and involve all stakeholders affected by the issue. Allow stakeholders to have some input into the identification and evaluation of strategy alternatives.
Transparency	Communicate and engage with all stakeholders. Strive to increase the awareness of the interest and preferences of all stakeholders.
Protect Core Values	Identify issues that can create real strife or moral outrage amongst stakeholders. Handle and attempt to resolve conflict in a way that aims to find compromise or consensus solutions.
Substance	Be clear about the role of researchers in collecting and disseminating information and to facilitate discussion. Ensure that research process provides multiple alternatives from which decision makers can choose.

Note: These points have been adapted from HarmoniCOP (2005) which is a key output of a major EU project into improving participation in water management.

Based on these principles, a research process to define a management strategy for urban household scale rainwater tanks would involve the following four steps.

(1) Develop an *understanding of the context.* Typical questions to answer are:
   (a) How many tanks are there and what type are they?
   (b) What were the different reasons why tanks were installed?
   (c) What are the regulations regarding design and installation?
   (d) What condition are the tanks thought to be in?
   (e) Are data available on tank condition?
(2) *Define the strategies* to be considered for further exploration. Typical questions to answer are:
   (a) What strategies can increase the likelihood of adequate tank design?
   (b) What strategies can increase the likelihood of adequate tank installation?
   (c) What strategies can increase the likelihood of adequate tank maintenance?
   (d) What are the community and stakeholder attitudes to different strategy features?
(3) *Assess strategies* using available data synthesized in a computational model to evaluate likely effectiveness of a strategy. Typical questions to answer are:

(a) What are the success criteria for a strategy (as defined by those key stakeholders who care a lot)?
(b) What combination of strategy features will help achieve satisfaction of the success criteria?
(c) Will a given combined strategy be acceptable to stakeholders?

(4) *Plan, implement and evaluate* a chosen strategy. Typical questions to answer are:
(a) How will the strategy be paid for? Does the strategy need to change to make it financially viable? If so, go back to step 3.
(b) Who should be responsible for each of the tasks in the strategy?

These activities, tasks and knowledge bases are further described in the following sections through examples based on research in SEQ, Australia. Figure 7.1 provides an overview of the research framework.

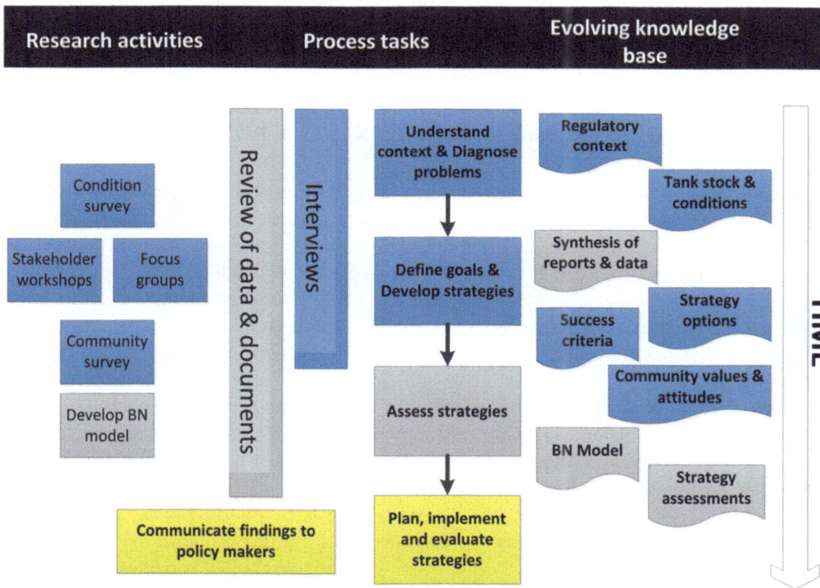

Figure 7.1 Research framework for identifying and assessing management strategies. Note: BN means to Bayesian Networks; Blue colour indicates generation of information/data; Grey colour indicates synthesis or analysis; Yellow colour indicates actions to change the system and/or implement new policy.

7.7 APPLICATION OF THE RESEARCH FRAMEWORK: THE SOUTH EAST QUEENSLAND CASE STUDY

7.7.1 South East Queensland context

In the South East Queensland study, the research process was initiated through meetings with the Queensland Water Commission which at the time (2011) provided independent policy advice to Government on South East Queensland regional water security, demand and supply options (thus their interest in rainwater tanks). In these meetings, concerns were raised about the condition of household tanks across the region, noting the need for a coherent plan to ensure the acceptable condition of these tanks. Through these meetings, an initial list of stakeholders was established: primarily state government departments with some role or responsibility relating to tanks, local governments with responsibilities relating to tanks, as well as experts within various professional organisations. It was also identified that the community is an important player in this issue. The researchers then set out to establish the context of tanks via interviews with stakeholders and experts as well as review of relevant reports and documentation.

7.7.1.1 Historical context

The history of urban rainwater tanks in South East Queensland, further described in Text Box 1, was established through discussions with experts. Broadly speaking, household rainwater tanks in South East Queensland were classified into three categories: those tanks installed without a rebate or financial incentive to promote tanks; those installed with householders receiving rebates for their installation; and many tanks installed to meet the water savings target which was in place for some time (see Text Box 7.1) for new dwellings required under the Queensland Development Code Mandatory Part 4.2 (QDC MP 4.2).

TEXT BOX 7.1 EXTRACT DESCRIBING THE HISTORY OF RAINWATER TANKS IN SOUTH EAST QUEENSLAND (MOGLIA *ET AL.* 2013, PP. 1–2):

'Rainwater tanks have always been part of the rural and urban landscape in Australia, ever since the days of early settlement. Growing concern about rainwater tanks as a potential breeding site for the *Aedes aegypti* mosquito and the spread of dengue fever saw the demise of rainwater tanks in urban areas across Queensland during the 1960s and early 1970s. With the onset of the millennium drought in Australia at the turn of the twenty-first century, rainwater tanks were widely promoted in urban areas as a supplementary water source for non-potable uses to alleviate the demand on centralised potable water supply and increase resilience to drought. Financial incentives such as rebates were provided by all levels of government to promote and encourage the uptake of rainwater tanks.

'Legislation enforcing installation of rainwater tanks in new housing has also been enacted in various Australian jurisdictions. Rainwater tanks in urban areas are currently mandated for new dwellings in South Australia and New South Wales. The SEQ Water Strategy advocates the reduction of household mains water consumption to increase the security of water supply, with alternative water resources (rainwater and stormwater) in new developments expected to reduce demand on bulk water supplies by nearly 7% by 2056 (QWC, 2010). The Queensland Development Code (QDC) Mandatory Part (MP) 4.2 (2007) required all new Class 1 residential dwellings in SEQ built after 1 January 2007 to meet a water savings target of 70 kilolitres/household/year (kL/hh/yr) (Queensland Government, 2009). A common approach to meet this water savings target was a 5 kL rainwater tank connected to at least 100 m^2 of roof area and internally plumbed to the toilet for toilet flushing, to the cold water tap in the washing machine and to an external tap(s) for outside water use and garden irrigation.

'In SEQ alone, approximately 59,000 homes with rainwater tanks have been built since 2007 (ABS, 2010). There are over 300,000 tanks in SEQ (Gardiner, 2009). An additional 745,000 new dwellings are projected to be built by 2031 (QWC, 2010). Following the repeal of laws mandating the installation of alternative water supply systems on 1 February 2013, buildings in Queensland no longer have to meet compulsory water savings targets (Department of Housing and Public Works, 2013). Local governments can now choose to opt-in to water savings requirements in recognition of Queensland's varying climatic conditions and regional circumstances. Builders in these local government areas will still need to comply with water savings requirements. Rainwater tanks can still be installed voluntarily by homeowners and builders in all areas of the state, but must comply with the health and safety standards set out in QDC MP 4.2.'

7.7.1.2 Regulatory and practical context

The regulatory and practical context involves both the formal legislative context, as well as the informal context, how things are done in practice. The formal legislative context was explored by means of relevant documentation, such as the QDC MP 4.2 (Queensland Government, 2008), the Public Health Act 2005 (Queensland Government, 2005), and guidelines to minimise mosquito and biting midge problems in new development areas (Queensland Health, 2002). Further details relating to standards and guidelines applicable to rainwater harvesting systems, and other relevant regulation can be found in Moglia *et al.* (2012a). The informal context was explored by means of interviews and discussions with experts and professionals. The type of information collected in these interviews related, for example, to how the approval process for tanks is or isn't carried out in practice, and the rich and colourful stories of times when people have had problems with tanks. Through these stories, a list of different ways that tanks fail (failure modes) was identified, as well as some initial indication as to causative chains leading to failure. Furthermore, through this process, the research team was also able to provide a more complete picture of all the stakeholders relating to rainwater tanks (see Moglia *et al.* 2012).

7.7.1.3 Tank stock and general condition

After interview discussions, the research team soon established that there was no reliable available data on the condition of tanks in the region, leaving the team unclear as to the extent of the problem with poorly maintained tanks. A handful of studies had explored the motivations and practices of householders (Gardiner, 2009; Gardiner, 2010; Gardiner *et al.* 2008; Tilbrook, 2009) but no data existed on the physical condition of tanks. Furthermore, it was widely argued that it would be difficult to fund and carry out a large scale survey of the physical condition of the 300,000 tanks in the region.

In a study related to our research, an inspection program of 200 tanks was undertaken by Biermann *et al.* (2012) (see Chapter 5), but these data were only available halfway through our study. It was therefore thought that in the absence of physical data, it would be appropriate to canvas water professionals including plumbers on their expectations and experiences relating to tanks; especially when these water professionals are also tank owners. Over 250 water professionals were surveyed via email, to elicit their experiences. This generated some data that could be used to establish a base estimate of the extent of the problems (Moglia *et al.* 2013a). This data provided expert judgments as to the likely time before failure of system components, as well as the likely causes of any failures. Using the data in combination with Monte Carlo computer modelling simulations, the team established a rough estimate of what percentage of tanks would be under inadequate condition if various frequencies of inspection and maintenance were occurring (see Table 7.5).

Table 7.5 Relationship between frequency of inspection and estimated rates of inadequate tank condition (or failure) in a simulated population of tanks.

Frequency of inspections	Proportion of tank systems with blocked gutters (overhanging trees)	Proportion of tanks with broken pumps	Proportion of tanks with broken meshing	Proportion of tanks that are structurally broken
1 month	17%	1%	1%	0%
3 months	39%	4%	4%	0%
6 months	54%	8%	10%	1%
1 year	67%	19%	21%	2%
2 years	75%	37%	38%	5%

7.7.2 Definition of strategies

With the results of the Monte Carlo simulations indicating that, in all likelihood, a good proportion of tanks were expected to have some problems, it was concluded that there is a need for a management strategy. Having established the need for a management strategy, there was a real need to further explore exactly how the issue of tanks in South East Queensland ought to be managed. It was also acknowledged that there was a remaining knowledge gap regarding the actual state of tanks, and that the topic was contentious. From the previous stakeholder discussions, many ideas as to what can be done to manage tanks had emerged, and these proposed ideas were not all consistent with each other. For example, industry stakeholders suggested that a register of existing tanks should be established whilst others suggested that this would aggravate home owners and make management difficult. The team recognised two key points regarding the situation:

(1) Without the cooperation from tank owners, it is unlikely that any strategy to manage tanks would succeed. Therefore, serious consideration will have to be given to the views and preferences of tank owners.
(2) Stakeholders, such as the Queensland Water Commission and Queensland Health have good reasons for making sure that tanks across the region are operated in a safe and efficient manner.

It was further recognised that to find solutions to this problem would require creative ideas, as well as cooperation between multiple stakeholders. To develop solutions, the research team designed a two-part process, engaging both with the community, and with government and industry stakeholders and experts. The first step was to engage with relevant industry and government stakeholders through a dedicated workshop to identify and explore possible options for improving the management of tanks, with follow-up interviews with individual industry professionals. The second step was to engage with community stakeholders using focus groups to understand public impressions and to investigate possible community responses to alternatives.

7.7.2.1 Industry and government stakeholder engagement: Workshop

A stakeholder workshop was conducted to identify the various ways in which the management of rainwater tanks could be improved. Participants were stakeholders from across six different sectors; state government; regulatory entities; water utilities; local council; academia, and industry representatives including rainwater tank designers and manufacturers, engineers, plumbers and the plumbing industry representative body. All stakeholders had involvement in some aspect of the life cycle of rain water tank systems.

The workshop process was designed to identify and explore a range of possible strategies for encouraging and ensuring rainwater tanks are well maintained and managed. The research team specifically wanted to understand the benefits and barriers associated with each potential strategy, assess participants' perceptions of the difficulty in implementation of a particular strategy, and the strategy's potential effectiveness. Following an initial short presentation of relevant background information, the workshop included three activities: a brainstorming session; an in-depth analysis of key strategies; and a voting task. The brainstorming session identified a diverse range of possible strategies, which were then grouped and key strategies prioritised for more in-depth analysis. The in-depth analysis focussed on discussion of benefits, barriers, and implementation issues. A perceptual map was created, which summarised each strategy in relation to the others in terms of the perceived impact on rainwater tank maintenance and ease of implementation. The voting task identified participants' preferences for each strategy (Figure 7.2).

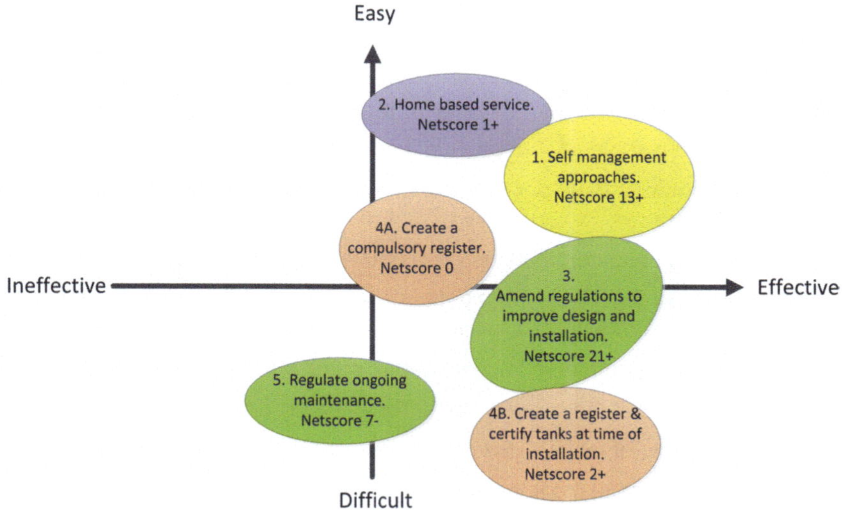

Figure 7.2 Perceptual map of groups of strategies and net voting preference scores. *Source*: Adapted from Walton *et al.* (2012). *Note*: A net preference score was calculated for each group of strategies, with the most preferred and least preferred votes added together. A positive net score indicated a strategy was liked overall, whereas a negative net score indicated a strategy was not liked overall.

Workshop participants generated a range of potential approaches, including improvements in installation practices, improvements in design to reduce maintenance requirements, and various methods of supporting/encouraging householders to manage tank maintenance by themselves. There was little support for setting up a register of tanks to help track and enforce maintenance, nor for taking the responsibility for tank maintenance away from householders (e.g., by making the local council responsible for maintenance). Participants believed that enabling options, such as education and awareness campaigns and the introduction of a dedicated and interactive website to provide tailored information to tank owners about tank maintenance, were preferable to regulatory processes. Interventions that focussed on improving tank design and ensuring correct installation were favoured because participants viewed such changes as capable of reducing potential maintenance problems. In addition, participants felt that regulatory processes would be both difficult and costly to implement and manage, and that compliance within the community would be low. Enabling strategies were considered easier to implement and likely to be more effective (Walton *et al.* 2012).

These views about likely community responses needed to be tested with community members themselves.

7.7.2.2 Community stakeholder engagement: Focus groups

A series of community focus groups were held to gain a deeper understanding of what the public thinks about rainwater tank maintenance and how people might respond to various strategies to promote or encourage maintenance. Focus group participants included tank owners (28 participants) and non-tank owners (14 participants) and the sessions were conducted in small groups (7 participants) with a research facilitator. A series of eleven different strategy ideas, developed from the industry and government stakeholder workshop, were put to the groups (Table 7.6), and each idea was discussed in terms of its potential effectiveness to achieve change and its likely acceptance by the community.

Table 7.6 List of strategy ideas presented to focus groups.

Number	Strategy ideas
1	Leave it to householders to manage for themselves; the status quo
2	Leave it to householders, but provide them with support • *Advice on what's to be done; fact sheets; helpline; directory of plumbers and tank cleaning services*
3	Leave it to householders, but increase householder awareness • *Promote benefits of keeping a tank maintained; highlight the consequences of not keeping a tank maintained; reminders and prompts*
4	Home service – you pay to have someone come and inspect your tank
5A	Create a register of tanks – rely on tank owners co-operation
5B	Create a register of tanks – make it compulsory to have tanks registered
6A	Inspect tanks – make it compulsory to have your tank inspected when it is first installed; to check it has been installed properly
6B	Inspect tanks – make it compulsory to have your tank inspected every couple of years
6C	Inspect tanks – make it compulsory to have your tank inspected when your house is sold
7	Tank design – make it compulsory to improve the design of the tank so that less things will need maintaining
8	Maintenance information – make it compulsory to be given information about tank maintenance when you have it installed

Open-ended questions were used to explore the benefits and barriers associated with each strategy idea. An output of these focus groups was a preliminary model to describe the socio-psychological factors that influence whether householders intend to undertake maintenance, and whether they are likely to put that intention into action. The community focus groups also reinforced the idea that registers of tanks would be highly unpopular and suggested that there was a strong mandate for householders' self-management approaches. Perceptions of each strategy, in terms of how effective and how acceptable the strategy would be at fostering tank maintenance, were mapped onto a perceptual diagram (Figure 7.3).

7.7.3 Assessing strategies

With a large array of strategies identified for managing rainwater tanks, the next step is to construct a framework for strategy evaluation. Key success criteria need to be established and a multi-criteria assessment model developed. A knowledge base capable of generating inferences of likely strategy success also needs to be developed for the assessment model. This knowledge base synthesises all existing information and formulates coherent judgments about the possible success of each strategy. To fulfil these requirements in the South East Queensland case, a research process involving three steps was undertaken:

(1) Establishing success criteria for strategies; that is, establish what a successful strategy would achieve.
(2) Undertake a survey that quantitatively measures the socio-psychological factors relating to householder tank-related behaviours.

(3) Synthesize the existing data and survey findings using computer simulations that can model different strategies for their chances of achieving the success criteria.

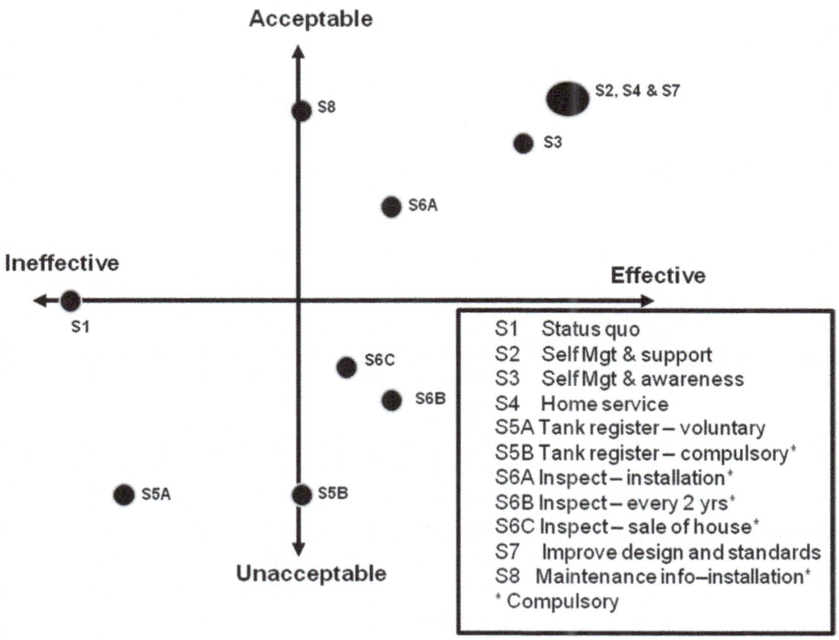

Figure 7.3 Perceptual map of strategy ideas from focus groups. *Source*: Adapted from Walton *et al.* (2012b).

7.7.3.1 Establishing success criteria for strategies

Interviews were undertaken with those stakeholders who had indicated a strong interest in tanks being adequately looked after. The research team interviewed staff members at the state government entities: Queensland Water Commission, Queensland Health, the Department of Natural Resources and Mines, as well as local engineering consulting firms with significant experience in designing tank systems for new developments. The focus of the discussions was to establish clear and unambiguous criteria that help evaluate whether a rainwater tank management strategy is successful. Not all of those interviewed stated any success criteria. Domestic rainwater tank owners are also considered as key stakeholders, but it is not practical or appropriate to allow them to define assessment criteria. However, their views and preferences about strategy options were measured through a household survey and incorporated into the modelling. The criteria for successful rainwater tank management developed from the stakeholder interviews are presented in Table 7.7.

7.7.3.2 Survey of householders

To quantitatively evaluate the behavioural response from the community to various management strategies, a survey was undertaken of 533 domestic tank owners from South East Queensland (Walton & Gardner, 2012; Walton & Gardner, 2014). The survey used a choice modelling approach to test tank owners' preferences for different types of intervention mechanisms that could be used to encourage them to maintain their tanks. Based on the strategies identified in the previous workshop, the interventions varied in the level of coerciveness embodied in the intervention, from a regulatory and 'push' approach through to an enabling and 'pull' approach. Six different strategies were evaluated: 1) introduction of tank registers; 2) introduction of tank inspections; 3) use of penalties (fines); 4) leaving tank owners to their own devices; 5) provision of information and regular awareness campaigns; and 6) use of an incentive (rebate). The strategies were combined in random order into eight different choice sets and each strategy evaluated for perceptions of fairness, effectiveness, and acceptability. The survey also investigated the links between socio-psychological factors and tank maintenance based on previous focus group findings (Walton *et al.* 2012).

Table 7.7 Criteria for successful rainwater tank management in the South East Queensland case study.

Criteria	Details
A. Adequate water savings	At least 90% of tanks need to provide the projected water savings from now and into the future. From the *water planning* perspective this also needs to be certain fact, backed up by sampling of the condition of tanks and statistical analysis of the condition data (see criterion D).
B. Acceptable low risk of mosquito breeding in tank systems	In the context of South East Queensland, at least 99% of tanks need to be protected by mosquito meshing. Furthermore, a general household requirement is that the amount of stagnant open water needs to be kept to a minimum.
C. Acceptable low risk of health risks related to poor drinking water quality	Stakeholders consider that it will be impossible to ensure drinking water quality at an adequate level with adequate certainty. Information campaigns need to ensure that the community are advised not to drink rainwater when safer options are available.
D. Knowledge of the condition of tanks	It was argued by the key stakeholders that the condition of tanks needs to be known. Not all tanks need to be inspected, but an adequate sample of randomly selected tanks needs to be inspected regularly. Adequate water savings need to be assessed every 3–5 years, and acceptable low health risks need to be assessed on an on-going basis every year. To support the inspection program, there is a need for knowledge of the tank stock (see criterion E).
E. Knowledge of the tank stock	In order to undertake a program of random inspections of rainwater tanks, there needs to be a database of tanks and their locations. Currently this type of information is managed by local councils, but there is a need to centralise this dataset into a single location. Whilst the community have indicated that they do not want a register of tanks, a database could be developed based on collection of existing available information about rainwater tanks from councils, and so on.

Source: Moglia *et al.* (2012b)

Survey results indicated that, although tank owners had low levels of awareness of the tasks involved in keeping a tank maintained, there were high levels of support among respondents for the introduction of measures that could foster tank maintenance within the community. Approximately 63% of survey respondents either didn't know that a tank had to be maintained or didn't know what needed to be done to maintain a rainwater tank. Only 37% of respondents reported being both aware of the need for tank maintenance and knowing what to do. Nonetheless, the mean score for supporting the introduction of measures to encourage tank maintenance was 5.13 (SD = 1.66) on a scale of 1–7 with higher scores indicating a more favourable view (Walton & Gardner, 2012). Clearly, information campaigns aimed at rainwater tank owners is a first step in order to raise awareness of tank maintenance issues but it is uncertain the level of success that can be expected from such a campaign.

However, community acceptance for the various interventions revolved around how fair and how effective the public viewed each intervention, with fairness twice as important as effectiveness. This meant that the more fair and effective the policy seemed to a person, the greater their acceptance of that policy. Interventions that were based on 'enabling' tank maintenance, and those that provided incentives for tank maintenance, were viewed as being more fair, effective and acceptable than interventions based on penalties and monitoring through registers and inspections. This is consistent with research into social acceptance of for water recycling and other alternative water supply systems (Nancarrow *et al.* 2010).

These preferences for incentives and enabling interventions were in line with the socio-psychological factors that seemed to underpin tank maintenance behaviour. The survey also found that favourable attitudes towards tank maintenance, a strong self-image as someone who likes to keep things well maintained, and perceptions of capability and self-efficacy were all important predictors of a person's intention to undertake tank maintenance.

These results suggested that policy makers could not only focus on policies that are based around 'enabling' factors and incentives, but also develop programs that address the socio-psychological issues. Based on the survey findings, strategies to foster a favourable attitude towards tanks maintenance could also include interventions which promote the benefits of tank maintenance and highlight the 'costs' of not keeping a tank maintained. Costs could include costs to both the individual, such as the loss of unrestricted water use for the garden, and to the wider community, for example, the risk to public health and the increased burden to the mains water supply. Similarly, solutions that link tank maintenance with other home maintenance activities could be appealing to those who see themselves as someone who keeps things well maintained.

Finally, interventions that support a person's belief in their own capability and provide confidence to undertake tank maintenance could be effective. Education initiatives, awareness campaigns and information services could result in improved self-efficacy beliefs, and ultimately influence tank maintenance behaviour. Furthermore, survey results demonstrated that even with a positive intention to undertake tank maintenance, if people don't feel capable or confident of their ability, then tank maintenance is less likely to occur. This finding further highlights the importance of providing enablers as a policy intervention to support tank maintenance behaviour.

7.7.3.3 Synthesising the data and using models for evaluation of strategies

It is necessary for policy makers to make choices about what type of management approach should be in place for domestic rainwater tanks. This choice is neither obvious nor straightforward, because there are many different factors to consider and a large body of information to incorporate into the decision making process. In fact, such factors also interact with each other, creating considerable complexity and placing this type of analysis beyond cognitive mental models. Furthermore, rigorous analysis of decision problems leads to better and more sustainable decisions and computer-based analytical tools can help to structure and analyse the decision problem. The likely effectiveness of strategies for rainwater tank management was evaluated using a Bayesian Network model, which can incorporate existing data, judgments and understanding. The model is described in detail in (Moglia *et al.* 2012b). Other models and/or decision analysis approaches may also be suitable for analysing this decision problem.

A Bayesian Network model is a probabilistic model describing a system of logical statements, albeit where the statements are probabilistic rather than absolute – thus more realistically reflecting the uncertainty that is ever-present in real-world decision problems (Pearl, 2000). As such, Bayesian Networks can incorporate (probabilistic) deductive, inductive and abductive reasoning – which are all key mechanisms by which humans and their computers can make inferences. Technically, Bayesian Networks are constructed as a system of equations. To define this system of equations, the modeller needs to define model factors, their possible states, as well as the interactions between factors. Interactions between factors are described through conditional probability tables, and the model is initiated by specifying the probability distributions of under-lying conditions.

As input into the model, the research team used the household survey results to describe socio-psychological factors contributing to behavioural response to strategy features, the judgments by engineering experts on what contributes to keeping a tank in good condition, and the stakeholder-identified success criteria. The model is shown in Figure 7.4.

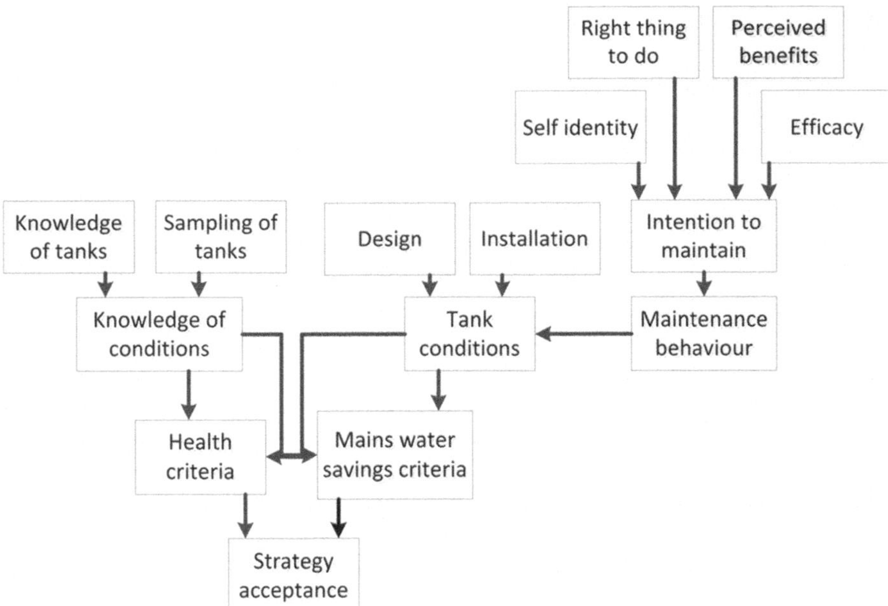

Figure 7.4 Bayesian Network model of the effectiveness of rainwater tank management strategies.

The way that the model is structured and populated with data is as follows:

- A domestic tank owner's motivation to maintain his/her tank is influenced by the perceived benefits, the self-identity, the efficacy, and whether the person thinks it is the right thing to do. The conditional probability tables describing this are defined through the householder survey.

- The motivation to maintain a tank does not always translate into action, that is, actual householder maintenance. The assumed likelihoods for this occurring for the different cases (motivated, unmotivated, or undecided) is set in a conditional probability table that can be subject to sensitivity analysis to explore various assumptions.
- A tank's condition is influenced by the design, installation and maintenance of the tank. This influence has been defined in conditional probability tables by an engineering consultant, and the assumptions can be explored using sensitivity analysis.
- The initial conditions describing what percentage of tanks have been adequately installed and designed, and so on, have been defined through professional judgments from email-based survey (Moglia *et al.* 2013a), but this should be defined through physical surveys of tank conditions, such as those reported by Biermann *et al.* (2012).
- The likelihood of adequate water savings and/or mosquito problems for a tank in poor condition has been defined by research team members, but again this should be established through physical surveys of tank conditions as per Biermann *et al.* (2012).
- The knowledge of tank conditions is influenced by the judged knowledge of the tank stock and whether regular inspections are being undertaken. The influence has been established on the basis of assumptions made by stakeholders in the interviews on defining clear and unambiguous criteria for strategy success. These are relatively straightforward and unambiguous, but these assumptions can also be explored using sensitivity analysis.

The model should not be viewed as an exact representation of reality, but rather an approach for synthesising the existing information in a single analysis. The model is a tool to help with structured logical reasoning and synthesis of existing information beyond the capabilities of human cognitive abilities. Problems with underlying data or assumptions will remain, but it is possible to explore sensitivity of conclusions to data and assumptions. A key premise for the modelling is that the model makes data issues and assumptions underlying the decision problem more transparent.

Figure 7.4 shows the nodes and arcs of the Bayesian Network model that was developed and encapsulates the main factors contributing to achieving the criteria in Table 7.7. The key factor is the condition of tanks, which is influenced by design, installation and maintenance behaviours. Maintenance behaviours are in turn dependent on the intention to maintain a tank, which in turn relates to a number of socio-psychological factors. However, the fact that tanks are in good condition is not sufficient by itself to meet the success criteria of stakeholders. For example, knowledge of tank condition is an important factor contributing to meeting health and water savings criteria, which in turn depends on both an understanding of where tanks are located (knowledge of tank stock) as well as a condition assessment program (sampling of tanks) to confirm that tanks are in good condition. State government departments also need some certainty that all key criteria are being met.

Behind the nodes and arcs of this diagram are conditional probability tables set up with available data and judgments by key informants (for detailed information see Moglia *et al.* 2012b). Using the Bayesian Network model, it is possible to input various conditions such as the distribution in the community of socio-psychological factors related to maintenance behaviour, as well as various design, installation and maintenance strategies and various levels of knowledge of tanks and condition assessment programs. The output of the model is a likelihood/probability of achieving the success criteria, allowing for scenario analysis as per Table 7.8, showing that success is dependent on a broad approach that considers all relevant factors.

Using the collected data synthesized into the Bayesian Network model, and running scenarios as per Table 7.8, a number of recommendations for a tank management program were made (Moglia *et al.* 2012b).

Table 7.8 Scenarios based on Bayesian Network modelling: running scenarios describing different situations of design, installation and maintenance for rainwater tanks in South East Queensland, and showing the resulting likelihoods of tanks being in good condition and success criteria fulfilment.

Scenario	% of tanks in good conditions	Knowledge of the tank stock (highest likelihood state)	% likelihood of health and potable water savings criteria fulfilment
1. **Worst case:** poor design & installation; poor knowledge of stock and too small sample of tanks. Status quo maintenance.	61%	Anecdotal	25%
2. **Benchmark design:** same as worst case but benchmark designs	74%	Anecdotal	29%
3. **Benchmark installation:** same as worst case but benchmark installations	72%	Anecdotal	28%
4. **Benchmark installation & design:** same as worst case but benchmark installations and design	86%	Anecdotal	32%
5. **Benchmark design & installation and rigorous knowledge:** same as Scenario 4 plus benchmark knowledge of stock and sampling of tanks	86%	Good knowledge	86%
6. **Best case:** like Scenario 5 but with socio-psychological factors significantly improved	93%	Good knowledge	91%

The recommendations centered on improving the three pillars of rainwater tank management – promoting good design, promoting good installation, and promoting good maintenance – in a three-phase process. The first steps relate to collecting information and better data management, as well as reviewing the need for legislative and other changes to the institutional framework. Some things that could be implemented in first phase at relatively low cost are the strategies to motivate, remind and enable owners to maintain tanks by addressing the underlying factors that contribute to the intention to maintain the tank. This would be done by providing information and developing the skills and ability of tank owners, as well as promoting the public and individual benefits of tank ownership. In the second phase, there is a move towards action including randomised inspections of tanks, changes to the design and installation guidelines and standards, and implementation of a training program for plumbers. The third phase is to review the efficacy of the chosen strategies, and to evaluate what is working and what is not working. It will also be important at this stage to make a fair assessment of the public health risk, and to evaluate whether such risk can be managed or whether further actions would be required.

7.8 DISCUSSION

Without adequate management of rainwater tanks at a regional/urban level, it is foreseen that there will be difficulties with achieving water savings targets (as discussed by Beal *et al.* in Chapter 3) and there are some public health risks that may materialise. The need to manage rainwater tanks would appear to be mainly driven by state government and local council needs and desires to ensure adequate water supply for the city/region and that health risks are adequately managed. An ongoing inspection program is required in order to fulfil these criteria of key stakeholders. This will come at a financial cost, regardless of whether householders have the key responsibility for maintenance. On the basis of experiences in a ongoing project in Melbourne, Victoria, it is now known that an inspection takes about one hour once it has been scheduled; and about 5 inspections a day a can be carried out (1.6 hours per inspection including travel time). Someone would need to coordinate the inspection process, and to have responsibility for the adequate management of tanks across the region. Fairness principles would dictate that this cost is carried by government and not householders.

However, government investment in rainwater tanks could perhaps be viewed as equivalent to investments in large dams. There are ongoing operations and maintenance costs associated with managing dams and these costs are passed on to householders via their water bills. Rainwater tanks can be viewed as 'small dams'. Therefore, it seems reasonable that the costs for the condition assessment and inspection programs for rainwater tanks, and the cost of any information campaigns, are also passed on to all householders via their water bills. On the basis that a suburb of mandated tanks can be likened to a small, central reservoir, all the community should pay for the tank inspection and condition assessment programs, not just the tank owners. The additional cost to the average water bill would be minimal and rainwater tanks would still be a cost-effective way to supplement water supplies for the city.

Related to the cost of the condition assessment and inspection programs is how the tank management requirements impact on the lifecycle costs of tanks, and thus on the competitiveness, or cost-effectiveness, of urban domestic rainwater tanks when compared with other options. The extent of the lifecycle costs depends on whether there is a cost attached to the labour for the maintenance tasks. By allocating some reasonable times to undertake the tasks in Table 7.2 (~17 hours per annum) and at a cost of AUS$50 per hour for labour, the annual cost of completing the maintenance tasks would be around AUS$830. This is equivalent to a net present value of just under AUS$9,500 over 15 years with a 5% discount rate. However, the reality is that much of the work is not charged for, as maintaining a rainwater tank can be considered as part of the regular upkeep of a property by the householder. This is a further reason why householders should have the responsibility for maintaining the domestic tanks, although considerations have to be made for the elderly and those who for various reasons do not have the confidence to undertake such tasks.

Due to the need for multiple perspectives when defining detailed intervention strategies, the process would be greatly assisted with input from a panel including representatives with deep knowledge and understanding of the various aspects of the rainwater tank management problem, including urban water planners, public health officials and representatives from the plumbing industry. It is unclear at this stage who should take on this responsibility. Furthermore, in the case of SEQ, the housing construction industry is likely to be affected by any policy regarding whether houses are required to have rainwater tanks installed, and is also often involved in the installation of tanks, and should therefore also be engaged in the process. Stakeholders need to coordinate amongst themselves to do something about this problem, or the future of domestic urban rainwater tanks looks questionable.

Whilst these issues appear to pose some questions about the viability of rainwater tanks, we argue that domestic urban rainwater tanks have a future in Australia and elsewhere. Domestic rainwater tanks complement other urban water supplies, especially at times of restricted mains supplies. However, there is a need to take the management of tanks seriously. The overall cost of devising a management strategy for tanks would not amount to much on a per-tank basis, but will require stakeholders to take control of the situation and move towards well-resourced action. Action needs to be implemented in acknowledgment that this is a new problem, with significant amounts of uncertainty attached to it, requiring significant buy-in from a range of stakeholders as well as the collection of adequate biophysical data through tank surveys and water savings assessments.

It is clear that there is a dilemma in the management of rainwater tanks in SEQ because on one hand householders want minimal interference from governments at any level, but on the other hand there is a need to ensure that public health risks are acceptable and water savings are achieved, plus key stakeholders need some certainty about such matters. Philosophically, rainwater tanks are an uncommon type of assets which provides key public good outcomes but that are in private ownership. Very few precedents exist for how to manage this type of infrastructure asset, although septic tanks appear to potentially fit into this category. It could however be argued that incentive structures are rather different.

From a theoretical perspective, if one assumes a *Homo Economicus* worldview where people are rational, self-interested and behaviour is dictated by personal incentives, the likelihood of contributing to the public good and thus maintaining private rainwater tanks depends on the presence or lack of incentives, that is, carrots and/or sticks (i.e., inherent incentives plus push or pull policies). The *Homo Economicus* perspective is largely discredited as unrealistic (Kahneman & Tversky, 2000; Gintis, 2000; Henrich *et al.* 2001) and we adopt the view that people are much more complex than what the *Homo Economicus* worldview dictates. In fact, even laboratory experiments may not adequately describe behaviour and case studies are now commonly the favoured approach for understanding behaviour (Levitt & Lis, 2008).

In the case of rainwater tank maintenance, we currently do not have data on what works and what doesn't work in terms of inciting rainwater tank maintenance behaviours. However, a reasonable assumption seems to be that it is probable that householders, to a large degree, want to do the right thing and to look after their own property. If that is the case, then providing information and promoting an understanding of the need to maintain tanks will have a significant positive outcome. It is important to remember that ultimately, householders are legally free to choose whether to maintain their rainwater tanks but are also liable to ensure that their assets do not pose health risks to themselves or their fellow citizens.

The alternative to a 'softly' approach to managing urban rainwater tanks also requires costly enforcement. A middle ground is recommended where a condition assessment program of private rainwater tanks is undertaken on an ongoing basis. Such a condition assessment program may be linked to penalties if rainwater tanks are thought to pose health risks to the community (a small stick); acknowledging that householders are legally liable for public health risks but not for water savings. Householders may also need to concede that a centralised database containing information about all rainwater tanks is necessary in order to manage public health risks associated with rainwater tanks.

The research framework to develop rainwater management strategies outlined in this Chapter is based on a participatory framework involving both qualitative and quantitative data collection through surveys and interviews as well as modelling using Bayesian Networks. The Bayesian Networks help synthesize the findings into a coherent system of probabilistic causation statements; and this model can be used to help develop and assess management strategies for their likelihood of achieving success. The case study and modelling undertaken in South East Queensland shows that any strategy to improve rainwater tank maintenance will need to consider multiple factors (i.e., not just behaviour, or design, or installation) and will have to deal with factors in combination. Furthermore, due to the uncertain and socially complex nature of this problem, an adaptive and participative approach will be necessary, reiterating previous calls by Moglia *et al.* (2010, 2011a, 2011b) where the efficiency of a strategy is assessed before moving on to any amendments. This then demands adoption of a systematic schedule of tactics, combined with random inspections of rainwater tanks to evaluate outcomes and promote learning.

7.9 CONCLUSIONS

This chapter has described the issues involved with rainwater tank management, and has provided a research framework to develop management strategies, exemplified with a case study application in South East Queensland, Australia. The research has shown that there is need for policy or interventions aimed at encouraging rainwater tank maintenance by tank owners. Rainwater tanks have a future in Australia and elsewhere. Domestic rainwater tanks complement other urban water supplies, especially at times of restricted mains supplies. However, there is a need to take the management of tanks seriously.

The long-term benefits from rainwater tanks in urban areas require the proper design, installation and maintenance of the tank assets. The risks associated with inadequate installation and maintenance can impact not only the security of future water supply, but also pose risks to public health by creating habitats for mosquitoes which can act as vectors for diseases such as dengue fever. Such risks can be mitigated by simple maintenance practices. But scarce information is available on the upkeep and condition of rainwater tanks post-installation in urban areas.

The undertaking of tank maintenance tasks by tank owners is driven by their motivation and their perception of competence (or self-efficacy) to undertake such tasks. Adequate knowledge of the maintenance needs and the ability to fulfil such tasks is required; however these vary markedly among urban rainwater tank owners. Any strategy to ensure tank maintenance will need to consider multiple factors (i.e., not just behaviour, or design, or installation) and will have to deal with factors in combination. The interventions strategies that are most likely to promote the operation and maintenance of rainwater tanks need to be perceived as being *fair* to generate acceptance by tank owners. In particular, the strategy needs to reinforce the benefits associated with tank use and to provide tank owners with knowledge on tank maintenance tasks.

Another major consideration for rainwater tank management is the associated cost. There is a cost involved with management of tanks, regardless of whether householders have the key responsibility for maintenance. The annual cost of completing the tank maintenance tasks is estimated to be approximately AUS$830 (~17 hours per annum, at a cost of AUS$50 per hour). However, much of this work is not charged for and is considered as part of the regular upkeep of a property by the householder. This is a further reason why householders should have the responsibility for maintaining the domestic tanks. An ongoing inspection program is required in order to fulfil the tank water supply security and public health obligations, and this also comes at a financial cost. It seems reasonable that the costs for condition assessment programs for rainwater tanks, and the costs of any information campaigns, are passed on to all householders via their water bills, not just the tank owners.

Due to the uncertain and socially complex nature of this problem, an adaptive and participative approach will be necessary, where the efficiency of a strategy is assessed before moving on to any amendments. A systematic schedule of tactics needs to be adopted, combined with random inspections of rainwater tanks, to evaluate outcomes and promote learning. The strategies will need to be periodically reviewed and re-assessed by field survey for effectiveness as tank owner motivation evolves.

Whilst these recommendations are those of the authors, it is also important to remember that this is a situation that requires cooperation from a broad alliance of stakeholders. In such situations, it is usually not constructive to enforce solutions that do not have wide support, but instead, stakeholders, including householders, need to take part in fair and transparent negotiations. However, that is a step which is outside the scope of our research framework.

7.10 REFERENCES

Baguma D., Loiskandl W. and Jung H. (2010). Water management, rainwater harvesting and predictive variables in rural households. *Water Resources Management*, **24**(13), 3333–3348.

Biermann S., Sharma A., Chong M. N. and Umapathi S. (2012). Assessment of the Physical Characteristics of Individual Household Rainwater Tank Systems in South East Queensland, Urban Water Security Research Alliance, Technical Report No. 66, Brisbane, Australia.

Gardiner A. (2009). Domestic rainwater tanks: usage and maintenance patterns in South East Queensland. *Water*, **36**(1), 151–156.

Gardiner A. (2010). Do rainwater tanks herald a cultural change in household water use? *Australasian Journal of Environmental Management*, **17**(2), 100–111.

Gardiner A., Skoien P. and Gardner T. (2008). Decentralised water supplies: south-east Queensland householders' experience and attitudes. *Water*, **36**(1), 53–58.

Garling T. and Schuitema G. (2007). Travel management targeting reduced private car use: effectiveness, public acceptability, and political feasibility. *Journal of Social Issues*, **63**, 139–153.

Gintis H. (2000). The human actor in ecological-economic models. beyond homo economicus: evidence from experimental economics. *Ecological Economics*, **35**, 311–322.

Gould J. E. (1996). Rainwater utilisation in botswana: problems and possibilities. *Science, Technology and Development*, **14**(3), 72–87.

HarmoniCOP (2005). Learning Together to Manage Together – Improving Participation in Water Management, D. Ridder, E. Mostert and H. A. Wolters (eds), European Commission, Osnabruck, Germany.

Henrich J., Boyd R., Bowles S., Camerer C., Fehr E., Gintis H. and McElreath R. (2001). In search of Homo economicus: behavioral experiments in 15 small-scale societies. *American Economic Review*, **91**(2), 73–84.

House of Lords Science and Technical Committee (2011). Behaviour Change, 2nd Report of Session 2010–12, Authority of the House of Lords, London, UK.

Hurlimann A. (2011). Household use of and satisfaction with alternative water sources in Victoria Australia. *Journal of Environmental Management*, **92**(10), 2691–2697.

Kahneman D. and Tversky A. (2000). Choices, Values and Frames. Cambridge University Press and the Russell Sage Foundation, New York.

Karim M. R. (2010). Assessment of rainwater harvesting for drinking water supply in Bangladesh. *Water Science and Technology: Water Supply*, **10**(2), 243–249.

Levitt S. D. and Lis J. A. (2008). Economics: homo economicus evolves. *Science*, **319**(5865), 909–910.

Mankad A. (2012). Decentralised water systems: emotional influences on resource decision making. *Environment International*, **44**(1), 128–140.

Mankad A. and Greenhill M. (2014). Motivational indictors predicting the engagement, frequency and adequacy of rainwater tank maintenance. *Water Resources Research*, **50**(1), 29–38.

Mankad A. and Tapsuwan S. (2011). Review of socio-economic drivers of community acceptance and adoption of decentralised water systems. *Journal of Environmental Management*, **92**(3), 380–391.

Mankad A., Chong M. N., Gardner T. and Sharma A. (2012). Examining biophysical and socio-demographic factors across mandated tank users in urban Australia: a linking step towards achieving best practices. *Water Resources Management*, **26**(7), 1983–1998.

Mankad A., Greenhill M., Tucker D. and Tapsuwan S. (2013). Motivational indicators of protective behaviour in response to urban water shortage threat. *Journal of Hydrology*, **491**(1), 100–107.

Moglia M., Cook S., Sharma A. and Burn S. (2010). Assessing decentralised water solutions: towards a framework for adaptive learning. *Water Resources Management*, **25**(1), 217–238.

Moglia M., Alexander K. S. and Sharma A. (2011a). Discussion of the enabling environments for decentralised water systems. *Water Science and Technology*, **63**(10), 2331–2339.

Moglia M., Sharma A., Alexander K. and Mankad A. (2011b). Perceived performance of decentralised water systems: a survey approach. *Water Science and Technology: Water Supply*, **11**(5), 516–526.

Moglia M., Tjandraatmadja G. and Sharma A. (2011c). How long will a rainwater tank last? Do you know? In: UWSRA 3rd Science Forum: Science Forum and Stakeholder Engagement: Building Linkages, Collaboration and Science Quality, D. K. Begbie and S. L. Wakem (eds), Urban Water Security Research Alliance, Brisbane, Australia.

Moglia M., Tjandraatmadja G. and Sharma A. (2012a). Initial investigation into the governance and management models for rainwater systems, Urban Water Security Research Alliance, Technical Report No. 50, Brisbane, Australia. http://www.urbanwateralliance.org.au/publications/UWSRA-tr50.pdf.

Moglia M., Tjandraatmadja G., Sharma A. K., Walton A., Umapathi S. and Gardner J. (2012b). Proposed Strategy Portfolio for the Management of Rainwater Tanks: The South East Queensland Case, Urban Water Security Research Alliance, Technical Report No. 75. Brisbane, Australia. http://www.urbanwateralliance.org.au/publications/UWSRA-tr75.pdf (accessed 24 December 2014).

Moglia M., Tjandraatmadja G. and Sharma A. K. (2013a). Exploring the need for rainwater tank maintenance: survey, review and simulations. *Water Science and Technology: Water Supply*, **13**(2), 191–201.

Moglia M., Walton A., Sharma A. K., Tjandraatmadja G., Gardner J. and Begbie D. (2013b). Management of rainwater tanks: lessons and findings from South East Queensland. *Water Journal – Australian Water Association*, **40**(5), 90–95.

Nancarrow B. E., Porter N. B. and Leviston Z. (2010). Predicting community acceptability of alternative urban water supply systems: a decision making model. *Urban Water Journal*, **7**(3), 197–210.

Osbaldiston R. and Schott J. P. (2012). Environmental sustainability and behavioural science: meta-analysis of pro-environmental behaviour experiments. *Environment and Behavior*, **44**, 257–299.

Pearl J. (2000). Causality. Models, Reasoning and Inference. Cambridge University Press, New York, USA.

Queensland Government (2005). Public Health Act 2005, Reprint No. 2B Last Revision: December 2010. State of Queensland, Brisbane, Australia.

Queensland Government (2008). MP4.2 – Water Savings Target, Department of Local Government and Planning, Brisbane, Australia.

Queensland Government (2011). Rainwater Tanks, Vol. 18/06/2011, Department of Environment and Resource Management, Brisbane, Australia.

Queensland Health (2002). Guidelines to Minimise Mosquito and Biting Midge Problems in New Development Areas. State of Queensland, Brisbane, Australia.

Queensland Health (2007). Rainwater Tanks: A Guide to Keeping Your Tank Safe. State of Queensland, Brisbane, Australia.

Rodrigo S., Sinclair M. and Leder K. (2010). A survey of the characteristics and maintenance of rainwater tanks in urban areas of South Australia. *Water Science and Technology*, **61**(6), 1569–1577.

Standards Australia (2004). ATS5200.466 2004. In: Technical Specification for Plumbing and Drainage Products – Rainwater Tank Connection Devices, Standards Australia.

Standards Australia (2008). HB230-2008, Rainwater Tank Design and Installation Handbook, Standards Australia.

Tilbrook R. (2009). Rainwater tanks in new housing development: a survey of the issues, knowledge and uses of residents in the Northlakes Estate, a study for Lake Macquarie City Council, NSW. In Unpublished Newcastle, University of Newcastle, Australia.

Walton A. and Gardner J. (2012). Rainwater Tank Maintenance: Community Perceptions of Policy Options, Urban Water Security Research Alliance, Technical Report No. 71, Brisbane, Australia. http://www.urbanwateralliance.org.au/publications/UWSRA-tr71.pdf.

Walton A. and Gardner J. (2014). Community acceptance of policy options for managing maintenance of rainwater tanks. *Local Environment*, **20**(5), 565–580, DOI: 10.1080/13549839.2014.884554.

Walton A., Gardner J., Sharma A., Moglia M. and Tjandraatmadja G. (2012). Exploring Interventions to Encourage Rainwater Tank Maintenance, Urban Water Security Research Alliance, Technical Report No. 59, Brisbane, Australia. http://www.urbanwateralliance.org.au/publications/UWSRA-tr59.pdf.

WHO (2008). Guidelines for Drinking-Water Quality, 2nd Addendum–3rd edn, World Health Organization, Geneva, Switzerland.

WSAA (2002). Integrated rainwater tank systems – a supplement to the water supply code of Australia. WSA03-2002, Water Services Association of Australia Inc., Melbourne, Australia.

Chapter 8

Public perceptions; motivational drivers, and maintenance behaviour for urban rainwater tanks

Aditi Mankad, Kelly Fielding and Sorada Tapsuwan

ABSTRACT

The purpose of this chapter is to present approaches to examine psychosocial factors influencing rainwater tank adoption and maintenance and their application in South East Queensland, Australia. The chapter reviews the limited social research on these issues that has been conducted to date and addresses three research areas: 1. Key dimensions of public acceptance for rainwater tanks, 2. Exploring relationships between key dimensions of public acceptance and water-related attitudes and perceptions, and 3. Investigating key psychological variables likely to influence rainwater use and tank maintenance beyond public acceptance. Psychological theories used to frame and interpret the psychosocial research findings are also introduced, namely, protection motivation theory and self-determination theory. Appropriate methodologies are discussed to address these research areas and a case study example provided to illustrate the methodological approach. The research presented explores attitudes of users and non-users of rainwater tanks, examines perceptions of water shortage threat, and helps to understand the management and maintenance of rainwater systems at the household level. Psychological and behavioural data is also compared across those with voluntary (retrofitted) rainwater tanks installed in existing homes and those implemented subject to a local development code (mandated). The role of willingness to pay for rainwater tanks is also considered. A general summary brings together all relevant research findings and considers the 'way forward' for social research on rainwater tanks.

Keywords: alternative water; amotivation; community attitudes; protection motivation; self-determination.

8.1 INTRODUCTION

Rapid population growth and changing climatic conditions around the world mean that there is increasing pressure on traditional water supplies to cope with high demand within urban environments (Brown *et al.* 2008). To address these water shortages there is a need to adopt alternative water sources, including household-level rainwater tanks, designed to supplement mains water for non-potable applications such as gardening, toilet-flushing, and laundry. Rainwater harvesting is certainly not a new option, and is one that has been voluntarily adopted in various other parts of the world, such as in Africa (e.g., Zambia; Handia *et al.* 2003), the USA (e.g., California, Colorado; Adrian *et al.* 2002), the UK (e.g., Ward *et al.* 2007) and Europe (e.g., Spain; Farreny *et al.* 2011). Although there is an extensive international literature on the viability of domestic rainwater harvesting, there has been very little attention paid to the social dimensions of this issue, in particular to the question of why people do or do not install rainwater tanks and to the motivational drivers of rainwater tank maintenance. This is surprising when considering the relatively low levels of uptake of rainwater harvesting systems in some countries. For example, only 26% of Australian households utilise a rainwater tank (ABS, 2010) and even in Queensland (Qld) where government rebates and regulation have been in place to encourage greater adoption, that proportion is only slightly higher with 36% of households using a rainwater tank (ABS, 2010). This suggests that the majority of Australian households, particularly in urban areas, have not adopted on-site rainwater harvesting as part of their domestic water supply.

In some urban regions the installation of rainwater tanks for non-potable uses has been legislated as part of building development codes to meet potable water savings (e.g., New South Wales, Queensland, India, urban areas in New Zealand). As an example, from January 2007 to February 2013, all new homes built in South East Queensland (SEQ) were required to install a water collection/conservation device that enabled households to save 70 kL of water annually (DHPW, 2013). This was typically achieved through a 5 kL rainwater tank plumbed into the internal fixtures within a home, such as the washing machine cold water tap and toilets, as well as an external tap (Queensland Development Code (QDC) MP 4.2; DPI, 2010). The recent repeal of QDC MP 4.2 still allows local councils to 'opt-in' to QDC MP 4.2 if they can demonstrate that introducing the requirements will deliver a benefit to the community (DHPW, 2013). Prior to the QDC MP 4.2 legislation coming into effect, state and local governments offered home owners of existing dwellings substantial rebates to assist in covering the costs associated with purchasing and installing (i.e., retrofitting) a rainwater tank. Despite these rebates, which varied between \$1000–\$1500, less than 20% of home-owners in the urbanised area of SEQ retrofitted their homes with a rainwater tank prior to the introduction of QDC MP 4.2 (RWIMU, 2009). Water administrators and researchers were not equipped with available knowledge to explain why the uptake of rainwater tanks was not higher, despite the obvious advantages of rainwater tanks in providing increased water supply to homes, minimising the impact of water restrictions and unlimited rainwater use outdoors for tanks not internally plumbed.

The example of SEQ, where there was low uptake of household rainwater collection systems despite financial incentives that helped to overcome any cost barriers to their implementation, suggests that social and psychological factors may be playing a role in residents' decisions to install a rainwater tank. Therefore, social research can help to identify underlying reasons why people do or do not install rainwater tanks. Until recently, there was little empirical investigation of the psychological underpinnings of intended or actual rainwater use and/or maintenance of rainwater tanks. Much of the general social research on rainwater tanks conducted in the past was carried out by Gardiner and colleagues in SEQ (Gardiner *et al.* 2007; Gardiner *et al.* 2008; Gardiner, 2009; Gardiner, 2010). However, more recent research conducted by Mankad and colleagues (Mankad *et al.* 2010; Mankad *et al.* 2011; Mankad & Tucker, 2013) examined psychosocial issues surrounding rainwater use and maintenance and this research will be presented in the case studies provided in this chapter.

In the remainder of the chapter we provide a review of the social literature on rainwater use. The research on this topic has been conducted in Australia and therefore the literature reviewed is focused on the Australian context. We then outline two psychological frameworks that may be particularly relevant to social research on rainwater tanks. We present methodological recommendations for addressing two key social research questions relating to rainwater tanks: 1) Investigating public acceptance and uptake of rainwater tanks; and 2) Identifying motivational drivers of rainwater tank maintenance behaviours. We also provide case studies to illustrate the use of these methodologies to address the two key research topics.

8.2 PAST SOCIAL RESEARCH ON RAINWATER USE IN URBAN AUSTRALIA

Historically, Australians have had a preference for using harvested rainwater in their homes, until local governments introduced mandatory charges for centralised mains water within urban areas, whether homes used the mains water or not (Coombes & Kuczera, 2002). As Coombes and Kuczera explain, this mandatory charge was to ensure the economic viability of a centralised water supply for councils and steer people away from their preferred, independent, rainwater supply. Thus, Australia's water culture shifted in the late 19th century from having a reliance on rainwater tanks to relying on a centralised water grid (Brown *et al.* 2008). Even as recently as the 1990s, rainwater tanks (mainly built from galvanised iron) were strictly discouraged in urban areas by local governments because they were considered to be hazardous to public health due to the risk of breeding mosquitoes, leading to the outbreak of arboviruses such as dengue fever (WHO, 2002). This historical context is likely to have had a prevailing and significant influence on present-day community concerns about the quality of rainwater harvested from tanks, and helps to explain why the adoption of rainwater tanks in urban areas remains lower than advocates would hope. An unrestricted supply of relatively inexpensive centralised urban water may also help to explain why rainwater tanks are more widely used in peri-urban areas (where connections to centralised supplies are not feasible due to technical and economic considerations) than in urban areas, where all homes are connected to a centralised town water supply.

In addition to a shift in water provision processes in urban Australia, the introduction of modern cultural innovations, such as hot water systems and water-intensive appliances like washing machines, meant that people could no longer rely on their rainwater tanks to adequately and reliably provide the amount of water required to support these modern water use habits (Moy, 2012; Shove, 2003). These modern changes were not designed with water conservation in mind. Rather, they reflected society's changing views on how water was used and its functional value in society; water was no longer seen as a basic need for survival. Instead, water had become a tool in creating new expectations about the ideal standards for quality of life (e.g., the expectation that hot water could be accessed immediately and that a washing machine will produce clean clothes and maintain personal hygiene). As a consequence, generations of Australians were losing touch with Australia's historical affinity with, and reliance on, rainwater and all that came along with that, such as frugal water use habits and a working knowledge of the water cycle.

There has been much social science research on the acceptance of alternative water sources in modern urban Australia, however, it has mainly been limited to public and industry acceptance of centralised forms of alternative water, such as recycled municipal wastewater and desalinated seawater. Specific social science research on rainwater tank acceptance has been limited. The research that does exist on rainwater suggests that the public are accepting of rainwater harvesting and, in some instances, indicate a preference for rainwater over other forms of alternative water (e.g., Hurlimann, 2007; Hurlimann & Dolnicar, 2010). In Marks *et al.*'s (2008) survey of public acceptance of alternative water sources, participants viewed rainwater as the cleanest form of water and a majority of respondents were willing to use it for drinking and domestic uses.

Gardiner's (2009) descriptive research on rainwater tanks demonstrated that rainwater tanks were highly valued in the community because of their benefits at the household level, and their perceived positive impact on the environment by reducing reliance on the mains water supply grid. Her research suggested that there were distinct groups of tank owners: environmentally committed tank owners (i.e., had a tank to reduce their environmental impact), garden-focused retrofitters (i.e., used their tank primarily for recreational gardening), and disinterested tank owners (i.e., saw tank as only useful during drought and were unlikely to persevere with a tank in the long term). These findings suggest that due to their differing perspectives on tank use, these groups would likely have differing motivational interests in learning more about appropriate tank maintenance, and they may embrace the lifestyle of independent water use in different ways.

Although the use of alternative water supply systems is increasing worldwide, it is still not widespread in many developed areas and the growth rate in rainwater tank use from 2007 to 2010 in Australia was 6% (WWAP, 2012; ABS, 2010). Further, the cost of rainwater from a tank is still high when compared to the cost of town water supply from a large, centralised system. Thus, cost savings may not necessarily be the main driver for the adoption of rainwater tanks as the literature (Marsden Jacob Associates, 2007; Tam *et al.* 2010; Hall, 2012) indicates that the cost of water supplied from rainwater tanks is higher than from centralised water supplies.

Arguably, cost-effectiveness may not be the main driving factor for the adoption of rainwater tanks because the pay-off period is long and the upfront cost is high. Other non-economic factors, such as quality of life, lifestyle and the freedom to avoid water restrictions may be the key to encourage rainwater tank adoption. It is clear that little research has focused on rainwater tank adoption or the role of psychological and economic factors in adoption, acceptance and use of this alternative water source. Given this dearth of existing knowledge, in this chapter we focus on how we can gain a better understanding of public perceptions of rainwater tanks as well as the motivational drivers for adopting and maintaining rainwater tanks. In the next section we provide a brief overview of two psychological theories that can help us to understand these issues. These theories provide the broad theoretical framework for the case study research presented later in the chapter.

8.3 THEORETICAL FRAMEWORKS FOR UNDERSTANDING PUBLIC ACCEPTANCE, ADOPTION AND MAINTENANCE OF RAINWATER TANKS

One way to think about the installation of rainwater tanks is as a way for householders to deal with the threat of drought and water scarcity. Protection Motivation theory (PM theory; Rogers, 1975; 1983), which is a framework that outlines the cognitive processes that underpin people's decisions to respond to threat (i.e., to take protective action), is therefore relevant to understanding rainwater tank installation decisions. According to this perspective, there are two processes that feed into decisions: threat appraisal and coping appraisal (see Figure 8.1). People will experience greater threat if they think they are vulnerable to a severe situation. In the context of rainwater tank adoption, residents are likely to perceive greater threat if they believe that drought conditions will directly affect them and that the effect of drought on their lives will be severe. The other process that feeds into decisions is the coping appraisal. This appraisal process involves perceptions of how effective the recommended response will be in overcoming the threat (response efficacy), one's ability to carry out these recommended responses (self-efficacy), as well as an assessment of the costs of responding (e.g., time, effort, money). So, if residents believe that rainwater tanks can help to reduce the threat of water scarcity, if they believe that they have the means and capability to install a rainwater tank, and that the time, effort or financial costs are not too onerous, then people will be more likely to intend to install a rainwater tank. Bringing these two processes together then, the decision to install a rainwater tank will be more likely to the extent that householders feel threatened by drought and water scarcity and that they believe that a rainwater tank is an effective way to address that threat.

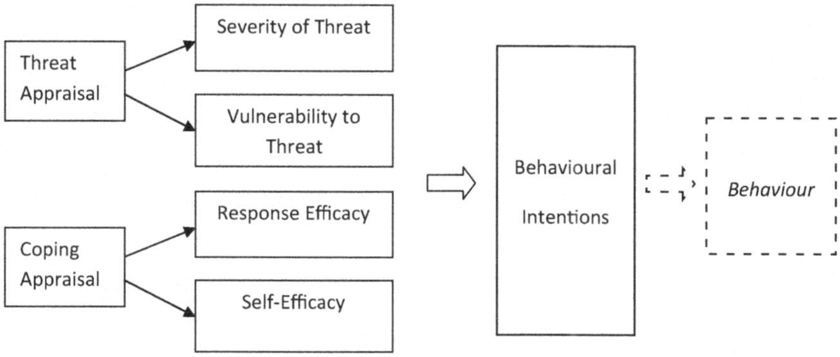

Figure 8.1 An abbreviated depiction of variables associated with protection motivation theory (Rogers, 1983).

When considering the issue of rainwater tank maintenance behaviour, the motivation that people experience to engage in this behaviour is particularly important. Self-determination theory (SDT) (Deci & Ryan, 2000), a meta-theory of human motivation, is therefore particularly relevant to this issue. Self-determination theory proposes that the social environment influences the quality of motivation that people experience. Better quality motivation (defined as more self-determined motivation) is more likely when the social environment satisfies three basic human needs, that is: for autonomy, competence and relatedness. Autonomy refers to people having choice and agency in their decisions, competence refers to a sense of self-efficacy in relation to specific actions, and relatedness refers to a sense of belonging and connectedness to others. It is hypothesized that social environments that foster greater autonomy, competence and relatedness result in more internalized, self-determined motivation and, therefore, better behavioural outcomes (Deci & Ryan, 2000).

Self-determination theory theorises about the quality of motivation by distinguishing between different types of motivation, which sit along a continuum from less self-determined to more self-determined motivation (Grolnick & Ryan, 1987). At one end of the self-determination continuum (Figure 8.2) is *amotivation,* which reflects a lack of action stemming from not valuing a particular behaviour, not feeling competent to engage in the behaviour or the belief that it will not yield the desired outcome. Any of these rationales could underpin residents' amotivation or apathy for engaging in rainwater tank maintenance. At the other end of the continuum is *intrinsic motivation* whereby individuals engage in behaviours for the inherent interest, enjoyment and satisfaction derived from the behaviour. As described in Figure 8.2, Ryan and Deci (2000) acknowledge that much of human activity is extrinsically (externally) motivated, rather than intrinsically (internally) motivated, although behaviours vary with regards to how much they are internally or externally motivated. As it seems unlikely that many people maintain their tank for the sheer enjoyment of it, the motivation for rainwater tank maintenance may be more likely to be externally motivated, for example, by government regulation or by neighbourhood norms of what other householders do to maintain their tanks. Self-determination theory is also relevant to the issue of retrofitting versus mandating rainwater tank installation. Although mandating installation of rainwater tanks has the advantage of encouraging greater uptake, the downside may be that the lack of choice reduces people's sense of autonomy and reduced autonomy has been shown to have negative impacts. One potential negative impact in relation to rainwater tanks is that people may be less likely to maintain their tanks.

Lowest---SELF-DETERMINATION-----------------------------------Highest					
Amotivation	External regulation	Introjected regulation	Identified regulation	Integrated regulation	Intrinsic motivation

Figure 8.2 The self-determination continuum (Grolnick & Ryan, 1987).

8.4 INVESTIGATING PUBLIC ACCEPTANCE OF RAINWATER TANKS: QUALITATIVE METHODS

An important social research question in relation to rainwater tanks is to understand what influences public acceptance and adoption of these systems. Results from past research have indicated that there is a high level of overall acceptance of rainwater for non-potable applications, however, adoption of domestic rainwater tank systems is low in urban areas. This high level of community acceptance coupled with low levels of adoption suggests a need to understand the barriers that may exist to uptake of these systems. The question of identifying barriers and facilitators is particularly amenable to qualitative research as this approach can provide a deeper understanding of the issue and allow researchers to gain insight into the issue from the perspective of the participants (Silverman, 2013). For research topics that have received little attention, it is helpful to begin with an exploratory qualitative approach, such as focus groups and semi-structured interviews, and, if appropriate, quantitative research can be conducted subsequently to investigate whether the qualitative findings generalise to the broader population.

Qualitative interviews can be unstructured and represent an interaction between interviewers and respondents that has a general plan but no specific questions. However, addressing the issue of barriers and facilitators to rainwater tank adoption lends itself more to a semi-structured interview approach whereby the interviewer has a set of questions that guide the interaction, but there is also scope to follow-up other issues that are raised during the interviews. Focus groups are another qualitative method that could be used to investigate this issue. Focus groups have the advantage of being quick and relatively low-cost and the group dynamic can sometime allow additional information to come to the fore. On the other hand, the researcher has less control than when conducting individual interviews and there are more complex logistics required to organise focus groups.

In Section 8.4.1 below we provide a case study of the use of semi-structured qualitative interviews to explore the facilitators and barriers to the adoption of rainwater tanks in SEQ. The case study represents a first phase of a broader research program undertaken as part of the Urban Water Security Research Alliance (UWSRA) seeking to identify determinants of public acceptance and adoption of rainwater tanks.

8.4.1 Identifying facilitators and barriers to rainwater tank adoption in South East Queensland

The qualitative interview case studyfocused on exploring attitudes and perceptions of household decentralised systems (rainwater tanks and greywater systems) among those with and without such systems on their property (Mankad *et al.* 2010; Mankad & Tucker, 2013). Throughout the study, it became evident that rainwater tanks were the most salient of decentralised systems, and therefore we limit our discussion here to the results on key dimensions of acceptance of rainwater tanks.

This case study comprised qualitative semi-structured interviews with homeowners who did and did not have functional rainwater tank systems on their property (i.e., Users and Non-Users respectively). The aim of this study was to identify key themes or drivers of acceptance that could be used in future quantitative research (e.g., assisting in survey development). Participants with rainwater tanks were sampled from three locations. The first was an eco-development located within Gold Coast local council, approximately 100 kilometres south of Brisbane, and 7 km inland from the coast. The location was chosen as it is the site of a relatively recent development with a strong focus on environmentally sustainable living and the development relied solely on several decentralised water supply systems working in tandem. The second location was within Brisbane local council, situated approximately 15 km west of Brisbane city. This location was chosen because it was the site of a new sustainable greenfield development that included decentralised water systems, which will eventually comprise 22 lots when completed. At the time of data collection, only six lots had been developed and residents had been living in their homes for a minimum of 12 months. The third location, within Moreton Bay local council, is a region 45 km north of the city of Brisbane designed for environmental sustainability with just over 100 lots in the development. Participants without rainwater tanks were recruited from the same localities in areas adjacent to the eco-developments. The Users and Non-users groups were matched on dwelling location and were comparable on demographic criteria such as age and gender.

Participants were recruited via:

- *Telephone recruitment,* using a specialised research recruitment company;
- *Door-knock recruitment,* where research assistants approached individual home owners and explained the study's purpose, as well as participation requirements; and
- *Emailed newsletter,* where housing developer restrictions meant that researchers were not permitted to access the community's residents. Instead, an email outlining the study and participation requirements was forwarded to residents via the community's intranet and residents were asked to contact researchers directly if they were interested in taking part.

Involvement in the study was voluntary and all respondents gave written consent for their participation. Participants were assured of confidentiality and were given a $100 gift voucher in recognition of their participation (Dillman, 2007). Participant selection was based on the following criteria:

(1) Respondents were homeowners or paying their mortgage; this was a delimitation applied to participant recruitment to exclude renters because, typically, renters do not have the authority to make decisions regarding modifications to the property, such as the installation of a rainwater tank.
(2) Users and Non-users were matched on locality of residence to ensure similarities in the surrounding environmental condition of both groups and to minimise demographic differences related to variables such as income and education.
(3) As noted above, those with rainwater tanks (Users) were recruited from purpose-built eco-developments and those without rainwater tanks (Non-users) were recruited from established suburbs adjacent to those eco-developments.
(4) It was necessary that the Non-users were not using any type of decentralised water system so as not to confound the interpretation of results.

Interviews took place in individual respondents' homes, with two researchers present. The interviews varied between 25–60 minutes in duration and were conducted within a single week. Participants were asked to respond openly and honestly to the questions and were informed that they were free to withdraw at any time without penalty. Individuals were also asked their permission for researchers to audiotape the interview to maintain accuracy of the data being obtained.

Given the semi-structured format, interviewers began the session by asking both groups of respondents what their general background knowledge was of the water supply being plumbed into their homes. This was to encourage participants to think more broadly about their household water and its use, before moving on to discuss more detailed aspects of their beliefs, values and knowledge regarding decentralised systems. After the initial question, the interview guidelines for the two groups varied, due to differences in exposure to, and experience with rainwater tanks:

- *Users of rainwater tanks* were asked to describe their current residential water systems and to identify the main reasons why their household had chosen to adopt this water supply. Respondents were encouraged to demonstrate their knowledge of the systems adopted and any related issues, such as maintenance and functioning. Advantages and disadvantages associated with a rainwater tank were discussed and respondents were asked to think about ideal changes or improvements they would make, if given the opportunity. Householders were asked to consider how others (i.e., Non-users) may feel about their rainwater tank systems and highlight any concerns they had as users of rainwater systems.
- *Non-users of rainwater tanks* were asked a number of preliminary questions, relating to their understanding of what an on-site rainwater tank system was, awareness of rainwater tank systems available for residential use, and general thoughts concerning these alternative water supplies. Interviewers then guided the topic of conversation towards gaining insight into the reasons why current Non-users would or would not consider adopting rainwater tanks in the future, as well as their thoughts about why others may choose to install rainwater tanks. Participants' perceptions of the associated advantages and disadvantages were explored, and respondents were encouraged to discuss concerns they had regarding alternative water supplies.

Interviews were transcribed verbatim and participants were not identified within the transcript at any stage. Data were then coded and content analysed using NVivo 8 (QSR International, 2008) by two researchers directly involved in the study, plus one team member who was not involved in data collection to limit interpretation biases. Analysis of the data included identifying and extracting recurrent themes and sub-themes within the interview text, to gain an understanding of participants' beliefs regarding rainwater tanks, the existing level of knowledge among both Users and Non-users of rainwater tanks, as well as personal and social values attributed to rainwater tanks. Inter-rater reliability was high amongst the three raters.

As a qualitative and exploratory study, it is important to understand that this research methodology was not concerned with a population sample which would be subject to statistical analysis. Therefore, the final number of participants in this study was determined based on *data saturation* being achieved (Marshall, 1996; Sandelowski, 1995). Saturation refers to a completion of the analysis process when no new concepts are identified in the data. The ultimate sample size is, therefore, determined by data saturation or peak, and traditional 'rules' about an appropriate sample size should not be applied to qualitative data.

8.4.2 Case study findings

Results indicated a high level of overall acceptance for the use of rainwater for non-potable applications. Both groups indicated that living sustainably was a desirable way of life (e.g., utilising a rainwater supply to supplement household water). However, participants acknowledged that the decision to live sustainably was usually influenced by external factors such as property regulations and financial constraints (e.g., rainwater tank infrastructure was significantly more expensive than 'normal' infrastructure). Both groups also cited the benefits (or potential benefits for the Non-users) of having increased water supply around the home, especially with regards to an increase in the quantity of water available to water gardens and lawns.

The majority of respondents from the Users group stated that property characteristics, such as locality, outlook and geographic orientation, were dominant factors for choosing to live in their current homes. Environmental benefits associated with using rainwater were of secondary importance. Users cited many advantages of having a rainwater supply; the primary advantage being the superior quality of drinking water produced from the rainwater tanks, which was perceived to be free of chemicals such as chlorine and fluoride, unlike council water. Cost savings from harvesting one's own water supply were also mentioned as another benefit. However, most Users conceded that the true financial benefits of using an alternative water source (rainwater) would probably only become apparent after at least 10 years of continuous use, usually because of high running costs.

Users cited the high amount of electricity required to pump water from a rainwater tank to the household taps as a major disadvantage of using a tank and, in some locations, participants raised concerns regarding poor installation of rainwater tanks, which led to problems such as leakage and back flow. This then led to tensions between water system providers and the residents. Amongst a number of other concerns raised, Users cited maintenance issues as a disadvantage of owning and using rainwater systems, as they required ongoing maintenance, which could be costly in some instances. Such problems were cited as a severe disadvantage of living in an area that had employed an innovative building approach where water efficient infrastructure was being trialled and was, thus, still experiencing teething problems. These problems caused residents to be less confident in the quality of their harvested water and, as a result, people cited being less likely to use the water for potable *or* non-potable purposes (Figure 8.3).

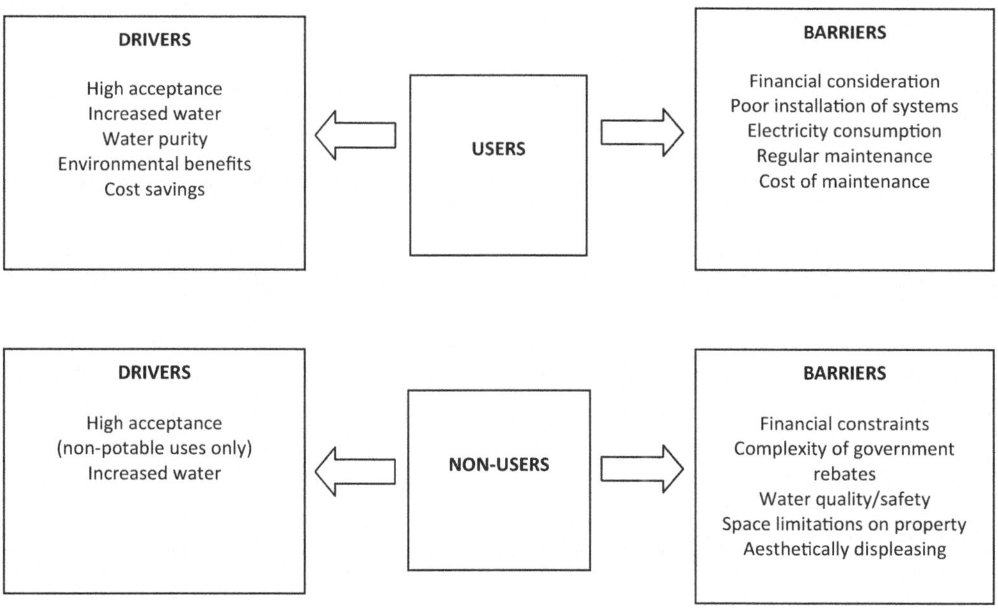

Figure 8.3 Drivers and barriers to living with rainwater tanks, as cited by users and non-users of the infrastructure (Mankad *et al.* 2010).

Non-users typically formed their attitudes and perceptions about rainwater tanks based on observations of rainwater tank use of friends and neighbours (i.e., a type of normative comparison). These observational experiences were reportedly positive, possibly explaining why the data highlighted such high acceptance of rainwater tanks among Non-users. However, there was consistent doubt among Non-users of the ability of rainwater tanks to adequately treat the water to drinking standards. This is likely to be because participants typically reported seeing acquaintances using their rainwater primarily for outdoor applications, such as watering the garden and washing vehicles. Consequently, Non-users stated their intentions for installing a rainwater tank would be for non-potable uses, as they were reluctant to use rainwater for indoor applications, particularly drinking and cooking. Non-users consistently mentioned three dominant barriers to adopting rainwater systems around their homes: 1) financial limitations (particularly in low-income areas), 2) perceived complexity of applying for government rebates, and 3) space constraints on their property (Figure 8.3). Most Non-users stated that, if building a new house, they would opt for installing an underground rainwater tank at the building stage, despite the greater capital costs, rather than installing a tank outside the new home. Non-users felt that their homes were not designed to accommodate a rainwater tank and there seemed to be a clear perception among this group that external rainwater tanks damaged the housing aesthetic (Figure 8.3).

Interestingly, when talking about water use patterns, a majority of participants, both Users and Non-users, explained that they were no longer as 'strict' in their water savings behaviour because they believed there was little threat of water shortages 'these days' following significant rainfall events and reported consequently having longer showers and watering their gardens more than they used to when the drought was 'on'. This attitude demonstrated a low perceived threat for future water shortages among all participants and was evidently influencing their current water use behaviours. This finding alluded to the importance of threat and risk perceptions in water use behaviours and intended adoption of rainwater tanks. Therefore, the data highlighted a need to understand the influence of perceived threat and vulnerability to threat on rainwater tank adoption and use.

8.5 INVESTIGATING PUBLIC ACCEPTANCE OF RAINWATER TANKS: QUANTITATIVE METHODS

While qualitative approaches can provide rich data that allows researchers to drill down into the experiences of specific groups within society, quantitative approaches such as surveys can reach representative samples to gain a broader understanding of issues. Quantitative surveys also allow researchers to test for statistical relationships between variables and can assess the utility of specific theoretical perspectives for understanding an issue. It also allows researchers to investigate whether themes that emerged from qualitative research can be generalised to the broader population of interest. In the following section we provide a case study of a quantitative survey that was conducted in SEQ as part of the UWSRA social research program to explore acceptance of rainwater tanks.

8.5.1 Predictors of rainwater tank adoption in South East Queensland

The quantitative case study was guided by Protection Motivation Theory as well as testing the importance of dominant themes emerging from the qualitative phase (discussed in Section 8.4). It aimed to examine the role of perceptual and motivational factors in predicting rainwater tank ownership. Protection Motivation Theory was outlined in Section 8.3; according to this perspective the likelihood of an individual carrying out risk-reduction behaviours, such as installing a rainwater tank in response to a perceived water shortage threat, is increased if there is: 1) belief in one's vulnerability to the water shortage threat; 2) belief in the severity of the water shortage threat; 3) belief that the suggested risk-reduction (protective) behaviours are effective ways to mitigate the risk of water shortage; and 4) belief that one is capable of carrying out protective behaviours to successfully avoid water restrictions associated with the water shortage threat. A quantitative survey that includes measures of these variables can help to identify whether these factors are key determinants of public acceptance of rainwater tanks. As cost had emerged in the qualitative interviews as an important barrier to rainwater tank adoption, we also incorporated a choice experiment in the survey that assessed willingness to pay for a rainwater tank.

An online survey method was chosen to collect quantitative data on 'protective' motivations surrounding water shortage threat (Mankad *et al.* 2011; Mankad *et al.* 2013). Participants were recruited from local government areas within SEQ: Brisbane city, Gold Coast, Logan, Moreton Bay, Redland city, and Sunshine Coast. Participants were verified as SEQ residents through their self-reported postcode. The survey comprised:

- *Standard demographic questions* (e.g., age, income, occupation), as well as other more specific descriptive items relating to rainwater (e.g., uses for rainwater, treatment process for rainwater at home, number of bedrooms/bathrooms in the house, actual household water consumption).
- *Perception and motivation* items adapted from qualitative findings (Mankad *et al.* 2010) and the structure of items adhered to existing psychological theory (Rogers, 1983; see relevant PM theory discussion in Section 8.3). Responses to the psychological items in the survey (i.e., threat severity, threat vulnerability, response costs, response efficacy, self-efficacy, subjective knowledge, norms) were typically made using a 5-point Likert format, where only the lowest and highest points were named (e.g., 1 = strongly disagree; 5 = strongly agree). Reliability coefficients were generated for each of the psychological sub-scales, as a measure of internal consistency and reliability between items within the same scale. For more detailed information regarding the items used in this online survey, please refer to Mankad *et al.* (2011) and Mankad *et al.* (2013).
- *Intentions to install a rainwater tank* was the main dependent variable and was measured on a scale from 1 (very unlikely) to 5 (very likely). Participants were also asked if they had already installed a rainwater tank on their property.
- *Willingness to pay* for rainwater tanks was evaluated using the choice experiment method (see Louviere & Hensher, 1982; Louviere & Woodworth, 1983) to determine how much households in SEQ were willing to pay for various rainwater tank sizes, when faced with choices of near perfect substitutes, which were greywater systems and bores or spear pumps. The amount of money that an individual is willing to pay is used to reflect his or her preferences for one system over another. The questionnaire and experimental design of the choice experiment study is detailed in Tapsuwan *et al.* (2014).

The study consisted of a cross-sectional survey targeting residents within SEQ, a region identified as hydrologically 'at risk' (QWC, 2009). The aim was to ensure that the sample population was comparable with the most recent Australian Bureau of Statistics population data available (ABS, 2007). Participants were recruited through an online research panel (database) from local government areas within SEQ; potential participants were verified as SEQ residents through their self-reported postcodes. People initially received an invitation email to take part in the survey and were then directed to the survey by following the survey hyperlink. Participants were also informed via the invitation email that if they completed the survey, they would receive 'reward points' through the online research company as incentive for participating in the study. An appropriately weighted incentive can enhance participation rates (Dillman, 2007).

A total of 406 participants fully completed the online survey; respondents were screened prior to participation to ensure that they were home owners or paying a mortgage of a free-standing dwelling, which was connected to the centralised town water supply. This was to exclude short-term renters, apartment owners and rural property dwellers from participating in the survey, ensuring that participants only comprised urban citizens who actively engaged in water supply-related decisions within their homes. The survey took approximately 20 minutes to complete. After completing the component of the survey that assessed psycho-social factors, participants completed the demographics section and then moved on to take part in the choice modelling component.

8.5.2 Case study findings

8.5.2.1 Psychological predictors of rainwater tank adoption

Results indicated that subjective knowledge, threat appraisal, response efficacy, response costs, subjective norms and social norms accounted for 43% of variance in intentions to install a rainwater tank, which is considered a moderate to high amount of explained variance in the psychological sciences (Cohen, 1988).

Table 8.1 lists the key unique contributors to the model of rainwater tank adoption intentions in order of relative contribution (Mankad *et al.* 2011; Mankad *et al.* 2013). Response efficacy was the strongest predictor of intentions to install a rainwater tank. That is, individuals who felt that they could manage these systems well were more likely to express intentions to install rainwater tanks in the future and more likely to already have a rainwater tank system at home. The next most important predictor was threat appraisal. This relationship shows that the more residents felt that they were threatened by future water shortages the more they intended to install a rainwater tank. Subjective knowledge, which refers to how much people thought that they knew about the water situation in the region was also positively related to rainwater tank adoption. Interestingly, participants' social awareness of what was expected of them as citizens (i.e., the extent to which they perceived that important others think they should install a rainwater tank) was also an influential factor in their intention to own a rainwater tank.

While the factors discussed to this point are positively related to rainwater tank adoption, response costs was negatively related. This means that the more that residents felt there were costs (e.g., time, affordability, space) related to rainwater tanks, the less likely they were to install a tank. It was also evident that those residents who had installed a rainwater tank perceived significantly lower costs associated with using a rainwater tank, as well as greater response efficacy (i.e., higher confidence that a rainwater tank can protect the householder from the negative effects of future water restrictions).

Table 8.1 Relative contribution of the five key variables that significantly predicted protective behavioural intentions towards rainwater tank adoption (Mankad *et al.* 2013).

Predictor	Beta coefficient
Response efficacy	0.253
Threat appraisal	0.237
Response costs	−0.228
Subjective knowledge	0.168
Subjective norms	0.141

Note: Higher beta scores indicate a greater weight (i.e., greater contribution to the explanatory model) and a negative sign indicates a negative relationship (i.e., greater perceived response costs is associated with poorer tank adoption).

The results suggested that motivational factors played a significant role in rainwater tank adoption. In particular, the strength of perceived effectiveness of rainwater tanks (i.e., response efficacy) as a driver of intentions to install a rainwater tank, suggests that it is critical to clearly communicate how the installation of rainwater tanks in homes can serve as a protective response to the threat posed by urban water stress. For example, if people believe that a rainwater tank is effective in protecting the individual, their lifestyle and their property from a water shortage threat, they may be more likely to report strong intentions to engage in rainwater tank adoption and actively seek out knowledge specific to rainwater tank installation and use.

8.5.2.2 Willingness to pay for a rainwater tank

The online survey also included a series of choice questions to evaluate the influence of rainwater tank cost and the cost of near perfect substitutes on the adoption of rainwater tanks (see Tapsuwan *et al.* 2014). Designing an efficient water supply management strategy requires an assessment of consumer preferences, as consumers contribute to the augmentation of supply through the adoption of rainwater tanks or other decentralised water systems (e.g., greywater or groundwater). Consumer preferences can be understood through their stated willingness to pay levels for various types of water supply sources. In the context of rainwater tank adoption, willingness to pay can help inform policy-makers whether rainwater tanks are preferred over other types of decentralised water systems or not.

In addition to ascertaining preferences through a series of choice questions, socio-economic variables are also collected in willingness to pay studies because there are many factors, in addition to the water system itself, that determine how much an individual is willing to pay. Theoretically, the most important economic factor is income because an individual can only pay as much as he or she can afford. More recently, sophisticated statistical software packages, such as MPlus (Muthén & Muthén, 2012), enable the estimation of structural models where psycho-social factors can be integrated into the willingness to pay models. Hence, we now have a better understanding of the underlying socio-economic and attitudinal factors that drive preferences and willingness to pay.

Although rainwater tanks are considered market goods, there is a lack of market data to reliably associate socio-economic and attitudinal information to people's purchasing behaviour, especially when consumers have to choose between installing a rainwater tank versus other types of decentralised water technology. Additionally, finding out how much people are willing to pay for future installations is in fact ascertaining preferences in the context of a hypothetical market. Given the current policy practice of intervening in markets and encouraging adoption through the use of consumer subsidies or mandated adoption, it is important to find out the gap between the willingness to pay value placed on these technologies and the actual market price.

Results from the willingness to pay study shows that the willingness to pay for rainwater tanks ranged from \$2100 to \$5000 for small rainwater tanks (<5000 ltr), \$2100 to \$7400 for medium-sized rainwater tanks (5000–25,000 ltr), and \$800 to \$3600 for large rainwater tanks (>25,000 ltr). For comparison, willingness to pay for greywater systems ranged from \$1700 for greywater diversion devices for outdoor use only to \$14,000 for greywater treatment devices for both indoor and outdoor non-potable uses. (Tapsuwan *et al.* 2014). The willingness to pay estimates for rainwater tanks show diminishing marginal return on tank size, as indicated by the low willingness to pay for large tanks relative to medium-sized tanks. As the sample consist mostly of urban households, large rainwater tanks would not be as desirable as small and medium size tanks because of space requirement. Households may even view the size as excess to their needs. When compared to actual market prices (Figure 8.4), willingness to pay was substantially lower and subsidies would be required to encourage people to buy rainwater tanks. Subsidies of \$1650, \$2650 and \$4950 for small, medium and large rainwater tanks, respectively, would increase the adoption level to 63% of the sample and a full price subsidy would be required to achieve 100% adoption. In order for the remaining 33% of the sample to install rainwater tanks, the technology would essentially have to be given out for free. Of course, considerations must be made about whether investments in government subsidies or hand-out programs are worth the cost savings associated with not having to augment a new source of supply. Having said that, at least 17% of the sample will buy small or medium size rainwater tanks at market price (i.e., no subsidies required).

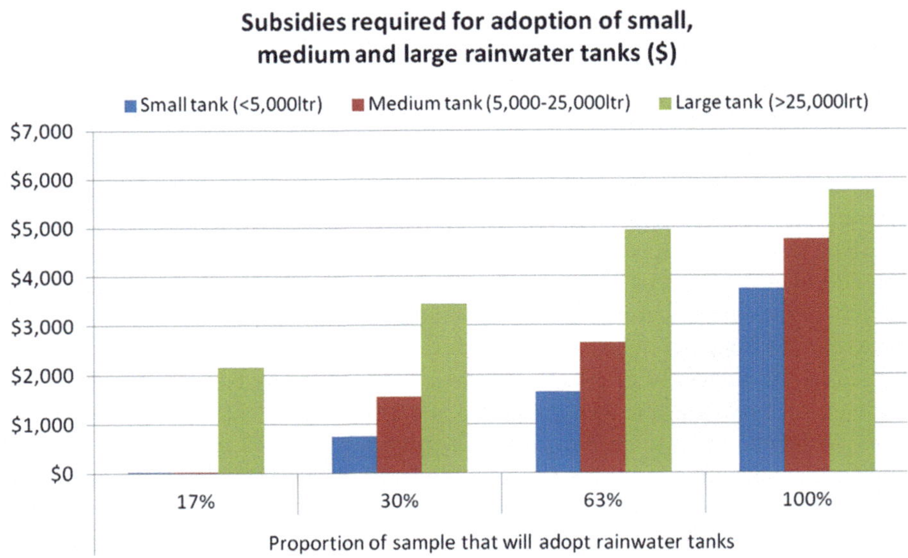

Figure 8.4 Subsidies required for the adoption of small, medium and large rainwater tanks.

8.6 IDENTIFYING MOTIVATIONAL DRIVERS OF RAINWATER TANK MAINTENANCE: QUANTITATIVE METHODS

A second important social research question relating to domestic rainwater tanks is to understand what motivates residents to maintain and manage their tanks. If rainwater tanks are to be an integral and effective part of the domestic water supply system in the long term it is important that they are maintained to ensure their safety and proper functioning. This question could fruitfully be addressed through qualitative and quantitative social research approaches. Conducting qualitative interviews or focus groups with householders who have rainwater tanks that have been retrofitted or that have been mandated in a new home would provide important insights into the facilitators and barriers of tank maintenance. Representative quantitative surveys of the same two groups (i.e., retrofitters and mandated installers) can describe the general level of tank maintenance being undertaken by households as well as the factors that predict householders' willingness to engage in maintenance activities. Quantitative surveys also allow researchers to statistically assess the size of the relationship between predictors and outcomes (i.e., rainwater tank maintenance).

In the following sections we provide two case studies that adopt quantitative research methods to identify the extent of householders' maintenance behaviour, whether maintenance activities vary between households that have retrofitted their tanks or been mandated to install them in a new dwelling, and what factors (including psychological factors) motivate householders to engage in maintenance activities. The two case studies are drawn from research conducted as part of the UWSRA in SEQ.

8.6.1 Drivers of rainwater tank maintenance behaviour for mandated tank owners

The first case study used a population-based mail out survey to examine the rainwater tank maintenance activities of residents with 'mandated' rainwater tanks in SEQ. The survey was broadly framed by self-determination theory and examined the basic motivational needs that are considered prerequisites of motivated behaviour (Mankad *et al.* 2012c). The mail out survey comprised a range of questions, designed to collect descriptive information about participants' rainwater tank set-up, as well as questions within the motivational orientations framework (Deci & Ryan, 2000), to provide greater insight into maintenance-related attitudes and behaviours of rainwater tank owners. Categories of items included:

- *Filtering questions* – to ensure the sample only included households with rainwater tanks and other sample requirements.
- *Demographics and general psychosocial descriptors* – age, gender, income, household composition, occupation, education, ethnicity, past history of living with a rainwater tank, householder satisfaction with rainwater, willingness to use rainwater for household uses, and perceptions of responsibility for maintaining their rainwater tank. Some items used a structured response format (i.e., male or female) and other items utilised the Likert format (e.g., 1 = Strongly disagree, 5 = Strongly agree), depending on which was most appropriate to measure participant responses.
- *Physical tank set-up* – tank volume, tank connections, plumbing, water level indicator, purposes for rainwater use (e.g., 'toilet flushing', 'clothes washing', 'garden irrigation'), estimate of how much of the household's total water use was from the rainwater supply, how often householders used their rainwater (e.g., daily, weekly, monthly), and a description of the types of water appliances and fixtures installed in the home (e.g., low flow taps and shower heads, dual flush toilets).
- *Psychological perception subscales* – perceived competence, perceived autonomy and perceived relatedness; responses were made using a 5-point Likert format (e.g., 1 = Not at all true, 5 = Very true) to indicate level of agreement to statements.
- *Attitudinal subscales* – perceptions of tank ownership, perceived water culture, perceived water rights; responses were once again made using a 5-point Likert scale as a measure of agreement (see Mankad *et al.* 2012a for a more detailed description of survey items and wording).
- *Regularity of rainwater tank maintenance (Dependent Variable)* – how frequently participants self-reported carrying out particular tasks on a rainwater tank maintenance schedule. The list of 13 maintenance tasks included checking and cleaning the first flush device, checking mosquito screens and repairing holes and other damage, inspecting and clearing gutters, checking pipes for structural integrity, and so on. All tasks were described using a specific time frame (e.g., 'check mosquito-proof screen ... *every 6 months*', 'check for evidence of bird or animal access... *every 3 months*'), so that participants could indicate whether their maintenance behaviour aligned with the correct maintenance guidelines, as outlined through the Australian Government (DHA, 2004; Queensland Health, 2007). The response scale was a 5-point Likert scale (1 = Never, 2 = Rarely, 3 = Sometimes, 4 = Almost always, 5 = Always).

Any householders who had been contacted to participate in any previous rainwater tank associated social research studies were not approached for the present study. Units and apartment blocks were also not sampled, as it was unclear if householders residing in these complexes would be responsible for maintaining their own rainwater tanks. Recruitment was limited to Local Government Areas (LGA) of SEQ where mandated lists were supplied. These LGA included: Gold Coast, Logan, Moreton Bay, Redland and Sunshine Coast. The target sample size for each LGA was determined based on an estimation of the proportion of residents living in that area. A conscious effort was made to avoid suburbs that were affected by the severe flooding events that occurred in the region in January 2011. In particular, Brisbane City, Ipswich and Lockyer Valley local government areas were excluded from the study for this reason. Surveys were mailed out to relevant homes in SEQ using a three-staged mail-out strategy, with the first stage involving an introductory postcard, to notify potential participants of the upcoming survey on rainwater tanks, followed by a mail-out questionnaire, reply paid envelope and eco-friendly shopping bag, intended to serve as a minor incentive to encourage participants to complete their surveys (Dillman, 2007). Two weeks later, a reminder postcard was sent to participants encouraging them to return their completed surveys. Participants were given approximately six weeks to return completed surveys.

8.6.2 Case study results

Descriptive analyses of participants' knowledge of their physical tank set-up showed that most (76%) participants knew the size of their rainwater tanks and the large majority (95%) knew that their tank was connected to a pump. There were two areas where there was less knowledge. First, as shown in Figure 8.5, only 61% of participants knew whether their rainwater tanks were connected to mains water for back-up supply, whereas the remainder (39%) either did not know or were not sure of such a technical aspect of their mandated tanks (Mankad *et al.* 2012c). Second, 44% of participants were not aware of any trickle top-up or automatic switching device installed on their tanks to allow mains water flows if the rainwater is not available, which is a compulsory feature of mandated tanks. The overall absence of knowledge among a high percentage of mandated homeowners indicates that these homeowners might not be maintaining their tanks to the required standard for the tank to be effectively saving mains water.

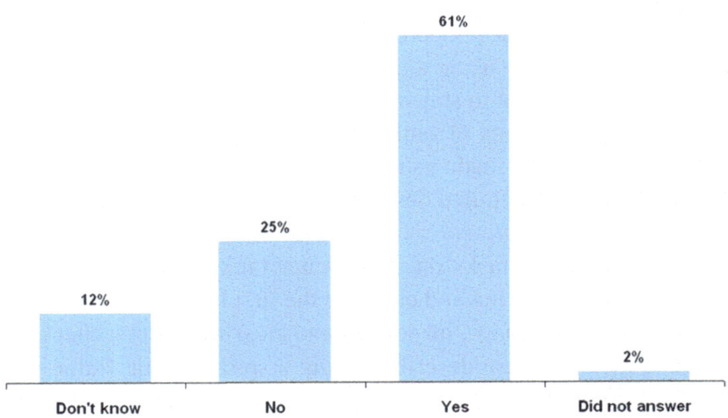

Figure 8.5 Homeowners' knowledge of the existence of mains water top-up in their mandated tanks.

Table 8.2 displays the descriptive statistics for the psychosocial factors (competence, autonomy, relatedness) as well as for frequency of tank maintenance behaviours. The table also shows the results of the multiple regression analysis. The mean score of 2.18 on a 5-point scale for frequency of tank maintenance demonstrates that, on average, homeowners were rarely engaging in tank maintenance behaviours. Mean scores on the perceived competence, autonomy, and relatedness could range from 4 to 20. The means on these variables therefore suggest moderate levels of perceived competence, autonomy, and relatedness in relation to tank maintenance. Multivariate analysis (i.e., multiple regression analysis) was conducted to investigate the extent to which the psychosocial factors (competence, autonomy, relatedness) predicted tank maintenance behaviour. Multiple regression analysis allows an examination of the relative contribution of each of the variables in predicting tank maintenance frequency. The beta weights represent the standardised regression weights for each of the predictor variables included in the regression model and the p values indicate the significance of the contribution of each of the predictor variables. As Table 8.2 shows, homeowners' perceptions of their own competence in maintaining their tank had the largest beta weight and it was therefore the most important predictor of regularity in tank maintenance behaviours (Mankad *et al.* 2012c). The positive relationship shows that the more perceived competence homeowners reported, the more tank maintenance they engaged in. When we consider that most mandated participants had limited knowledge of their tank set-up (as noted above), and that participants reported only moderate perceptions of competence with respect to tank maintenance, it is perhaps not surprising that there is also poor frequency of tank maintenance behaviour (Table 8.2). The regression results suggest, however, that if householders perceived competence increased, that this might also increase the frequency of their tank maintenance behaviour. This finding is further supported by self-determination theory, which posits that people will be more motivated when they feel more competent although we also acknowledge that because our data is correlational we cannot establish a causal link between competence and maintenance behaviour in this study. Autonomy was also found to be a significant factor in determining tank maintenance behaviours, however, as the beta weight shows, it is a weaker predictor of tank maintenance behaviour than competence. In this study, autonomy was conceptualised as the government supporting homeowner independence with respect to water supply (e.g., '*the government gives me freedom to make my own decision about my rainwater tank*'). Results showed that higher perceptions of autonomy support were related to a greater regularity in reported maintenance behaviours. That is, participants who perceived the government as supportive in allowing citizens the freedom to make their own decisions about installing a rainwater tank tended to engage more when maintaining their tank. However, results indicate that this form of autonomy is a significant, but weak positive predictor of maintenance behaviour. Relatedness was only very weakly related to frequency of maintenance behaviour.

Table 8.2 Descriptive and regression results for the influence of basic psychological needs.

Predictor	*M*	*SD*	β	*p*
Competence	13.697	4.827	.467	<0.001
Autonomy	12.92	3.20	.110	0.003
Relatedness	13.18	4.14	−.070	0.038
Dependent variable			*F* (3, 750)	*p*
Frequency of maintenance	2.18	1.48	79.999	<0.001

Note: Scores on competence, autonomy and relatedness could range from 4–20 with higher scores indicating higher perceived competence, autonomy and relatedness; scores on frequency of maintenance could range from 1–5 with higher scores indicating more frequent maintenance.

Results also showed that feelings of personal choice with respect to tank adoption were important to homeowners and encouraged *more regular* tank maintenance behaviours. Participants said they were happy to maintain their rainwater tanks themselves (57%), or with assistance from the local council (25%), rather than paying someone to maintain their tanks (Figure 8.6). An overwhelming majority of participants were not in favour of water utilities being responsible for tank maintenance. Therefore, enabling householders by increasing their levels of competence (e.g., applied education about maintenance) could go some way in increasing rates of tank maintenance.

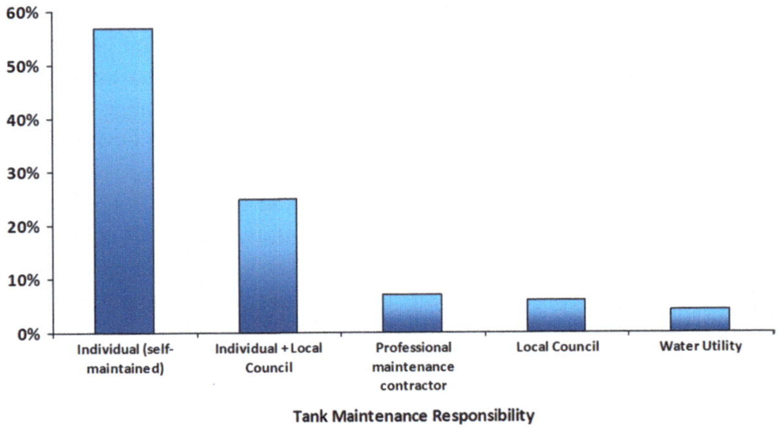

Figure 8.6 Perceptions of tank maintenance responsibility among householders with mandated rainwater tanks (Mankad *et al.* 2012c).

8.6.3 Drivers of rainwater tank maintenance behaviour for retrofitted versus mandated tanks

In this second quantitative case study in SEQ, a survey was conducted to compare tank maintenance behaviour among retrofitted and mandated rainwater tank owners (see Mankad & Greenhill, 2014). Participants were SEQ residents from local government areas (Gold Coast, Logan, Moreton Bay, Redland and Sunshine Coast) who had a rainwater tank installed on their property, either voluntarily (retrofitted) or because of development guidelines (mandated). Participants for the mail out survey were randomly selected from a database of households known to have rainwater tanks. The selection of participants was defined by two criteria:

1. *Households with* retrofitted *rainwater tanks* – these households applied for government rebates to install rainwater tanks at home. They were not required by legislation to have rainwater tanks at home but chose to have them (in conjunction with government rebates).
2. *Households with* mandated *rainwater tanks* – these households applied for new water accounts from 2007 onwards. New legislation was introduced in 2007 mandating that all new residential homes require internally plumbed rainwater tanks. Therefore, any new water accounts opened from 2007 include the mandated rainwater tanks.

The main categories of question included in the survey were (see Mankad & Greenhill, 2014):

- *Self-determined motivation* was measured with the 23-item Motivation toward the Environment Scale (Pelletier *et al.* 1998). Householders responded on a 7-point Likert scale (1 = strongly disagree, 7 = strongly agree) to statements that tapped into each of the type of motivation. Consistent with recommendations, a self-determination index was created.
- *Amotivation* was measured with the Amotivation toward the Environment Scale (Pelletier *et al.* 1999). Participants responded to 16 items that assessed capacity, strategy, effort and helplessness in relation to rainwater tank maintenance on a 7-point Likert scale (1 = strong disagree, 7 = strongly agree). The items were combined into an overall amotivation scale.

- *Tank maintenance behaviour including*:
  - *Engagement in tank maintenance behaviours* was measured by asking householders whether they engaged in 11 maintenance behaviours (e.g., check and clean first flush devices, check tank mosquito-proof screens and flap valves for rips, holes and defects, check inside of tank for accumulated sediment, check pipes for leaks or damage). Response options included: [3] already undertaken the behaviour, [2] planned on undertaking the behaviour, [1] had not undertaken the behaviour.
  - *Frequency of tank maintenance behaviours* was assessed by asking householders how often they undertook each of the 11 maintenance behaviours. Respondents rated their frequency of behaviour on a scale that ranged from: after it rains (considered the most frequent behaviour and coded as 5), through to every 2 to 3 years (least frequent and coded as 1); householders were assigned 0 for those behaviours they had not engaged in at all.
  - *Adequacy of tank maintenance behaviours* was measured through matching responses to the frequency questions described above with the frequency recommendations of the Australian Government guidelines (DHA, 2004; Queensland Health, 2007). Householders' responses to the frequency questions were recoded so that answers matching recommended frequencies were scored 2, and those not congruent were scored 1.

SEQ residents were randomly selected to participate in this study, from a database of households known to have rainwater tanks. The final sample comprised 570 mandated and 1443 retrofitted householders (*see* Mankad & Greenhill, 2014 for more details on the study's methodology). The study showed that mandated and retrofitted households had an average tank volume of approximately 8,500 Litres (8.5 kL) and, interestingly, only 87.6% mandated householders said their tank was plumbed into their home (compared to 20.9% of retrofitted tanks). Given the development code required all mandated tanks to be internally plumbed, this suggests some lack of knowledge or adherence on the part of householders.

8.6.4 Case study results

Comparisons were made using self-determination theory, specifically, the self-determination continuum presented in Section 8.3 and Figure 8.2 (Grolnick & Ryan, 1987; Deci & Ryan, 2000). Basic descriptive information on participants with mandated and retrofitted tanks and their engagement, frequency and adequacy of rainwater tank maintenance showed that retrofitted tank owners consistently reported higher scores (e.g., a mean score of 23.87 for engagement in tank maintenance) than mandated tanks owners (mean score of 20.31), and the t-tests confirmed that the differences were statistically significant (see Table 8.3).

More complex hierarchical multiple regression analyses were conducted to explore the relationship between motivation and tank maintenance behaviours, where type of tank installation (i.e., mandated vs. retrofitted) was included at the first step of the regression. The expected differences between Retrofit and Mandated householders emerged on motivation, and participants who voluntarily retrofitted their rainwater tank were more likely to engage in maintenance than mandated rainwater tank owners. Specific results showed that:

- Amotivation was the strongest predictor of poor engagement in tank maintenance.
- Amotivation was also a strong predictor of infrequent maintenance behaviours and less adequacy in the frequency of tank maintenance.
- Mandated participants reported higher amotivation and thus had a lower self-determined motivation than retrofitted participants.

- Retrofitters reported higher intrinsic, self-determined motivation than mandated participants.
- Retrofitters demonstrated a higher sense of tank ownership compared to mandated participants.
- Higher self-determination scores were aligned with greater engagement in tank maintenance, higher frequency of maintenance behaviours and greater adequacy in the frequency of tank maintenance.

Table 8.3 A comparison of mandated and retrofitted rainwater tank owners on the three tank maintenance dependent variables (Mankad & Greenhill, 2014).

Subscales	TYPE	Mean	SD	*t*	df	Sig. (*p*)	Cohen's *d*[a]
Engagement in tank maintenance[b]	Mandated	20.31	6.95	−10.632	1986	0.000	0.48
	Retrofitted	23.87	6.62				
Tank maintenance frequency[c]	Mandated	1.28	1.16	−12.061	1986	0.000	0.54
	Retrofitted	1.97	1.15				
Adequacy of maintenance frequency[d]	Mandated	1.23	1.39	− 9.475	1986	0.000	0.43
	Retrofitted	1.91	1.48				

Note: for simplified explanations of statistical terms used in social psychology (t, df, p, d), please see Pallant, 2007.
[a] Interpretation guidelines proposed by Cohen (1988): .2 = small effect size, .5 = moderate effect size.
[b] Score range 1–33
[c] Score range 0–11
[d] Score range 0–5

Although the differences in motivation between the two groups was found to have a relatively small effect size (Cohen, 1992), the relationship between motivation and maintenance is consistent across the three measures (see Figure 8.7). More in-depth analysis of participants' motivational drivers showed that people who engaged in greater tank maintenance were those who likely identified themselves as 'water savers' or 'environmentally friendly', where engaging in maintenance behaviour was part of the householder's self-concept and was aligned with their other personal values (e.g., sustainability). These individuals were less amotivated in that they had a purpose to their maintenance behaviour; this was also the case for those who engaged in more *frequent* and *adequate* tank maintenance. Results showed that people who engaged in more frequent tank maintenance were also likely to do so because of an inherent pleasure or satisfaction derived from the activity itself. Finally, individuals who engaged in maintenance behaviours that were aligned with frequency recommendations (i.e., *adequacy*) were found to also be motivated by the value they placed on the maintenance behaviour. That is, they may have viewed their rainwater as a valuable resource for recreational gardening and, therefore, placed more value on tank maintenance as it allowed them to reach their goal of a productive garden.

In summary, consistent with self-determination theory, when residents had not chosen to install their rainwater tanks, they felt more amotivated and less intrinsically motivated to maintain their tanks than residents who had voluntarily installed their tank. Not surprisingly, the mandated tank owners also engaged in less tank maintenance than the retrofitted tank owners. A reason for their greater amotivation was perhaps a lack of understanding of what needed to be done and a lack of willingness to put in the effort because they had not necessarily made the choice to install the tanks.

8.7 CONCLUSION

In this chapter we have outlined key social research questions that relate to understanding public acceptance and adoption of domestic rainwater tanks and the predictors of householders' rainwater tank maintenance behaviour. There has been very little research addressing these topics and the research that does exist has been conducted within the Australian context. We have outlined methodological approaches and case study examples that can help to develop an evidence base to answer these social research questions. In the remainder of this chapter we summarise the lessons that have been learned from the research conducted to date.

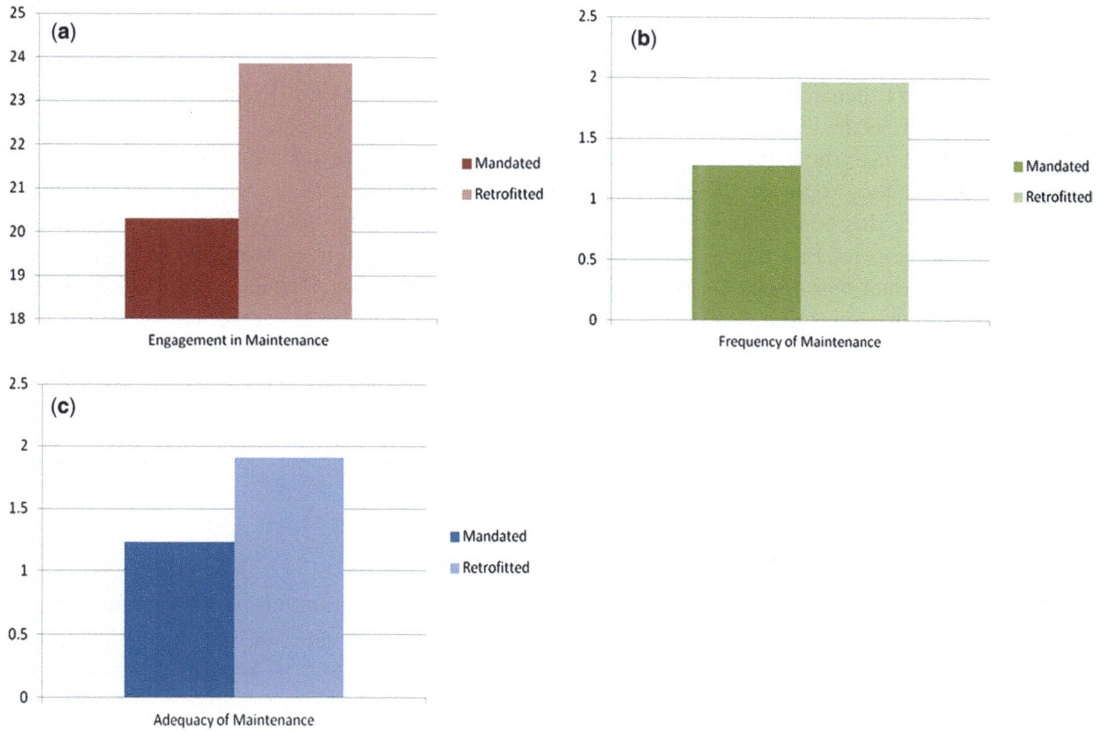

Figure 8.7 A comparison of mandated and retrofitted mean scores on the three maintenance dependent variables: (a) engagement, (b) frequency, and (c) adequacy (Mankad & Greenhill, 2014; Mankad *et al.* 2012a).

8.7.1 How can we influence public acceptance and adoption of rainwater tanks?

The findings to date suggest that the perceived costs, for example, financial costs or the time, effort, knowledge and capability associated with installing rainwater tanks can act as a barrier to installation. While rebate programs can help to overcome the financial burden on households, the bureaucratic process involved in applying for rebates may be a further hurdle that some people cannot get past. This finding suggests the need to streamline rebate programs to ensure the process is as easy as possible. The findings also suggest the importance of providing easy access to step-by-step information that helps householders navigate the installation of a rainwater tank. This has the added benefit of raising householders' self-efficacy in relation to tank installation – a factor that was shown to be an important predictor of householders' intentions to install a tank in the case study research. The space constraints that some householders experience on their property are a real barrier that cannot always be overcome, but the aesthetic considerations that were mentioned by some householders in the qualitative interviews can be addressed through a wide range of tank sizes, shapes and colours and also by installation of underground tanks, but at an increased cost. There is also the likelihood that aesthetic responses to tanks may be influenced by the prevailing norms: the more ubiquitous that tanks are, the more likely they will be to seem 'normal' and an integral part of a home.

Knowledge of the water context in the region and the potential for individual householders to be threatened by water scarcity and drought may be an important part of the decision to install a rainwater tank. It makes sense that people need to feel some sense of vulnerability in order to be motivated to take action to protect themselves against this threat. What needs to go hand in hand with this sense of vulnerability, though, is the belief that rainwater tanks can be an effective way to protect against the threat of water scarcity. Hence, the effectiveness of programs that promote the installation of rainwater tanks may be increased if they focus on the benefits of rainwater tanks for householders and the community more broadly. While this conclusion seems straightforward, research has shown that people may value water for different reasons. Past research suggests that people typically retrofit rainwater tanks when they have an interest in water-based activities where a greater supply of 'personal' water is required, such as for gardening, topping up swimming pools, or washing large vehicles (e.g., boats, caravans) (Gardiner, 2009; 2010; Gardiner *et al.* 2008; Mankad *et al.* 2010). Hence, although it makes sense to focus on the benefits of rainwater tanks for protecting the individual, their lifestyle and their property, it is also important to communicate that in times of extreme drought the chief contribution of rainwater tanks could be to reduce the pressure on drinking water supplies and thereby increase water security in urban areas.

8.7.2 How can we encourage more effective maintenance of domestic rainwater tanks?

The research outlined in the case studies in this chapter provides some important preliminary insights into householders' current rainwater tank maintenance behaviours. It appears that householders have relatively low levels of knowledge about their tanks and are not engaging in appropriate maintenance activities. Householders are not supportive of outsourcing tank maintenance to professionals so programs are needed that help people to build the knowledge and skills that allow them to effectively maintain their tanks. Perceived competence emerged as a significant predictor of tank maintenance behaviour: the more competent householders felt, the more likely they were to maintain their tanks. This finding emphasises the importance of providing rainwater tank owners with information that helps them to the build competence in this area. Of course, there is widespread recognition that information alone does not always lead to behaviour change but accompanying the 'how to' information with 'why to' information can provide people with important rationales for engaging in new behaviours (see Osbaldiston & Schott, 2012 for an evaluation of other effective environmental behaviour change strategies). Agencies seeking to change behaviour sometimes try to achieve this end through highlighting that people in the community are not engaging in particular actions (like rainwater tank maintenance). But research suggests that this may have paradoxical effects whereby people take their cues from others and this may lead them to think: 'if no one else is doing it, how important/effective can it be?' Therefore messages that emphasise that other rainwater tank owners think that rainwater tank maintenance is important may be an appropriate norm to communicate to rainwater tank owners.

Another important consideration for agencies seeking to promote rainwater tank adoption and appropriate ongoing maintenance of these systems are the findings in relation to installation of retrofitted versus mandated tank systems. It is clear that where rainwater tanks have been mandated, householders have less sense of ownership of the system, have less knowledge about it, and engage in less maintenance of their rainwater tanks than householders who have voluntarily chosen to retrofit them. Householders with mandated tanks appear to feel greater apathy and less intrinsic satisfaction for tank maintenance. These findings raise important issues for government agencies seeking to address water security through rainwater tank adoption. On the one hand, legislation that mandates installation of rainwater tanks could result in much broader uptake of these systems than measures that encourage voluntary uptake. On the other hand, the lack of choice that legislation introduces may undermine householders' willingness to maintain their systems. These findings suggest the need for agencies to pay particular attention to households with mandated tanks and to encourage greater engagement with the tank as well as greater knowledge of its functioning. Strategies that help to overcome the resentment that some may feel about having to maintain a system that they didn't choose to install may be helpful (e.g., by communicating the benefits of the system that could offset the costs).

Future research is needed to develop the evidence base for strategies to promote greater rainwater tank adoption and maintenance. The current research has been conducted in Australia and there is clearly a need to examine whether the findings generalise to other national, legislative, cultural, and environmental contexts. Qualitative research that explores householders' experiences of managing their rainwater tanks could also be beneficial as it may provide a deeper understanding of the barriers that people face in relation to these activities. Finally, qualitative (e.g., focus groups) and quantitative (e.g., field experiments) research that examines the efficacy of different informational and engagement strategies to encourage rainwater tank maintenance would also provide useful data for agencies with oversight of domestic rainwater tanks.

8.8 REFERENCES

ABC (Australian Broadcasting Corporation) (2009). Drought over in SE Qld. ABC News Online. http://www.abc.net.au/news/2009-05-20/drought-over-in-se-qld/1688796 (accessed 14 January 2014).

ABS (Australian Bureau of Statistics) (2007). 2006 Census Tables: South East Queensland. http://www.censusdata.abs.gov.au/ABSNavigation/prenav/TopicList?newmethod=Place+of+Usual+Residenceandnewtopic=Educational+Attainmentandcollection=Censusandperiod=2006andareacode=3andgeography=andmethod=andproductlabel=andproducttype=Census+Tablesandtopic=andnavmapdisplayed=trueandjavascript=trueandbreadcrumb=LPTandtopholder=0andleftholder=0andcurrentaction=301andaction=301andtextversion=false (accessed 17 September 2012).

ABS (Australian Bureau of Statistics) (2010). Environmental Issues: Water Use and Conservation. http://www.abs.gov.au/ausstats/abs@.nsf/MediaRealesesByCatalogue/629A13C5A1CFAC3CCA2577DF00155272?OpenDocument (accessed 2 September 2013).

Beal C., Gardner T., Sharma A., Barton R. and Chong M. (2011). A Desktop Analysis of Potable Water Savings from Internally Plumbed Rainwater Tanks in South East Qld, Urban Water Security Research Alliance, Technical Report No. 26, Brisbane, QLD, Australia.

Beal C., Sharma A., Gardner T. and Chong M. (2012). A desktop analysis of potable water savings from internally plumbed rainwater tanks in South East Queensland. *Water Resouce Management*, **26**, 1577–1590.

Biermann S., Sharma A., Chong M. N., Umapathi S. and Cook S. (2012). Assessment of the Physical Characteristics of Individual Household Rainwater Tank Systems in South East Queensland, Urban Water Security Research Alliance, Technical Report No. 66, Brisbane, QLD, Australia.

Brown R., Keath N. and Wong T. (2008). Transitioning to water sensitive cities: historical, current and future transition states. In *11th International Conference on Urban Drainage*, Edinburgh, Scotland, pp. 1–10.

Chong M. N., Mankad A., Gardner T. and Sharma A. (2011). Baseline Characteristics of Mandated Rainwater Tank Users in South East Queensland (Phase 1), Urban Water Security Research Alliance, Technical Report No. 48, Brisbane, QLD, Australia.

Cohen J. (1988). Statistical Power Analysis for the Behavioural Sciences. 2nd edn. Lawrence Erlbaum Associates, NJ, USA.

Cohen J. (1992). A power primer. *Psychological Bulletin*, **112**(1), 155–159.

Coombes P. J. and Kuczera G. (2002). Integrated urban water cycle management: moving towards systems understanding. In: *2nd National Conference on Water Sensitive Urban Design*, Brisbane, pp. 1–8.

Coombes P. J., Kuczera G. and Kalma J. D. (2003). Economic, water quantity and quality results from a house with a rainwater tank in the inner city. *Australian Journal of Water Resources*, **7**(2), 111–120.

Deci E. L. and Ryan R. M. (2000). The "what" and "why" of goal pursuits: human needs and the self-determination of behavior. *Psychological Inquiry*, **11**, 227–268.

DEWS (Department of Energy and Water Supply) (2013). Queensland Water Commission: Water Restrictions and Efficiency Management Plans. Queensland Government. http://www.dews.qld.gov.au/policies-initiatives/water-sector-reform/queensland-water-commission (accessed 20 January 2014).

DHA (Department of Health and Ageing) (2004). Guidance of the use of rainwater tanks. enHealth Council, National Public Health Partnership. Australian Government Department of Health and Ageing. http://enhealth.nphp.gov.au/council/pubs/documents/rainwater_tanks.pdf (accessed 7 December 2010).

DHPW (Department of Housing and Public Works) (2013). Construction: Water Supply Systems. http://www.hpw.qld.gov.au/construction/BuildingPlumbing/Building/WaterSupplySystems/Pages/default.aspx (accessed 30 October 2013).

DPI (Department of Planning and Infrastructure) (2010). Water Savings Target. http://www.dip.qld.gov.au/sustainable-housing/water-savings-tergets.html (accessed 3 December 2010).

Dillman D. A. (2007). Mail and Internet Surveys: The Tailored Design Method, 2nd edn, Wiley, Hoboken, NJ.

Dolnicar S. and Hurlimann A. (2009). Drinking water from alternative water sources: differences in beliefs, social norms and factors of perceived behavioural control across eight Australian locations. *Water Science and Technology*, **60**(6), 1433–44. doi:10.2166/wst.2009.325.

Gardiner A. (2009). Domestic rainwater tanks: usage and maintenance patterns in South East Queensland. *Water Supply*, (February), **36**(1), 73–77.

Gardiner A. (2010). Do rainwater tanks herald a cultural change in household water use? *Australasian Journal of Environmental Management*, **17**(2), 100–111.

Gardiner A., Skoien P. and Gardner T. (2007). Social dimensions of decentralised water supply systems. In: *Water Policy and Coordination Forum 2*. 15th November, Riverglenn, Qld Department of Natural Resources and Water, Brisbane, Australia.

Gardiner A., Skoien P. and Gardner T. (2008). Decentralised water supplies: South East Queensland householders' experience and attitudes. *Water: Journal of the Australian Water Association*, **35**(1), 17–22.

Grolnick W. S. and Ryan R. M. (1987). Autonomy in children's learning: an experimental and individual difference investigation. *Journal of Personality and Social Psychology*, **52**, 890–898.

Hunter L. M., Hatch A. and Johnson A. (2004). Cross-national gender variation in environmental behaviors. *Social Science Quarterly*, **85**(3), 677–694. doi:10.1111/j.0038-4941.2004.00239.x.

Hurlimann A. C. (2007). Is recycled water use risky? An Urban Australian community's perspective. *Environmentalist*, **27**(1), 83–94. doi:10.1007/s10669-007-9019-6.

Hurlimann A. and Dolnicar S. (2010). Acceptance of water alternatives in Australia. *Water Science and Technology*, **61**(8), 2137–42. doi:10.2166/wst.2010.007.

Kollmuss A. and Agyeman J. (2002). Mind the Gap: why do people act environmentally and what are the barriers to pro-environmental behavior? *Environmental Education Research*, **8**(3), 239–260. doi:10.1080/1350462022014540.

Louviere J. and Hensher D. (1982). On the design and analysis of simulated choice or allocation experiments in travel choice modelling. *Transportation Research Record*, **890**, 11–17.

Louviere J. and Woodworth G. (1983). Design and analysis of simulate consumer choice or allocation experiments: an approach based on aggregate data. *Journal of Marketing Research*, **20**, 350–367.

Mankad A. and Gardner M. J. (2014). The role of personal experience in frequency of rainwater tank maintenance and policy implications. *Local Environment: The International Journal of Justice and Sustainability*. doi: 10.1080/13549839.2014.959907.

Mankad A. and Greenhill M. (2014). Motivational indicators predicting the engagement, frequency and adequacy of rainwater tank maintenance. *Water Resources Research*, **50**, 29–38.

Mankad A. and Tapsuwan S. (2011). Review of socio-economic drivers of community acceptance and adoption of decentralised water systems. *Journal of Environmental Management*, **92**(3), 380–391. doi:10.1016/j.jenvman.2010.10.037.

Mankad A. and Tucker D. (2013). Alternative household water systems: perceptions of knowledge and trust among residents of South East Queensland. *Ecopscyhology*, **4**(4), 296–307. doi:10.1089/ECO.2012.0050.

Mankad A., Tucker D., Tapsuwan S. and Greenhill M. P. (2010). Qualitative Exploration of Beliefs, Values and Knowledge Associated with Decentralised Water Supplies in South East Queensland Communities. Urban Water Security Research Alliance, Technical Report No. 25, Brisbane, QLD, Australia.

Mankad A., Tapsuwan S., Greenhill M. P., Tucker D. and Malkin S. (2011). Motivational Indicators of Decentralised Systems Use among Householders in South East Queensland. Urban Water Security Research Alliance, Technical Report No. 44, Brisbane, QLD, Australia.

Mankad A., Tucker D. and Greenhill M. P. (2012a). Mandated versus Retrofitted Tank Owners: Psychological Factors Predicting Maintenance and Management. Urban Water Security Research Alliance, Technical Report No. 51, Brisbane, QLD, Australia.

Mankad A., Chong M., Gardner T. and Sharma A. (2012b). Examining biophysical and socio-demographic factors across mandated tank users in urban Australia: a linking step towards achieving best practices. *Water Resources Management*, **26**, 1983–1998.

Mankad A., Chong M. N., Umapathi S. and Sharma A. (2012c). The Role of Physical and Psychological Factors in Understanding Rainwater Tank Maintenance. Urban Water Security Research Alliance, Technical Report No. 64, Brisbane, QLD, Australia.

Mankad A., Greenhill M., Tucker D. and Tapsuwan S. (2013). Motivational indicators of protective behaviour in response to urban water shortage threat. *Journal of Hydrology*, **491**, 100–107. doi:10.1016/j.jhydrol.2013.04.002.

Marks J., Martin B. and Zadoroznyj M. (2008). How Australians order acceptance of recycled water: national baseline data. *Journal of Sociology*, **44**(1), 83–99. doi:10.1177/1440783307085844.

Marsden Jacob Associates (2007). The Cost-Effectiveness of Rainwater Tanks in Urban Australia. National Water Commission, Canberra.

Marshall M. N. (1996). Sampling for qualitative research. *Family Practice*, **13**(6), 522–5.

Martin I. M., Bender H. and Raish C. (2007). What motivates individuals to protect themselves from risks: the case of wildland fires. *Risk Analysis*, **27**(4), 887–900. doi:10.1111/j.1539-6924.2007.00930.x.

Moy C. (2012). Rainwater tank households: water savers or water users? *Geographical Research*, **50**(2), 204–216. doi:10.1111/j.1745-5871.2011.00720.x.

Muthén L. K. and Muthén B. O. (2012). Mplus User's Guide, 7th edn, Muthén and Muthén, Los Angeles, CA.

Osbaldiston R., and Schott J. P. (2012). Environmental sustainability and behavioral science: meta-analysis of proenvironmental behavior experiments. *Environment and Behavior*, **44**(2), 257–299.

Pallant J. (2007). SPSS Survival Manual, 3rd edn, Open University Press, New York, NY.

Queensland Health (2007). Rainwater tanks: A Guide to Keeping Your Tank Safe. Queensland Government, Queensland Health. http://www.health.qld.gov.au/ph/Documents/ehu/32922.pdf (accessed 7 December 2010).

QSR International. (2008). NVivo. QSR International Pvt Ltd, Melbourne, VIC, Australia.

QWC (Queensland Water Commission) (2008). Water Restrictions. Queensland Government, Brisbane.

QWC (Queensland Water Commission) (2009). South East Queensland Water Strategy – Revised Draft November 2009. The State of Queensland (Queensland Water Commission) November, 2009, Brisbane, QLD, Australia.

RWIMU (Regional Water Information Management Unit) (2009). Rebate Program Uptake Rates. Queensland Water Commission, Brisbane, QLD, Australia.

Rogers R. W. (1975). A protection motivation theory of fear appeals and attitude change. *Journal of Psychology*, **91**, 93–114.

Rogers R. W. (1983). Cognitive and physiological processes in attitude change: a revised theory of protection motivation. In: *Social Psychophysiology*, J. Cacioppo and R. Petty (eds), Guilford Press, New York, pp. 153–176.

Ryan R. M. and Deci E. L. (2000). Self-determination theory and the facilitation of intrinsic motivation, social development, and well-being. *American Psychologist*, **55**, 68–78.

Sandelowski M. (1995). Focus on qualitative methods: sample size in qualitative research. *Research in Nursing and Health*, **18**(2), 179–183.

Sharma A., Tjandraatmadja G., Cook S. and Gardner T. (2013). Decentralised systems: definition and drivers in the current context. *Water Science and Technology*, **67**, 2091–2101.

Shove E. (2003). Converging conventions of comfort, cleanliness and convenience. *Journal of Consumer Policy*, **26**, 395–418.

Silverman D. (2010). Doing Qualitative Research: A Practical Handbook. Sage, London.

Tam V. W. Y., Tam L. and Zeng S. X. (2010). Cost effectiveness and tradeoff on the use of rainwater tank: an empirical study in Australian residential decision-making. *Resources, Conservation and Recycling*, **54**(3), 178–186. doi:10.1016/j.resconrec.2009.07.014.

Tapsuwan S., Burton M., Mankad A., Tucker D. and Greenhill M. (2014). Adapting to less water: household willingness to pay for decentralised water systems in urban Australia. *Water Resources Management*, **28**(4), 1111–1125.

Tjandraatmadja G., Cook S., Ho A., Sharma A. and Gardner T. (2009). Drivers for Decentralised Systems in South East Queensland. Urban Water Security Research Alliance, Technical Report No. 13, Brisbane, QLD, Australia.

WWAP (World Water Assessment Program) (2012). The United Nations World Water Development Report 4: Managing Water Under Uncertainty and Risk. UNESCO, Paris.

Zelezny L. C., Chua P.-P. and Aldrich C. (2000). Elaborating on gender differences in environmentalism. *Journal of Social Issues*, **56**(3), 443–457.

Chapter 9

Chemical quality of rainwater in rain tanks

Mirela I. Magyar and Anthony R. Ladson

ABSTRACT

The number of rain tanks in urban areas is increasing rapidly and although most tank water is for non-potable use there are a large number of households where tanks supply water for drinking. Comparing tank water with drinking water guidelines shows that high lead concentrations and low pH are common issues. We identified 32 studies where elevated metals concentrations in tank water were of concern with lead being an issue in 31 of these studies. A meta-analysis suggests that, in urban areas in Australia, about 22% of tanks can be expected to provide water with lead concentrations that do not meet drinking water standards. The risk of lead contamination needs to be taken seriously and there are opportunities to improve water quality through treatment, by better maintenance and improved tank design.

Keywords: water quality; rain tanks; heavy metals; lead; water supply.

9.1 INTRODUCTION

This chapter reviews the chemical quality of water captured and stored in rain tanks and supplied to various end uses. Any discussion of water quality involves a comparison; we need to compare the quality of water supplied by a rain tank with that required by its intended use. Water that is acceptable for drinking may not be of sufficient quality for say, electronic fabrication processes, whilst water that doesn't meet drinking water standards may be fine for garden irrigation.

Water captured in rain tanks will be appropriate for many uses. Problems can arise when the quality of water supplied by a tank is lower than required for its intended use. There is now sufficient information to indicate when problems with the quality of tank water are likely and where intervention is appropriate. We will highlight key issues in this chapter.

Generally, our focus is on domestic tanks in urban areas. This is justified for two reasons. First, pollutants in dust and in the air in urban areas may compromise water quality. Second, use of rainwater tanks in many urban areas is increasing rapidly. For example, in Brisbane, for dwellings less than 1 year old, the proportion with a rain tank increased from 27% in 2007 to 57% in 2010 (White, 2014). The rapid expansion of rainwater tanks in cities is new and there are some important water quality issues which, we will show, are not being taken seriously enough.

9.1.1 The uses of tank water

What can tank water be used for in an urban domestic setting? Wong (2006) discusses the concept of fit-for-purpose water supply for domestic water uses. He suggests that tank water is preferred or compatible with all domestic uses although drinking mains water is recommended if it is available (Table 9.1). The potential compatibility of rainwater with many domestic demands is consistent with the use of rainwater tanks to augment mains water supply (see Chapters 2 and 3 of this book).

Table 9.1 Suggested domestic uses for tank water from Wong, 2006.

	Preferred use	Compatible use	Notes
Garden		✓	Lower quality sources are preferred
Kitchen – cold water		✓	Treatment may be required to improve microbial quality
Kitchen- hot water	✓		
Laundry	✓		
Toilet		✓	
Bathroom – cold water		✓	Treatment may be required to improve microbial quality
Bathroom – hot water	✓		

A significant difference between water supplied from a tank compared to that from the mains is the lack of centralised control. In developed countries there is substantial background effort to keep mains water to a consistent high standard. Infrastructure is maintained, water is treated, quality is monitored and people are warned if the standard drops. There is an organised response to extreme events including boil-water advisories if there is a risk to health (Fitzgerald *et al.* 2014).

The use of rain tanks means that ownership, operation, and maintenance of part of the water supply infrastructure is transferred from urban water authorities to property owners and occupants. Therefore, tank maintenance, and more importantly, the quality of water supplied from the tanks is the householders' responsibility. The lack of monitoring and control and the limited expertise of householders suggest we should be conservative when selecting uses for tank water. It is likely this happens in practice. We know that in urban areas the majority of rain tanks supply water for non-potable uses, such as for garden watering because, in capital cities, 67% of tanks are not plumbed into dwellings (ABS, 2013). Guidelines recommend against the consumption of tank water if there is an alternative potable supply available (Coombes & Mitchell, 2006; DHS, 2007).

There would be little interest in the quality of tank water if it all just ended up on gardens. Significantly in a large number of houses (but small overall percentage), tank water is supplied for uses involving human contact and occasional ingestion. Rain water supplied to hot water systems has the potential to be used for cooking (Chapman *et al.* 2006) and there are also many cases in urban areas where people drink rain tank water even when mains water supply is available. Table 9.2 shows that around 127,000 households in Australian capital cities use tank water for drinking.

Table 9.2 Number of households in capital cities where rain tanks are the main source of water for drinking (ABS, 2010).

Capital City[1] (State)	Number of households using rainwater tanks for drinking	Households (%)
Sydney (NSW)	23,600*	1.4*
Melbourne (VIC)	17,300*	1.2*
Brisbane (QLD)	29,500*	4.0*
Adelaide (SA)	33,400	7.0
Perth (WA)	17,900*	2.7*
Hobart (TAS)	5300*	6.3*
Total	**127,100**	**2.5**

[1]Estimates for Darwin and Canberra were not available; *estimate has a relative standard error of 25% to 50% and should be used with caution (ABS, 2010)

9.1.2 Water quality guidelines

There are no specific recommendations for water quality from rainwater tanks, but as a substantial number of tanks are used to provide water for drinking, it is appropriate to compare tank water quality with drinking water guidelines. In this chapter we mainly use the Australian Drinking Water Guidelines (ADWG) as a reference (NHMRC & NRMMC, 2014). There are also water quality recommendations for recreational water use and the use of water for agriculture (ANZECC & ARMCANZ, 2000). A selection of guideline values are provided in Table 9.3, mainly focussing on metals which are the chemical water quality constituents of most concern in tank water (as we discuss below). Lead is one chemical water quality parameter of particular interest because of the risk to health from lead exposure. Lead is also a common contaminant of tank water, as we will show in this chapter.

Table 9.3 Recommended concentration values (units of μg/L) for metals in water for different human exposures and end uses.

Guidelines	Cd	Cr[1]	Cu	Fe	Mn	Ni	Pb	Zn
Australian Drinking Water Guideline (ADWG) **health** concentration upper limits (NHMRC & NRMMC, 2014)	2	50	2000	–	500	20	10	–
ADWG **aesthetic** concentrations upper limits (NHMRC & NRMMC, 2014)	–	–	1000	300	100	–	–	3000
Agricultural **irrigation** long-term trigger value (ANZECC & ARMCANZ, 2000)	10	100	200	200	200	200	2000	2000
Recreation guideline maximum value for primary contact (ANZECC & ARMCANZ, 2000)	5	50	–	–	–	100	50	

[1]Guidelines are for hexavalent chromium. If total chromium exceeds 0.05 mg/L in drinking water a separate analysis for hexavalent chromium is required.

The potential adverse health effects of lead contamination of drinking water are both serious, and well known. A maximum lead concentration of 10 μg/L is specified in the ADWG and in guidelines published by the World Health Organisation (WHO, 2008). Exposure to high concentrations of lead in drinking water has been associated with spontaneous abortion, still birth and high rates of infant mortality (Edwards, 2014). Lead exposure also has implications for children's intellectual functioning (Lanphear *et al.* 2005) and premature adult mortality (Nawrot & Staessen, 2006). Lead can also cause harmful effects on the gastrointestinal tract, joints, reproductive system, kidneys and haemoglobin synthesis (Goyer, 1993) and may increase risk of hip fractures (Dahl *et al.* 2013). Lead is a cumulative toxin with long term chronic effects from the accumulation of lead in the human skeleton, as well as short term acute effects from episodic high doses (Shih *et al.* 2007). Drinking lead contaminated tank water has been linked to elevated blood lead levels (Body, 1986; Maynard *et al.* 2003). Substantial adverse health effects have been reported for lead concentrations in drinking water of 40 to 80 μg/L (Edwards, 2014).

Another important water quality guideline is pH. The ADWG state that the pH should be between 6.5 and 8.5 to minimise the risk of corrosion and encrustation of pipes and fittings. The degree of saturation of calcium carbonate in water, the Langelier Index, can also be a useful indicator of the risks of corrosion or the formation of scale. Low pH can increase the concentration of metals dissolved in tank water.

9.2 WATER QUALITY FROM RAINFALL TO ROOF TO TANK

The quality of water provided by a rain tank depends on the quality of water that enters the tank, the processes within the tank, and any contamination as water passes to a supply point. In this section we review rainwater quality, the influence of roof materials, and ways to improve quality of tank inflows.

9.2.1 Rainwater and atmospheric influences on inflow quality

The quality of water entering a tank is influenced by the quality of rain and the runoff from the roof. In the literature, distinctions are made between the following processes.

- Wet deposition is the material carried in by rain. Its composition depends on anthropogenic, marine or natural sources.
- Dry deposition refers to particulate contaminants in the air that settle on the roof surface between rain events. Dry deposition is influenced by the sources of pollution and their proximity, the number of days between rain events, and the aspect of the roof (Huston *et al.* 2009). Generally, the longer the time between rain events, the greater the deposition and the higher the contaminant loads that will be washed into tanks.
- Bulk deposition is the term used when both dry and wet deposition concentrations are combined.

Another important process is the chemical interaction of rainwater with roof materials which may enhance the concentration of constituents such as metals. We discuss this in the next section.

Before providing a statistical overview of the roof runoff quality, we first mention some specific results from two studies which aimed to relate contamination of stored rainwater to atmospheric sources. In a study in Sydney, Australia (average annual rainfall 1200 mm), wet deposition was found to have high concentrations of some metals (Kus *et al.* 2010a; Kus *et al.* 2010b). In particular, lead exceeded 10 μg/L which is the limiting value for the Australia Drinking Water Guideline (Table 9.3). The source of these high lead values, as suggested by earlier work in Sydney (Birch & Taylor, 1999; Birch & Scollen, 2003) is from the dust near roads with heavy traffic. Copper and Manganese were also detected in rainfall but at concentrations below drinking water guidelines.

In Brisbane, Australia (average annual rainfall 1000 mm/year), bulk deposition (dry + wet) was investigated on a monthly basis between April 2007 and March 2008 at 16 sites (Huston *et al.* 2009). The samples were analysed for 30 metals and 155 organic compounds. The main water quality concerns related to the concentrations of lead and cadmium which exceeded the ADWG recommended values in 10.3% and 1.7% of samples respectively.

Daily flux values for metals in Brisbane were also measured by Huston *et al.* (2009); a selection of average flux values is shown in Table 9.4. These are comparable to fluxes in other cities (Venice, Paris, Los Angeles) but lower than Taiwan and Tokyo. For the Brisbane data there were substantial differences between monitoring sites depending on land use. Fluxes (μg/m^2/d) increased from outer suburban to inner suburban to city/heavy traffic/light industrial. The main water quality concern in the Brisbane data relates to lead.

Table 9.4 Concentrations of metals in wet and bulk deposition in Sydney (Kus *et al.* 2010b) and Brisbane (Huston *et al.* 2009).

Metal	Sydney (Wet deposition) (μg/L)	Brisbane (Bulk deposition) Mean annual concentration (μg/L)	Brisbane (Average daily fluxes) (μg/m^2/d)
Al	108	–	–
Cd	–	0.1	0.32
Cr	–	0.7	1.8
Cu	6	2	5.5
Fe	72	–	–
Mn	17	–	–
Ni	–	0.4	–
Pb	17	2.1	5.9
Zn	77	16.4	45.5

We can put these lead fluxes into perspective as follows. Consider a tank of rainwater which is contaminated with lead at a concentration that just exceeds the ADWG value of 10 μg/L. For a 5 kL tank, a typical size for an urban allotment, this means the tank water contains 0.05 g of lead. The average daily lead flux in Brisbane is 5.9 μg/m^2/d so 100 m^2 of connected roof area could supply about 0.2 g of lead per year to the tank. This is 4 times that required to contaminate the water.

9.2.2 The effect of roof material on the quality of tank inflows

Along with atmospheric wet and dry deposition, metals can be supplied from roof materials which are commonly a source of zinc, lead and copper. Zinc is a widely used roofing metal, with galvanised steel roofs common in Australia. Over time, particularly in coastal environments, solid zinc is converted to soluble zinc chloride and will be washed off the roof and into a rain tank. Chang *et al.* (2004) also found high zinc concentrations from timber shingle roofs which they attributed to chemical treatments for fungi protection.

Lead flashing has also been widely used for weather proofing joints on roofs, although this is less common in new developments. Guidelines such as AS/NZS 3500.3 (Standards Australia, 2003), Building Code of Australia (ABCB, 2012) and manufacturers' web sites advise that lead flashing is not recommended, and should be avoided especially on any roof that is part of a potable water catchment area (ABCB, 2012). Nonetheless there is lead flashing on many existing roofs in Australia.

Where researchers have sampled water on both roofs and in gutters, they have found that roof materials contribute metals loads to tanks (Yaziz *et al.* 1989; Forster, 1996; Zobrist *et al.* 2000; Gromaire *et al.* 2001; Polkowska *et al.* 2002; Chang *et al.* 2004; Huston *et al.* 2012). Roof type has a significant influence on runoff quality. There are the obvious links between galvanised roofs and zinc, lead flashing and lead concentrations, and copper flashing and copper concentrations. There are also less obvious associations. Timber roofs may contribute zinc, asphalt roofs are a source of lead, and gravel roofs can retain and release metals (Zobrist *et al.* 2000; Van Metre & Mahler, 2003; Chang *et al.* 2004).

9.2.3 A statistical overview of roof runoff quality

The runoff quality to any individual tank will be influenced by a larger number of factors. Examining individual roofs and roof runoff provides useful information on sources of contaminants and processes of contamination. However it is difficult to generalise these results to the large number of rain tanks that are used in urban areas. A statistical description of roof runoff quality can provide a broader view and flag likely water quality issues.

A number of reviews of roof runoff quality area available (Meera & Ahammed, 2006; Lye, 2009; Abbasi & Abbasi, 2011), but here we focus on the work by Duncan (1999, 2006) which summarised the results from a large number of studies providing information on mean concentration values and variability for a range of water quality constituents. Where literature data was sufficient, Duncan (1999, 2006) also related roof runoff quality to broad classes of land use for example, urban and rural. Information assembled by Duncan is the basis of pollutant generation in the widely used stormwater management model MUSIC (eWater, 2009).

Roof runoff concentrations (mean and ±1 standard deviation) for lead and cadmium are shown in Figure 9.1 along with allowable concentrations from the Australian Drinking Water Guidelines. Pollutant concentrations are considered on a log scale because they approximate a log normal distribution (Duncan, 1999, 2006).

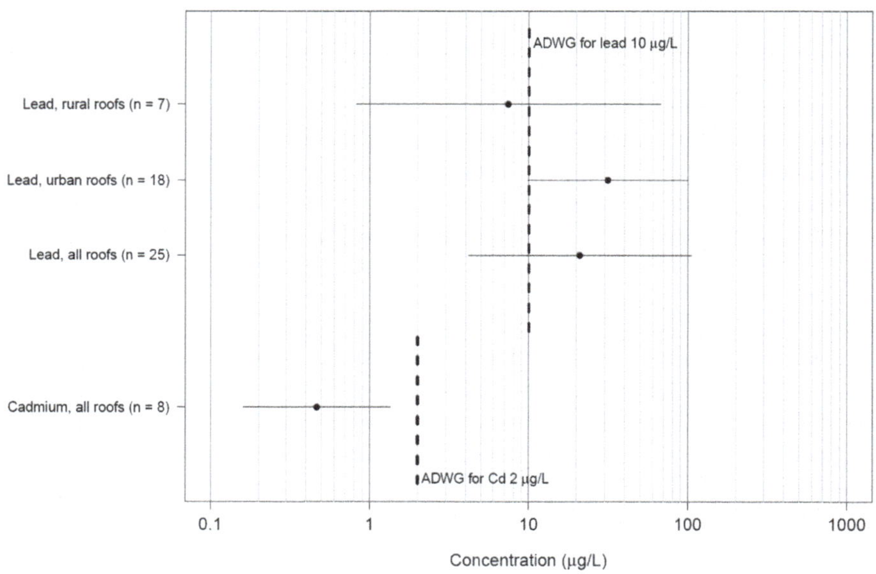

Figure 9.1 Concentration of lead and cadmium in roof runoff compared to the Australian drinking water guidelines. Data show mean concentrations ±1 standard deviation. The number of studies is shown in brackets (Duncan, 1999, 2006).

These data suggest that the lead concentration of roof runoff is likely to be high, with the potential to contribute to poor water quality in a rain tank. The mean lead concentration for runoff from urban roofs, is greater than the drinking guideline values. Even runoff from rural roofs can supply water with high lead concentrations. In contrast, the mean concentration of cadmium in roof runoff is substantially less than the recommended drinking water concentration. Hence it is much less likely for tank inflows to exceed guidelines for cadmium than for lead.

The pH of roof runoff has also been measured in a number of studies (Figure 9.2). Roof runoff is usually acidic and generally has a pH well below drinking water guidelines although there are exceptions. For example naturally occurring limestone dust may moderate pH (Al-Khashman, 2009). There can also be temporary periods of high pH following extreme weather (Vialle *et al.* 2011)

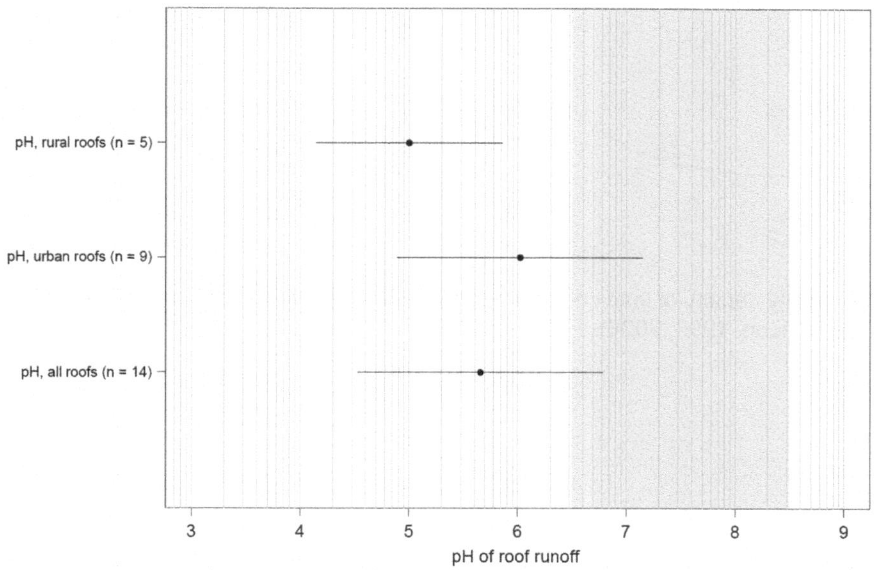

Figure 9.2 pH of roof runoff. Data show mean values ±1 standard deviation. The shaded area corresponds to drinking water guidelines for pH. The number of studies is shown in brackets (Duncan, 1999, 2006).

The data from Duncan (1999, 2006) can also be given a probabilistic interpretation. Probability densities for lead concentrations for three roof types are shown in Figure 9.3. These are based on the mean and standard deviation of lead concentrations in rainwater as documented by Duncan (1999, 2006). The substantial likelihood of roof runoff exceeding drinking water guideline is shown with a large proportion of the probability mass to the right of the ADWG value. It is also possible to quantify the probability that roof runoff concentration will exceed guidelines using standard probability theory (Walpole & Myers, 1978). For example, based on data provided by Duncan (1999, 2006) the probability that the lead concentration in runoff from an urban roof will exceed 10 μg/L is 84%.

Similar probability calculations were undertaken for a range of other water quality constituents using data compiled by Duncan (1999, 2006). Those chemical constituents where roof runoff is least likely to meet drinking water guidelines can be flagged as requiring further consideration (Figure 9.4). These results show that roof runoff in both rural and urban areas is likely to fail to meet drinking water guidelines because of low values of pH and high concentrations of lead.

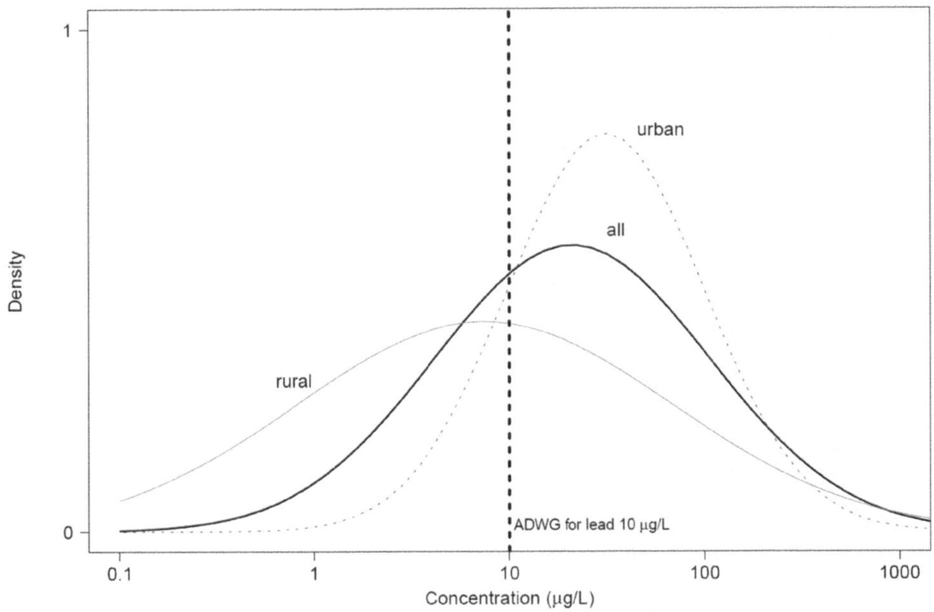

Figure 9.3 Probability density of lead concentration in roof runoff from roofs in urban and rural areas (and combined data) (Duncan, 1999, 2006).

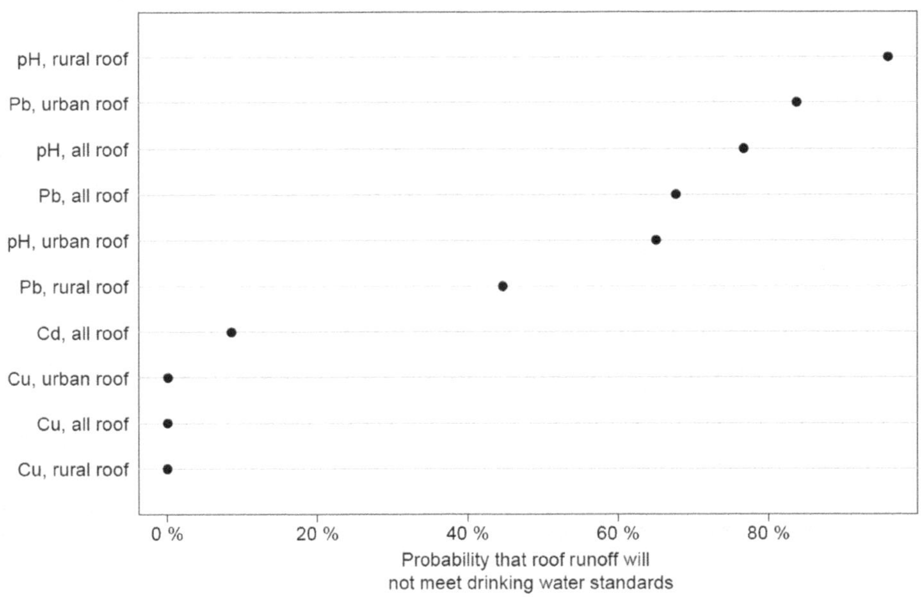

Figure 9.4 Probability that roof runoff will not meet drinking water standards for a range of water quality constituents in runoff from roofs in urban and rural areas (and combined data). Lead and pH stand out as the major issues (analysis based on Duncan, 1999, 2006).

9.2.4 Improving tank inflow quality

We have suggested three sources of contamination for inflows to tanks, pollutants in rain, dry deposition, and interaction with roof materials. From the point of view of a householder, there is little that can be done about the quality of rain or dry deposition. Several studies suggest that lead flashing on roofs can be a source of lead in tank water, so painting or removal of lead flashing may improve water quality (Huston *et al.* 2012; Magyar *et al.* 2014). The Building Code of Australia (Volume 2, Clause 3.5.1) (ABCB, 2012) states that lead flashing should not be used on any roof that is part of a potable water catchment.

There are several strategies that can be undertaken to reduce wash off of materials that deposit on the roof. Gutters can be cleaned to reduce the amount of material entering the tank. Installation of leaf guards may also be advantageous as lead dissolution is increased in the presence of leaf litter (Body, 1986). The other common strategy is to install a first flush device to catch and discard the first proportion of roof runoff which usually has the highest concentration of contaminants (Bach *et al.* 2010; Kus *et al.* 2010a).

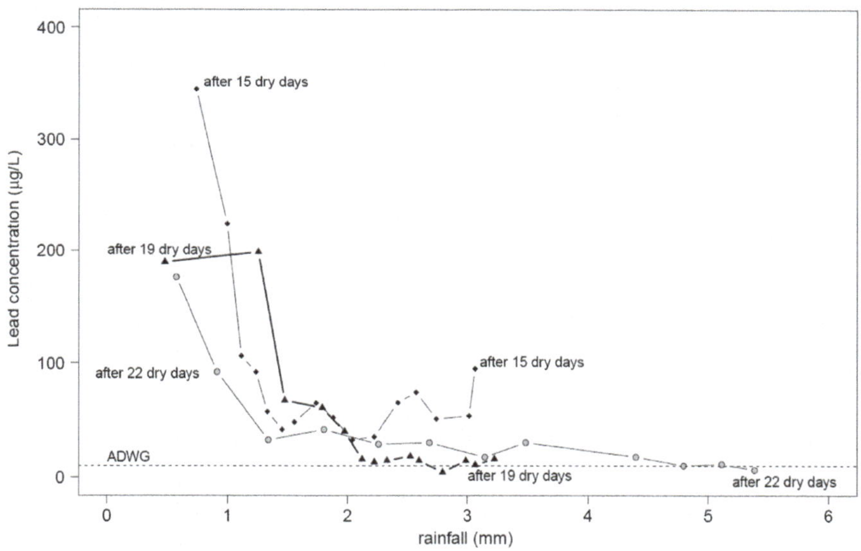

Figure 9.5 Lead concentrations in roof runoff for three events following dry periods. The high initial concentrations were attributed to a first flush effect (Kus *et al.* 2010a). There was no lead flashing on this roof.

First flush devices aim to reduce the loads from material washed off in the early part of a rain storm. They will be most effective in situations where dry deposition contributes a relatively high proportion of contaminants. This could include circumstances where there are long dry periods between intense rain events, or where the air-shed above the roof contributes contaminants that settle on a roof. Monitoring by Kus *et al.* (2010a) in Sydney showed a first flush effect for suspended solids, turbidity, lead and other constituents. The results for lead are shown in Figure 9.5 for three events following extended dry periods. For the first millimetre or two of rain, the concentration of lead in roof runoff was 10 to 30 times the concentration in rainfall. After 5 mm, the concentration had decreased to a similar value to that in rain which was about the drinking water guideline value of 10 μg/L. Most domestic first flush devices only discard about 20 L because they are limited by the storage in the first flush device; commonly a 2–3 m length of 100 mm pipe. For a connected roof area of 100 m^2, this is equivalent to only 0.2 mm. Gardner *et al.* (2004) showed that discarding the first 1 mm of roof runoff had very little effect on the event mean concentration of lead and other contaminants. Where a first flush device discards a significant amount of runoff they will substantially reduce the proportion of rainfall harvested by a rainwater tank (Gardner *et al.* 2004).

The proportion of houses that use first flush devices was measured in a number of studies and is highly variable, ranging from 3% to 40% with highly inconsistent results even in the same city (Table 9.5)

Table 9.5 Number of tanks with first flush devices as reported by various studies.

City	No. Tanks with a First Flush device (%)	Reference
Adelaide	7 (36.8%)	Rodrigo *et al.* (2012)
Adelaide	(8%)	Heyworth *et al.* (2006)[1]
Adelaide	100 (30.8%)	Rodrigo *et al.* (2010)
Brisbane	12 (39%)	Huston *et al.* (2012)
Melbourne	2 (3%)	Magyar (2010)
Northeast Victoria	9 (18%)	Spinks *et al.* (2006)
Sydney	40 (40%)	Ferguson (2011)
Tamborine Mountain	14 (5%)	Mark Rigby and Assoc. (2002)

[1]The number of tanks in the study by Heyworth et al. (2006) is not recorded but is likely to be at least 500.

9.2.5 The rain tank as a water treatment device

The above discussion suggests it will be common for water with low pH and high lead concentration to enter a rain tank. Ideally, processes in the tank, such as settling of sediment, should mean that the water supplied to users is cleaner than the inflows into the tank. However commercial tank design is not focussed on achieving the delivery of the cleanest water.

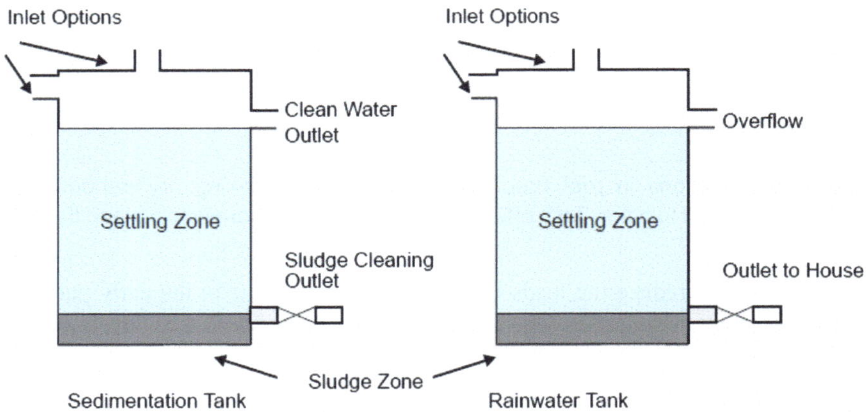

Figure 9.6 Comparison of a sedimentation tank and a rainwater tank. Adapted from Kiely (1997) and Magyar (2010).

Generally, water from a tank is supplied through an outlet situated close to the bottom of the tank, usually between 50 mm to 200 mm above the base (Figure 9.6). The overflow of the tank is situated at the top of the tank, just below the inlet and is connected to the urban stormwater system Magyar (2010). This arrangement of inlet/outlet/overflow maximises the volume of water stored in the tank. Unfortunately it also means that solid pollutants that settle at the bottom of the tank are potentially re-entrained and hence are exported in the supplied water, unless they are removed. The highest quality water in a rain tank is in the layer close to the tank surface and it will be lost via the overflow when the tank spills.

The performance of a rain tank can be contrasted with that of a sedimentation tank used for water treatment. In a sedimentation tank, high quality water is delivered via overflow of surface water while contaminants settle on the base and are periodically removed. There have been suggestions that delivered tank water could be improved in quality by, for example, using a floating inlet. Whilst these are commercially available, they are not widely used. There are also a range of other potential design improvements such as the position of the entry compared to the outlet and the shape of the tank base (Magyar *et al.* 2011a).

Another factor that militates against the delivery of high quality water from tanks is the pH of the tank water. Water that enters a tank will likely have low pH, and except for concrete tanks, there are no processes that will result in the pH increasing to the desired range. pH below 6.5 can lead to metals being present in dissolved form in tank water and may contribute to corrosion of plumbing.

9.3 MEASUREMENTS OF RAIN TANK WATER QUALITY

9.3.1 Introduction

The discussion so far has highlighted possible water quality issues for rain tanks. It is clear that inflows are likely to have lead concentrations that exceed drinking water guidelines and low pH. Sediments will settle in the tank but these will be vulnerable to disturbance and re-entrainment in the extracted water. Sampling of tank water can test the extent of these problems. In this section we systematically review the literature on tank water quality and key processes within tanks.

9.3.2 Studies of tank water quality: A summary

There have been many studies of tank water quality including: assessment of water delivered from a tank tap; from hot and cold water taps inside a house which are supplied from a tank; and sampling of the bulk water within the tank. In terms of chemical water quality, most studies test for metals. Far fewer studies consider organic contaminants although some results are available (Bannister *et al.* 1997; Spinks *et al.* 2006; Huston, 2009; Huston *et al.* 2009).

Lead has been found to exceed drinking water standards in many tanks. Our literature review identified 32 studies where chemical water quality issues associated with tank water were mentioned. In all but one, lead was key concern (Table 9.6). Across the range of studies, cadmium concentrations exceeding drinking water standards was the second most common chemical contaminant, followed by a much smaller number of papers that reported issues with metals such as arsenic and nickel. High zinc concentrations were also common, but as there is no health related guideline for zinc concentration we will not consider this further. In some studies metal contamination can be linked to a specific source. For example in Brisbane, most occurrences of high cadmium concentrations in tank water were linked to emissions from a glass factory, which is the largest emitter of cadmium to the atmosphere in Brisbane (56 kg during 2003–2004) (Chapman *et al.* 2008). In contrast, lead contamination of tanks in Melbourne could not be linked to an individual source (O'Conner *et al.* 2009).

Where water has been tested for organic pollutants, they have not been found to be a health issue. Huston (2009) tested water in 26 tanks in Brisbane for 122 pesticides, 19 polycyclic aromatic hydrocarbons and 16 phenolic compounds, and found that there were no concentrations that would present a chronic or acute health hazard. Similarly Spinks *et al.* (2006) undertook a study of water quality in 49 tanks following a large bushfire in north east Victoria, Australia. This study was motived by concerns about contamination of tank water by polycyclic aromatic hydrocarbons (PAHs) from incomplete combustion of organic matter, and arsenic from burnt, copper chrome arsenate (CCA) treated wood. However these compounds were found not to be a significant issue. Rather, the main concerns related to cadmium which was found at levels above ADWG in tanks at two properties. In both cases, the roofs were galvanised iron over 50 years old and Spinks *et al.* (2006) suggest the corrosion of the roof was the likely source.

Table 9.6 Studies of tank water quality where metals have been identified as a water quality issue.

No	Study	City/region	No. tanks in study	Metals exceeding health drinking water standards in some tanks[1]
1	Fuller (1981)	South Australia	80	Pb
2	MWAWA (1983)	Perth	92	Pb
3	Sharpe and Young (1984), Young and Sharpe (1984)	Ohio	30–40	Pb, Cd
4	Gumbs and Dierbert (1984)	Antilles	46	Pb
5	Body (1986)[2]	Port Pirie	31	Pb
6	Haebler and Waller (1987)	Caribbean	11	Pb
7	Olem and Berthouex (1989)	Kentucky-Tennessee	25	Pb
8	Duncan and Wight (1991)	Melbourne	25	Pb, Cd
9	Bannister *et al.* (1997)	Victoria	21	Pb
10	Simmons *et al.* (2001)	Auckland	125	Pb
11	Zhu *et al.* (2004)	Northern China	12	Pb
12	Spinks *et al.* (2005)	Urban areas, East Coast Australia	6	Pb
13	Handia (2005)	Zambia	5	Pb
14	Spinks *et al.* (2006)	NE Victoria, Australia	49	Cd
15	Hart and White (2006)	Alaska	50	Pb, Cu
16	Martin *et al.* (2007)	Newcastle	1	Pb
17	Morrow *et al.* (2007)[3]	NSW, Qld, Vic, SA	40	Pb, Ar, Ni
18	Chapman *et al.* (2006, 2008)	Brisbane	30	Pb, Cd
19	Chapman *et al.* (2008)[4]	Australia	69	Pb
20	Magyar *et al.* (2008)	Melbourne	64	Pb, Cd
21	Peters *et al.* (2008)	Bermuda	112	Pb, Se
22	Huston *et al.* (2009)	Brisbane	26	Pb
23	Radaideh *et al.* (2009)	Jordan	92	Pb, Cr
24	Morrow *et al.* (2010)	East Coast, Australia	10	Pb
24	Kus *et al.* (2010b)	Sydney	11	Pb
25	Daoud *et al.* (2011)	Palestine	44	Pb

Table 9.6 Studies of tank water quality where metals have been identified as a water quality issue (Continued).

No	Study	City/region	No. tanks in study	Metals exceeding health drinking water standards in some tanks[1]
26	Huston *et al.* (2012)	Brisbane	31	Pb
27	Rodrigo *et al.* (2012)	Adelaide	19	Pb
28	Morrow and Dunstan (2012)	Brisbane, Sydney, Melbourne	30	Pb
29	Stump *et al.* (2012)	Texas	36	Pb
30	van der Sterren *et al.* (2013)	Sydney	5	Pb
31	Magyar *et al.* (2014)	Melbourne	6	Pb
32	Malassa *et al.* (2014)	Hebron, Palestine	44	Pb, Cr, Mn, Ni, Ag

[1]Although many studies report high concentrations of zinc, there is no health-based guideline value for zinc in the Australian Drinking Water Guidelines so zinc concentration is not considered in this chapter. The aesthetic (taste) guideline value is a maximum of 3000 μg/L.
[2]The study by Body (1986) was specifically focussed on tanks with high lead concentrations in Port Pirie, SA, a town with a lead smelter.
[3]High arsenic concentrations were recorded in a single sample.
[4]Combined results from Adelaide (6 tanks), Brisbane (6), Broken Hill (6), Canberra (5), Sydney (6), Wollongong (6).

Although we found 31 studies that identified lead concentration as a concern, it should also be pointed out that not all studies of tank water found concentrations of lead that exceeded drinking water guidelines. In particular, work undertaken by Coombes in Newcastle, Australia did not find high lead concentrations in tank water (Coombes *et al.* 2000a, 2000b, 2002, 2005, 2006). There are also studies in Tamborine Mountain Queensland (Mark Rigby & Associates, 2002) and Kefalonia Island, Greece (Sazakli *et al.* 2007) where lead levels were not elevated. Huston (2009) speculates that there may be particular characteristics of these studies that explain these results. For example, the study by Sazakli *et al.* (2007) sampled large (>300 kL) concrete tanks and samples were drawn from the middle of the water column. Concrete tanks generally have high pH water so are less likely to have metals in dissolved form. The pH of rainfall in the study by Sazakli *et al.* (2007) is also unusually high which is possibly related to $CaCO_3$ dust from the limestone geology of the island (*sensu* Conlan & Longhurst, 1993; Al-Kashman, 2009). In addition, there is limited industrial land use near the tanks. The studies by Coombes *et al.* (2000a, 2000b, 2002) included tanks that had mains water top up so any contaminants in tank water would have been diluted. The Coombes *et al.* (2002) study was of a concrete tank. At Figtree place, central Newcastle, Coombes *et al.* (2000a, 2000b, 2005) found that rainwater exceeded drinking water standards for lead but tank water did not; suggesting it was processes within the tank that resulted in cleaner water for example, settling of particles (Spinks *et al.* 2003b). Sludge in these tanks was found to contain high concentrations of lead (Spinks *et al.* 2005). In the Tamborine Mountain study, 79% of tanks were concrete and 28% contained a mixture of rainwater and other water – imported and bore water (Mark Rigby & Associates, 2002).

9.3.3 Meta-analysis: Proportion of tanks with high lead concentrations

A systematic review of the literature in Table 9.6 was undertaken to estimate the proportion of rain tanks that supply water with lead levels greater than drinking water guidelines. Our focus is on Australian capital cities so we choose recent Australian studies that included a random sample of tanks. Only recent studies are relevant as there have been interventions to reduce the amount of lead in the environment over time. For example, lead-free petrol was introduced into Australia in the 1980s, and mandated in 2002; lead concentration in paint has been decreased from up to 50% before 1950, to a maximum of 0.1% from 1997 (enHealth, 2004). The selected studies are summarised in Table 9.7. The proportion of tanks with lead concentrations exceeding drinking water guidelines ranged from 9% to 45% (Figure 9.7).

Table 9.7 Selected studies that estimate the proportion of tanks that exceed drinking water guidelines for lead.

Study	City	No. tanks in study	No. tanks exceeding drinking water guidelines for lead (percentage)	95% confidence limits on percentage of tanks exceeding drinking water guidelines
Chapman *et al.* 2008	Brisbane	30	9 (30%)	14.7%–49.4%
Chapman *et al.* 2008	National	69	6 (9%)	3.3%–18%
Magyar *et al.* 2008	Melbourne	49	16 (33%)	19.9%–47.5%
Huston *et al.* 2009	Brisbane	38	5 (14%)	4.4%–28.1%
Kus *et al.* 2010b	Sydney	11	5 (45%)	16.7%–76.6%
Rodrigo *et al.* 2012	Adelaide	19	3 (16%)	3.4%–39.6%
Average (meta-analysis)			**21.8%**	**12.8%–34.7%**

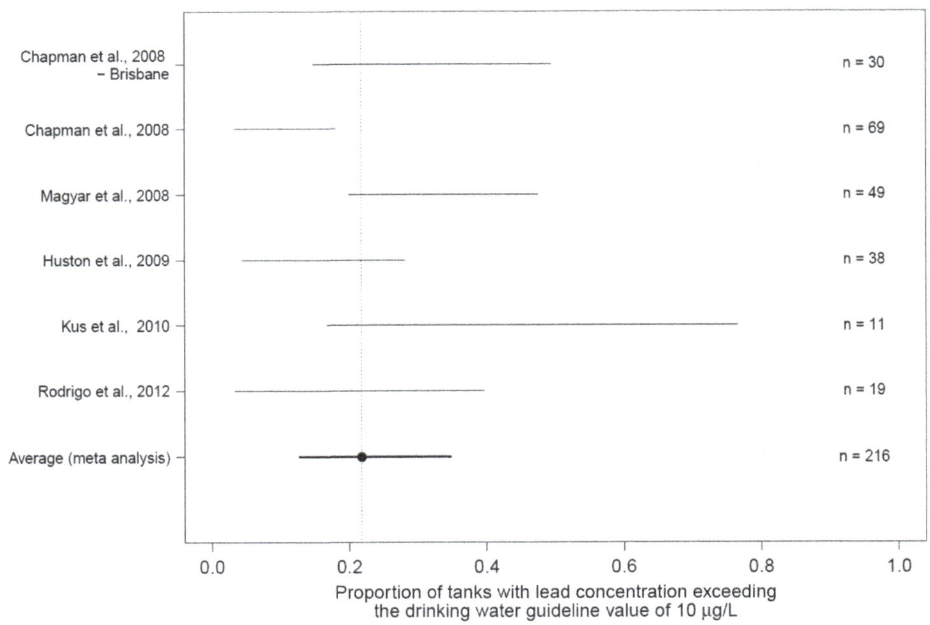

Figure 9.7 Proportion of tanks from recent studies in Australian capital cities where lead concentration exceeded drinking water guidelines.

We used meta-analysis to estimate the average proportion of tanks with high lead concentration. Our analysis followed the recommendations of Trinquart and Touze (2009) and used the meta package in R (Schwarzer, 2014; R Core Team, 2014). Using a random effects model, the average proportion of tanks predicted to exceed drinking water guidelines for lead is 21.8%, with the 95% confidence limits of 12.8% to 34.7%.

As shown in Table 9.2, it is estimated that 127,000 households in Australian capital cities drink tank water. We can now estimate that 21.8% of these tanks will supply water with lead concentrations greater than drinking water guidelines. This is a total of 27,700 tanks (95% CI 16,300–44,000 tanks).

9.3.4 pH of tank water and relationship with lead concentration

As discussed in section 9.2.3 roof runoff is usually acidic. For water supplied by rain tanks, the main issue associated with low pH is enhanced metal solubility with main concern being lead which, as we have seen, is commonly found in roof runoff and tank water. Although the solubility behaviour of lead is complex and depends on other factors including water temperature and the concentrations of other solutes in water, Hem (1976) notes that lead may be soluble in dilute natural water below pH 6.5, with solubility increasing as rapidly as pH decreases (Rickard & Nriagu, 1978; Bodek *et al.* 1998). This behaviour was broadly confirmed in rain tanks sampled by Magyar *et al.* (2011b) who measured the partitioning between soluble & particulate lead as the pH of the water changed.

Tank water has been found to be acidic except where there is a strong influence of tank material. For example, Huston (2009) in a study of 29 tanks in Brisbane found a mean value of pH of all tank water samples of 6.1 (SD = 1, n = 325) (Figure 9.8). This is much lower, and more variable than Brisbane mains water, reported by Chapman *et al.* (2008) as 7.9 (SD = 0.14, n = 607). In Huston's (2009) study, tank water pH was generally outside the guideline values for drinking water (Figure 9.8) and in the range where metallic lead is increasing in solubility. The high lead concentrations in tank water samples measured by Huston (2009) were strongly linked to low pH (Figure 9.9). Where pH exceeded 7, lead concentration met drinking water standards. Similar results were reported by Simmons *et al.* (2001). They found that systems with pH < 6.5 were 4 times more likely to have elevated lead levels than those with pH > 6.5.

In Huston's study, the water from concrete tanks had significantly higher pH than plastic or steel tanks. Other studies have also found that concrete tanks can lead to higher pH (Thomas & Greene, 1993; Simmons *et al.* 2001; Handia, 2005) although the effect decreases with tank age (Handia, 2005; NHMRC & NRMMC, 2014).

There is also an interaction between pH in tank water and sediments that settle at the bottom of the tank. Low pH water can liberate lead and other metals from these sediments. High concentrations of lead have been reported in tank sediments in a number of studies and reviews (Sharpe & Young, 1984; Michaelides & Young, 1984; Gumbs & Dierbert, 1984; Scott & Waller, 1987; Spinks *et al.* 2005; Coombes *et al.* 2006; Magyar *et al.* 2007; Abbasi & Abbasi, 2011; Magyar *et al.* 2014). Tank sediment can also be re-suspended and supplied to end uses (Magyar *et al.* 2011b). If tank water is acidic then we can expect lead to be present in dissolved form and delivered in dissolved form to end uses.

9.3.5 Quality aspects of tank supply to hot water systems

The supply of tank water to domestic hot water systems has been suggested as one way to reduce microbial contamination (Spinks *et al.* 2003a; enHealth, 2004). However, Chapman *et al.* (2008) found that the use of rainwater in hot water systems can lead to increases in lead concentrations. In their analysis, 16% of hot water samples had lead concentrations greater than drinking water guidelines compared to 4% of samples collected prior to the hot water system. They recommended that where hot water systems are supplied from a rainwater tank, hot water should not be used for drinking or cooking. Although Chapman *et al.* (2008) did not propose a process for the observed increase in lead concentration in hot water, a potential mechanism is the increase in solubility of lead salts that occurs with increases in water temperature, especially when pH is low (Clever & Johnston, 1980). Lead contamination of water can also occur from brass fittings inside hot water systems or in house plumbing (McCafferty *et al.* 1995). This will be exacerbated when water is hot and has low pH.

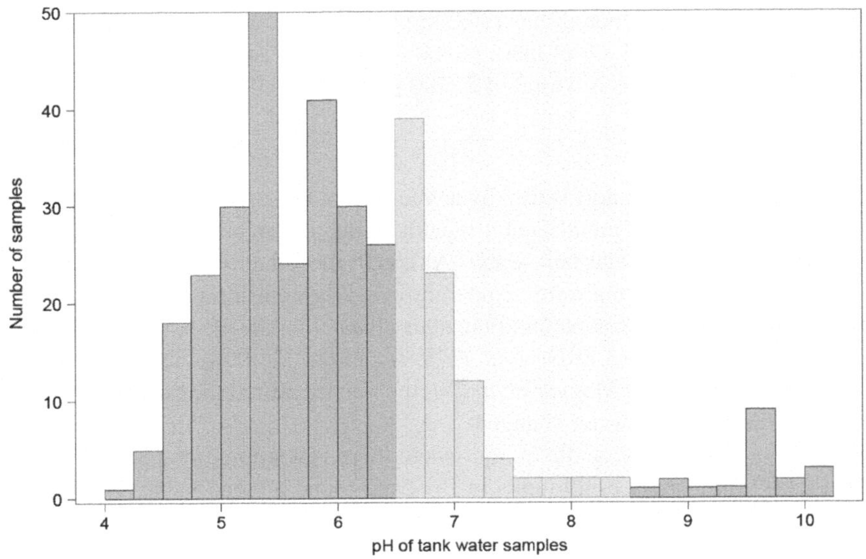

Figure 9.8 The distribution of pH from all the water samples in 29 Brisbane tanks ($n = 352$, Huston, 2009). Lightly shaded region is acceptable pH range specified in the Australian drinking water guidelines.

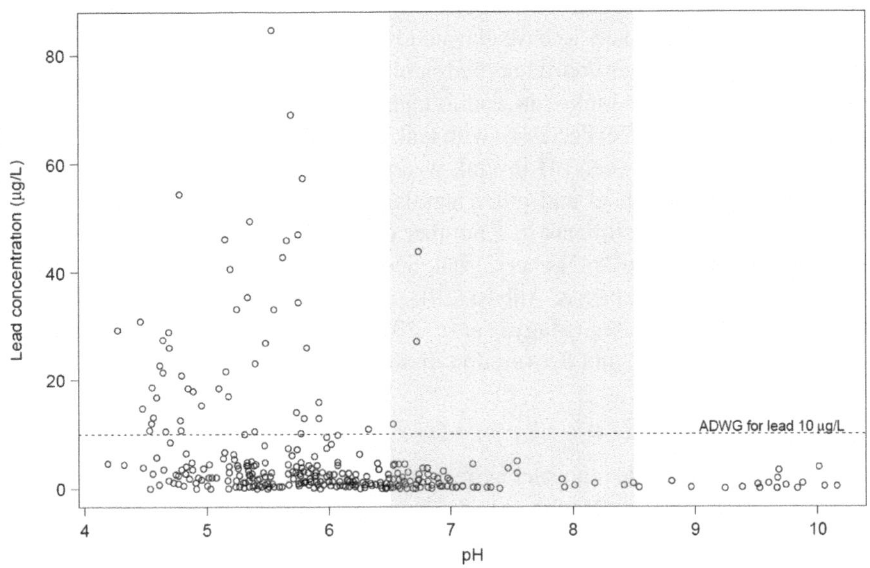

Figure 9.9 In a study of tank water quality in Brisbane, water samples with high lead concentration (>10 ug/L) occurred when pH was less than 7 (Huston, 2009). The shaded region corresponds to pH values as recommended in the Australian drinking water guidelines.

9.4 IMPROVING THE QUALITY OF WATER SUPPLIED BY RAIN TANKS

Although not the focus of this chapter, we can suggest some ways to improve the quality of water delivered from rain tanks. The first step must be to take the risk of lead contamination seriously. We have identified 31 studies where lead concentration in tank water exceeded drinking water standards and only a few studies where it did not. Once the problem is acknowledged, researchers and designers can work to mitigated it. Currently, many guides aimed at practitioners who design and install tanks, place little emphasis on lead contamination and how it can be avoided (e.g., Melbourne Water, 2005; Coombes & Mitchell, 2006; MPMSAA, 2008).

A particular issue is tank maintenance, which if carried out appropriately, has the potential to improve water quality. However recommended procedures are seldom undertaken by householders and even if they were, water quality issues would not necessarily be avoided. For example tank inspection for accumulated sediments is recommended only every 2–3 years (Cunliffe, 1998; enHealth Council, 2004; NRMMC *et al.* 2009) yet Magyar *et al.* (2014) found that sediments in tanks were already heavily polluted after only one year post installation. Several researchers have found it is common for householders to undertake little or no maintenance. In a survey of 10,000 households in Brisbane, almost all (95%) had never had their rainwater tank professionally cleaned (Mankad & Greenhill, 2014). In their study of the health effects of the consumption of tank rainwater in South Australia, Heyworth *et al.* (2006) described maintenance procedures as rudimentary. Sludge had never been removed in at least 42% of tanks (26% of respondents were unaware of whether sludge had ever been removed). Other studies confirm the lack of maintenance (Bannister *et al.* 1997; Spinks *et al.* 2006; Rodrigo *et al.* 2010; Van der Sterran *et al.* 2013; Moglia *et al.* 2013). An ABS survey of 2558 households with rain tanks found that 58% had done no maintenance in the last 12 months that is, they had not cleaned gutters, checked or repaired inlets, insect screens or pipe work, nor had they checked or cleaned tank sediments (ABS, 2013). If tank water is to be used for drinking then clearly tank maintenance must become a priority. A suggested list of maintenance tasks is provided by Moglia *et al.* (2013).

Other ways to improve tank water quality include pH adjustment, better tank design (Michaelides & Young, 1984; Magyar *et al.* 2011a), floating outlets and water treatment for example, filtration.

9.5 CONCLUSION

Our conclusion is that the chemical quality of rainwater in rain tanks is of concern if tank water is to be used for drinking. Lead concentration exceeding the drinking water guideline of 10 ug/L is common. Based on a meta-analysis of 6 studies, we estimate that 22% of tanks in Australian capital cities will have high lead levels. Low pH is also widely reported which can increase the proportion of lead in dissolved form. High lead concentrations in rain tank water are widely reported in the literature, with 31 studies identifying lead as a water quality issue. Only a few studies have found lead concentrations that always meet drinking water standards.

Other than lead, the chemical quality of tank water generally meets water quality guidelines. A few studies mention excursive concentrations of other metals such as cadmium, nickel, selenium, copper and arsenic. In some cases; these can be linked to specific sources. However organic pollutants such as polycyclic aromatic hydrocarbons (PAHs), pesticides and herbicides do not seem to be a common or widespread issue. Even when high concentrations of PAHs were expected, such as after a bushfire or tanks located close to a freeway, they were not detected in rain tanks.

If tank water is not be used for drinking then there is less concern. The trigger value for lead in agricultural water supply (2000 μg/L) is much larger than that for drinking water (10 μg/L) and is unlikely to be exceeded unless there is a specific contamination issue.

The potential for high lead concentration in tank water needs to be taken seriously with better advice required for tank system designers and the general public. Tank water quality can be improved by regular maintenance, but this is currently not widely practiced. There is also a need to reconsider rain tank design to focus on delivery of high quality water.

9.6 REFERENCES

Abbasi T. and Abbasi S. A. (2011). Sources of pollution in rooftop rainwater harvesting systems and their control. *Critical Reviews in Environmental Science and Technology*, **41**(23), 2097–2167.

ABCB (2012). Building Code of Australia. Australian Building Codes Board.

ABS (2013). Environmental issues: water use and conservation. Catalogue No. 4602.0.55.003. Australian Bureau of Statistics.

Al-Khashman O. A. (2009). Chemical characteristics of rainwater collected at a western site of Jordan. *Atmospheric Research*, **91**, 53–61.

ANZECC and ARMCANZ (2000). Australian and New Zealand guidelines for fresh and marine water quality. Paper No. 4, Volume 1, Australian and New Zealand Environment and Conservation Council (ANZECC) and Agriculture and Resource Management Council of Australia and New Zealand (ARMCANZ).

Bach P., McCarthy D. T. and Deletic A. (2010) Redefining the stormwater first flush phenomenon. *Water Research*, **44**(8), 2487–2498.

Bannister R., Westwood J. and McNeill A. (1997). Investigation of microbiological and chemical water quality in rainwater tanks in Victoria. Water Ecoscience for the Department of Natural Resources and Environment, Victoria.

Birch G. and Taylor S. (1999). Source of heavy metals in sediments of the Port Jackson estuary, Australia. *The Science of the Total Environment*, **227**, 123–138.

Birch G. F. and Scollen A. (2003). Heavy metals in road gust, gully pots and parkland soils in a highly urbanised sub-cathchment of Port jackson, Australia. *Australian Journal of Soil Research*, **41**, 1329–1342.

Bodek I., Lyman W. J., Reehl W. F. and Rosenblatt D. H. (eds) (1988). Environmental Inorganic Chemistry. Properties, Processes and Estimation Methods. Pergamon Press. A special publication by SETAC.

Body P. (1986). The Contamination of Rainwater Tanks at Port Pirie, South Australian Health Commission, Adelaide.

Chang M., McBroom M. W. and Beasley R. S. (2004). Roofing as a source of nonpoint water pollution. *Journal of Environmental Management*, **73**(4), 307–315.

Chapman H., Gardner T., Huston R., Chan A. and Shaw G. (2006). Chemical Water Quality and Health Risk Assessment of Urban Rainwater Tanks. 7th International conference on Urban Drainage Modelling and the 4th International Conference on Water Sensitive urban Design. Monash University, Melbourne, Australia.

Chapman H., Cartwright T., Huston R. and O'Toole J. (2008). Water Quality and Health Risks from Urban Rainwater Tanks. Cooperative Research Centre for Water Quality and Treatment, Salisbury, South Australia.

Conlan D. E. and Longhurst J. W. S. (1993). Spatial variability in urban acid deposition, 1990: results from the Greater Manchester Acid Deposition Survey (GMADS) network in the UK. *Science of the Total Environment*, **128**(2–3), 101–120.

Coombes P. and Mitchell G. (2006). Urban water harvesting and reuse. In: Australian Runoff Quality: A Guide to Water Sensitive Urban Design, T. H. F. Wong (ed.), Engineers, Australia.

Coombes P. J., Argue J. R. and Kuczera G. (2000a). Figtree Place: a case study in water sensitive urban development (WSUD). *Urban Water*, **1**(4), 335–343.

Coombes P. J., Kuczera G., Kalma J. D. and Dunstan, R. H. (2000b). Rainwater quality from roofs, tanks and hotwater systems at figtree place. In: Hydro 2000: Interactive Hydrology Proceedings. Institution of Engineers, Australia, pp. 1042–1047.

Coombes P. J., Kuczera G. and Kalma J. D. (2002). Economic, Water Quantity and Quality Rresults from a House with Rainwater Tank in the Inner City. 27th Hydrology and Water Resources Symposium, 20–23 May 2002, Institution of Engineers Australia, Melbourne.

Coombes P. J., Kuczera G. and Kalma J. D. (2003). Economic water quantity and quality impacts from the use of a rainwater tank in the inner city. *Australian Journal of Water Resources*, **7**(2), 111–120.

Coombes P. J., Dunstan H., Spinks A. T., Evans C. and Harrison T. (2005). An Overview of a Decade of Research into the Quality of Rainwater Supplies Collected from Roofs. 12th International Rainwater Catchment Systems Conference: Mainstreaming Rainwater Harvesting New Delhi, India.

Coombes P. J., Dunstan H., Spinks A., Evans C. and Harrison T. (2006). Key Messages from a Decade of Water Quality Research into Roof Collected Rainwater Supplies. 1st National Hydropolis Conference, Stormwater Industry Association, Perth, WA.

Dahl C., Søgaard A. J., Tell G. S., Flaten T. P., Hongve D., Omsland T. K., Holvik K., Meyer H. E. and Aamodt G. (2013). Do Cadmium, Lead, and Aluminum in drinking water increase the risk of hip fractures? *Biological Trace Element Research*, **157**, 14–23.

Daoud A. K., Swaileh K. M., Hussein R. M. and Matani M. (2011). Quality assessment of roof-harvested rainwater in the West Bank, Palestinian Authority. *Journal of Water and Health*, **9**(3), 525–533.

DHS (2007). Rainwater use in urban communities. Department of Human Services, State of Victoria.

Duncan H. P. (1999). Urban stormwater quality: a statistical overview. Cooperative Research Centre for Catchment Hydrology, Melbourne, pp. 134.

Duncan H. P. (2006). Urban stormwater pollutant characteristics. In: Australian Runoff Quality: a guide to Water Sensitive Urban Design. T. H. F. Wong (ed.), Engineers, Australia.

Duncan H. P. and Wight D. J. (1991). Rainwater Tanks for Domestic Supply in the Melbourne Area. Melbourne and Metropolitan Board of Works, Melbourne, Australia.

Edwards M. (2014). Fetal death and reduced birth rates associated with exposure to lead-contaminated drinking water. *Environmental Science and Technology*, **48**(1), 739–746.

enHealth Council (2004). Guidance on use of rainwater tanks, Australian Government, Department of Health and Ageing.

eWater (2009). MUSIC - Model for Urban Stormwater Improvement Conceptualisation – User Guide 6. http://www.toolkit.net.au/Tools/MUSIC, eWater Cooperative Research Centre Canberra, Australia.

Ferguson M. (2011). Rainwater Tank Monitoring Report. Sydney Water.

Fitzgerald S. K., Stanford B. D. and Khan S. J. (2014). Lessons from a decade of extreme weather events for Australian drinking water supplies. *Journal of the Australian Water Association*, **41**(2), 144–149.

Forster J. (1996). Patterns of roof runoff contamination and their potential implications on practice and regulation of treatment and local infiltration. *Water Science and Technology*, **33**(6), 39–48.

Fuller C. O., Martin T. J. and Walters R. P. (1981). Quality aspects of water stored in domestic rainwater tanks: a preliminary study. Domestic Rainwater Tanks Working Party, Department of Environment, South Australia.

Gardner T., Baisden J. and Millar G. (2004). Rainwater First Flush Devices – Are They Effective? Sustainable Water in the Urban Environment. 2004 Conference Brisbane, Aug 30–31.

Goyer R. A. (1993). Lead toxicity – current concerns. *Environmental Health Perspectives*, **100**, 177–187.

Gromaire M. C., Garnaud S., Saad M. and Chebbo G. (2001). Contribution of different sources to the pollution of wet weather flows in combined sewers. *Water Research*, **35**(2), 521–533.

Gumbs A. F. and Dierbert F. E. (1984). Heavy metals in the drinking water from cisterns supplying single-family dwellings on St Maarten, Netherlands Antilles. In: 2nd International Conference on Rain Water Cistern Systems. International Rainwater Catchment Systems Association, St Thomas, Virgin Islands.

Haelber R. H. and Waller D. H. (1987). Water Quality of Rain Water Collection Systems in the Eastern Caribbean. 3rd International Rainwater Cistern Systems Conference, International Rainwater Catchment Systems Association, Khon Kaen, Thailand.

Handia L. (2005). Comparative study of rainwater quality in urban Zambia. *Journal of Water Supply Research and Technology – Aqua*, **54**(1), 55–64.

Hart C. and White D. (2006). Water quality and construction materials in rainwater catchments across Alaska. *Journal of Environmental and Engineering Science*, **5**, 19–25.

Hem J. D. (1976). Inorganic chemistry of lead in water. In: Lead in the Environment. T. G. Lovering (ed.), Geological Survey Professional Paper 957 United States Geological Survey, pp. 5–12.

Heyworth J. S., Glonek G., Maynard E. J., Baghurst P. A. and Finlay-Jones J. (2006). Consumption of untreated tank rainwater and gastroenteritis among young children in South Australia. *International Journal of Epidemiology*, **35**, 1051–1058.

Huston R. (2009). Chemical Contaminants in Urban Rainwater Ranks. PhD Thesis, Griffith University, Australia.

Huston R., Chan Y. C., Gardner T., Shaw G. and Chapman H. (2009). Characterisation of atmospheric deposition as a source of contaminants in urban rainwater tanks. *Water Research*, **43**, 1630–1640.

Huston R., Chan Y. C., Chapman H., Gardner T. and Shaw G. (2012). Source apportionment of heavy metals and ionic contaminants in rainwater tanks in a subtropical urban area in Australia. *Water Research*, **46**(4), 1121–1132.

Kiely G. (1997). Environmental Engineering, McGraw-Hill, London, New York.

Kus B., Kandasamy J., Vigneswaran S. and Shon H. K. (2010a). Analysis of first flush to improve the water quality in rainwater tanks. *Water Science and Technology*, **61**(2), 421–428.

Kus B., Kandasamy J., Vigneswaran S. and Shon H. K. (2010b). Water quality characterisation of rainwater tanks at different times and locations. *Water Science & Technology*, **61**(2), 429–439.

Lanphear B. P., Hornung R., Khoury J., Yolton K., Baghurst P., Bellinger D. C., Canfield R. L., Dietrich K. N., Bornschein R., Greene T., Rothenberg S. J., Needleman H. L., Schnaas L., Wasserman G., Graziano G. J. and Roberts R. (2005). Low-level environmental lead exposure and children's intellectual function: An international pooled analysis. *Environmental Health Perspectives*, **113**, 894–899.

Lye D. J. (2009). Rooftop runoff as a source of contamination: a review. *Science of the Total Environment*, **407**(21), 5429–5434.

Magyar M. I. (2010). Investigating Factors that Affect Water Quality From Rainwater Tanks in Australia. PhD Thesis, Monash University.

Magyar M. I., Mitchell V. G., Ladson A. R. and Diaper C. (2007). An investigation of rainwater tanks quality and sediment dynamics. *Water Science and Technology*, **56**(9), 21–28.

Magyar M. I., Mitchell V. G., Ladson A. R. and Diaper, C. (2008). Lead and other heavy metals: common contaminants of rainwater tanks in Melbourne. Water Down Under 2008 Conference, incorporating the Hydrology and Water resources Symposium and the 4th International Conference on Water Resources and Environmental Research, Adelaide, Australia. Institution of Engineers, Australia.

Magyar M. I., Ladson A. R., Mitchell V. G. and Diaper C. (2011a). The effect of rainwater tanks design on sediment re-suspension and subsequent outlet water quality. *Australian Journal of Water Resources*, **15**(1), 71–84.

Magyar M. I., Ladson A. R. and Diaper C. (2011b). Sediment transport in rainwater tanks and implications for water quality. In: Sediment transport in aquatic environments. A. J. Manning (ed.), InTech, Rijeka, Croatia, pp. 65–88. http://www.intechopen.com/books/sediment-transport-in-aquatic-environments (accessed 28 December 2014).

Magyar M. I., Ladson A. R., Diaper C. and Mitchell V. G. (2014). Influence of roofing materials and lead flashing on rainwater tank contamination by metals. *Australian Journal of Water Resources*, **18**(1), 71–83.

Malassa H., Al-Rimawi F., Al-Khatib M. and Al-Qutob M. (2014). Determination of trace heavy metals in harvested rainwater used for drinking in Hebron (south West Bank, Palestine) by ICP-MS. Environmental Monitoring and Assessment, **186**(10), 6985–6992. doi:10.1007/s10661-014-3904-5.

Mankad A. and Greenhill M. (2014). Motivational indicators predicting the engagement, frequency and adequacy of rainwater tank maintenance. *Water Resources Research*, (10.1002/2013WR014338f).

Mark Rigby and Associates (2002). Tamborine Mountain Water Quality Project: a Report on the Sampling and Assessment of Potable Tank Water From Ninety-Five Dwellings on Tamborine Mountain, Queensland. Tamborine Shire, Queensland.

Martin A. R., Coombes P. J. Dunstan R. H., Evans C. and Morrow A. C. (2007). The Passage of Direct Precipitation to Rainwater Storage: A Case Study; Newcastle, Australia. 13th International Rainwater Catchment Systems Conference: Rainwater and Urban Design 2007, Sydney, Australia. International Rainwater Catchment Systems Association.

Maynard E., Thomas R., Simon D., Phipps C., Ward C. and Calder I. (2003). An evaluation of recent blood lead levels in Port Pirie, South Australia. *Science of the Total Environment*, **303**(1–2), 25–33.

Meera V. and Ahammed M. M. (2006). Water quality of rooftop rainwater harvesting systems: a review. *Journal of Water Supply: Research & Technology*, **55**(4), 257–268.

Melbourne Water (2005). WSUD Engineering Procedures: Stormwater. CSIRO publishing, Melbourne.

Moglia M. G., Tjandraatmadja G. and Sharma A. K. (2013). Exploring the need for rainwater tank maintenance: survey, review and simulations. *Water Science and Technology: Water Supply*, **13**, 191–201.

Morrow A. C. and Dunstan R. H. (2012). Effects of atmospheric deposition, the rainwater delivery system and in-tank processes on roof harvested rainwater quality. *Air Quality and Climate Change*, **46**(3), 22–27.

Morrow A. C., Coombes P. J., Dunstan H., Evans C. and Martin A. (2007). Elements in Tank Water – Comparisons with Mains Water & Effects of Locality & Roofing Materials 13th International Rainwater Catchment Systems Conference: Rainwater and Urban Design 2007, Sydney, Australia. International Rainwater Catchment Systems Association.

Morrow A. C., Dunstan R. H. and Coombes P. J. (2010). Elemental composition at different points of the rainwater harvesting system. *Science of the Total Environment*, **408**(20), 4542–4548.

MPMSAA (2008). Rainwater Tank Design and Installation Handbook. Master Plumbers and Mechanical Services Association of Australia. National Water Commission and Australian Rainwater Industry Development Group.

MWAWA (1983). Metropolitan Water Authority, Western Australia. Personal Communication.

Nawrot T. S. and Staessen J. A. (2006). Low-level environmental exposure to lead unmasked as silent killer. *Circulation*, **1388**(114), 1347–1349.

NHMRC and NRMMC (2014). Australian Drinking Water Guidelines. National Health and Medical Research Council and Natural Resource Management Ministerial Council, Australian Government, Canberra.

NRMMC, EPHC and NHMRC (2009). Australian Guidelines for water Recycling: Managing Health and Environmental Risks (phase 2). Stormwater Harvesting and Reuse. National Water Quality Management Strategy. Natural Resources and Management Ministerial Council, Environmental Protection and Heritage Council, National Health and Medical Research Council.

O'Connor J. B., Mitchell V. G., Magyar M. I., Ladson A. R. and Diaper C. (2009). Where is the Lead in Rainwater Tanks Coming From? H2009, 32nd Hydrology and Water Resources Symposium, Newcastle, Australia.

Olem H. and Berthouex P. M. (1989). Acidic deposition and cistern drinking water supplies. *Environmental Science and Technology*, **23**(3), 333–340.

Peters A. J., Weidner K. L. and Howley C. L. (2008). The chemical water quality in roof- harvested water cisterns in Bermuda. *Journal of Water Supply: Research & Technology*, **57**(3), 153–163.

Polkowska Z., Gorecki T. and Namiesnik J. (2002). Quality of roof runoff water from an urban region. *Chemosphere*, **49**, 1275–1283.

R Core Team (2014). R: A language and environment for statistical computing. R Foundation for Statistical Computing, Vienna, Austria. http://www.R-project.org/. (accessed 4 May 2014).

Radaideh J., Al-Zboon K., Al-Harahshed A. and Al-Adamat R. (2009). Quality assessment of harvested rainwater for domesitic uses. *Jordan Journal of Earth and Environmental Sciences*, **2**(1), 26–31.

Rickard D. T. and Nriagu J. O. (1978). Aqueous environmental chemistry of lead. In: The Biogeochemistry of Lead in the Environment, J. O. Nriagu (ed.), Amsterdam; New York: Elsevier/North-Holland Biomedical Press; New York: sole distributors for the USA and Canada, Elsevier/North-Holland, pp. 219–284.

Rodrigo S., Sinclair M. and Leder K. (2010). A survey of the characteristics and maintenance of rainwater tanks in urban areas of South Australia. *Water Science & Technology*, **61**(6), 1569–1577.

Rodrigo S., Sinclair M. and Leder K. (2012). Water and sediment quality of rainwater tanks. *Journal of the Australian Water Association*, **39**(1), 83–87.

Sazakli E., Alexopoulos A. and Leotsinidis M. (2007). Rainwater harvesting, quality assessment and utilization in Kefalonia Island, Greece. *Water Research*, **41**(9), 2039–2047.

Schwarzer G. (2014). meta: Meta-Analysis with R. R package version 3.5-0. http://CRAN.R-project.org/package=meta (accessed 4 May 2014).

Scott R. S. and Waller D. H. (1987). Water Quality Aspects of a Rain Water Cistern System. 3rd International Rainwater Cistern Systems Conference, International Rainwater Catchment Systems Association, Khon Kaen, Thailand.

Sharpe W. E. and Young E. S. (1984). Occurrence of Selected Heavy Metals in Rural Roof-Catchment Cistern Systems. 1st International Conference on Rain Water Cistern Systems. June, International Rainwater Catchment Systems Association, Honolulu, Hawaii, USA.

Shih R. A., Hu H., Weisskopf M. G. and Schwartz B. S. (2007). Cumulative lead dose and cognitive function in adults: a review of studies that measured both blood lead and bone lead. *Environmental Health Perspectives*, **115**(3), 483–492.

Simmons G., Hope V., Lewis G., Whitmore J. and Gao W. (2001). Contamination of potable roof-collected rainwater in Auckland, New Zealand. *Water Research*, **35**(6), 1518–1524.

Spinks A. T., Coombes P. J., Dunstan R. H. and Kuczera G. (2003a). Water Quality Treatment Processes in Domestic Rainwater Harvesting Systems. 28th Hydrology and Water Resources Symposium. Wollongong, The Institution of Engineers, Australia, Vol. 2, pp. 227–234.

Spinks A. T., Dunstan R. H., Coombes P. J. and Kuczera G. (2003b). Thermal Destruction Analysis of Water Related Pathogens at Domestic Hot Water System Temperatures. 28th International Hydrology and Water Resources Symposium. 10–14 Nov, 2003. Wollongong, NSW. Institution of Engineers, Australia.

Spinks A. T., Dunstan R. H., Coombes P., Berghout B., and Kuczera G. (2005). Tank Sludge as a Sink for Bacterial and Heavy Metal Contaminants and Its Capacity for Settlement, Resuspension and Flocculation Enhancement. 12th International Rainwater Catchment Systems Conference: Mainsteaming Rainwater Harvesting, New Delhi, India. International Rainwater Catchment Systems Association.

Spinks J., Phillips S. Robinson P. and Van Buynder P. (2006). Bushfires and tank rainwater quality: a cause for concern? *Journal of Water and Health*, **4**(1), 21–28.

Standards Australia (2003). AS/NZS 3500.3 (2003) Plumbing and Drainage. Part 3: Stormwater drainage.

Stump B., McBroom M. and Darville R. (2012). Demographics, practices and water quality from domestic potable rainwater harvesting systems. *Journal of Water Supply Research and Technology-Aqua*, **61**(5); 261–271.

Thomas P. R. and Greene G. R. (1993). Rainwater quality from different roof catchments. *Water Science and Technology*, **28**(3–5), 291–299.

Trinquart L. and Touze E. (2009). Pitfalls in meta-analysis of observational studies: lessons from a systematic review of the risks of stenting for intracranial atherosclerosis. *Stroke*, **40**, e486–e587, doi: 10.1161/STROKEAHA.109.556290.

Van der Sterren M., Rahman A. and Dennis G. R. (2013). Quality and quantity monitoring of five rainwater tanks in Western Sydney. *Journal of Environmental Engineering* (ASCE), **139**(3), 332–340.

Van Metre, P. C. and Mahler, B. J. (2003). The contribution of particles washed from rooftops to contaminant loading to urban streams. *Chemosphere*, **52**(10), 1727–1741.

Vialle C., Sablayrolles C., Lovera M., Jacob S., Huau M. C. and Montrejaud-Vignoles M. (2011). Monitoring of water quality from roof runoff: Interpretation using multivariate analysis. *Water Research*, **45**(12), 3765–3775.

Walpole R. E. and Myers R. H. (1978). Probability and statistics for engineers and scientist. Macmillan.

White I. (2014). Rainwater tanks – reign no more? *Journal of the Australian Water Association*, **41**(2), 163–170.

WHO (2008). Guidelines for Drinking Water Quality. World Health Organisation, Geneva.

Wong T. H. F. (ed.) (2006). Australian Runoff Quality: a guide to Water Sensitive Urban Design. Engineers, Australia.

Yaziz M. I., Gunting H., Sapari N. and Ghazali A. W. (1989). Variations in rainwater quality from roof catchments. *Water Research*, **23**(6), 761–765.

Young E. S. and Sharpe W. E. (1984). Atmospheric deposition and roof-catchment cistern water quality. *Journal of Environmental Quality*, **13**(1), 38–43.

Zhu K., Zhang L., Hart W., Liu M. and Chen H. (2004). Quality issues in harvested rainwater in arid and semi-arid Loess Plateau of northern China. *Journal of Arid Environments*, **57**(4), 487–505.

Zobrist J., Müller S. R., Ammann A., Bucheli T. D., Mottier V., Ochs M., Schoenenberger R., Eugster J. and Boller M. (2000). Quality of roof runoff for groundwater infiltration. *Water Research*, **34**(5), 1455–1462.

Chapter 10

Microbiological quality and associated health risks with the use of roof-captured rainwater

Warish Ahmed and Simon Toze

ABSTRACT

This chapter reviews the available research reporting on the microbial quality of roof-captured rainwater (RCR) and provides insight on the potential health risks associated with the consumption of untreated tank water. The chapter also highlights a series of studies that were undertaken in South East Queensland to determine the microbial quality of tank water; identify the possible sources of faecal pollution in tank water; and provide information on the inactivation of faecal indicator bacteria in a RCR system. The studies indicated that the quality of the tank water is highly variable in terms of the presence of faecal indicator bacteria and zoonotic bacterial and protozoa pathogens. One of the studies identified possum and birds as the likely sources of faecal contamination in tank water. It remains prudent to disinfect RCR before using it as a potable water source.

Keywords: Faecal indicator bacteria; Zoonotic pathogens; Quantitative PCR; Health risks; Quantitative microbial risk assessment; Roof-captured rainwater.

10.1 INTRODUCTION

Water authorities worldwide are exploring alternative water sources to meet ever-increasing demands for potable water due to the adverse impacts of climate change on water supply. Roof-captured rainwater (RCR) has been considered as an alternative water source for both potable (drinking) and non-potable uses (irrigation, toilet flushing, car washing, showering, and clothes laundering). Australia, Canada, Denmark, Germany, India, Japan, New Zealand, Thailand and the United States have investigated the potential benefits of RCR (Despins *et al.* 2009; Evans *et al.* 2006; Uba & Aghogho, 2000). The numbers of rainwater tanks as a source of water for Australian urban and rural households are increasing. For example, 26% of Australian households used a rainwater tank as a source of water in 2010 compared with 19% in 2007 and 17% in 2004 (ABS, 2010). South Australia continues to have the highest proportion of households with a rainwater tank (48% and 49% in 2004 and 2010, respectively) but there was a marked increase in the proportion of households with a rainwater tank in Queensland (17% in 2004 to 36% in 2010) and

Victoria (16% in 2004 to 30% in 2010). There are several advantages to using rainwater tanks, including: (i) reducing the demand on the town water supply; (ii) providing an alternative water supply during times of water restrictions; and (iii) reducing stormwater runoff that can often degrade creek ecosystem health (see Chapter 13). Despite rainwater tanks being widely used for potable water supply in rural and peri-urban areas in Australia, tank water has not been widely used for drinking in urban settings due to a lack of information on the presence and risk from microbiological and chemical pollutants. Another shortcoming is the lack of appropriate guidelines specifying the use of tank water for both drinking and non-potable uses, and how the human health risks from microbiological and chemical pollutants can be managed. The most significant issue in relation to untreated tank water for drinking is the health risks associated with microbial pathogens (Ahmed *et al.* 2008; Crabtree *et al.* 1996; Simmons *et al.* 2001). A wide array of pathogens could be present in the faeces of birds, insects, mammals, and reptiles that have access to the roofs. Consequently, following rain events, animal droppings and other organic debris deposited on the roofs and gutters can be transported into the tanks via runoff. In this scenario, if the untreated water collected from the roof were used for drinking, there are potential disease risks for people consuming this water. Only limited information is available, however, regarding the actual health risks associated with the uses of tank water.

Many countries have provided subsidies to encourage the installation of RCR systems for drinking and non-potable uses with the specific aim of decreasing the use of mains water. For instance, in 2006, the Queensland Government initiated the 'Home Water Wise Rebate Scheme' which provided subsidies to residents who used rainwater for non-potable domestic uses. More than 260,000 householders were granted subsidies by December 2008 when the scheme was concluded (Beal *et al.* 2011). The Queensland Department of Health does not recommended using RCR for drinking where municipal (town) water is available. However, Queensland regulations do not prohibit the plumbing of rainwater tanks to supply drinking water. Thus, if a person chooses to use rainwater for drinking or any other purpose, then that person is responsible for ensuring that the quality of the water is good enough for its intended use.

A general community perception is that tank water is safe to drink without having to undergo prior treatment. For example, there are 22% of rural households in Australia that are completely dependent on rainwater for their domestic supply (ABS, 2010). The safety of drinking rainwater was further supported by the epidemiological surveys of gastroenteritis in South Australia that concluded tank water poses no increased risk of gastroenteritis when compared with town water (Heyworth *et al.* 2006; Rodrigo *et al.* 2010). In contrast, a number of studies on the microbial quality of tank water have reported the presence of specific zoonotic pathogens (i.e., capable of being transmitted from animals to humans) in individual or communal tank systems, suggesting that a real health risk exists (Ahmed *et al.* 2008; Birks *et al.* 2004; Crabtree *et al.* 1996; Lye, 2002; Simmons *et al.* 2001; Uba & Aghogho, 2000).

The purpose of this chapter is to: (i) highlight international research studies investigating the microbiological quality of tank water; and (ii) discuss disease outbreaks linked to the consumption of untreated tank water. The chapter includes a review of a series of studies associated with microbial pollution of tank water undertaken in South East Queensland (SEQ), Australia.

10.2 FAECAL INDICATORS AND PATHOGENS IN ROOF-CAPTURED RAINWATER

10.2.1 Faecal indicators

Drinking water guidelines have been used to assess the microbial quality of the tank water. For most guidelines, this entails the non-detection of *Escherichia coli* in 100 mL of water (NHMRC–NRMMC,

2004; WHO, 2004). Even when tank water is not used for drinking, the assessment of the microbial quality is usually undertaken by monitoring the presence of faecal indicator bacteria (Ahmed *et al.* 2008, 2010; Appan, 1997; CRC WQT, 2006; Dillaha & Zolan, 1985). The World Health Organization (WHO) recommends that for drinking water, the total coliform numbers should be <10 colony forming units (CFU)/100 mL in 95% of samples collected from a particular water source (WHO, 2004). These guidelines also indicate that if the numbers of total coliforms are >20 CFU/100 mL water, further treatment should be undertaken prior to drinking. It should be noted that total coliforms may not be a suitable indicator of faecal contamination in RCR. Total coliforms in water can comprise microorganisms that could be faecal in origin but also could be from other sources such as soil and vegetation. Thus, their presence in rainwater tanks could be through contamination from dust and plant materials. *E. coli* and enterococci are considered much better indicators of faecal contamination, and most guidelines stipulate that the numbers of these organisms should be zero CFU/100 mL for drinking water. Such stringent guideline values have been established for these indicators as they are commonly found in high numbers in human and animal faeces. Their presence in drinking water, therefore, indicates a strong likelihood that faecal sourced pathogens also occur, and that the water is not suitable for drinking (Baudisöva, 1997).

Most research studies on tank water reported to date used faecal indicator bacteria to assess the microbiological quality of the water. Figure 10.1 shows the percentage of positive samples for faecal indicator bacteria in tank water samples reported in 17 international research studies. In Micronesia, 176 tank water samples were surveyed for faecal coliforms. Of these, 68% of the samples contained measurable numbers of faecal coliforms (Dillaha & Zolan, 1985). Despite the high numbers of samples not complying with the guidelines, the authors suggested that the tank water could be reasonable for drinking. Lye (1987) also reported that the microbiological quality of tank water was consistently good in a study of the occurrence of faecal coliforms in tank water samples in Kentucky, USA, where only one sample out of 30 had faecal coliforms numbers >10 CFU/100 mL.

In contrast, several studies reported higher numbers of faecal indicators in tank water samples, which therefore, did not comply with acceptable drinking water guideline values (Figure 10.1). A recent study in South Korea reported that 72% of tank water samples were positive for *E. coli* and the numbers were above the WHO drinking water guideline value of zero CFU/100 mL (Lee *et al.* 2010). High numbers of *E. coli* were also found in tank water samples in Denmark, where *E. coli* was observed to be present in 11 out of 14 systems, with numbers ranging from 4 to 900 CFU/100 mL. The conclusion made in this study was that the presence of *E. coli* indicated that the water may not be suitable for drinking (Albrechtsen, 2002).

In Victoria, Australia, 49 rainwater tanks surveyed for the presence of *E. coli* and enterococci found that 33% were positive for *E. coli* and 73% positive for enterococci, therefore exceeding the Australian Drinking Water Guidelines of zero CFU/100 mL (ADWG, 2004; Spinks *et al.* 2006). High numbers of *E. coli* (ranging from 4 to 800 CFU/100 mL) and enterococci (5 to 200 CFU/100 mL) were also reported in tank water samples tested in South East Queensland, Australia (Ahmed *et al.* 2008). In this study, 63% and 78% of the 27 samples were positive for *E. coli* and faecal streptococci, respectively. Ahmed *et al.* (2008) concluded that, as *E. coli* could not be detected in a number of the water samples that were positive for enterococci (a sub-group of faecal streptococci), tank water should be tested for both *E. coli* and enterococci, as well as relevant indicators (where possible) to obtain multiple lines of evidence on the occurrence of faecal contamination. On the basis of these results, they concluded that *E. coli* by itself as an indicator may be of limited use to assess the microbial quality of tank water samples.

Several studies also reported that enterococci are more prevalent in tank water samples compared to *E. coli* (Ahmed *et al.* 2008, 2010; CRC WQT, 2006; Spinks *et al.* 2006), and thus may be a better indicator for assessing faecal contamination. The greater prevalence of enterococci in tank water may be due to the fact that enterococci persist in the water longer than *E. coli* (Anderson *et al.* 2005; McFeters *et al.* 1974).

It has also been reported that enterococci are better indicators of faecal contamination in environmental waters compared to *E. coli* (Kinzelman *et al.* 2003). Nonetheless, more studies on the potential sources and the relative persistence are needed that compare the usefulness of *E. coli* versus enterococci in tank water samples before any recommendations can be made concerning which indicators may be the most suitable.

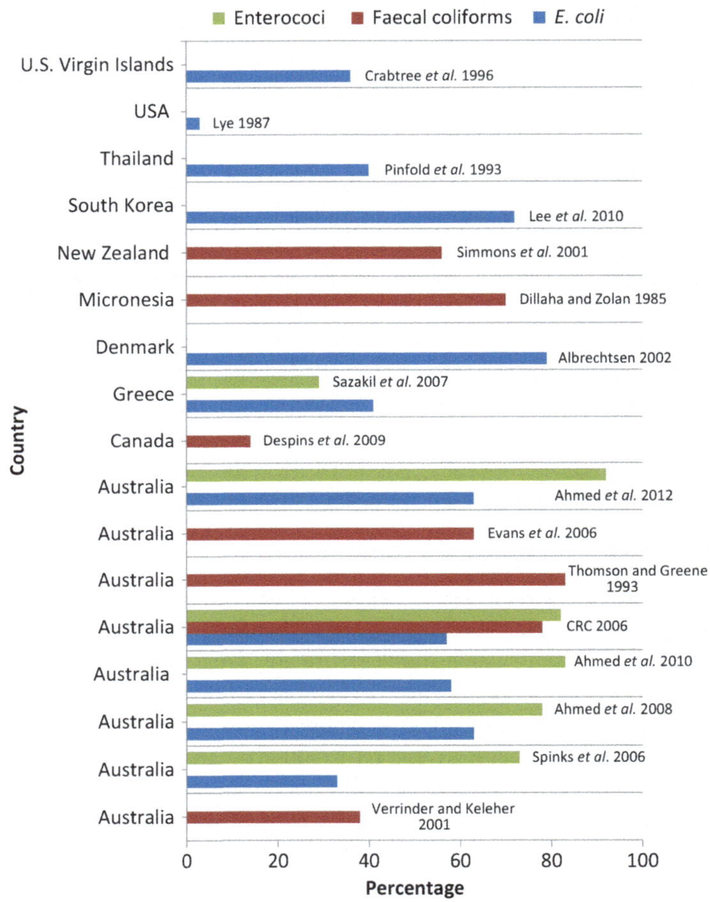

Figure 10.1 Percentage of samples positive for faecal indicators in tank water samples (Ahmed *et al.* 2011).

10.2.2 Bacterial pathogens

Faecal indicator bacteria such as *E. coli* and enterococci generally used to detect and estimate the level of faecal contamination in water. These are relatively harmless to human health but are used to indicate the presence of potential health risks. To date, only a relatively small number of studies have investigated the presence of bacterial pathogens in tank water samples (Table 10.1). In one of these studies, Simmons *et al.* (2001) reported the presence of *Salmonella* spp. in 1% of the 115 tank water samples tested in Auckland, New Zealand. *Campylobacter* spp. has also been detected in tank water samples in New Zealand using polymerase chain reaction (PCR) (Savill *et al.* 2001). In all, 37.5% of the samples tested were positive for *Campylobacter* spp., with numbers ranging from <0.06 to 0.56 most probable number (MPN)/100 mL tank water. *C. jejuni* have also been detected in tank water samples in Denmark (Albrechtsen, 2002). On the basis of these findings, the author concluded that connecting rainwater tanks to the drinking water systems would increase the level of risk of gastroenteritis and respiratory illness. A recent study that used PCR to detect evidence of bacterial pathogens in tank water in South East Queensland, Australia, found that approximately 20% of tank water samples were positive for *Campylobacter* spp. and *Salmonella* spp. that can cause gastrointestinal illness (Ahmed *et al.* 2010a). The authors suggested that tank water samples tend to decrease in microbiological quality after rain events, although the impact of duration of the intervening dry periods between rain events on runoff water quality remained unknown. Similarly, the tank storage time required after rain events for improvement in microbial quality also remained unknown.

Table 10.1 Percentage and number of samples positive for bacterial pathogens in rainwater samples.

Country	Percent of samples (number of samples in brackets) positive for bacterial pathogens				References
	***Campylobacter* spp.**	***Salmonella* spp.**	***Shigella* spp.**	***Vibrio* spp.**	
Australia	45 (27)	11 (27)	–	–	Ahmed *et al.* (2008)
Australia	20 (100)	17 (100)	–	–	Ahmed *et al.* (2010)
Australia	2 (67)	3 (67)	–	–	CRC WQT (2006)
Denmark	12 (17)	–	–	–	Albrechtsen (2002)
New Zealand	37 (24)	–	–	–	Savill *et al.* (2001)
New Zealand	ND (125)	1 (125)	–	–	Simmons *et al.* (2001)
Nigeria	–	67 (6)	67 (6)	67 (6)	Uba and Aghogho (2000)

ND: Not detected; –: Not tested.

10.2.3 Opportunistic bacterial pathogens

To date, only a small number of studies have investigated the presence of opportunistic bacterial pathogens in tank water samples (Table 10.2). Opportunistic pathogens such as *Legionella* spp. and *Aeromonas* spp. are considered to be of concern for respiratory and gastrointestinal aspects of human health. In addition, these opportunistic pathogens are also responsible for urinary tract infection, skin and soft tissue infections, pneumonia and wound infections. The links between water and *Legionella* infections is well known, and a recent study by Khajanchi *et al.* (2010) demonstrated a link between *Aeromonas* spp. isolated from clinical and water samples, indicating transmission from water. Simmons *et al.* (2001) reported the presence of *Aeromonas* spp. in RCR samples collected in Auckland, New Zealand, with 20% of the 125 samples tested being positive for *Aeromonas* spp.

Based on the positive detections obtained, the authors concluded that tank water samples were not suitable for drinking, and recommended further research on the *Aeromonas* spp. because of its high prevalence and association with gastroenteritis in both adults and children. *Aeromonas* spp., along with other opportunistic pathogens such as *Pseudomonas* spp., *Legionella* spp., and *Mycobacterium* spp., have also been detected in tank water samples in Denmark (Albrechtsen, 2002). The author concluded that if collected rainwater is used for drinking, it would increase the risk of illness from opportunistic pathogens.

Another study that used PCR to detect evidence of opportunistic bacterial pathogens in tank water in South East Queensland, Australia, found that around 7% of the 100 samples were positive for *Aeromonas hydrophila* (Ahmed *et al.* 2010a). In addition, 8% of the samples were positive for the respiratory opportunistic pathogen *L. pneumophila*.

Table 10.2 Detection of opportunistic bacterial pathogens in roof-captured rainwater.

Country	Percent of samples (number of samples in brackets) positive for opportunistic bacterial pathogens				References
	***Aeromonas* spp.**	***Pseudomonas* spp.**	***Legionella* spp.**	***Mycobacterium spp.***	
Australia	15 (27)	–	26 (27)	–	Ahmed *et al.* (2008)
Australia	7 (100)	–	8 (100)	–	Ahmed *et al.* (2010)
Australia	32 (56)	–	15 (67)	–	CRC WQT (2006)
Denmark	14 (14)	7 (14)	71 (7)	7 (14)	Albrechtsen (2002)
New Zealand	20 (125)	–	ND (125)	–	Simmons *et al.* (2001)
Nigeria	–	83 (6)	–	–	Uba and Aghogho (2000)
U.S. Virgin islands	–	–	80 (10)	–	Broadhead *et al.* (1988)

ND: Not detected; -: Not tested.

10.2.4 Protozoa pathogens

Despite a well-established zoonotic (i.e., transmission of diseases from animals to humans) link, the presence of protozoa pathogens in rainwater tanks has not been extensively investigated, with only a few studies examining tank water samples for the presence of *Giardia* spp. and *Cryptosporidium* spp. (Table 10.3). Crabtree *et al.* (1996) reported that 45% and 23% of 45 water samples tested from private and public rainwater systems in the U.S. Virgin Islands were positive for *Cryptosporidium* oocysts and *Giardia* cysts, respectively. The levels of cysts and oocysts were found to range from 1 to 10 organisms/100 L, with one sample containing 70 oocysts/100 L. The numbers of *Cryptosporidium* oocysts in all the positive samples were well above the acceptable guidelines of zero tolerance of protozoa pathogens. Simmons *et al.* (2001) also reported the presence of *Cryptosporidium* spp. in 4% of 50 tank water samples in Auckland, New Zealand. However, unlike Crabtree *et al.* (1996), Simmons *et al.* (2001) were unable to detect any *Giardia* cysts. Albrechtsen (2002) similarly reported the presence of *Cryptosporidium* spp. in Danish tank water samples. They tested 17 rainwater samples, of which 6 were positive for *Cryptosporidium* spp. The numbers of *Cryoptosporidium* spp. were as high as 50 oocysts/L. However, as reported in Simmons *et al.* (2001), *Giardia* spp. was not detected in any of the samples.

In contrast, another study in South East Queensland, Australia, reported the presence of *G. lamblia* in 19% of 21 tank water samples tested, yet none of the samples were positive for *C. parvum* (Ahmed *et al.* 2008, 2010). One possible reason for the differences in the type of protozoa pathogen detected in the Australian study and those from overseas may be the different wildlife that has access to roofs. In South East Queensland, for example, marsupial possums are a common animal in and around urban dwellings and frequently traverse roof tops in their nocturnal movements. Possums are known to be carriers of *Giardia* cysts (Ahmed *et al.* 2012a; Marino *et al.* 1992). Thus, possums could be a major source of *Giardia* spp. in tank water samples in South East Queensland, whilst birds or other wild animals are the likely source of the *Cryptosporidium* spp. in the overseas studies.

Table 10.3 Detection of protozoa pathogens in roof-captured rainwater.

Country	Percent of samples (number of samples in brackets) testing positive for protozoa pathogens		References
	***Cryptosporidium* spp.**	***Giardia* spp.**	
Australia	–	19 (21)	Ahmed *et al.* (2008)
Australia	ND	15 (100)	Ahmed *et al.* (2010)
Denmark	35 (17)	ND (17)	Albrechtsen (2002)
New Zealand	4 (50)	ND (125)	Simmons *et al.* (2001)
U.S. Virgin islands	45 (45)	23 (45)	Crabtree *et al.* (1996)

ND: Not detected; -: Not tested.

10.2.5 Likely sources of *Escherichia coli* harboring toxin genes in rainwater tanks

E. coli is often characterized as a *commensal* or harmless bacterium (Hartl *et al.* 1984). However, certain strains of *E. coli* can be pathogenic and responsible for both intestinal and extraintestinal infections (Kaper *et al.* 2004). It has been reported that faeces of warm-blooded animals may contain high numbers of *E. coli* bacteria carrying virulence genes (Ishii *et al.* 2007). These virulence genes allow pathogenic *E. coli* to cause a wide array of infections, such as diarrhoea, urinary tract infections (UTIs), neonatal meningitis, soft-tissue infections, and bacteremia. Pathogenic *E. coli* strains that are capable of causing diseases in humans and animals can be categorised as intestinal pathogenic *E. coli* (InPEC) or extraintestinal pathogenic *E. coli* (ExPEC) (Russo & Johnson, 2000). These pathotypes contain various combinations of virulence genes for the attachment and elaboration of hemolysins and enterotoxins (Bertin *et al.* 2001). ExPEC strains have the special ability to cause extraintestinal infections such as UTIs, neonatal meningitis and sepsis, and wound infections which can lead to serious complications and death. ExPEC strains possess virulence genes combinations that are distinct from those of strains that cause intestinal infections. Despite increasing evidence that *E. coli* strains from several animal hosts contain virulence genes, and the fact that some have been shown to cause intestinal and extraintestinal diseases in humans, none of the studies have determined whether *E. coli* bacteria found in rainwater tanks carry virulence genes and are potentially able to cause intestinal or extraintestinal infections in humans.

Ahmed *et al.* (2012b) report on a study to identify the likely sources of *E. coli* strains harbouring toxin genes in rainwater tanks. Understanding the contamination sources could enable better control of the sources, thus minimising the potential public health risks. In their study, 200 *Escherichia coli* isolates from 22 rainwater tank samples in SEQ, Australia were tested for the presence of ten toxin genes (stx_1, stx_2, *hlyA*, *ehxA*, LT1, ST1, *cdtB*, *eastl*, *cnfl*, and *cvaC*) associated with intestinal and extraintestinal pathotypes. 428 *E. coli* isolates from bird and possum faecal samples were also tested for the ten toxin genes as these animals had been previously identified as potential sources of faecal contamination of RCR (Ahmed *et al.* 2012a).

Among the 10 toxin genes tested, 4 genes (i.e., ST1, *eastl*, *cdtB*, and *cvaC*) were detected in 43 of 200 *E. coli* strains. The remaining toxin genes stx_1, stx_2, *hlyA*, *exhA*, LT1, and *cnfl* could not be detected in any of isolates tested. Among the 200 isolates tested, 8 (4%), 25 (13%), 19 (10%), and 1 (0.5%) strains were positive for the ST1, *east1*, *cdtB*, and *cvaC*, respectively (Table 10.4). Among the 214 *E. coli* isolates tested from birds, 30 (14%), 11 (5%) and 18 (8%) strains contained the *east1*, *cdtB*, and *cvaC* toxin genes, respectively. Similarly, among the 214 possum *E. coli* isolates, 74 (35%) contained only the *east1* toxin gene (Table 10.4). Biochemical phenotypes of 14 (33%) *E. coli* strains from seven rainwater tanks and nine (21%) *E. coli* strains from six rainwater tanks were identical to a number of biochemical phenotypes of *E. coli* strains isolated from bird and possum faeces, respectively, suggesting that these animals are probably the sources of these *E. coli* strains in rainwater tanks.

Table 10.4 Occurrence of *Escherichia coli* harbouring toxin gene in rainwater tanks, bird and possum faecal samples.

Samples	No. of *E. coli* tested	No. of *E. coli* isolates harbouring toxin genes (%)	Distribution of *E. coli* harbouring toxin genes into intestinal and extraintestinal pathotypes (%)			
			ETEC	EaggEC	EPEC/ExPEC	ExPEC
			ST1	*east1*	*cdtB*	*cvaC*
Rainwater tanks	200	43 (22)	8 (4)	25 (13)	19 (10)	1 (0.5)
Birds	214	55 (23)	ND	30 (14)	11 (5)	18 (8)
Possums	214	74 (35)	ND	74 (35)	ND	ND

ND: Not detected.

The presence of these toxin genes in *E. coli* strains from rainwater tanks is of concern as toxins are the most obvious virulence factors in pathogenic *E. coli*. For example, *E. coli* strains harbouring *east1* toxin gene alone have been reported to show a clear association with diarrhoea. Similarly, *E. coli* strains harbouring the *cdtB* or ST1 toxin genes are known to cause extraintestinal as well as intestinal infections. The authors acknowledged however that in the absence of any *in vivo* study, it was not possible to determine whether strains harbouring toxin genes in rainwater tanks were in fact capable of expressing pathogenicity, and because of that, these strains can be considered only as potential pathogenic strains.

To identify the likely sources of these potential clinically significant *E. coli* in rainwater tanks, a source-tracking approach was undertaken. All *E. coli* strains isolated from bird and possum faecal samples harbouring toxin genes were biochemically fingerprinted. The principle of the biochemical fingerprinting with the PhPlate system has been described previously (Möllby *et al.* 2003). This method uses quantitative measurements of the kinetics of several biochemical reactions of bacteria in microtiter plates with dehydrated substrates (Möllby *et al.* 2003). Biochemical fingerprints of *E. coli* obtained from rainwater tanks were compared with those found in bird and possum faeces cal to establish a potential link.

Of the 43 strains from rainwater tank samples, 14 (from seven tanks) and nine (from six tanks) had identical biochemical phenotypes to those found in bird and possum faecal samples, respectively. Five strains from 4 rainwater tanks were identical to those isolated from both bird and possum faecal samples. Based on the results, the authors concluded that the presence of potential clinically significant *E. coli* in rainwater tanks most likely originated from bird and possum faeces.

Based on these results, Ahmed *et al.* 2012 a,b recommended that rainwater be treated with effective disinfection procedures such as filtration, ultraviolet disinfection or simply boiling the water prior to drinking. Maintenance of good roof and gutter hygiene and elimination of overhanging tree branches and other structures where possible to prevent the flocking of possums and birds should be considered for improved water quality.

10.2.6 Presence and source of faecal indicators and zoonotic pathogens in household drinking water taps fed from rainwater tanks in South East Queensland

Many householders who drink tank water use an under-sink filtration (USF) system to reduce exposure to pathogenic microorganisms, suspended solids and harmful chemicals. Little is known regarding the prevalence of zoonotic pathogens in household drinking water taps fed by rainwater tanks. Little is also known of the prevalence of zoonotic pathogens in wild animals, such as birds and mammals, which are most likely contaminating rainwater tanks. Mammals can get access to the roof via overhanging trees or electricity cables, or by climbing to the roof via walls or other structures attached to the house. Birds are attracted to a roof by overhanging trees or structures mounted on the roof, such as television aerials and solar panels. Knowing the source of pathogenic microorganisms is important in order to design management strategies to reduce public health risks.

Ahmed *et al.* (2012a) investigated the prevalence and numbers of faecal indicators (*E. coli* and enterococci) and zoonotic bacterial (*Campylobacter* spp. and *Salmonella* spp.) and protozoa (*Cryptosporidium parvum* and *Giardia lamblia*) pathogens in water samples from rainwater tanks and corresponding connected household taps in South East Queensland. They also investigated the prevalence of the above-mentioned pathogens in faecal samples from possums and various species of wild birds. Conventional culture-based methods were used to enumerate *E. coli* and enterococci, and qPCR was used to obtain the numbers of zoonotic pathogens in roof captured rainwater (RCR), connected household tap water (CHTW), and animal faecal samples.

Their study area, Currumbin Ecovillage, is located on the southern end of the Gold Coast, South East Queensland, Australia. The Ecovillage is known for its sustainable residential developments and is often viewed as a blueprint for future urban development. Twenty-four households participated in the study. All the households use RCR for drinking, as well as other non-potable water uses such as car washing, clothes laundering, showering, and gardening. Two water samples were collected from each household (i.e., one from the rainwater tank and one from the connected household tap), giving a total of 48 samples from the 24 households. Samples were collected within 1 to 4 days after a rain event. Brush tail possum faecal samples ($n = 40$) were obtained from the possum removal service. Bird faecal samples ($n = 38$) were collected from botanical gardens, bird sanctuaries, and a veterinary hospital. DNA was extracted from all water and animal faecal samples and tested with qPCR.

Amongst the 24 households, *E. coli* was cultured from 15 (62%) RCR and 14 (58%) connected household tap water (CHTW) samples. Similarly, 22 (92%) RCR and 20 (83%) CHTW samples contained cultured enterococci. The numbers of *E. coli* bacteria in these samples ranged from 1 to 230/100 mL (for RCR) and 1 to 300 CFU/100 mL of water for CHTW. For enterococci, the numbers were 2 to 110 CFU/100 mL (for RCR) and 1 to 110 CFU/100 mL (for CHTW). Among the 24 households, 5 (21%), 1 (4%), and 3 (13%) RCR samples contained *Campylobacter* spp., *Salmonella* spp., and *G. lamblia*, respectively (Table 10.5). Similarly, 5 (21%) and 3 (13%) of the CHTW samples contained *Campylobacter* spp. and *G. lamblia*, respectively. The *Salmonella* spp. could not be detected in CHTW samples. The numbers of *Campylobacter* cells in RCR and household tap water samples ranged from 5 to 110 (in RCR) and 12 to 19 (in CHTW) cells/L of water. Similarly the estimated number of *Salmonella* cells was 7,300 (in RCR)/L of water. The numbers of *G. lamblia* cysts ranged from 120 to 580 (in RCR) and 110 to 140 (in CHTW)/L of water. The numbers of faecal indicators and pathogens were pooled for all RCR and CHTW samples to determine whether the numbers were correlated between RCR and CHTW samples.

Table 10.5 Numbers of zoonotic pathogens in roof captured rainwater and connected household tap water samples in four houses at Currumbin Ecovillage.

Household ID	Numbers (mean) of bacterial and protozoa pathogens/L of water			
	Campylobacter spp.		*G. lamblia*	
	RCR	CHTW	RCR	CHTW
H1	ND	ND	120	140
H7	ND	ND	160	140
H14	5	12	580	110
H15	30	19	ND	ND

ND: Not detected; RCR: Roof-captured rainwater; CHTW: Connected household tap water.

The number of *E. coli* ($P > 0.78$), enterococci ($P > 0.64$), *Campylobacter* ($P > 0.44$), and *G. lamblia* ($P > 0.50$) in RCR did not differ significantly from the numbers observed in the CHTW samples. Table 10.5 shows the numbers of *Campylobacter* spp. and *G. lamblia* detected in RCR and CHTW samples in four households.

Around 58% of households in this study did not use any filtration methods; therefore, the presence of faecal indicators and zoonotic pathogens in the CHTW samples was not unexpected. Ten of the 24 households had USF installed; however, these systems did not appear to be effective in removing faecal indicators and zoonotic pathogens.

G. lamblia was detected in three of the 24 tanks tested in this study. All three corresponding CHTW samples also contained *G. lamblia*. It should be noted that these households did not apply any filtration methods prior to drinking. The high numbers of *G. lamblia* cells in both tank water (120 to 580 cysts/L) and connected household tap water samples (110 to 140 cysts/L) may pose health risks to the consumers because of the low (1–10 cysts) infectious dose of *G. lamblia*. A little is known regarding the occurrence of Giardiasis in the community. Giardiasis is not notifiable in Queensland by clinicians but several medical doctors have expressed their concerns that rainwater may have increased the levels of Giardiasis in the community. It is also possible that the qPCR method used in the study overestimated the levels of *G. lamblia* in tank water samples because PCR cannot distinguish between viable and non-viable cysts.

Among the 40 possum faecal samples tested, *Campylobacter* spp., *Cryptosporidium parvum*, and *G. lamblia* were detected in 60%, 13%, and 30% of samples, respectively. Among the 38 bird faecal samples tested, *Campylobacter* spp., *Salmonella* spp., *C. parvum*, and *G. lamblia* were detected in 24%, 11%, 5%, and 13% of the samples, respectively. In all, 60% of possum and 24% of bird faecal samples contained *Campylobacter* spp. All bird faecal samples contained *C. jejuni* but none of the possum faecal samples contained *C. jejuni*. Possum and bird faecal samples also contained *G. lamblia*, with the numbers of cysts ranging from 21 to 1600 (for possums) and 1.3 to 120 (for birds) per gram of faeces suggesting these animals are the likely sources of *G. lamblia* in rainwater tanks.

10.2.7 Inactivation of faecal indicator bacteria in a roof-captured rainwater system

The presence of clinically significant *E. coli* in rainwater tank samples in South East Queensland (Ahmed *et al.* 2012b) raises questions regarding the persistence of faecal indicator bacteria and pathogens in rainwater tank samples, as well as in faecal deposits on the roof and in the gutter. A range of climatic and biological factors have been shown to influence the inactivation of faecal indicator bacteria and pathogens. These factors include temperature, moisture content, solar radiation, relative humidity, pathogen type, presence of biodegradable organic matter and interaction with other micro-organisms. The faecal indicator bacteria and pathogens from bird and animal droppings deposited on the roofs and in the gutters are expected to inactivate rapidly due to harsh environmental conditions such as temperature, UV radiations and loss of moisture. However, certain conditions such as the shaded portion of the roof, precipitation, availability of biodegradable organic matter in the gutter may prolong the inactivation of faecal indicator bacteria and pathogens. However, very little has been documented about the inactivation of bacteria and pathogens on the roofs, in the gutters, and in the tank water.

Ahmed *et al.* (2014) undertook experiments in South East Queensland, Australia to obtain information on the inactivation of faecal indicator bacteria in those time periods between faeces being deposited on the roof and washing off into the tank. The study investigated the persistence of *E. coli* and enterococci on the roof and in the gutter of a ‘model’ RCR system to determine: (i) the time required to achieve a one $\log_{10}$ (i.e., 90% or T_{90}) reduction of *E. coli* and enterococci on the corrugated iron roof and in the gutters; and (ii) the inactivation time (T_{90}) of faecal indicator bacteria in the tank water.

A model RCR system (similar set up to RCR systems commonly in use in Australian domestic dwellings) was built comprising a 5000 L polyethylene tank, 2 m^2 roof constructed of corrugated iron sheets with steel guttering and plastic downpipe leading water into the tank. The tank was placed in direct sunlight (received minimum 5 h sunlight per day during the experiment). Faecal indicator bacteria contamination of roof and gutter was simulated using homogenized possum faecal slurry spiked with known numbers of *E. coli* and enterococci.

For the roof and gutter inactivation experiments, 5 mL of faecal slurry was poured into a series of 50 mm petri dishes and placed on the corrugated iron roof and in the gutter of the experimental structure. The petri dishes were exposed to diurnal cycles of insolation. Moist sediment containing vegetation and organic debris was added to the gutters to simulate typical unclean, urban household gutters. The petri dishes were kept under the vegetation and organic debris and the numbers of surviving faecal indicator bacteria were enumerated. The inactivation experiment in tank water was undertaken using diffusion chambers as described by Toze *et al.* (2004).

For each faecal indicator bacteria all determined numbers in each replicate at each sampling occasion were converted to $\log_{10}$ values and plotted over time. One $\log_{10}$ reduction time (T_{90}) for each fecal indicator bacteria was determined from each plot using the following equation as previously described (Gordon & Toze, 2003).

$$T_{90} = -t/(\log_{10} C_t/C_0)$$

where C_0 is the number (CFU/mL) at day 0, C_t is the final number (CFU/mL) at day t. A linear regression was fitted to each plot and the slope was taken as the inactivation rate. The inverse of these calculated inactivation rates was then used as the determination of the one $\log_{10}$ reduction time (T_{90}). The average T_{90} on each sampling occasion was determined from replicates of each faecal indicator bacteria. Where the inactivation in some experiments was biphasic, two T_{90} values were calculated, one for the initial inactivation (first phase) and the other for the second stage of the inactivation (second phase). An analysis of variance (ANOVA) was performed on the T_{90} values on the roof and in the gutter under different conditions. For statistical comparison, T_{90} values derived from the first phases of various experimental conditions were used. Details of the statistical analysis can be found in Ahmed *et al.* (2014).

The inactivation rates of faecal indicator bacteria in possum faecal slurry placed on the roof were evaluated under sunlight and shade conditions and are shown in Figure 10.2. Under direct sunlight, *E. coli* is rapidly inactivated ($T_{90} = 2$ h) compared to shade, where a slow non-linear (biphasic) inactivation rate [$T_{90} = 53$ h (first phase) and 9 h (second phase)] was observed (Table 10.6). Similar results were also obtained for enterococci. No significant ($P > 0.05$) difference was observed in T_{90} value of *E. coli* compared to enterococci under sunlight conditions. Significant ($P < 0.001$) difference, however, was observed in T_{90} value of *E. coli* compared to enterococci for shade conditions (Table 10.6).

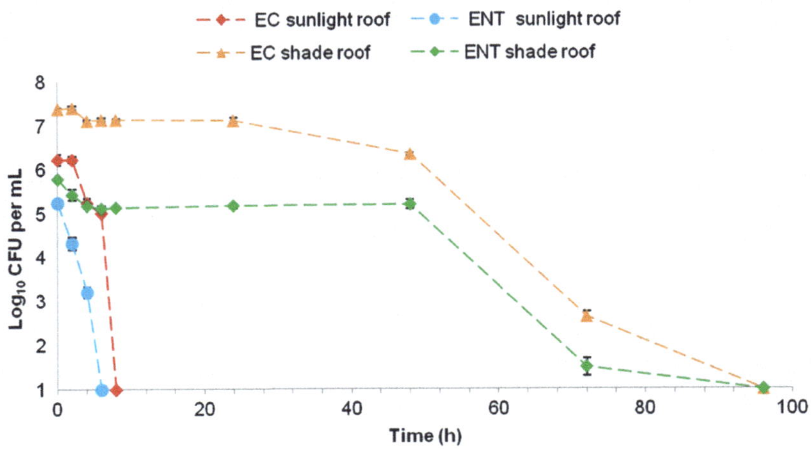

Figure 10.2 Log_{10} (mean ± standard deviation) colony forming units (CFU) of culturable *Escherichia coli* (EC) and *enterococci.* (ENT) during the course of inactivation on the roof under direct sunlight and shade conditions.

Table 10.6 T_{90} inactivation time of *Escherichia coli* and enterococci on the roof, in the gutter and tank water. R^2 of the regression line used to calculate T_{90} is shown in brackets.

Faecal indicators	Experiments	Conditions	T_{90} (h) (R^2)	
			First phase	Second phase
E. coli	Roof	Sunlight	2 (0.72)	
		Shade	53 (0.85)	9 (0.95)[a]
	Gutter	Clean	22 (0.66)	3 (0.99)[a]
		Unclean	20 (0.73)	6 (0.88)[a]
	Tank water		72 (0.86)	273 (0.84)[c]
Enterococci	Roof	Sunlight	2 (0.95)	
		Shade	9 (0.92)	18 (0.80)[c]
	Gutter	Clean	2 (0.80)	
		Unclean	6 (0.97)	
	Tank water		38 (0.99)	195 (0.94)[d]

[a]Three data points were used to calculate T_{90} inactivation times.
[b]Four data points were used to calculate T_{90} inactivation times.
[c]Five data points were used to calculate T_{90} inactivation times.
[d]Six data points were used to calculate T_{90} inactivation times.
R^2: Coefficient of determination.

The inactivation rates of faecal indicator bacteria were evaluated in the clean and unclean gutters under direct sunlight (Figure 10.3). The organic matter and vegetation in the unclean gutter shaded the faecal slurry from sunlight. *E. coli* showed biphasic inactivation rates under both clean and unclean gutter conditions [T_{90} = 22 h (first phase) and 3 h (second phase)] for the clean gutter conditions and [T_{90} = 20 h (first phase) and 6 h (second phase) for the unclean gutter conditions. Enterococci showed much more rapid initial inactivation (T_{90} = 2 h) for the clean gutter and the unclean gutter (T_{90} = 6 h) compared to *E. coli*. The T_{90} values for *E. coli* and enterococci were significantly different ($P < 0.001$) from each other for both clean and dirty gutters. Significant ($P < 0.001$) differences in T_{90} values occurred for enterococci between clean and unclean gutter conditions.

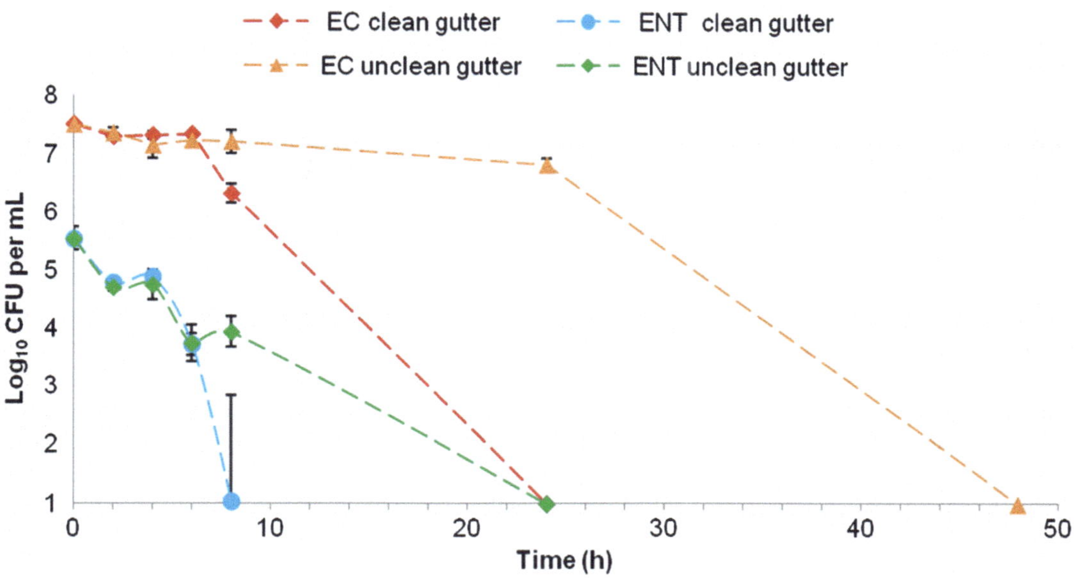

Figure 10.3 $\log_{10}$ (mean ± standard deviation) colony forming units (CFU) of culturable *Escherichia coli* (EC) and *Enterococcus* spp. (ENT) during the course of inactivation in the clean and unclean gutters.

The inactivation rates of faecal indicator bacteria were determined under *in-situ* conditions in the rainwater tank exposed to natural sunlight (Figure 10.4). Here *E. coli* fell below detection limit after 576 h whereas enterococci were detected up to 816 h. Both *E. coli* [T_{90} = 72 h (first phase) and 273 h (second phase)] and enterococci [T_{90} = 38 h (first phase) and 195 h (second phase)] showed non-linear biphasic inactivation. Significant difference was observed between the T_{90} value of *E. coli* inactivation compared to enterococci (paired t-test, $P = 0.0003$) in the tank water.

In conclusion, faecal indicator bacteria, especially *E. coli*, can survive longer (T_{90} = 53 h) on the roof under shade conditions compared to sunlight conditions. This could have an impact on health risks associated with tank water use. If there is a rainfall event after the deposition of faecal matter on a shaded roof, it is highly likely that faecal indicator bacteria and other faecal pathogens could be transported to the tank water. When introduced to the tank, a slower inactivation process may take place (T_{90} = 38–72 h). Further research is required to understand the persistence of bacterial and protozoa pathogens on the roof and in tank water in comparison to faecal indicator bacteria because certain pathogens are known to be more persistent in the environment than faecal indicator bacteria.

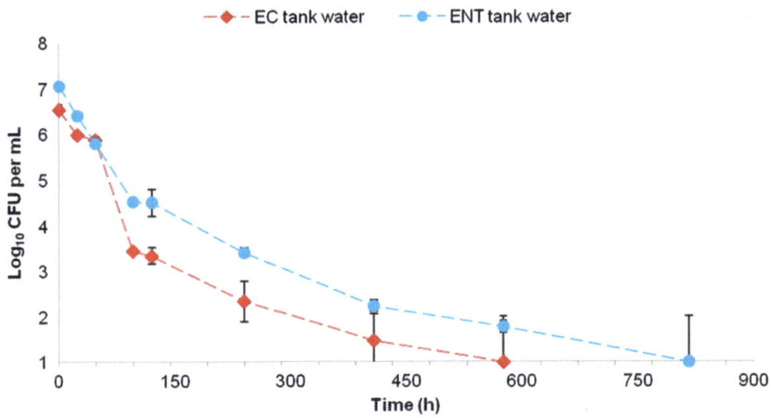

Figure 10.4 Log_{10} (mean ± standard deviation) colony forming units (CFU) of culturable *Escherichia coli* (EC) and enterococci (ENT) during the course of inactivation in the tank water.

10.3 HEALTH RISKS ASSOCIATED WITH ROOF CAPTURED RAINWATER

While most studies relating to RCR have focused on the detection of microorganisms in the tank water samples, there have also been eight case control studies reporting on sporadic gastroenteritis associated with the consumption of untreated tank water (Figure 10.5). The factors that can influence the level of actual risk include the type and numbers of pathogen carried by the infected animals, the time between deposition of faecal matter on the roof and pathogens being flushed into the tank, the size (kL) of the tank, the form of human exposure (consumption from drinking vs. exposure to droplets in the shower or toilet flushing), and the relative inactivation of the different pathogens in the roof to tank ecosystem.

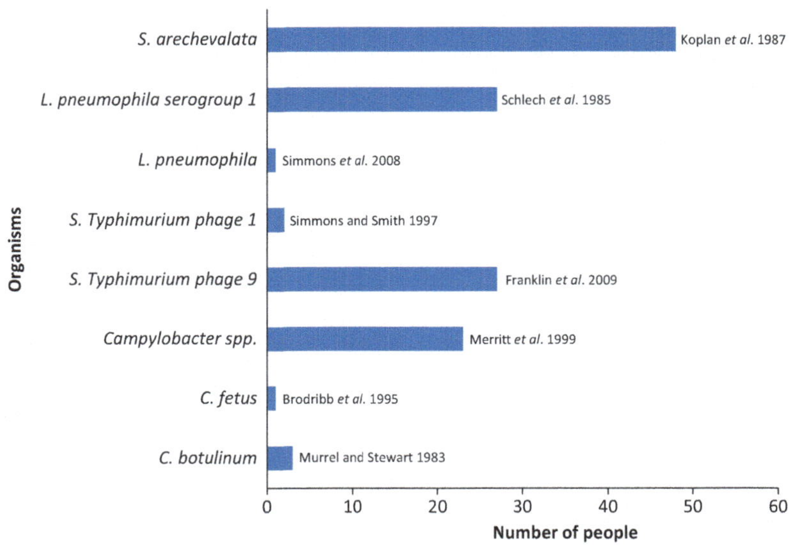

Figure 10.5 Reported cases of illnesses associated with the use of roof-captured rainwater.

An example of how all these factors can create a human health risk is a *Salmonella arechevalata*-related gastroenteritis outbreak reported by Koplan *et al.* (1978) that occurred among 83 campers (48 positive cases) in Trinidad, West Indies. Epidemiological and bacteriological studies were performed to identify the source of *S. arechevalata* infection. Through patient surveys, food items were ruled out as the source of the gastroenteritis. However, water samples collected from two kitchen taps connected to a RCR system were found to be positive for *S. arechevalata*, although the bacterium could not be isolated from the tank water sample itself. A sanitary survey suggested that rainwater washed faecal matter containing *S. arechevalata* from the roof into the tank, leading to the gastroenteritis outbreak. To further test this hypothesis, a number of intestinal samples were collected from local birds, rodents, and reptiles that were assumed to be the source of the contamination. Somewhat surprisingly, *S. arechevalata* could not be isolated from the faeces of these animals.

Another description of a potential link between pathogens and tank water was an outbreak of three cases of infant botulism in New South Wales, Australia, where *Clostridium botulinum* type B was isolated from soil around one house and in the rainwater tank from another house (Murrell & Stewart, 1983). *Clostridium botulinum* type A was also present in soil, dust from a vacuum cleaner, and the rainwater tanks. The presence of *C. botulinum* spores in the rainwater tanks was suggested as contributing to the occurrence of the infant botulism cases. As a result of this study, consumers with infant children were advised to disinfect water for the first 6 months of the infants' lives.

Another study reported the isolation of *Campylobacter fetus* from a 64 year old patient, which was subsequently linked to the tank water (Brodribb *et al.* 1995). Three sets of blood cultures from the patient were positive for *C. fetus* using PCR. The rainwater was the only source of water supply for this particular household. To confirm the source of infection, a sample of the tank water was tested and *C. fetus* was isolated from as little as 200 mL of water sample. The patient was advised to boil the tank water before consumption, and no further incidence of the illness was reported.

Merritt *et al.* (1999) reported an outbreak of *Campylobacter* enteritis among 23 resort staff in Queensland, Australia where untreated tank water was used as a potable supply. Food was initially suspected as the possible source of infection. However, none of the food samples were positive for the *Campylobacter* spp., but four rainwater tanks were positive for total coliforms and *E. coli*, with particularly high total coliform numbers found in one tank. The authors reported a strong association between gastroenteritis and consumption of water from a dispenser that had probably been filled from one of the contaminated tanks. It was hypothesized that the *Campylobacter* spp. that caused the outbreak may have been introduced into one or more of the tanks by contamination with the faeces of wild animals.

Another reported outbreak of gastroenteritis with a strong link to contaminated tank water occurred in Melbourne, Australia. An outbreak of *Salmonella* Typhimurium occurred at a rural school camp supplied by a rainwater tank (Franklin *et al.* 2008). Twenty seven of 55 students at the camp showed the disease symptoms (gastroenteritis). *Salmonella* (phage type strain DT9) was found in both the faecal specimens of patients and water taps supplying the untreated rainwater used for drinking, indicating a direct link between the tank water and the disease outbreak. Simmons and Smith (1997) also reported the isolation of *S.* Typhimurium (phage type I) from two of four family members who sought medical attention due to gastrointestinal symptoms. The family lived in a beach-side house and used tank water for all household uses. *S.* Typhimurium phage type I and faecal coliforms were isolated from the house's tap water, leading the authors to conclude that tank water was the likely source of infection, although the original contamination source in the tank water could not be confirmed.

An outbreak of Legionnaires' disease in an isolated suburb of Auckland, New Zealand, was linked to tank water. Isolates of *L. pneumophila* serogroup 1 (Lpl) from patient's clinical specimens (one out of four) were identical to the same serogroup of *L. pneumophila* present in the nozzle of a local marina water blaster used to clean boats. Sampling of nearby rainwater collection systems revealed that spray from the water blaster had been deposited on the roofs of houses in the local area. The *L. pneumophila* was washed into rainwater tanks, and residents were then exposed through bathroom showers (Simmons *et al.* 2008).

In a reported outbreak of 27 cases of Legionnaire's disease amongst tourists visiting the U.S. Virgin Islands, the infections were thought to have originated from inhalation of *L. pneumophila* detected in the drinking water system of a local hotel (Schlech *et al.* 1985). Case control and microbiological studies were undertaken to identify the mode of transmission. The exact mode of transmission during this outbreak could not be determined, but the potable water was strongly implicated as the most probable cause in the absence of other sources. The hotel obtained its potable water from a RCR system, and the identical serogroup of the *L. pneumophila* isolated from the infected patients was found in the stored tank water, as well as in hot and cold water taps. No further cases of Legionellosis were identified after the hotel water system was chlorinated.

The most recent incidence of RCR-acquired infections was reported in a case control study in Tasmania, Australia, where single variable associations were found between drinking untreated water and cases of infection with *Salmonella mississippi* (Ashbolt & Kirk, 2006). The second highest risk was found to be associated with exposure to untreated tank water.

The above studies strongly suggest that the untreated rainwater may be a causal or contributing factor for gastroenteritis and other illnesses. Several authors highlighted the need for evaluating the actual health risks from drinking untreated tank water (Ahmed *et al.* 2008; Lye, 1992; Simmons *et al.* 2001). An epidemiological study to identify the risk of gastroenteritis among 4 to 6 year old children who drank tank water compared with children who ingested treated mains water in South Australia noted that the consumption of tank water did not increase the level of self reported gastroenteritis relative to mains water consumption (Heyworth *et al.* 2006). The authors concluded, however, that their data could also have reflected a level of acquired immunity among regular users of tank water, and therefore may not reflect the actual risk to the new users. A later study investigated whether drinking untreated rainwater contributed to the incidence of gastroenteritis in the community in Adelaide, Australia (Rodrigo *et al.* 2010). The authors conducted a double-blinded, randomised controlled trial installing sham water treatment units and recorded incidences of gastroenteritis for 12 months in a total of 300 households. The results clearly indicated that consumption of untreated rainwater did not contribute appreciably to community gastroenteritis. However, the generality of the results must be qualified as susceptible and immunocompromised persons were not included in the study.

Dean and Hunter (2012), in a search of relevant bibliographic databases, also reported that there was no significant difference in health risks between RCR and water supplies. However, their analysis only included a small number of studies. In contrast, in New Zealand, numerous cases of campylobacteriosis have been associated with rainwater collection systems (Eberhart-Phillips *et al.* 1997). A case control study found a strong association between gastrointestinal diseases caused by *Campylobacter* spp. and the consumption of tank water. It was hypothesised that the RCR systems in their study had been contaminated by birds roosting on the roofs.

It is also probable that incidences of gastrointestinal diseases associated with tank water are underreported. It is generally accepted that only a portion of people with gastroenteritis seek medical attention, and most faecal specimens (if collected) from patients are not tested for the presence of pathogens. In Australia, it has been estimated that only 8% to 11% of *Campylobacter*- and *Salmonella*-related food-borne gastroenteritis cases are reported (Hall *et al.* 2006), whereas in the United States, it has been estimated that only 10% to 33% of water-related gastroenteritis is reported (Frost *et al.* 1996). These statistics suggest that any gastroenteritis cases from ingesting tank water would have at least similarly low reporting levels, if not even lower.

Another limitation is that cases of gastroenteritis from drinking untreated rainwater could also be masked by the background levels of gastroenteritis from other sources, such as consumption of food and community-based infections. The most credible epidemiological study to date reported that the consumption of tank water did not increase the risk of gastroenteritis compared to consuming mains water (Heyworth *et al.* 2006; Rodrigo *et al.* 2010). However, such results should be interpreted with care due to the lack of sensitivity of the epidemiological tool to detect gastroenteritis (Craun *et al.* 2004; Hrudey & Hrudey, 2004). It has been estimated that to detect illness at an annual rate of 100 cases/10,000 people/year would require samples of 416,000 participants (Eisenberg *et al.* 2006). Considering the high costs and time required, epidemiological studies of infection from sources such as RCR may not be practical for the sensitive detection of gastroenteritis in the community.

10.3.1 Quantitative Microbial Risk Assessment to determine health risk from the use of roof-captured rainwater

Risks associated with contamination of RCR may be difficult to document using epidemiological tools since infections are likely to affect a small number of individuals. More comprehensive approaches are needed to examine relative risks and risk-reduction strategies. One such approach is Quantitative Microbial Risk Assessment (QMRA), which can examine and account for various environmental biases associated with environmental trials (Enger *et al.* 2012). QMRA is a four-step probabilistic tool for estimating the human health risk associated with defined scenarios from exposure to specified pathogens (NRC, 1983). The four steps are: (i) hazard identification; (ii) exposure assessment; (iii) dose – response assessment; and (iv) risk characterisation. Although, QMRA models have been used to quantify disease risk in many contexts, only a few studies have attempted to identify the inherent risk of infection associated with the drinking and non-potable uses of tank water. One such study was undertaken by Fewtrell and Kay (2007), who investigated the risk of infection of *Campylobacter* from toilet flushing with tank water in homes in the United Kingdom. A QMRA estimate was performed on *Campylobacter* infection via ingestion of aerosols. The outcomes of this QMRA estimate concluded that any risk from flushing the toilet with captured rainwater would be well within the acceptable range of 1×10^{-6} DALYs (disability-adjusted life years) per person per year.

QMRA was also used to quantify the risk of infection associated with the exposure to zoonotic pathogens from drinking and non-potable uses of tank water in South East Queensland, Australia. Ahmed *et al.* (2010b) applied QMRA analysis to estimate the health risk from the use of roof-captured rainwater as potable or non-potable water. The aims of this research study were to: (i) to quantify the number and frequency of occurrence of *Salmonella*, *G. lamblia*, and *L. pneumophila* microorganisms in 214 tank water samples from 82 rainwater tanks in South East Queensland by using qPCR based methods; and (ii) to apply QMRA analysis in order to estimate the risk of infection from exposure to these pathogens found in RCR. The uniqueness of this study stems from the fact that instead of measuring faecal indicators, pathogens capable of causing illness were quantified and this ingested dose information was combined with QMRA to assess the human health risk of using RCR for potable and non-potable end uses.

The authors used qPCR to obtain the numbers of potential pathogens in rainwater tank samples. A subset of samples were also analysed to identify the occurrence of pathogens in tank water samples. The authors used a four-step QMRA analysis for estimating the human health risk associated with defined scenarios involving exposure to specified pathogens. The pathogen number in tank water and the volume ingested/inhaled by a person are estimated. Because of a lack of information regarding the proportion of PCR-detected cells and cysts that are viable, it was assumed all the PCR-detected cells and cysts were viable and capable of causing infections (conservative approach). To estimate the possible pathogen dose received by an individual, the likely infection routes appropriate to each pathogen must be considered. Considering possible routes, the infection risk associated with each of a total of six scenarios was estimated.

For salmonellosis and giardiasis risk, the scenarios were (i) liquid ingestion due to drinking of rainwater on a daily basis, (ii) accidental liquid ingestion due to garden hosing twice a week, (iii) aerosol ingestion due to showering on a daily basis, and (iv) aerosol ingestion due to hosing twice a week. For legionellosis risk, the scenarios were (i) aerosol inhalation due to showering on a daily basis and (ii) aerosol inhalation due to hosing twice a week. The volumes ingested and inhaled were extracted from relevant literature. An exponential dose-response model was used for *G. lamblia*, while for *Salmonella*, a beta-Poisson dose response relationship was used. Finally the risks were characterized by combining dose-response assessment and the probability of infection (expressed as likely numbers of infections per 10,000 persons per year) for the urban South East Queensland community, and comparison with an arbitrary but commonly accepted risk level of one extra infection per 10,000 persons per year.

Of the 214 samples tested during phases one and two, the *Salmonella* spp., *G. lamblia* and *L. pneumophila* were detected in 23 (10.7%), 21 (9.8%), and 12 (5.6%) rainwater samples, respectively. The numbers of *Salmonella* spp., *G. lamblia* and *L. pneumophila* in quantifiable samples ranged from 65 to 380 cells/L, 0.6 to 3.6 cysts/L, and 60 to 170 cells/L of water, respectively. The risk of infection from *Salmonella* spp., *G. lamblia*, and *L. pneumophila* associated with the use of rainwater for showering and garden hosing was calculated to be well below the threshold value of one extra infection/10,000 persons/year in urban South East Queensland. However, the calculations for infection based on the 'worst-case' assumptions indicated that if undisinfected rainwater were ingested by drinking, then the infection incidence is expected to range from 9.8 to 54 (*Salmonella* spp.) and from 20 to 130 (*G. lamblia*) cases per 10,000 persons/year in urban South East Queensland per year (Table 10.7).

Table 10.7 Infection risks for individuals exposed to contaminated tank water for risk scenarios determined using a QMRA analysis of 214 water samples from 82 rainwater tanks in South East Queensland.

Pathogen exposure and risk scenario	Range of infection risk/10,000 people in South East Queensland from a single event	Range of infection risk/year (no./10,000 people/year)
***Salmonella* spp.**		
Liquid ingestion via drinking	5.4×10^{-1}–2.9×10^{0}	9.8×10^{0}–5.3×10^{1}
Liquid ingestion via hosing	3.0×10^{-3}–1.5×10^{-2}	1.0×10^{-2}–8.0×10^{-2}
Aerosol ingestion via showering	1.0×10^{-3}–5.0×10^{-3}	1.9×10^{-2}–1.0×10^{-1}
Aerosol ingestion via hosing	5.0×10^{-6}–2.9×10^{-5}	2.6×10^{-5}–1.5×10^{-4}
G. lamblia		
Liquid ingestion via drinking	1.1×10^{0}–6.9×10^{0}	2.0×10^{1}–1.3×10^{2}
Liquid ingestion via hosing	5.0×10^{-3}–3.4×10^{-2}	3.0×10^{-2}–1.8×10^{-1}
Aerosol ingestion via showering	2.1×10^{-3}–1.3×10^{-2}	3.9×10^{-2}–2.4×10^{-1}
Aerosol ingestion via hosing	1.0×10^{-5}–6.5×10^{-5}	5.3×10^{-5}–3.3×10^{-4}
L. pneumophila		
Aerosol inhalation via showering	1.4×10^{-4}–4.0×10^{-4}	2.6×10^{-3}–7.3×10^{-3}
Aerosol inhalation via hosing	4.0×10^{-4}–1.1×10^{-3}	2.1×10^{-3}–5.8×10^{-3}

One major limitation of the health risk estimates presented in the study is that PCR method does not provide information regarding the viability and infectivity of detected pathogens, which is critical for QMRA analysis. In the absence of such data, the authors assumed that 100% of the PCR-detected cells and cysts are viable and infective. This assumption is likely to overestimate the risk of infection, as 100% of the PCR-detected cells and cysts may not be viable. However, the overestimation of risk could be preferable to an underestimation obtained via culture-based methods.

L. pneumophila at the levels detected in the RCR samples, did not present a threat for uses of tank water as potable water. Uses of the tank water as non-potable water also presented no unacceptable threat to human health at the pathogen numbers detected.

This study concluded that the risk of infection from *G. lamblia* and *Salmonella* spp. associated with the use of rainwater for bi-weekly garden hosing was below the acceptable threshold value of one extra infection/10,000 persons/year. However, the estimated risk of infection from drinking the rainwater daily was 20 to 130 (for *G. lamblia*) and 9.8 to 54 (for *Salmonella* spp.) infections/10,000 persons/year. Both of these risk assessments are well above the acceptable guideline levels outlined in the Potable Reuse section of the Australian Guidelines for Water Reuse (NRMMC–EPHC–NHMRC, 2008).

Nonetheless, the overall predicted health risk appeared to be much higher than the reported incidences of giardiasis in the community where the study was undertaken. The incidences of these diseases are 5.7 cases of salmonellosis per 10,000 people in Queensland and up to 5 cases of giardiasis per 10,000 people in other Australian states over the past 10 years (noting that giardiasis is not a notifiable disease in South East Queensland). A number of explanations for this discrepancy are possible. There is a naturally high incidence of gastroenteritis in the community (e.g., 8000 cases per 10,000 people per year), which may mask the actual disease (Hellard *et al.* 2001). Before the disease can be reported in the Notifiable Diseases Surveillance System Database, it must first be identified, and not every individual will seek medical attention if the illness is mild and lasts only for a few days. Consequently, the incidence of disease indicated in the Notifiable Diseases Surveillance System Database is at best a minimum value and may be substantially underestimating actual disease incidence.

The methodology used to estimate health risk could have inflated the risk calculated due to the assumption of 100% of pathogen cells or cysts being viable and infective. In fact, a significant proportion of PCR detected cells or cysts may have originated from nonviable organisms. The QMRA also did not take into account households that used effective disinfection treatment of rainwater before using it as potable water. Another factor is the possibility of individuals acquiring immunity to certain pathogens due to frequent exposure. However, to counterbalance this, no attempt was made to include the greater infection risk to the elderly or immunocompromised for a given dose, since the dose-response relationships were based on healthy adults, and these relationships were applied uniformly across the population. There are also uncertainties about the dose-infection response relationship and its relationship to illness response. But perhaps more importantly, there were also uncertainties in the proportion of time that pathogens occurred in the tank due to the limited frequency of the bimonthly sampling regime.

The authors noted that further work is needed to improve the assumptions made in the analysis, such as the proportion of gene copies that represent both viable and infective organisms, since quantitative PCR (qPCR) does not provide information regarding viability or infectivity. In addition, longitudinal monitoring of the pathogens concentrations over about a year are needed to confirm or otherwise the assumption of continuous pathogen presence used in the model. Nevertheless, until more data become available to reduce some of these uncertainties, the results indicate that it would be prudent to disinfect RCR, such as by the installation of a UV disinfection unit, boiling, or other forms of disinfection.

Lim and Jiang (2013) challenged the current U.S. EPA acceptable risk benchmark (<1 case/10,000 persons/year) by quantifying the microbial risks associated with the consumption of domestic tank water-irrigated produce (lettuce, cucumbers and tomatoes). Their results showed that the lettuce presents the highest risk followed by tomato and cucumber. They found that the annual infection risks are very likely to exceed the U.S. EPA risk benchmark (1 infection/10,000 persons) by one order of magnitude. de Man *et al.* (2014) performed a QMRA using *L. pneumophila* as a target pathogen to quantify the risk of infection for using RCR as source water for splash parks. The results indicated that using rainwater as source water for splash parks may pose a health risk.

These studies show that health risk analyses is an important step in aiding the use of RCR, but they are currently restricted by the lack of comprehensive work on the prevalence of pathogens in tank water. Improved levels of available data would enable more accurate calculations of the level of risk and allow an assessment of the required levels of reductions in pathogen numbers for different end uses (drinking vs. non-potable uses). This can then enable appropriate treatment measures such as filtration and ultraviolet disinfection to be undertaken to reduce the risk of infection from tank water (Jordan *et al.* 2008).

10.4 CONCLUDING REMARKS AND RECOMMENDATIONS

The review of the published literature combined with specific case studies in South East Queensland suggest that the microbial quality of RCR may be of much lower quality than commonly perceived. On the basis of the reported case studies, it appears that the microbial quality of RCR is strongly influenced by the season, the number of dry days preceding a rainfall event, animal activities on and around the roof, geographical location, and other exposure factors (Lye, 2009; Kus *et al.* 2010). Microbial assessment should involve the analysis of tank water for actual pathogens, not just the faecal indicator bacteria. The limited data in the literature have predominantly indicated that the commonly used indicators such as faecal coliforms and *E. coli* may not be suitable indicators of the public health risk from a microbial point of view due to their often poor correlation with pathogens (Ahmed *et al.* 2010).

Little information is currently available on the number of enteric and opportunistic microbial pathogens that can be present in RCR. The majority of the studies reported in the literature assessed the quality of the tank water on the basis of the presence or absence of specific microbial pathogens, with little information available regarding the actual numbers of enteric and opportunistic pathogens. Detection and quantification of pathogens using culture based methods and PCR-based assays both have their limitations. qPCR-based assays quantify DNA from both viable and non-viable cells of a target organism and therefore yield higher numbers compared to culture-based methods. On the other hand, culture-based methods may underestimate the numbers of a target microorganism due to the presence of viable but non-culturable cells. However, the overestimation of risk could be preferable than the underestimation via culture based methods for scenarios where the tank water is used for drinking, as it indicates that a contamination event has occurred. Further studies, including longitudinal surveys of rainwater tanks would be required to shed some light on pathogen frequency of occurrence in tank water samples.

Several case control studies have established links between gastroenteritis and consumption of untreated tank water. However, these reported outbreaks tended to involve small numbers of individuals, and the reported illnesses were often related to use of a common tank (Brodribb *et al.* 1995; Franklin *et al.* 2008; Murrell & Stewart, 1983). On the other hand, other studies could not identify tank water as a source of infection, and therefore, could only hypothesize that tank water was the likely source of infection via circumstantial evidence (Koplan *et al.* 1978; Merritt *et al.* 1999; Simmons & Smith, 1997). It should be noted that most of these case control studies used culture-based methods to establish a link between tap water and faecal specimens from patients.

Additional studies that focus on the collection and matching of pathogenic strains from faecal specimens from self-reported incidences of gastroenteritis, and from potential sources such as tap water and tank water using sensitive molecular typing methods would provide valuable information to determine if there is a direct link between gastroenteritis and consumption of tank water. This could be more practical than a comparable epidemiological study due to the complexity and costs of such epidemiological studies, and the fact that the most incidences of gastroenteritis remain unreported. The use of a QMRA is a critical tool to assess overall health risks associated with tank water uses, particularly drinking.

The public health risks are likely to be higher if the untreated tank water is used for drinking compared to other non-potable uses. Our research findings indicate that certain householders had under-sink filtration installed. However, these systems did not appear to be effective in removing faecal indicator bacteria and potential zoonotic bacterial pathogens, because of poor maintenance or inappropriate choice of treatment device for the pathogens of concern.

Very little has been documented about the inactivation rates of faecal indicator bacteria and pathogens on the roofs, in the gutters and in the tank water. If there is a rainfall event within a week after the deposition of faecal matter on the roof, it is highly likely that faecal indicator bacteria would be transported to the tank water. When introduced into the tank, their 90% reduction may take 2–10 days. Further research is required to understand the persistence of bacterial and protozoa pathogens in tank water because certain pathogens are known to be more persistent than faecal indicator bacteria. The faecal contamination of RCR appears to be more common in improperly designed systems, as well as systems that are not well maintained. We recommend that all RCR systems should be appropriately maintained, including the cleanliness of the systems before rainfall events. Roofs and gutters especially should be cleaned frequently, whilst the receiving tanks should be cleaned at least two times per year to improve the quality of the water. The roof should be kept clear of overhanging trees, which may provide access for wild animals which may harbour high numbers of bacterial and protozoa pathogens. It is recommended that rainwater should be treated with effective treatment procedures such as filtration, ultraviolet disinfection or simply boiling the water prior drinking.

10.5 REFERENCES

ABS (2010). Environmental issues: Water Use and Conservation No. 4602.0.55.003. Australian Bureau of Statistics, Canberra.

ADWG (2004). Guidelines for Drinking Water Quality in Australia. National Health and Medical Research Council/Australian Water Resources Council.

Ahmed W., Huygens F., Goonetilleke A. and Gardner T. (2008). Real-time PCR detection of pathogenic microorganisms in roof-harvested rainwater in Southeast Queensland, Australia. *Appl. Environ. Microbiol.* **74**(17), 5490–5496.

Ahmed W., Goonetilleke A. and Gardner T. (2010a). Implications of faecal indicator bacteria for the microbiological assessment of roof-harvested rainwater quality in Southeast Queensland, Australia. *Can. J. Microbiol.* **56**(6), 471–479.

Ahmed W., Vieritz A., Goonetilleke A. and Gardner T. (2010b). Health risks from the use of roof-harvested rainwater in Southeast Queensland, Australia, as potable or non-potable water, determined using quantitative microbial risk assessment. *Appl. Environ. Microbiol.* **76**(22), 7382–7391.

Ahmed W., Gardner T. and Toze S. (2011). Microbiological quality of roof-harvested rainwater and health risks: a review. *J. Environ. Qual.* **40**(1), 13–21.

Ahmed W., Hodgers L., Sidhu J. P. S. and Toze S. (2012a). Fecal indicators and zoonotic pathogens in household drinking water taps fed from rainwater tanks in Southeast Queensland, Australia. *Appl. Environ. Microbiol.* **78**(1), 219–226.

Ahmed W., Sidhu J. P. S. and Toze S. (2012b). An attempt to identify the likely sources of *Escherichia coli* harboring toxin genes in rainwater tanks. *Environ. Sci. Technol.* **46**(9), 5193–5197.

Ahmed W., Richardson K., Sidhu J. P. S. and Toze S. (2012c). *Escherichia coli* and *Enterococcus* spp. in rainwater tank samples: comparison of culture-based methods and 23S rRNA gene quantitative PCR assays. *Environ. Sci. Technol.* **46**(20), 11370–11376.

Ahmed W., Richardson K., Sidhu J. P. S., Jagals P. and Toze S. (2014). Inactivation of faecal indicator bacteria in a roof captured rainwater system under ambient meteorological conditions. *J. Appl. Microbiol.,* **116**(1), 199–207.

Albrechtsen H-J. (2002). Microbiological investigations of rainwater and graywater collected for toilet flushing. *Water Sci. Technol.,* **46**(6–7), 311–316.

Anderson K. L., Whitlock J. E. and Harwood V. J. (2005). Persistence and differential survival of fecal indicator bacteria in subtropical waters and sediments. A*ppl. Environ. Microbiol.,* **71**(6), 3041–3048.

Appan A. (1997). Roof water collection systems in some Southeast Asian countries: status and water quality levels. *J. R. Soc. Health,* **117**(5), 319–323.

Ashbolt R. and Kirk M. D. (2006). *Salmonella mississippi* infections in Tasmania: The role of native Australian animals and untreated drinking water. *Epidemiol. Infect.,* **134**(6), 1257–1265.

Baudisöva D. (1997). Evaluation of *Escherichia coli* as the main indicator of fecal pollution. *Water Sci. Technol.,* **35**(11–12), 333–356.

Beal C., Gardner T., Sharma A., Barton R. and Chong M. (2011). A Desktop Analysis of Potable Water Savings From Internally Plumbed Rainwater Tanks in South East Queensland. Urban Water Security Research Alliance Technical Report No. 26, Brisbane, Australia.

Bertin Y., Boukhors K., Pradel N., Livrelli V. and Martin C. (2001). stx_2 subtyping of Shiga toxin-producing *Escherichia coli* isolated from cattle in France: detection of a new stx_2 subtype and correlation with additional virulence factors. *J. Clin. Microbiol.,* **39**(9), 3060–3065.

Birks R., Colbourne J., Hills S. and Hobson R. (2004). Microbiological water quality in a large in-building water recycling facility. *Water Sci. Technol.,* **50**(2), 165–172.

Broadhead A. N., Negron-Alvira A., Baez A., Hazen T. C. and Canoy M. J. (1998). Occurrence of *Legionella* species in tropical rainwater cisterns. *Caribb. J. Sci.,* **24**(1–2), 71–73.

Brodribb R., Webster P. and Farrell D. (1995). Recurrent *Campylobacter fetus* subspecies bacteraemia in a febrile neutropaenic patient linked to tank water. *Commun. Dis. Intell.,* **19**(13), 312–313.

Crabtree K. D., Ruskin R. H., Shaw S. B. and Rose J. B. (1996). The detection of *Cryptosporidium* oocysts and *Giardia* cysts in cistern water in the U.S. Virgin Islands. *Water Res.,* **30**(1), 208–216.

Craun G. F., Till D. G. and McBride G. (2004). Epidemiological studies and surveillance. In: Water zoonoses. WHO Emerging Issues in Water and Infectious Diseases Series, J. A. Cotruvo, A. Dufour, G. Rees, J. Bartram, R. Carr, D. O. Clive, F. Craun, R. Fayer and V. P. J. Gannon (eds), IWA Publishing, London, pp. 154–166.

CRC WQT (2006). Water Quality and Health Risks From Urban Rainwater Tanks. Research Rep. 42. Project 202604. CRC Water Quality and Treatment.

de Man H., Bouwknegt M., van Heijnsbergen E., Leenen E. J. T. M., van Knapen F. and de Roda Husman A. M. (2014). Health risk assessment for splash parks that use rainwater as source water. *Water Res.,* **54,** 254–261.

Dean J. and Hunter P. R. (2012). Risk of gastrointestinal illness associated with the consumption of rainwater: a systematic review. *Environ Sci Technol.,* **46**(5), 2501–2507.

Despins C., Farahbakhsh K. and Leidl C. (2009). Assessment of rainwater quality from rainwater harvesting systems in Ontario, Canada. *J. Water Supply Res. Technol. AQUA.,* **58**(2), 117–134.

Dillaha T. A. and Zolan W. Z. (1985). Rainwater catchment water quality in Micronesia. *Water Res.,* **19**(6), 741–746.

Eberhart-Phillips J., Walker N., Garrett N., Ball D., Sinclair D., Rainger W. and Bates M. (1997). *Campylobacteriosis* in New Zealand: results of a case-control study. *J. Epidemiol. Community Health,* **51**(6), 686–691.

Eisenberg J. N. S., Hubbard A., Wade T. J., Sylvester M. D., LeChevallier M. W., Levy D. A. and Colford J. M. (2006). Inferences drawn from a risk assessment compared directly with a randomized trial of a home drinking water intervention. *Environ. Health Perspect,* **114**(8), 1199–1204.

Enger K. S., Nelson K. L., Clasen T., Rose J. B., Eisenberg J. N. S. (2012). Linking quantitative microbial risk assessment and epidemiological data: informing safe drinking water trials in developing countries. *Environ. Sci. Technol.,* **46**(9), 5160–5167.

Evans C. A., Coombes P. J. and Dunstan R. H. (2006). Wind, rain and bacteria: the effect of weather on the microbial composition of roof-harvested rainwater. *Water Res.,* **40**(1), 37–44.

Fewtrell L. and Kay D. (2007). Quantitative microbial risk assessment with respect to *Campylobacter* spp. in toilets flushed with harvested rainwater. *Water Environ. J.,* **21**(4), 275–280.

Franklin L. J., Fielding J. E., Gregory J., Gullan L., Lightfood D., Poznaski S. Y. and Vally H. (2009). An outbreak of *Salmonella* Typhimurium 9 at a school camp linked to contamination of rainwater tanks. *Epidemiol. Infect.* **137**(3), 434–440.

Frost F. J., Craun G. F. and Calderon R. L. (1996). Waterborne disease surveillance. *J. Am. Water Works Assoc.,* **88**(9), 66–75.

Gordon C. and Toze S. (2003). Influence of groundwater characteristics on the survival of enteric viruses. *J Appl Microbiol.* **95**(3), 536–544.

Hall G., Raupach J. and Yohannes K. (2006). An Estimate of Under-Reporting of Food Borne Notifiable Diseases: *Salmonella, Campylobacter,* Shiga Toxin Producing *E. coli* (STEC). Working Paper 52. National Centre for Epidemiology and Population Health, Canberra, ACT, Australia.

Hartl D. L. and Dykhuizen D. E. (1984). The population genetics of *Escherichia coli. Annu Rev Genet.,* **18**(1), 31–68.

Hellard M. E., Sinclair M. I., Forbes A. B., and Fairley C. K. (2001). A randomised blinded, controlled trial investigating the gastrointestinal health effects of drinking water quality. *Environ. Health Pers.,* **109**, 773–777

Heyworth J. S., Glonek G., Maynard E. J., Baghurst P. A. and Finlay-Jones J. (2006). Consumption of untreated tank rainwater and gastroenteritis among young children in South Australia. *Int. J. Epidemiol.,* **35**(4), 1051–1058.

Hrudey S. E. and Hrudey E. J. (2004). Safe Drinking Water: Lessons From Recent Outbreaks in Affluent Countries. IWA Publishing, London.

Ishii S., Hansen D. L., Hicks R. E. and Sadowsky M. J. (2007). Beach sand and sediments are temporal sinks and sources of *Escherichia coli* in Lake Superior. *Environ. Sci. Technol.,* **41**(7), 2203–2209.

Jordan F. L., Seaman R., Riley J. J. and Yoklic M. R. (2008). Effective removal of microbial contamination from harvested rainwater using a simple point of use filtration and UV-disinfection device. *Urban Water J.,* **5**(3), 209–218.

Kaper J. B., Nataro J. P. and Mobley H. L. T. (2004). Pathogenic *Escherichia coli. Nat. Rev. Microbiol.,* **2**(2), 123–140.

Khajanchi B. K., Fadl A. A., Borchardt M. A., Berg R. L., Horneman A. J., Stemper M. E., Joseph S. W., Moyer N. P., Sha J. and Chopra A. K. (2010). Distribution of virulence factors and molecular fingerprinting of *Aeromonas* species isolates from water and clinical samples: Suggestive evidence of water-to-human transmission. *Appl. Environ. Microbiol.,* **76**(7), 2313–2325.

Kinzelman J., Ng C., Jackson E., Grauds S. and Bagley R. (2003). Enterococci as indicators of Lake Michigan recreational water quality: comparison of two methodologies and their impacts on public health regulatory events. *Appl. Environ. Microbiol.,* **69**(1), 92–96.

Koplan J. P., Deen R. D., Swanston W. H. and Tota B. (1978). Contaminated roof-collected rainwater as a possible cause of an outbreak of salmonellosis. *J. Hyg.,* **81**(2), 303–309.

Kus B., Kandasamy J., Vigneswaran S. and Shon H. K. (2010). Analysis of first flush to improve the water quality in rainwater tanks. *Water Sci. Technol.,* **61**(2), 421–428.

Lee J. Y., Yang J-S., Han M. and Choi J. (2010). Comparison of the microbiological and chemical characterization of harvested rainwater and reservoir water as alternative water sources. *Sci. Total Environ.,* **408**(4), 896–905.

Lim K-Y. and Jiang S. C. (2013). Reevaluation of health risk benchmark for sustainable water practice through risk analysis of rooftop-harvested rainwater. *Water Res.,* **47**(20), 7273–7286.

Lye D. (1987). Bacterial levels in cistern water systems of northern Kentucky. *Water Resour. Bull.,* **23**(6), 1063–1068.

Lye D. (1992). Microbiology of rainwater cistern systems: a review: (Cistern, Rainwater, Microorganisms). *J. Environ. Sci. Health A.,* **27**(8), 2123–2166.

Lye D. J. (2002). Health risks associated with consumption of untreated water from household roof catchment systems. *J. Am. Water Res. Assoc.,* **38**(5), 1301–1306.

Lye D. J. (2009). Rooftop runoff as a source of contamination: a review. *Sci. Tot. Environ.,* **407**(21), 5429–5434.

Marino M. R., Brown T. J., Waddington D. C., Brockie R. E. and Kelly P. J. (1992). *Giardia intestinalis* in North Island possums, house mice, and ship rats. *N. Z. Vet. J.,* **40**(1), 24–27.

McFeters G. A., Bissonnette G. K., Jezeski J. J., Thomson C. A. and Stuart D. G. (1974). Comparative survival of indicator bacteria and enteric pathogens in well water. *Appl. Environ. Microbiol.,* **27**(5), 823–829.

Merritt A., Miles R. and Bates J. (1999). An outbreak of *Campylobacter* enteritis on an island resort, north Queensland. *Commun. Dis. Intell,* **23**(8), 215–219.

Möllby R., I. Kühn and M. Katouli. (1993). Computerized biochemical fingerprinting-a new tool for typing of bacteria. *Rev. Med. Microbiol.*, **4**:231–241.

Murrell W. G. and Stewart B. J. (1983). Botulism in New South Wales, 1980–1981. *Med. J. Aust.*, **1**(1), 13–17.

NRC (1983). Risk Assessment in the Federal Government: Managing the Process. National Academies Press, Washington, D.C.

NRMMC–EPHC–NHMRC (2008). Australian Guidelines for Water Recycling: Managing Health and Environmental Risks (Phase 2). Augmentation of drinking water supplies. http://www.ephc.gov.au/taxonomy/term/39 (verified 18 November 2010).

Pinfold J. V., Horan N. J., Wirojanagud W. and Mara D. (1993). The bacteriological quality of rainjar water in rural northeast Thailand. *Water Res.*, **27**(2), 297–302.

Rodrigo S., Sinclair M., Forbes A., Cunliffe D. and Leder K. (2010). Drinking rainwater: a double-blinded, randomized controlled study of water treatment filters and gastroenteritis incidence. *Am. J. Public Health,* **101**(5), 842–847.

Russo T. A. and Johnson J. R. (2000). A proposal for an inclusive designation for extraintestinal pathogenic *Escherichia coli*: ExPEC. *J. Infect. Dis.,* **181**(5), 1753–1754.

Savill M. G., Hudson J. A., Ball A., Klena J. D., Scholes P., White R. J., McCormack R. E. and Jankovic D. (2001). Enumeration of *Campylobacter* in New Zealand recreational and drinking waters. *J. Appl. Microbiol.,* **91**(1), 38–46.

Sazakil E., Alexopoulos A. and Leotsinidis M. (2007). Rainwater harvesting quality assessment and utilization in Kelafonia Island, Greece. *Water Res.,* **41**(9), 2039–2047.

Schlech W. F., Gorman G. W., Payne M. C. and Broome C. V. (1985). Legionnaires' disease in Caribbean: An outbreak associated with a resort hotel. *Arch. Intern. Med.,* **145**(11), 2076–2079.

Simmons G. and Smith J. (1997). Roof Water Potable Source of *Salmonella* Infections. *N. Z. Public Health Rep.,* **4**, 5.

Simmons G., Hope V., Lewis G., Whitmore J. and Wanzhen G. (2001). Contamination of potable roof-collected rainwater in Auckland, New Zealand. *Water Res.,* **35**(6), 1518–1524.

Simmons G., Jusry S., Thornley C., Harte D., Mohiuddin J. and Taylor M. (2008). A Legionnaires' disease outbreak: A water blaster and roof-collected rainwater systems. *Water Res.,* **42**(6–7), 1449–1458.

Spinks J., Phillips S., Robinson P. and Van Buynder P. (2006). Bushfires and tank rainwater quality: a cause for concern? *J. Water Health,* **4**(1), 21–28.

Thomas P. R. and Greene G. R. (1993). Rainwater quality from different roof catchments. *Water Sci. Technol.,* **28**(3–5), 291–299.

Toze S., Hanna J., Smith T., Edmonds L. and McCrow A. (2004). Determination of water quality improvements due to the artificial recharge of treated effluent. In: Wastewater Reuse and Groundwater Quality, IAHS Publication, **285**, pp. 53–60.

Uba B. N. and Aghogho O. (2000). Rainwater quality from different roof catchments in the Port Harcourt District, Rivers State, Nigeria. *J. Water Supply Res. Technol.* AQUA., **49**, 281–288.

Verrinder G. and Keleher H. (2001). Domestic drinking water in rural areas: are water tanks on farms a health hazard? *Environ. Health,* **1**(3), 51–56.

WHO (2004). Guidelines for Drinking Water Quality, 3rd edn, World Health Organization, Geneva, Switzerland.

Chapter 11

Cluster-scale rainwater harvesting

Stephen Cook, Ashok K. Sharma, Thulo Ram Gurung, Luis E. Neumann, Magnus Moglia and Priya Chacko

ABSTRACT

Rainwater tanks have been adopted in cities as a non-potable water source that complements mains drinking water supply. The uptake of rainwater tanks in countries such as Australia has been driven by the need to reduce demand for drinking water, whilst also taking a more integrated approach to urban water management. Rainwater harvesting can provide a fit-for-purpose water source that minimises the impact of the urban water cycle on the local environment and receiving waters. Rainwater tanks are typically installed at the household scale for non-potable water source uses such as toilet flushing, laundry and garden irrigation in urban areas. However, this chapter reports on the design principles, case studies and literature of cluster-scale rainwater harvesting. The cluster-scale rainwater harvesting system collects rainwater from multiple household roofs, and then transports it by gravity to a communal storage tank. After treatment and separate storage, a dedicated pressurised water distribution system supplies treated rainwater to the individual homes for uses that could include potable. Cluster-scale rainwater harvesting can offer multiple benefits, including: economies of scale for capital costs, reduced land footprint on individual allotments, centralised treatment and disinfection, and flexibility in matching supply and demand for different sized households. Cluster-scale harvesting also offers the opportunity to formalise management and maintenance of local water sources, which address some of the serious limitations of household-scale harvesting.

Keywords: rainwater harvesting; decentralised systems; community infrastructure; integrated urban water management.

11.1 INTRODUCTION

Centralised water supply systems in cities have traditionally relied upon surface water catchments and groundwater to meet demand, however a burgeoning urban population has created the need to diversify the water supply source mix. These diversified water supply systems, which include rainwater harvesting systems, have the purpose of reducing pressure on drinking water supplies, as well as developing resilience to climate change and variability (Sharma *et al.* 2005; Ruth *et al.* 2007; Moglia *et al.* 2011). In Australia, governments have encouraged the adoption of rainwater harvesting systems through financial incentives and regulations. These household rainwater systems are secondary to mains drinking water supply, and are mostly used for non-potable purposes such as toilet flushing, laundry, and external uses including garden irrigation. Rainwater harvesting is part of a strategy to diversify water supply sources by reducing reliance on traditional water catchments and centralised infrastructure. They can also offer benefits to other parts of the water cycle, including: moderating peak stormwater runoff, reducing discharge of nutrients to receiving waters, and improving ecological health (Villarreal & Dixon, 2005; Farreny *et al.* 2011; Kim & Furumai, 2012), as outlined in more detail in Chapter 13.

In Australian cities, the most common form for rainwater harvesting has been household-scale rainwater tank systems. However, there are a number of limitations with this type of system including a lack of householder understanding of health risks (Domènech & Saurí, 2011) and inadequate maintenance that can result in supply failure (Moglia *et al.* 2013). Furthermore, cities around the world are moving to a more compact urban form (Chhetri *et al.* 2013). This urban densification has implications for rainwater harvesting as the land area for storage on individual allotment becomes limiting. This raises the need to consider alternative configurations for local water harvesting, such as communal storages on public land. The concept being tested in this chapter is that cluster-scale rainwater and stormwater systems system can provide improve reliability, performance and system management, relative to household-scale harvesting, in many development contexts.

Geisinger and Chartier (2005) referred to cluster-scale systems (in discussing decentralised wastewater) as servicing a group of dwelling or businesses that are relatively close together. The distinction between a cluster-scale harvesting scheme and a traditional large-scale regional scheme is ill-defined. We defined a Cluster-scale Rainwater Harvesting (CSRH) system as any system that supplies more than one user. This may be a few households or businesses that have a communal system, but it also includes systems at a scale of a small suburb. For example, rainwater runoff could be collected from the roof area of a few dozen households, stored centrally, and then reticulated back to households. Figure 11.1 provides a schematic of a cluster-scale rainwater harvesting system.

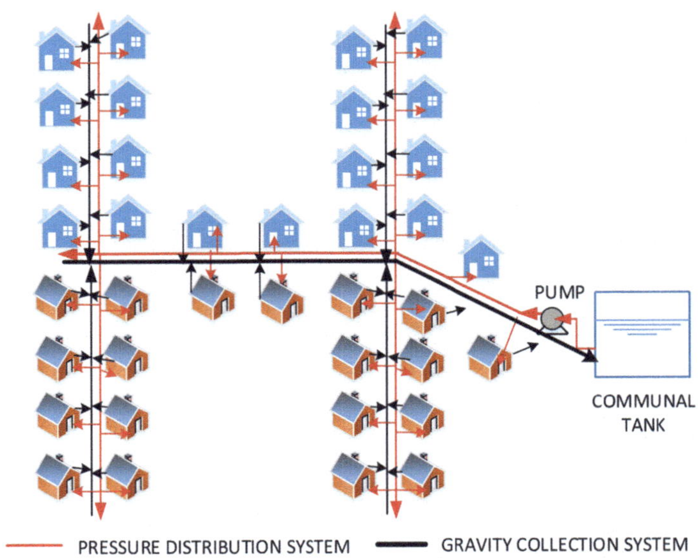

Figure 11.1 Schematic of a cluster-scale rainwater harvesting system.

The objectives of this chapter are to explore the potential benefits of harvesting, using and managing a rainwater system amongst a group of households relative to a single household. Following this, the chapter will also identify the development contexts where a CSRH scheme may be more suited compared to a household scheme. Suitability criteria include: rainwater yield, reliability of supply, water quality that is fit-for-purpose, impact on stormwater management, community acceptance, suitability to urban form, and appropriateness to management capacity and regulatory framework.

CSRH systems can be used in two development contexts. They can be used in areas that don't have access to reticulated mains water supply. CSRH systems that include treatment and back-up supply, can allow medium density developments in peri-urban and rural communities that don't have access to mains water supply. CSRH systems can also be used to provide a non-potable water supply in urban areas with traditional mains water supply. This will reduce demand for potable water as well as provide a water source for uses such as irrigation that is not subject to water restrictions during drought conditions.

In this chapter, firstly we present a literature review that highlights Australian and international experiences with CSRH. This literature review provides insights on the development contexts where CSRH has been applied, the drivers for their adoption, and the impediments for greater mainstream adoption. Four Australian case studies of CSRH are then presented, followed by a modelling study that compares cluster-scale approaches to household-scale rainwater systems. The chapter then explores the principles for the design of CSRH systems and the management implications, including identifying suitable management models.

11.2 LITERATURE REVIEW

11.2.1 Examples of cluster-scale rainwater harvesting

While literature on the application of rainwater harvesting at the scale of a single household is extensive, there is a dearth of reported studies on CSRH schemes. Makropoulos and Butler (2010) noted that the paucity of documented studies has lead to knowledge gaps such as context-independent characteristics that can be used to predict likely system performance and cost of CSRH systems. Sharma *et al.* (2009) presented a framework that can be used to assess the feasibility of alternative water servicing configurations, such as CSRH systems. This literature review focuses on the few reported example of CSRH systems.

Farreny *et al.* (2011) compared the performance and cost-effectiveness of four different configurations for rainwater harvesting in a dense urban area of Barcelona. The four different configurations included household-scale rainwater systems for both a retrofit and new development, and *neighbourhood-scale* rainwater systems for both a retrofit and new development. Their results showed that a neighbourhood-scale rainwater system for new developments was the most cost-effective as it did not require pavement disturbance and re-installation (Farreny *et al.* 2011). The economic feasibility of the neighbourhood-scale system was also strongly influenced by the housing density of the development. This supports the findings from Roebuck *et al.* (2011) who noted that rainwater systems at the scale of a single household are rarely cost-effective when all costs are taken into account.

Figtree Place is a medium density (45 dwellings per hectare) development in Newcastle, Australia. Figtree Place was designed to be a water sensitive development, which included CSRH systems that were used to satisfy demand for hot water and toilet flushing. Each CSRH system connected between four and eight dwellings. Coombes *et al.* (2000) described a detailed monitoring program undertaken to quantify the performance of the CSRH system in terms of water quality, mains water savings, stormwater management, costs and community acceptance. The study found that household hot water systems combined with other passive treatment processes in the rainwater tank (such as settling), effectively treated harvested rainwater to the microbiological standard required by the Australian Drinking Water Guidelines (National Health and Medical Research Council, 2011). The CSRH system also reduced demand for mains drinking water by around 60% (Coombes *et al.* 2000), and rainwater tanks as part of a water sensitive urban design approach reduced infrastructure costs by around 20% when compared to traditional stormwater practices.

Cook *et al.* (2012) presented a case study of a commercial high-rise building in Brisbane, Australia, where a CSRH system was used for toilet flushing and amenity garden irrigation. Monitoring studies showed that the system provided non-potable water with moderate reliability, but that the system yield was constrained by installation faults. Cook *et al.* (2012) highlighted the fact that because the rainwater system was not part of the building management system, faults were not identified and fixed in a timely manner, resulting in the system being offline for extended periods of time. Wannon Water provides an example of CSRH system that is integrated with the existing mains water supply system. This example and other case studies are presented in more detail in Section 11.3.

11.2.2 Impediments and benefits

The impediments faced by developers in achieving mainstream adoption of CSRH schemes are similar to those for decentralised wastewater systems. However there are some impediments that are specific to CSRH schemes.

The main impediments for mainstream adoption of alternative urban water systems are related to social and institutional factors, rather than technical factors (Brown *et al.* 2007). In particular, the inconsistency and fragmented nature of governance, regulations, and guidelines associated with water sensitive urban design (WSUD) and integrated urban water management (IUWM) are significant impediments (Tjandraatmadja *et al.* 2008; Sharma *et al.* 2010; Brown & Farrelly, 2009). This fragmentation has been attributed to the 'silo' administrative/governance structures of water authorities, local government, government departments and private industries that have evolved to deliver conventional water services (Mitchell, 2004). Brown and Clarke (2007) summarised the main institutional impediments to adopting WSUD (of which CSRH is a sub-component) as: insufficient skills and knowledge, organisational resistance, lack of political will, limited regulatory incentives, and unsuitable institutional capacity.

The skills and knowledge base required to adequately maintain and operate a CSRH scheme differ from those needed for conventional urban water systems. Moreover, management of decentralised water assets provides a challenge to the centralised management paradigm. There is also the need for the development of appropriate standards and guidelines to assist understanding by professions such as engineers and contractors and trades such as plumbers (Roy *et al.* 2008; US EPA, 2013).

As a cluster-scale approach to harvesting runoff is relatively novel in Australian cities, there is a lack of documented experience as to how a community-based harvesting scheme is likely to perform (Makropoulos & Butler, 2010). Well-studied projects such as Figtree Place (Coombes *et al.* 2000) provided a rich source of data, but there is the need for other studies in different development contexts. The paucity of demonstration projects can impede the development of suitable guidelines and effective regulatory frameworks (Brown & Farrelly, 2009). Demonstration projects that incorporate validation through ongoing monitoring can provide relevant data on environmental benefits (water quality improvement, flow management), associated risks, life cycle costs, externalities, adequacy of engineering standards and guidelines, and energy consumption (Fletcher *et al.* 2008; Domènech & Saurí, 2011).

The broader adoption of CSRH can also be impeded by a perceived lack of direct economic benefits (US EPA, 2013). Fane and Mitchell (2006) highlighted that decentralised water supply systems can be considered more expensive due to 'unfair' assessments of sunk and avoided infrastructure costs. Whilst comprehensive studies have been conducted on the cost-effectiveness of individual rain water tanks, see for example Coombes (2002) and Hall (2013), very few economic studies have been conducted on CSRH systems (Gurung *et al.* 2012; Gurung & Sharma, 2014). An exception is the work of Gurung *et al.* (2012) who undertook a conceptual analysis to determine the relative economies of individual rainwater tanks versus cluster-scale schemes, assuming a flat topography and a density of 20 dwellings per hectare. This analysis used Net Present Value (NPV), to assess the servicing from four to 576 homes with a CSRH scheme, and found around 200 homes as the most economic number on a per house basis. Nonetheless, Gurung *et al.* (2012) concluded that individual tanks cost less per household ($8,568) than a 200 home communal rainwater tank system ($10,180). The analysis also identified that the collection and distribution network costs of the CSRH system had dis-economies of scale; with the more houses connected, the higher the costs. However, this was slightly offset by economies of scale for the storage tanks. Changing the scenario from a steep slope to a gently sloping topography substantially reduced the NPV of the communal system. This was due to reduced pipe trenching and collection pumping costs due to shallower collection pipes depth and the use of a gravity fed system. We argue that although the cost of communal rainwater tank systems can be higher than the individual rainwater tanks, the communal systems have operational and management advantages, discussed later in this chapter. Chapter 12 contains more details on the economics of rainwater tank systems.

The case for the adoption of cluster-scale harvesting systems, relative to household-scale rainwater harvesting, is based on the following suppositions:

- That cluster-scale rainwater harvesting offers economies of scale by sharing the capital and operational expenses of the system among multiple users;
- That compared to household-scale rainwater tanks, there are better opportunities to balance supply and demand among households with different demand patterns; and
- The scale of cluster-scale rainwater harvesting can make it financially feasible to have formal arrangements in place for maintenance and operation, thus over-coming one of the major difficulties of maintaining household-scale rainwater harvesting systems.

The above factors are explored in more detail in this chapter.

11.3 CASE STUDIES OF CLUSTER-SCALE HARVESTING

The following section discusses four examples of cluster-scale rainwater harvesting and reuse. These Australian examples are situated in a range of development contexts and highlight the range of drivers for the adoption of CSRH schemes.

11.3.1 Capo di Monte

Capo di Monte (CDM) is a 46 home development at Mount Tambourine located in the hinterland ranges of the Gold Coast, South East Queensland. The development lies outside of the area serviced by municipal water and wastewater services, so for the development to proceed, it had to build a decentralised water and wastewater system. The communal rainwater system was designed to meet internal household uses that usually require potable water quality: kitchen, bathroom and laundry. A wastewater recycling scheme (described in Sharma *et al.* 2012; 2013) is used to satisfy non-potable demands: toilet flushing and garden irrigation. A groundwater bore is used to supplement both systems in times when demand is greater than can be supplied from harvested rainwater. The layout of the rainwater collection system and the location of communal rainwater storage tanks are shown in Figure 11.2.

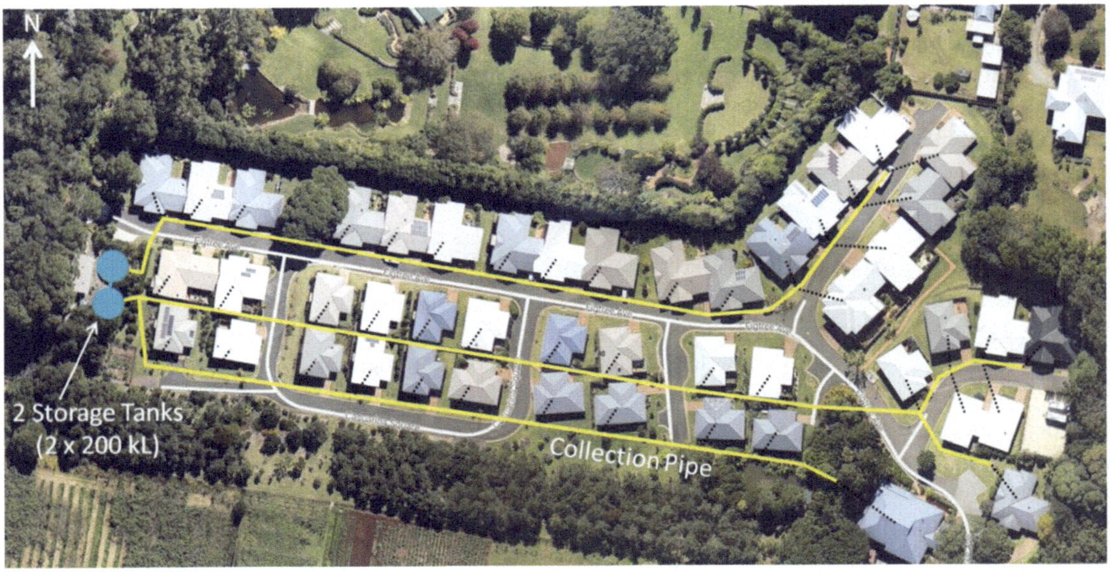

Figure 11.2 Capo di Monte cluster-scale rainwater harvesting system (Courtesy: Bligh Tanner Consulting Engineers, Brisbane, Australia).

CDM was planned as a retirement village, therefore residents are mostly 60 years or older and retired. The CDM population is 75 people, with an average household occupancy of 1.65 persons compared to a Brisbane average of 2.6 persons per dwelling. The Mount Tambourine weather station recorded an average annual rainfall of 1,318 mm/year over the period 1982 to 2005. Analysis of the rainfall record showed a pattern of relatively wet years with up to 2,000 mm, interspersed with drier years with rainfall of around 1,000 mm.

The communal rainwater system collects roof runoff through a network of household downpipes that feed into two main collector pipes which transfer the water by gravity into two 200 kL storage tanks. The 200 kL communal tanks are operated to have an active volume of 100 kL each, with 100 kL reserved for emergency fire-fighting capability, in accordance with state regulations. The total connected roof area is around 10,700 m^2, with houses having an average roof area of 222 m^2, and the community centre providing 488 m^2 of connected roof area.

A water treatment plant, comprising sand filtration, UV sterilisation and chlorination sends water to a 40 kL balance tank for subsequent pressurisation and distribution of *potable water* to each house and the small community centre which includes a swimming pool. A local bore provides supplemental water in times of insufficient rainfall or excess demand. The water supply components of the communal rainwater system are shown in Figure 11.3. The system is managed by an appropriately trained person who reports to the Body Corporate entity, which is legally responsible for management of all communal components of the development including sewage treatment, public open space irrigation, gardens and so on.

Monitoring studies detailed in Cook *et al.* (2012) demonstrated that the cluster-scale harvesting provided a reliable water source, albeit with an energy penalty. The rainwater system met around 80% of the potable demand (43 kL/hh/year) during the 4-year monitoring period. However, the energy required to provide this water was around 4 kWh/kL, with 75% of the energy use required for pressurising the reticulation system. Whilst the steeply sloping topography of CDM influenced this high specific energy use, Cook *et al.* (2012) found that the major cause was the oversized pump installed. A reduction in the pump size, while still meeting head and flow requirements, could reduce the specific energy for pumping by around 50%.

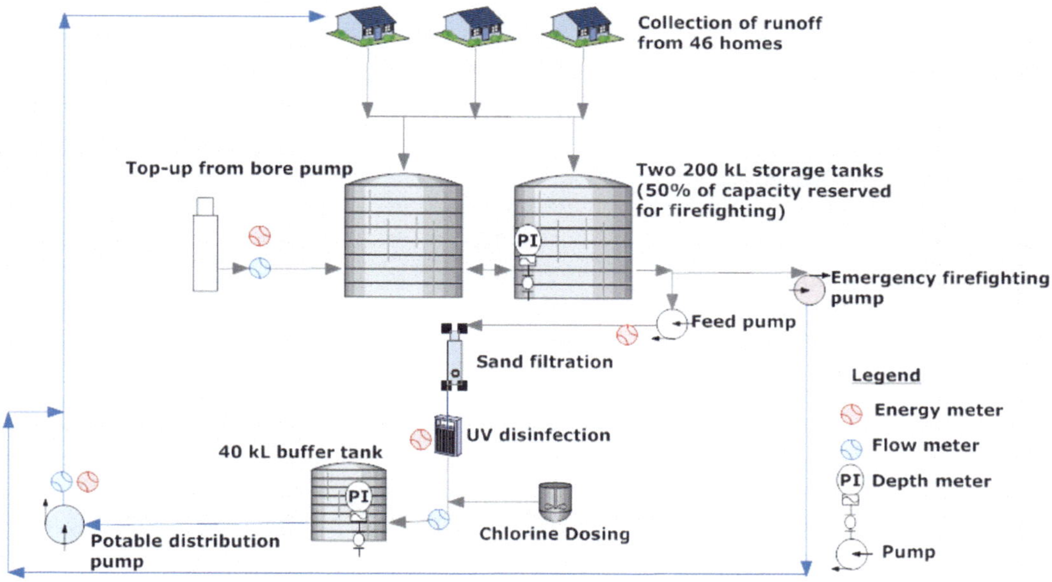

Figure 11.3 Capo di Monte hydraulic circuit, and water and energy meters.
Source: Cook *et al.* 2012, pg. 7.

11.3.2 Wannon water's roof water harvesting project

Wannon Water, a water utility in western Victoria, initiated an innovative project to harvest roof runoff from a housing estate to augment Warrnambool's drinking water supply. The harvesting scheme is located in an urban growth corridor of Warrnambool. The harvesting scheme was piloted for a 142 lot subdivision and it is envisaged that 3,000 homes will connected at full development.

The roof water harvesting scheme involved the construction of a dedicated network of roof water collection pipes. These collector pipes feed into a 2,000 metre trunk main, which transfers the water by gravity to the town's raw water storage. This impoundment primarily stores water collected from surface water sources such as the nearby Gellibrand River. The harvested roof water mixes with untreated stormwater in the basin (sourced by rural catchment runoff) prior to treatment at the Warrnambool Water Treatment Plant. The treated water is then supplied to the city's potable water reticulation system. Initial estimates indicated that (treated) rainwater harvested from the 142 homes would meet 75% of the annual water demand of these properties (Wannon Water, 2011).

The roof water harvesting scheme was identified after Wannon Water prepared a water supply and demand strategy, with a planning horizon of 50 years. This strategy identified that projected growth in demand for water could not be met by existing water supply sources, which were also sensitive to likely climate change impacts. A high level analysis was undertaken to estimate maximum yield and compare relative costs of water supply alternatives. Based on this analysis, the preferred options were the roof water harvesting scheme and the development of new groundwater resources (Table 11.1).

Table 11.1 Comparison of wannon water's selected augmentation options from the water supply and demand strategy.

Description of option	Estimated max yield (ML/yr)	CO_2 emissions (kg/ML)	Cost ($/ML)
Desalination of seawater	5000	2.66	$837
Desalination of groundwater	2000	2.29	$773
New groundwater resources	250–3400	0.26–3.47	$105–$246
Recycled water scheme from the wastewater treatment plant delivered by a dual reticulation network	488	Not calculated	$1,770
Winter flow harvesting from local river	1600	5.73	$487
Roof water harvesting form subdivisions via a separate pipe network	300	0	$240

Source: Wannon Water, 2012

Individual rainwater tanks had much higher operating costs than the regional roof water harvesting scheme due primarily to pumping costs from pumping energy, and pump maintenance and replacement. Major advantages at Wannon Water were that gravity could be used to transfer collected rainwater water to the storage basin, and existing infrastructure (storage basin, treatment plant and reticulation pipe network) used to reduce capital costs compared to schemes operating a dual pipe reticulation network (Wannon Water, 2012).

Also, it was considered that regional roof water harvesting would have a lower risk profile than individual rainwater tanks. This was because water harvested from the communal roof water scheme undergoes rigorous treatment at the regional centralised water treatment plant prior to being supplied to residents. This does not occur on individual rainwater tanks, even those with disinfection units installed (Ahmed *et al.* 2010). Another advantage is water supply back up from the other water sources as individual rainwater tanks do not have 100% supply reliability.

The most significant challenge faced by the Wannon scheme was that some developers were reluctant to implement the scheme due to the perceived cost burden. The cost per lot of connecting houses to the roof water harvesting scheme varied between $2,325 and $4,450. However, this cost was offset by removing the mandated requirement of installing a rainwater tank in each house, plumbed to the toilet. This avoided cost was estimated at around $4,000.

The initial capital investment needed to construct the scheme, was made possible by a Federal Government grant. This grant has meant the system is well documented and is supported by a toolkit that enables others to estimate the expected yield and costs associated with constructing and operating their own roof water harvesting scheme (Wannon Water, 2011).

11.3.3 Christie walk

Christie Walk is a brownfield medium density residential development, located in Adelaide's central business district. The development was initiated in 1999 and completed in 2006 (Downton, 2005). The development is situated on a 2,000 m^2 lot and contains 27 dwellings of varying types (detached, duplex, apartments etc.) housing a population of 50 people. Christie Walk was designed to demonstrate a sustainable approach to urban development and living using the concept of an eco-city. The water supply, wastewater and stormwater services were designed to be water efficient, minimise environmental impacts, and enhance the local ecological processes and liveability of the development. A feature of the development is a cluster-scale rainwater and stormwater harvesting scheme that supplies non-potable water (Downton, 2005).

Runoff from impervious surfaces at Christie Walk (roofs, paving, balconies, etc.) is collected via downpipes and grates to a stormwater collection pipe that drains by gravity to two underground concrete storage tanks (40 kL total capacity). The impervious collection area is approximately 1,300 m^2, which is around 65% of the total site area. The harvested runoff is reticulated back to the development via a ring main, providing non-potable water for toilet flushing and garden irrigation. The intended uses means there is no need for a treatment process. The non-potable supply is guaranteed with a mains back-up supply to the tanks, controlled by water level sensors. The system is fitted with non-return valves to prevent backflow and cross-contamination of mains water pipelines.

Excess rainwater and stormwater flow (i.e., overflow) that cannot be captured in the storage tanks is directed to a surface drain at the rear of the development. Monitoring of the harvesting and reuse scheme showed it reduced mains water demand by 45% compared to similar traditional households in South Australia. In fact, water use for Christie Walk households is actually significantly lower than the average water use for one person households in SA Water's service area. This is despite the average household size at Christie Walk being closer to two people (the average Adelaide household size is 1.85 people). The following may influence the lower water demand at Christie Walk:

- The built form and landscaping was designed to be water efficient. This included efficient household appliances and also the extensive use of native plants that are adapted to dry conditions; and
- Residents are highly motivated by a sustainable behaviour ethic, which is consistent with their choice of living in a development such as Christie Walk.

Christie Walk is managed as a community title development, so it has a standard body corporate arrangement comprised of owners. The ongoing maintenance is conducted by a team of volunteers from the residents. The management of the rainwater and stormwater harvesting system was formalised in a users' guide. This guide provides the technical details of the systems and also trouble-shooting steps in case of system failure.

11.3.4 Fitzgibbon chase potable roof water (PotaRoo) scheme

Fitzgibbon Chase is a residential development of the Urban Land Development Authority (ULDA) located in the northern suburbs of Brisbane, Australia. The development covers an area of approximately 295 hectares, and had the purpose of providing affordable, sustainable housing. This is demonstrated in the area's water management systems which include the reuse of rainwater and stormwater as alternative water sources; the former for potable uses, and the latter for non-potable purposes. The following information in this section has been taken from Dark and Hamlyn-Harris (2009).

11.3.4.1 Rainwater collection

The potable roof water project (PotaRoo) harvests rainwater from 1,230 homes, with a total roof catchment of about 110,000 m^2. Rainwater from the roofs is diverted into collection systems and collected in a series of satellite, subsurface communal rainwater tanks, from where it is pumped to a 0.8 ML centralised storage and treatment facility. Excess runoff overflows into the stormwater drainage system. The collected rainwater is treated to potable standards, and initially will be injected into the non-potable, stormwater reticulation line until the system is validated for potable quality, at which time the water will be connected into the main potable water supply network. A schematic layout of the roof water collection system depicting local and central rainwater storage tanks and their connections is shown in Figure 11.4.

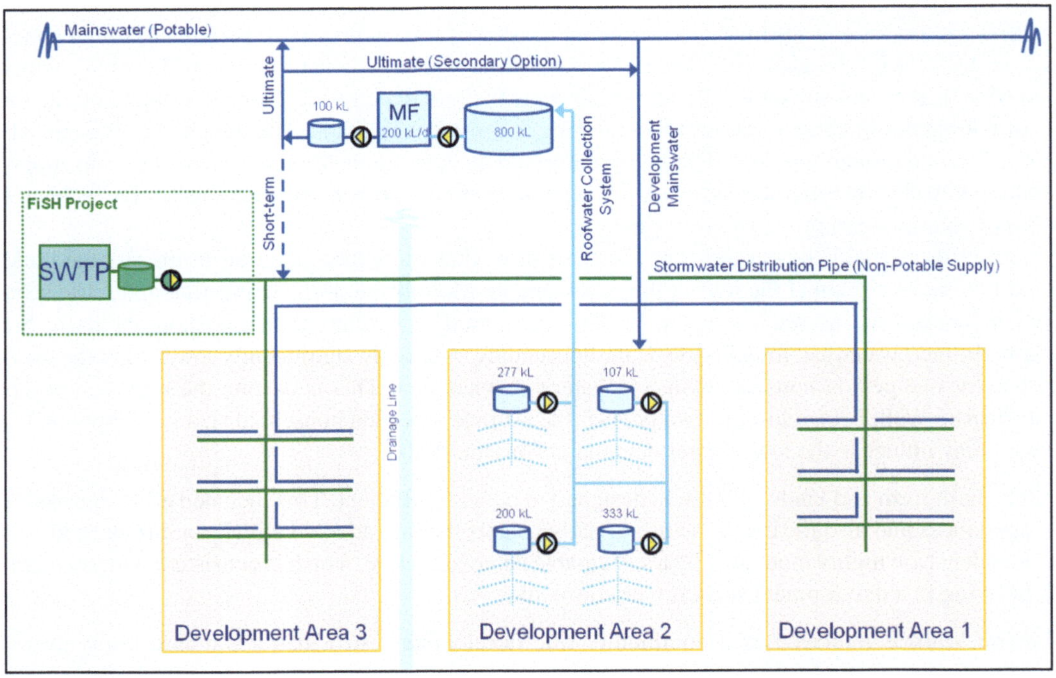

Figure 11.4 Schematic of fitzgibbon chase potable roof water systems (Courtesy Bligh Tanner Consulting Engineers, Brisbane, Australia).

11.3.4.2 Water balance

A daily water balance analysis was carried out over a 30-year period with various water treatment capacities (160 kL/day–500 kL/day). This study determined the optimum storage tank size for treated water to be 2 ML. Preliminary cost estimates indicated a treatment capacity of 200 kL/day was the most cost-effective. The water balance analysis showed that the scheme had the potential to supply approximately 44 ML/year (121 kL/day) of treated rainwater, approximately one third of the projected total potable demand.

11.3.4.3 Collection system

Analysis of rainfall intensity records showed that the majority (75%) of rainfall runoff occurred for events less than 30 mm/hr. Further analysis determined that limiting rainfall capture to events <30 mm/hour had only minor (<0.5 ML/year) impact on harvesting potential. The hydraulics of the gravity based collection system was therefore capped at 30 mm/hour. Each dwelling, with an average roof size of 100 m^2, has the potential to harvest a maximum flow of 0.83 L/sec, based on the capped flow capacity of the communal collection system. Communal collection pipes ranged between 100 mm to 375 mm uPVC pipe, with the majority no more than 225 mm in diameter (Dark & Hamlyn-Harris, 2009).

11.3.4.4 Storage systems

As shown in Figure 11.4, the flat topography of the area required the rainwater from the roofs to be first collected by gravity into four underground concrete tanks and then pumped to a centralised above-ground tank of 800 kL capacity for subsequent treatment, storage and distribution The combination of a gravity and pumping system for rainwater collection and transportation substantially reduced project costs.

11.3.4.5 Distribution and treatment system

The treatment plant has the ability to process 200 kL/day of rainwater and consists of a multi-barrier approach including pre-filtration, coagulation, pH adjustment, microfiltration, activated carbon, ion exchange and UV disinfection and fluoridation.

11.3.4.6 Supply of potable water

Initially, the treated rainwater will be distributed through the non-potable stormwater recirculation system, for use in toilets, cold water laundry washing and outdoor uses. After validation and regulatory approval (over a period of approximately three years) to ensure the treated water's suitability as a potable source, the water will be supplied into the mains water system, which incidentally also supplies houses outside of the Fitzgibbon development.

11.4 MODELLING THE PERFORMANCE OF INDIVIDUAL VS. CLUSTER RAINWATER HARVESTING

The performance of cluster-scale rainwater harvesting systems was explored using Urban Developer®, an integrated urban water cycle modelling tool which simulates urban water systems including water supply, stormwater, and wastewater at scales ranging from allotment up to clusters or subdivisions (eWater Cooperative Research Centre, 2011). The model scenario we chose simulated a hypothetical 42-house greenfield development in Melbourne, Australia, (average annual rainfall of 649 mm) where rainwater was used to augment non-potable uses. The performance of a CSRH scheme of different storage sizes was compared to the traditional approach of installing a rainwater tank of equivalent *pro rata* storage capacity in each allotment. The model scenarios included harvested rainwater being used for indoor non-potable uses only (toilet flushing and laundry), or indoor and outdoor uses, for different household occupancy rates of between one and four persons. Based on local census data, assumed model occupancy of one person, two, three and four persons represented 15%, 25%, 14% and 46% of Melbourne households, respectively. Estimates of the demand for rainwater were based on recent Australian end-use studies, in particular Beal *et al.* (2010). Indoor demand was estimated on a per capita basis, whilst outdoor demand was considered on a per household basis. For example, for a three person household, estimated indoor demand for rainwater was 50 kL/year, with an additional 6 kL/year for garden irrigation per household.

The impervious area (catchment) for the development was estimated from digitizing the layout of a representative greenfield development on the outskirts of Melbourne. The average lot area was 366 m^2 with 53% of this area estimated as roof catchment and a further 23% as other impervious surfaces (driveways, paved paths, etc.). The modelling considered two possible configurations for rainwater harvesting at the development:

(1) Individual household rainwater tanks with 100% of dwelling roof area connected to the tank. The simulations considered 3 tank sizes – 1 kL, 2 kL and 5 kL.
(2) Cluster-scale rainwater harvesting system with 100% of roofs in each household connected to the cluster-scale rainwater tank of the communal harvesting system.

We note that it may not be possible to connect 100% roof area in large houses due to problems with down pipes configuration and their connectivity, including aesthetic considerations.

Figure 11.5 shows that volumetric reliabilities increased as storage size increased, as expected. The simulated cluster-scale harvesting scheme used a storage volume equivalent to the combined storage of the individual houses. The CSRH showed a higher volumetric reliability compared to the average for households' volumetric reliability when non-potable demand was supplied from individual rainwater tanks. However, Figure 11.5 depicts via the vertical bars that reliability differed depending upon household occupancy (which was used as a surrogate for demand). Households with a lower occupancy rate are likely to achieve higher volumetric reliability with an individual tank than they would when connected to a CSRH scheme. The inverse is true for higher occupancy rate households.

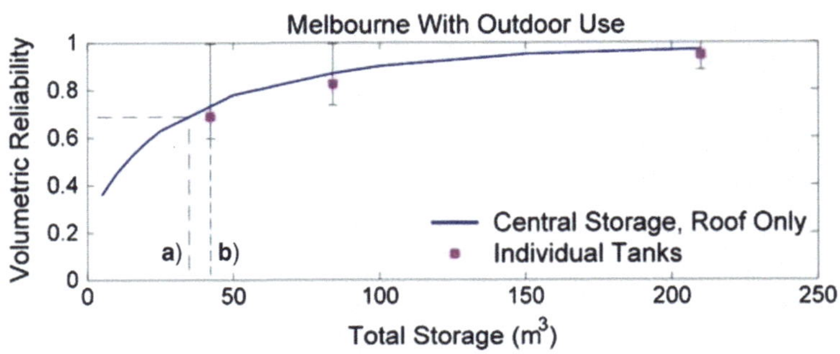

Figure 11.5 Volumetric reliability for a CSRH system compared to the average, maximum and minimum volumetric reliability when storage volume is supplied from individual household rainwater systems.

Also, it was possible to estimate the storage required for a cluster-scale system to achieve the same average reliability as that provided by the combined storage of individual rainwater systems. For the scenario where all 42 households have a 1 kL rainwater tank, the average volumetric reliability is 0.73 for a total combined storage of 42 kL. However, the same reliability can be achieved by using a cluster-scale harvesting scheme with a storage volume of 34 kL, a 19% reduction in total storage volume. This reduction in storage volume for cluster-scale systems has implication for both capital costs for storage tanks and the land footprint of rainwater harvesting systems. The *a* and *b* dotted lines in Figure 11.5 indicate that less storage volume is required for a CSRH system: (a) than the aggregate storage volume of individual rainwater systems; and (b) to achieve the same reliability.

11.5 METHOD FOR THE DESIGN OF CSRH SYSTEMS

The method for the design of CSRH systems presented in this section serves as the basis for the economic calculations of cluster-scale rainwater systems reported in Chapter 12.

11.5.1 Designing the CSRH system

The design of CSRH system is described in detail by Gurung and Sharma (2014). The main steps are summarised below:

(1) Develop a typical housing layout for use in a greenfield site, using collected information such as average lot size, roof area, housing density, street width and historical rainfall data.

(2) Use the collated information to develop various scales of housing layouts (4, 8, 16, 24, etc. houses).
(3) Assign the location of the rainwater storage tank within the development considering the overall topography.
  – For a flat topography, the communal rainwater tank should be situated in the centre to minimise costs of the rainwater collection and supply networks (Figure 11.6a);
  – In a sloping topography, the communal rainwater tank should be situated in the lower elevation side of the development to capture water easily by gravity (Figure 11.6b).

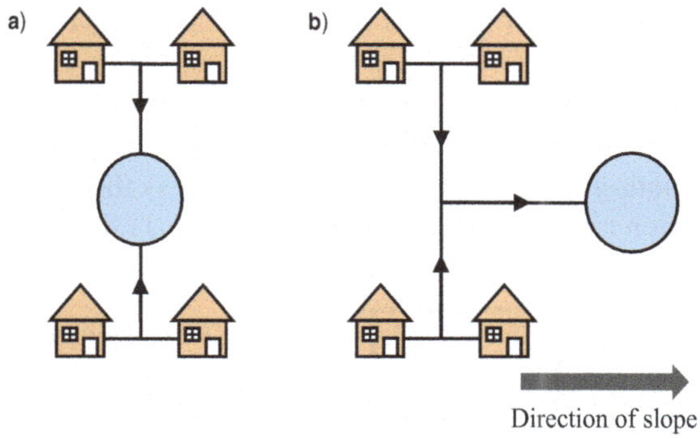

Figure 11.6 (a) Rainwater tank in the middle for flat topography; (b) Rainwater tank at lower elevation side for sloping topography.

(4) Plan the layout of the collection and distribution systems for each housing layout.
(5) Decide on the use of the rainwater; potable or non-potable purposes, and collate information on the consumption rates for various end uses.
(6) Estimate peak flows in rainwater collection and distribution systems for each housing layout (Figure 11.7).
(7) For each layout, estimate the size of main rainwater storage tanks, satellite tanks (interim storage tanks before the main storage tanks) if required, pipe sizes of collection and recirculation networks, and pump capacity. If satellite tanks are not considered, rainwater tanks will have to be placed underground, which may be very deep and thus uneconomical to construct and operate.

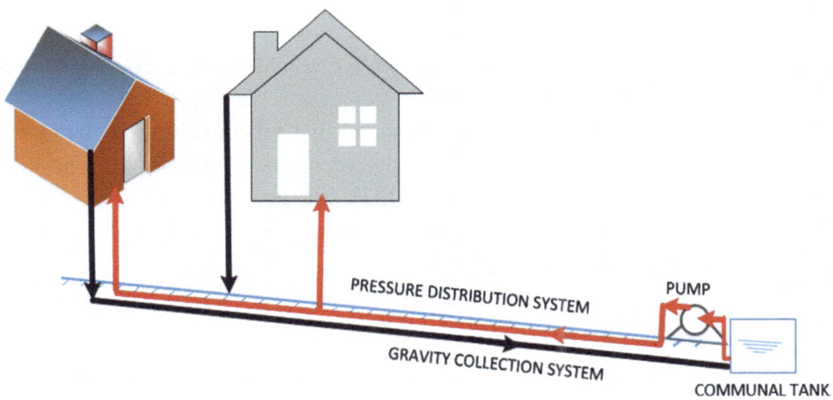

Figure 11.7 Communal rainwater tank systems – collection and distribution systems.

The method is illustrated here using the Capo di Monte (CDM) development example, which is used as the case study described in a previous section.

11.5.1.1 Rainwater collection pipes

Collection pipes transport the rainwater from roofs to the storage tanks and need to be adequately sized (Figure 11.7). Peak flow contributions from each household's roof can be calculated using the Rational Method (DERM, 2007). The size of the connection (pipe connectors) from the household downpipes to the communal rainwater collection pipe determines the runoff flow rate that can be harvested by the collection system.

At CDM, hydraulic analysis demonstrated that the size of the household's connector limited the peak rainfall intensity entering the communal system to <2 mm/5 min. In-ground pipe installed around the dwelling is connected to the household downpipes. The connector is then used to join the in ground pipe to the main communal collection pipe network, as shown in Figure 11.8.

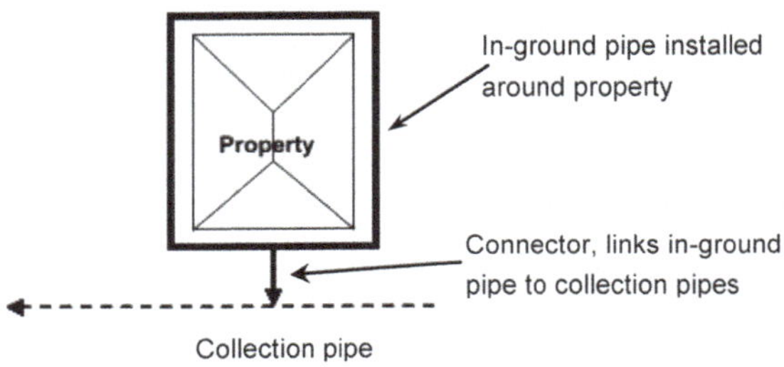

Figure 11.8 Schematic of a typical connection to the CRWH collection system.

Using this intensity with the Rational Method, the peak flow able to enter the system was found to be 1.28 L/s per dwelling assuming a roof size of 220 m^2 with 100% connectivity, and fractional rainfall catch losses of 0.875 mm. Manning's equation can be rearranged and used in the sizing of the gravity flow collection pipes (Figure 11.7), depending on the pipe material chosen (Rossman, 2010).

$$d^{8/3} = \frac{4^{5/3} \cdot Q \cdot n}{\pi \cdot S^{1/2}}$$

Where: d = diameter of pipe (m), Q = flow in pipe (m^3/s), n = Manning's roughness coefficient, and S = slope.

A minimum pipe-full flow velocity is required to ensure that self-cleansing of the pipe is maintained and scour does not occur (DERM, 2007). Local guidance recommends minimum pipe-full flow velocity for self-cleansing purposes to be 0.6 m/s (DERM, 2007). For polyvinyl chloride (PVC) pipe with an estimated roughness of 0.01 (Rossman, 2010) the minimum slope required to produce this velocity within the study area was 0.5% for a 100 mm diameter pipe.

11.5.1.2 Rainwater storage tanks

The sizing of storage tank for a CSRH system needs to consider the demand of intended end uses, rainfall pattern and reliability of supply. For the CDM case study, the harvested water is treated and used for all indoor potable demand. This implies a water demand pattern which has a typical diurnal, bimodal distribution, with peaks during the morning and early evening, but little seasonal variation (Cook *et al.* 2013). In contrast, harvested water used for irrigation is likely to have a pronounced seasonal variation in demand. Studies that partition household water consumption into specific end uses, such as those conducted in South East Queensland by Beal *et al.* (2010) are very useful in estimating the likely demand for a CSRH system.

For CDM, the urban water balance tool UVQ (Mitchell & Diaper, 2006), was used in the sizing of rainwater tanks. Historic rainfall data spanning 20 years was obtained from the BOM weather station and then 'censored' to exclude rainfall greater than 2 mm/5 min intensity, as per the flow limitation of the connector pipes. We found that the demand for rainwater could be satisfied on 98% of days over the 20-year period using a storage volume of 160 kL (Figure 11.9). To increase the reliability to 99.5% would require an impractically large storage of 280 kL. This is a large increase in storage volume for only a marginal gain in reliability (Cook *et al.* 2013). However, at CDM there was back-up supply from a groundwater bore to ensure reliable service, so rainfall did not have to meet all of the water demand. It was found that 90% of the daily demand could be met with an 80 kL tank, half the tank volume required to meet demand on 98% of the days. This analysis highlights that the availability of a back-up supply can minimise the need for large storage tanks whilst still maintaining a high level of service.

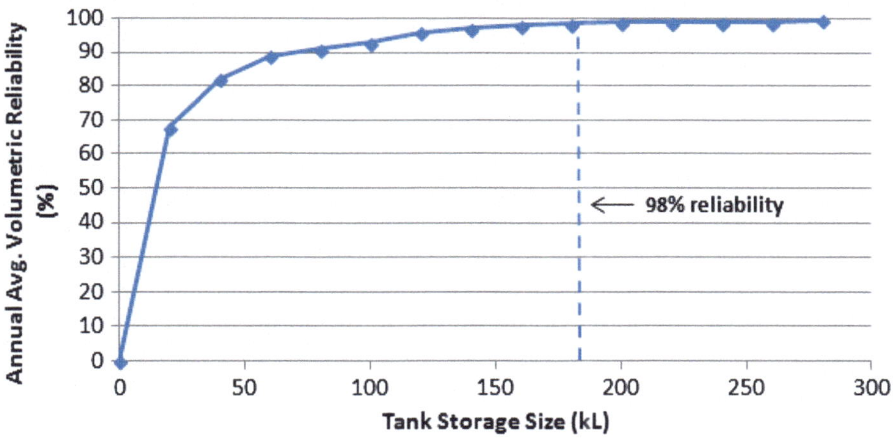

Figure 11.9 Volumetric reliability of a CSRH system at Capo Di Monte for different tank storage sizes. *Source*: Cook *et al.* 2012.

11.5.1.3 Rainwater distribution systems

Water distribution pipe sizes were estimated considering local planning guidelines and hydraulic methods for pipe sizing (Swamee & Sharma, 2008). Under local guidelines (GCCC, 2008), the minimum diameter of potable mains water is 100 mm with velocity not exceeding 2.5 m/s to prevent pipe failures. Checks for maximum velocity within the network can be calculated by dividing peak supply flow by the area of the pipe.

11.5.1.4 Pump capacities

The Darcy-Weisbach formula (Swamee & Sharma, 2008) is used to calculate head loss in pipes to ensure the network design meets the minimum pressure required at the property boundary for potable water, in this case 22 m of head (GCCC, 2008). The head required for distribution pumps is calculated using

Bernoulli's equation (Swamee & Sharma, 2008). The power required to pump the water to residents is then calculated using methods described in Swamee and Sharma (2008). A manufacturer's pump curve can then be used to choose the appropriate sizes of the pumps.

The energy usage by pumps is based on pump duties, sump well and sump capacities, and life cycle costing of the entire system can be conducted using methods described in Swamee and Sharma (2008) and Gurung and Sharma (2014). The economics of scale analysis of CSRH is described in Chapter 12.

11.6 WATER TREATMENT

CSRH offers the opportunity to use rainwater to meet potable water demands, such as drinking, as it provides economies of scale for the treatment plant. Another advantage of cluster-scale treatment of rainwater is that it enables the use of appropriate monitoring and management practices to ensure that treatment processes are operating to specification. Rainwater can be used for potable uses with adequate preventive risk management, including proper system design and maintenance and adequate treatment. The chemical and microbiological rainwater qualities for allotment-scale rainwater tanks are discussed in Chapters 9 and 10 respectively, and a similar quality can be expected in communal rainwater tanks as the roof area is still the rainwater catchment. The main water treatment objectives for potable use is reducing the turbidity and killing the pathogens. Hence, filtration (sand or MF membranes) and UV disinfection to ensure protozoan destruction is generally sufficient. Chlorine also needs to be added to prevent bacterial regrowth in the reticulation system, and this may cause disinfection by-product (DBP) problems if organic carbon content is high in the rainwater. This issue may require further research. More sophisticated treatments (e.g., ozone, RO, ion exchange, activated carbon) should only be required if stormwater is part of the raw water supply. Real-time monitoring of various parameters (eg. UV transmissivity, turbidity, residual chlorine, membrane integrity, etc.) will provide a high quality assurance that the treatment systems are working to specification.

The common treatment options provided for potable/non-potable use of harvested rainwater are illustrated in Figure 11.10 and summarised in the following sections.

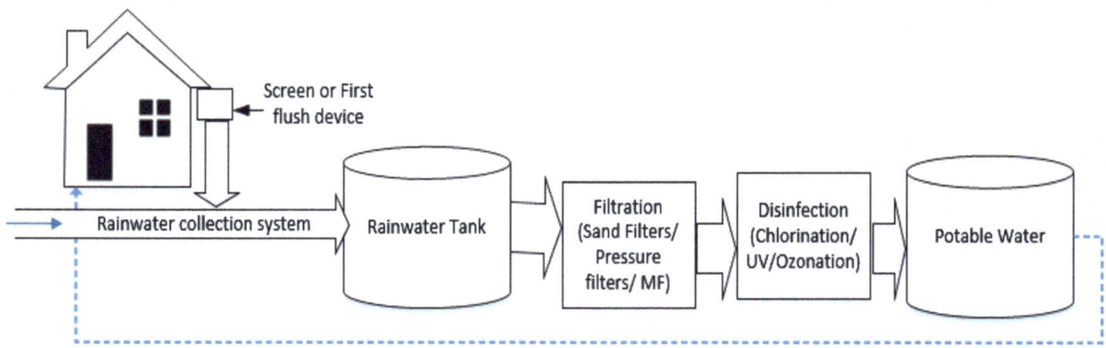

Figure 11.10 Arrangement of various rainwater treatment systems.

11.6.1 Preliminary treatment and filtration

Screening and first flush is considered as a preliminary treatment for improving harvested rainwater quality, followed by a filtration process which aids the effective removal of suspended impurities. Sand filtration using pressure filters is the most commonly used methods, where graded sand layers remove the fine sediments from the roof runoff. Regular backwashing is required for preventing clogging. Alternatively, membrane microfiltration can be used to reduce turbidity, organic matter and some microorganisms.

11.6.2 Disinfection

The most common technologies used for rainwater disinfection include chlorination, UV disinfection, ozonation and hydrogen peroxide disinfection. Liquid chlorine is generally dosed into a batch system or in-line system suited to the local conditions (Mintz *et al.* 1995; Salahuddin *et al.* 2011). However, chlorination is not suitable for water highly contaminated with organic matter due to the risk of DBP formation.

UV disinfection, on the other hand, is the most commonly used disinfection system for community supplies due to its ease of use, chemical free application, shorter retention time, small potential to form DBPs, but provides no residual disinfection in the distribution system (Roebuck *et al.* 2011; Keithley, 2012). Such systems can be installed in the pipe work delivering water either to the storage tank (as in the case of Capo di Monte), or from the tank to the receiver, or selectively to specific taps. To ensure the effectiveness of UV disinfection there is the need for upstream treatment that removes particles (through clarification and/or filtration). Even effluent with low measured turbidity (≤2 NTU) can still be difficult to disinfect due to large, undetected particles (Metcalf & Eddy, 2003). Other options for disinfection include ozone and hydrogen peroxide.

11.6.3 Post-treatment methods

Membrane filtration methods such as microfiltration, ultrafiltration (microfiltration uses 0.2-μ pores, while ultrafiltration excludes particles >0.02 μ), nanofiltration and reverse osmosis can be used as post-treatment steps to reduce trace contaminants if required. The influent is usually forced through the membrane by applying pressure, (Varbanets *et al.* 2009).

11.7 MANAGEMENT OF CSRH SYSTEMS

The broader adoption of CSRH systems requires consideration of business case and management models for the maintenance and operation of these systems. CSRH systems require cooperation amongst neighbouring households in a formal manner. For this reason, we have drawn upon experiences with decentralised wastewater systems in indentifying potentially suitable management models. Body corporate or community title arrangements can be applied when the CSRH system is part of a shared land title. However, there is the need to consider other management models where ownership of land is not shared.

CSRH systems offer a number of advantages for ensuring effective management when compared to household-scale rainwater systems. A CSRH system provides economy of scale and the option of sub-contracting operation and maintenance to industry specialists. This means that individual householders can adopt a more passive role thereby reducing the burden on householders who may lack the knowledge, time or motivation to maintain their rainwater system correctly. In this manner, CSRH systems offer the opportunity to formalise management arrangements. Using terminology coined by the US EPA (2005) in relation to decentralised wastewater treatment systems, CSRH require setting up a *Responsible Management Entity* (RME) The management of CSRH involves two phases: setup and ongoing management. The individual tasks are identified in Figure 11.11.

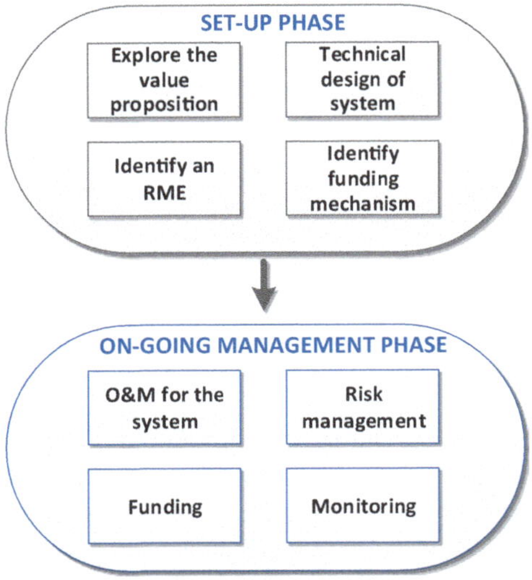

Figure 11.11 Responsible management entity tasks for a CSRH system.

Table 11.2 Management related decisions for CSRH systems.

Decision type	Choices	Considerations
Type of Responsible Management Entity (RME)	– Water utility – Local council – Home owners association – Body corporate or community title corporations – Privately owned businesses – Cooperative – Sub–contracting services – State agencies, urban land development authorities	– Experience in being an RME – Financing capacity of the RME – Long term continuity and stability of RME – Legal rights to enforce payments – Accountability and transparency of RME – Tax rules – Ease of set–up
Roles of RME	– Collect bill payments – Turn off services – Pay bills – Operate systems – Maintain systems – Monitor performance	– Technical capacity of the RME – Awareness of regulatory requirements – Motivation of the RME
Utilising O&M service providers	– Operate systems – Maintain systems – Monitor performance	– Contract arrangements – Continuity of services – Liability issues – Insurances – Accreditation and regulation
Payment approach	– Taxes – Water bills – Council rates – Body corporate fees – Bills to private company	– Can the RME enforce the payment of bills? – Can the RME turn off services if non–payment? – Motivation of householders/water users

In the early stage of planning, there are a number of decisions to be made in relation to RME activities (Table 11.2). This table has been developed based on experience with decentralised wastewater treatment systems which we believe can be adapted to CSRH systems. The decisions are:

- type of Responsible Management Entity;
- roles of the Responsible Management Entity;
- outsourcing Operation and Maintenance to service providers; and
- Choosing a payment approach that ensures financial sustainability, fairness, and adequate resourcing to cope with unexpected problems.

In addition to the decisions described in Table 11.2, it is also important to undertake *risk management planning* so that risks are identified and mitigated through adequate risk management strategies. Whilst the major risks relate to water quality and health issues, other key risks include financial viability and the capacity to ensure the ongoing operation of the system.

11.8 CONCLUSIONS

A cluster-scale approach to rainwater harvesting can provide a local, decentralised water supply in non-reticulated peri-urban areas, as well as reducing demand on drinking water in traditional urban areas. Household-scale rainwater harvesting systems have been the dominant mode of providing a rainwater source in established urban areas. However, this chapter has shown that cluster-scale systems *may* provide a better option for local water harvesting than household-scale systems. Specific findings included:

- There are a number of examples in different development contexts where cluster-scale approaches have been successfully implemented as a local, decentralised water source;
- Cluster-scale rainwater systems can allow for urban development in peri-urban areas not serviced by mains water supply;
- Cluster-scale rainwater harvesting can also be used in cities to reduce demand for mains water and remove the need for individual rainwater tanks as a non-potable water source;
- Cluster-scale systems can improve water quality management and reduce the land footprint required for storage, which can free some land for other purposes such as house extensions and so on;
- Cluster-scale harvesting can increase the overall yield and average reliability of rainwater harvesting, relative to household-scale systems;
- A back-up water supply is required in cases where the system is used for non-discretionary uses, such as potable demand or toilet flushing, to ensure 100% supply reliability;
- A cluster-scale approach can be used to extend the uses of harvested rainwater to potable uses by providing economies of scale for the more sophisticated treatment systems;
- There can also be economies of scale for cluster-scale collection and storage systems; and
- Cluster-scale harvesting offers the opportunity to formalise management of local water sources, which address some of the limitations of household-scale rainwater harvesting due to lack of householder skills or motivation to undertake adequate maintenance to ensure water quality and supply.

11.9 REFERENCES

Ahmed W., Vieritz A., Goonetilleke A. and Gardner T. (2010). Health risk from the use of roof-harvested rainwater in Southeast Queensland, Australia, as potable or nonpotable water, determined using quantitative microbial risk assessment. *Applied and Environment Microbiology*, **76**(22), 7382–7391.

Beal C. D., Stewart R. A. and Huang T. (2010). South East Queensland Residential End Use Study: Baseline Results – Winter 2010, Urban Water Security Research Alliance Technical Report 31.

Brown M., Liebman M. and Wiese R. (2008). Water Constraints Driving Advanced WSUD in Regents Park, Sydney. Sustainability Workshop, Stormwater Industry Association.

Brown, R. and Clarke J. (2007). Transition to Water Sensitive Urban Design: The Story of Melbourne, Australia, Monash University, Report No. 07/1.

Brown, R. R., Farrelly M. and Keath N. (2007). Summary Report: Perceptions of Institutional Drivers and Barriers to Sustainable Urban Water Management in Australia. Survey results of urban water professionals across Brisbane, Melbourne and Perth. Monash University National Urban Water Governance Program, Melbourne, Australia.

Brown R. R. and Farrelly M. A. (2009). Challenges ahead: Social and institutional factors influencing sustainable urban stormwater management in Australia. *Water Science and Technology*, **59**(4), 653–660.

Chhetri P., Han J. H., Chandra S. and Corcoran J. (2013). Mapping urban residential density patterns: compact city model in Melbourne, Australia. *City, Culture and Society*, **4**(2), 77–85.

Cook S., Sharma A. and Chong M. (2013). Performance analysis of a communal residential rainwater system for potable supply: a case study in Brisbane, Australia. *Water Resources Management*, 1–12.

Cook S., Sharma A., Gurung T. R., Chong M., Umapathi S., Gardner T., Palmer A. and Carlin G. (2012). Performance of Cluster Scale Rainwater Harvesting Systems: Analysis of Residential and Commercial Development Case Studies, Urban Water Security Research Alliance, Technical Report No. 68, p. 56.

Coombes P. J. (2002). An evaluation of the benefits of source control measures at the regional scale. *Urban Water*, **4**, 307–320.

Coombes P. J., Argue J. R. and Kuczera G. (2000). Figtree place: a case study in water sensitive urban development (WSUD). *Urban Water*, **1**(4), 335–343.

Dark L. and Hamlyn-Harris D. (2009). Feasibility Report for the Fitzgibbon Potable Roofwater Reuse Project. Bligh Tanner Consulting Engineers, Brisbane.

DERM (2007). Queensland Urban Drainage Manual Department of Environmental and Resource Management, Queensland Government, p. 1.

Domènech L. and Saurí D. (2011). A comparative appraisal of the use of rainwater harvesting in single and multi-family buildings of the Metropolitan Area of Barcelona (Spain): social experience, drinking water savings and economic costs. *Journal of Cleaner Production*, **19**(6–7), 598–608.

Downton P. (2005). Urban revolution. *Sanctuary*, **1**, 57–60.

eWater Cooperative Research Centre (2011). Urban Developer User Guide. eWater Cooperative Research Centre, Canberra.

Fane, S. A. and Mitchell C. (2006). Appropriate Cost Analysis for Decentralised Systems. Enviro '06, Melbourne.

Farreny R., Gabarrell X. and Rieradevall J. (2011). Cost-efficiency of rainwater harvesting strategies in dense Mediterranean neighbourhoods. *Resources, Conservation and Recycling*, **55**(7), 686–694.

Fletcher T. D., Deletic A., Mitchell V. G. and Hatt B. E. (2008). Reuse of urban runoff in Australia: a review of recent advances and remaining challenges. *Journal of Environmental Quality*, **37**(5), S116–S127.

GCCC (2008). Gold Coast Planning Scheme Policies, Policy 11: Land Development Guidelines, Section 4: Water Reticulation – Design Requirements, Gold Coast, Queensland, Gold Coast City Council.

Geisinger D. and Chartier G. (2005). Managed onsite/decentralised wastewater systems as long-term solutions. *Clearwaters*, **35**, 6–11.

Gurung T., Sharma A. and Umapathi S. (2012). Economics of Scale Analysis of Communal Rain Water Tanks, Urban Water Security Research Alliance, Technical Report No. 67.

Gurung T. R. and Sharma A. (2014). Communal rainwater tank systems design and economies of scale. *Journal of Cleaner Production*, **67**(0), 26–36.

Hall M. (2013). Review of Rainwater Tank Cost-Effectiveness in South East Queensland, Urban Water Security Research Alliance, Technical Report No. 105.

Huston R., Chan Y. C., Chapman H., Gardner T. and Shaw G. (2012). Source apportionment of heavy metals and ionic contaminants in rainwater tanks in a subtropical urban area in Australia. *Water Research* **46**(4), 1121–1132.

Keithley S. E. (2012). The Effect of Treatment on the Quality of Harvested Rainwater. Environmental and Water Resources Engineering, Cockrell School of Engineering, The University of Texas.

Kim J. and Furumai H. (2012). Assessment of rainwater availability by building type and water Use through GIS-based scenario analysis. *Water Resources Management*, **26**(6), 1499–1511.

Magyar M. I., Mitchell V. G., Ladson A. R. and Diaper C. (2007). An investigation of rainwater tanks quality and sediment dynamics. *Water Science & Technology*, **56**(9), 21–28.

Makropoulos C. and Butler D. (2010). Distributed water infrastructure for sustainable communities. *Water Resources Management*, **24**(11), 2795–2816.

Metcalf and Eddy, Tchobanoglous G., Burton F. L. and Stensel H. D. (2003). Wastewater Engineering, Treatment and Reuse, 4th edn, McGraw Hill Education.

Mintz E. D., Reiff F. M. and Tauxe R. V. (1995) Safe water treatment and storage in the home: a practical new strategy to prevent waterborne disease. *Journal of the American Medical Association*, **273**, 948–953.

Mitchell V., Taylor A., Fletcher T. and Deletic A. (2005). Stormwater Reuse-Potable Water Substitution for Melbourne. Institute for Sustainable Water Resources, Monash University, Melbourne, Australia. ISWR Report No 05/12.

Mitchell V. G. (2004). Integrated Urban Water Management: A review of current Australian practice. Australian Water Conservation and Reuse Research Program.

Mitchell V. G. and Diaper C. (2006). Simulating the urban water and contaminant cycle. *Environmental Modelling and Software*, **21**(1), 129–134.

Moglia M., Cook S., Sharma A. and Burn S. (2011). Assessing decentralised water solutions: towards a framework for adaptive learning. *Water Resources Management*, **25**(1), 217–238.

Moglia M., Tjandraatmadja G. and Sharma A. K. (2013). Exploring the need for rainwater tank maintenance: survey, review and simulations. *Water Science & Technology: Water Supply*, **13**, 191–201.

National Health and Medical Research Council (2011). Australian Drinking Water Guidelines, National Health and Medical Research Council, Canberra.

Rodrigo S., Sinclair M., Forbes A., Cunliffe D. and Leder K. (2011). Drinking rainwater: a double-blinded, randomized controlled study of water treatment filters and gastroenteritis incidence. *American Journal of Public Health*, **101**(5), 842–847.

Roebuck R., Oltean-Dumbrava C. and Tait S. (2011). Whole life cost performance of domestic rainwater harvesting systems in the United Kingdom. *Water and Environment Journal*, **25**, 355–365.

Rossman, A. L. (2010). Storm Water Management Model User's Manual. Water Supply and Water Resources Division, US Environmental Protection Agency, Cincinnati, OH.

Roy A., Wenger S., Fletcher T., Walsh C., Ladson A., Shuster W., Thurston H. and Brown R. (2008). Impediments and solutions to sustainable, watershed-scale urban stormwater management: lessons from Australia and the United States. *Environmental Management*, **42**(2), 344–359.

Ruth M., Bernier C., Jollands N. and Golubiewski N. (2007). Adaptation of urban water supply infrastructure to impacts from climate and socioeconomic changes: the case of Hamilton, New Zealand. *Water Resources Management*, **21**(6), 1031–1045.

Salahuddin M., Jeyagapal M. R., Al-Tameemi M. and Mogan V. (2011). An investigation into cistern for rainwater storage and sterilization for the new UBC Student Union Building. Sustainability Office, University of British Columbia, pp. 1–27.

Sharma A., Cook S., Tjandraatmadja G. and Gregory A. (2012). Impediments and constraints in the uptake of water sensitive urban design measures in greenfield and infill developments. *Water Science and Technology*, **65**(2), 340–352.

Sharma A., Cook S. and Chong M. (2013). Monitoring and validation of decentralised water and wastewater systems for increased uptake. *Water Science and Technology*, **67**(11), 2576–2581.

Sharma A., Grant A., Grant T., Pamminger F. and Opray L. (2009). Environmental and economic assessment of urban water services for a greenfield development. *Environmental Engineering Science*, **26**(5), 921–934.

Sharma A., Grant A., Gray S., Mitchell G., Grant T. and Opray L. (2005). Sustainability of Alternative Water and Sewerage Servicing Options. Prepared by CSIRO for Yarra Valley Water, Melbourne.

Swamee P. K. and Sharma A. K. (2008). Design of Water Supply Pipe Networks. John Wiley & Sons.

Tjandraatmadja G., Cook S., Sharma A., Diaper C., Grant A., Toifl M., Barron O., Burn S. and Gregory A. (2008). ICON water sensitive urban developments. CSIRO, Melbourne.

US EPA (2005). Handbook for Managing Onsite and Clustered (Decentralized) Wastewater treatment systems. An Introduction to Management Tools and Information for Implementing EPA's Management Guidelines. Cincinatti, U.S.

US EPA (2013). Rainwater Harvesting Conservation, Credit, Codes, and Cost Literature Review and Case Studies. United States Environmental Protection Agency, Washington, DC.

Varbanets M.-P., Zurbrügg C., Swartz C. and Pronk W. (2009). Decentralized systems for potable water and the potential of membrane technology. *Water Research*, **43**, 245–265.

Villarreal, E. L. and Dixon A. (2005). Analysis of a rainwater collection system for domestic water supply in Ringdansen, Norrköping, Sweden. *Building and Environment*, **40**(9), 1174–1184.

Wannon Water (2012). Roof water harvesting requirement: Expert report. VCAT Hearing. Wannon Water, Melbourne.

Wannon Water (2011). Roofwater Harvesting Evaluation Toolkit. Clearwater Resource Library, http://www.clearwater.asn.au/content/roofwater-harvesting-evaluation-toolkit (accessed March 2014).

Chapter 12

Economics of individual and communal rainwater tank systems

Murray R. Hall, Thulo Ram Gurung and Kym Whiteoak

ABSTRACT

This chapter considers the economic performance of rainwater tanks. The chapter starts at the scale of individual rainwater tanks and then considers communal rainwater tank systems. The chapter also progresses in scope by expanding analysis from cost-effectiveness to cost benefit analysis. The chapter presents general methods and demonstrates the approach by drawing upon recent research in South East Queensland, Australia where several hundred thousand rainwater tanks have been installed over the past decade.

The economics of individual rainwater tanks focuses on the variation and uncertainty in cost-effectiveness calculations. It illustrates that variation in assumptions such as the discount rate and the maintenance assumptions can have as large an impact as a standard deviation increase in the yield. When a number of modelling assumptions are changed to form a new scenario the mean cost-effectiveness can change by 50% even though the yield and basic cost data remain the same.

The economics of communal rainwater tanks focuses on design considerations such as scale and the effect on cost-effectiveness. The results suggest that there is an optimal scale for communal rainwater tanks and under or oversized systems will be less cost-effective.

The cost benefit analysis of rainwater tanks provides a focus on the importance of other benefits from rainwater tanks such as delayed infrastructure upgrades and reduced stormwater flows to receiving waters. The case study suggested that additional benefits for rainwater tanks were relatively small. Nonetheless, in some circumstances, rainwater tanks may provide a greater net benefit than options such as augmentation of the existing supply.

Keywords: Cost benefit analysis; cost-effectiveness; rain water tanks and systems.

12.1 INTRODUCTION

Over the past decade a number of articles have reviewed the financial performance of rainwater tanks for individual houses. These articles come from countries across the world including Australia, India, Bangladesh, China, Spain, UK and Italy. Some articles suggest rainwater tanks are very cost-effective, have short payback periods and are competitive with mains supplies (Alam *et al.* 2012; Coombes, Kuczera & Kalma, 2003; Ghisi & Schondermark, 2013; Islam, Chou & Kabir, 2011; Knights & Wong, 2008; Tam, Tam & Zeng, 2010). Other articles suggest that rainwater tanks are likely to have long payback periods and will be more costly than mains supplies for many years to come (Domenech & Sauri, 2011; Farreny, Gabarrell & Rieradevall, 2011; Morales-Pinzon, Luruena, Rieradevall, Gasol & Gabarrell, 2012; Roebuck, Oltean-Dumbrava & Tait, 2011).

Part of the difference may be due to the local context and differing water yields, social and institutional factors. However, there also appears to be a large variation in the scope of costs considered, the rationale for defining financial parameters such as the discount rate, the type of data used and the uncertainty, the system design as well as the consideration of additional benefits. Each section of this chapter presents a method and considerations for undertaking an analysis of the economic performance of rainwater tanks. The analysis begins at the level of an individual rainwater tank and considers the variation and uncertainty in cost-effectiveness. The section on communal rainwater tanks investigates the effect of scale on the cost-effectiveness. Finally, the third section of the chapter broadens the scope of benefits and considers Cost Benefit Analysis of rainwater tanks.

12.2 COST-EFFECTIVENESS OF INDIVIDUAL RAINWATER TANKS

12.2.1 Method

The method draws upon Life Cycle Costing and Cost-effectiveness Analysis (CEA). Cost-effectiveness Analysis is an established economic method for evaluating the cost of an option to achieve an objective (Pearce, Atkinson, and Mourato, 2006). Life Cycle Costing methods, which aim to capture costs over the entire life cycle, are also well established and standardised, for example StandardsAustralia (1999). The following steps provide general guidance for undertaking cost-effectiveness analysis and are demonstrated using a case study in South East Queensland (SEQ), Australia.

12.2.1.1 Step 1: Define the objectives and limitations

The cost-effectiveness of a rainwater tank will consider the cost to provide water over a period of time and can support decisions about whether or not to invest in a rainwater tank. However, the objective, costs and decision may be different for a homeowner, water utility or government entity. This means that the user group, and the decision to be made, needs to be clearly defined to justify the costs included in the calculation. For example, the cost-effectiveness of a rainwater tank for:

- a household may consider rebates, a market value for the water saved, and the market cost of money used to finance the tank system. Financial measures such as the payback period may be important for the household to decide upon investing in a rainwater tank.
- a water utility may limit costs to the rebates that it provides (and not the remaining costs incurred by the household) and use a Weighted Average Cost of Capital (WACC) for its access to finance. The WACC provides an average cost of capital that reflects the asset base and sources of finance for an organisation. The water utility may be interested in the levelised cost of rainwater tanks compared to other marginal supplies to decide whether or not to provide rebates. The levelised cost is the ratio between total life cycle cost and the total life time rainwater usage. The levelised cost is further defined in the following case study.
- a government policy perspective may consider all the capital and operating costs for a rainwater tank and the cost of money based upon government bond rates. The levelised cost of rainwater tanks may be an important consideration for a 'prudent and efficient' government policy for rainwater tanks.

An advantage of cost-effectiveness analysis is the focus on the costs to achieve a single objective. The focus on a single objective is justified where other benefits are small and will not affect the evaluation. These assumptions should be defined as limitations of the study. A broader framework for capturing benefits is discussed in the Section 12.4 dealing with Cost Benefit Analysis of rainwater tanks.

12.2.1.2 Step 2: Define the data variation and uncertainty

The calculation of cost-effectiveness can draw upon a number of different types of data. This has implications for the consideration of variability and uncertainty in the results. For example, capital costs for tanks and pumps may be based upon a survey of quotes from suppliers. The variation in results can be calculated and may be expected to be relatively small in a competitive market. Similarly, the water yield from rainwater tanks can be measured and a distribution developed. The yield may also be compared to modelled yield which may be based upon measured inputs such as rainfall, roof area, tank size and water consumption patterns. The information on the measured yield is generally unknown and thus the estimations are based on modelled yield, which is generally higher than the measured yield.

Thus, measured or modelled data may not be available for all variables and there will be uncertainty in the results beyond the variation in the data. For example, future costs such as interest rates, electricity and water as well as the life span of tanks, future yields and the future maintenance practices of households. The uncertainty can be considered using scenarios and sensitivity of the results. These variables need to be clearly identified to assess the rationale for the data selected and the sensitivity of the results. For example, the cost of maintenance practices for cleaning tanks could be estimated using a typical or 'nominal' cost to provide a scenario for future practice (Domenech & Sauri, 2011; Khastagir & Jayasuriya, 2011; Marsden Jacob, 2007; Tam *et al.* 2010). However, an alternative scenario may also be justified if current maintenance practices do not meet health guidelines. This is discussed in more detail in the following case study, in Section 12.12.2.2.

12.2.1.3 Step 3: Calculate the cost-effectiveness and test the sensitivity

There are a number of ways of expressing the cost-effectiveness (Pearce *et al.* 2006); the estimation of levelised cost is demonstrated in the following case study to understand cost-effectiveness. A sensitivity analysis can provide a way to understand the importance of variables for the cost-effectiveness results. This provides a way of understanding the effect of assumptions upon the results, and may also highlight the need for further data collection. The sensitivity analysis may begin with an analysis of the effect of each variable upon the results. This may involve changing each variable by a standard deviation if the distribution is known, or by some other factor if not. Variables that have a large effect on the results can be investigated to check the validity of the assumptions.

12.2.2 Case study – mandated rainwater tanks in South East Queensland, Australia

12.2.2.1 Define the objectives and limitations

The objective of the study was to consider the cost-effectiveness of rainwater supply for an internally plumbed five kilolitre (kL) rainwater tank. This tank configuration (size) was common in the region because it's performance was assumed to meet Queensland Development Code MP 4.2 requirements for a potable water saving of 70 kL per annum. These mandatory requirements were subsequently repealed in 2013. The analysis aimed to support a water company or government policy investment decision. In

2008, it was estimated that there were 313 000 rainwater tanks in the SEQ region of Australia (Gardiner, 2009). This was projected to increase to a total of 1.11 million tanks by 2056 due to building regulation requirements (QWC, 2008). Figure 12.1 shows the SEQ region and its local government areas.

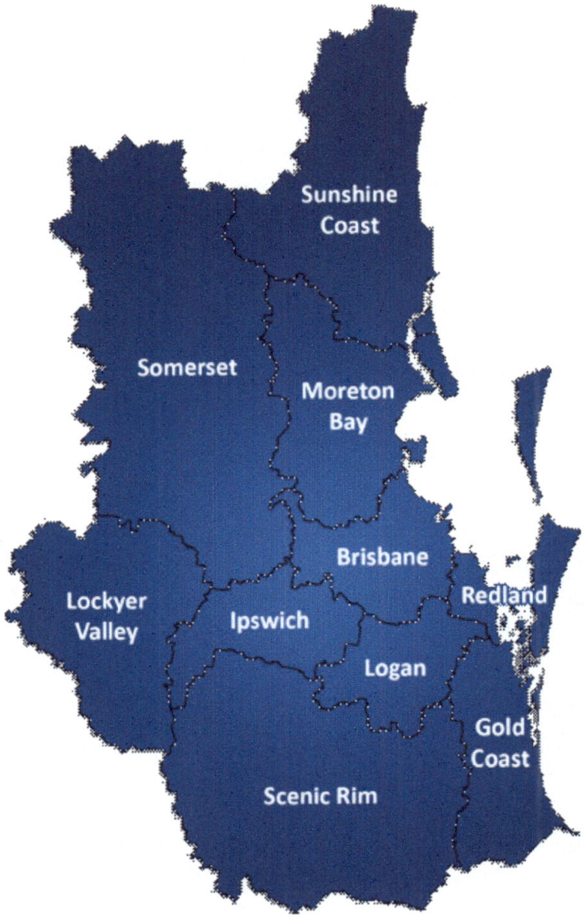

Figure 12.1 Local government areas in South East Queensland, Australia (http://en.wikipedia.org/wiki/South_East_Queensland).

Cost-effectiveness was expressed as a levelised cost as outlined in Equation 1. This approach considers the physical flow of water as a revenue stream and assumes that the unit cost of water (the levelised cost) as well as the discount rate is constant over the period of analysis. The analysis of levelised cost follows the approach outlined for the water and energy sectors (Fane, Robinson, and White, 2002; OECD, 2010).

$$\text{Levelised Cost} = \frac{C + \sum_1^n A_n/(1+i)^n}{\sum_1^n Y_n/(1+i)^n} \tag{12.1}$$

where C is the capital cost (\$), n is the year in the period of analysis, A is the annual payment (\$), i is the discount rate, Y is the annual yield (kL). These parameters are further described in the following section.

Each term in Equation 12.1 has a number of input variables which can be defined following local Life Cycle Costing (LCC) standards as defined in StandardsAustralia (1999). Capital may include the cost of the rainwater tank, the pump, laying of a concrete slab for the tank foundation, as well as plumbing and installation of both the tank and the pump. Annual costs may include operating costs such as electricity use by the pump, maintenance costs such as pump repair, roof gutter and screen maintenance and desludging, as well as capital replacement cycles for pumps and tanks over the period of analysis. The calculation of energy costs includes the specific energy of pumps as well as assumed price paths for electricity. Residual values of tanks and pumps were estimated as the fraction of the useful life of an asset remaining at the end of the period of analysis. For example, if the period of analysis was 30 years and the tank was replaced after 25 years, then only 20% of the value of the replacement tank was considered in the analysis.

The discount rate reflects the time cost of money and the risk profile of the borrower. From this perspective, an organisation seeking to finance an investment adopts a discount rate based on the interest rate of their borrowings. However, the discount rate also expresses how much *less* something is worth in the future compared to the present. This can raise a number of ethical and intergenerational issues (Ruth, 1993). The discount rate can also be the largest uncertainty in an economic analysis for environmental impacts over long time periods (Weitzman, 2007). For this reason, a number of authors have suggested that long-term analysis of costs and benefits should have lower discount rates than the market cost for money (Garnaut, 2008; Weitzman, 2001).

12.2.2.2 Define the data variation and uncertainty

Table 12.1 provides a summary of the data, scenario and sensitivity assumptions for the study. Probability distributions were developed for input variables to facilitate a Monte Carlo simulation for the cost-effectiveness. The differing availability of data meant that a number of approaches were possible for developing the input distributions. In some cases, detailed monitoring and modelling data was available which defined the probability distribution. In other cases, the amount of data was limited and there was insufficient information to define the distribution. A simplifying assumption was made to assume a triangular distribution for all input variables based upon IPCC recommendations for addressing uncertainty in data (IPCC, 2000). The distribution was defined by the most likely and upper and lower values for a 95% confidence interval. The effect of the choice of distribution was considered in the sensitivity analysis.

Other variables such as the discount rate, period of analysis and maintenance were defined in the scenario and sensitivity analysis. A 3% discount rate was assumed to capture a government perspective of investment as well as long periods of analysis (Weitzman, 2001, 2007). A 3% discount was adopted based on the benchmark Weighted Average Cost of Capital (WACC) for the 2013–15 period for price monitoring of SEQ water and wastewater retail activities (PWC, 2013). This discount rate is considered to be 'risk free' because it was based on the government bond rate (PWC, 2013). A 6% discount rate was also considered in the sensitivity analysis and reflects the current benchmark WACC which considers the cost of equity and debt for SEQ water utilities.

The maintenance of tanks was also an important consideration due to the apparent difference between current and recommended practice to manage health risks (Moglia, Tjandraatmadja & Sharma, 2011). For example, a survey of rainwater tank maintenance in SEQ indicated that relatively few respondents checked for the build-up of sediment in the tank and even less removed it as required (Mankad, Tucker & Greenhill, 2012) despite health guidelines. Tanks installed as part of mandatory building requirements in

Table 12.1 Summary of data, scenario and sensitivity assumptions for individual rainwater tank cost-effectiveness analysis.

	'Basic Scenario'			'Alternative Scenario'			Sensitivity	Note
	Triangular distribution			Triangular distribution				
	Min	Most likely	Max	Min	Most likely	Max		
Financial								
Discount rate (%)		3			6		6	Note 1
Period of analysis (years)		50			25		Same as 'alternative scenario'.	Note 2
Capital and installation ($AUD2012)								
Rainwater tank (5 kL)	1401	1544	1657					Note 3
Pump	722	790	962					
Plumbing	759	900	1017	Same as 'basic scenario'				
Tank installation	334	350	400					
Laying concrete slab	597	700	803					
Pump installation	200	250	300					
Operating, maintenance and replacement								
Energy								
Specific energy (kWh/kL)	1.1	1.48	1.9					Note 4
Unit cost for energy in 2012 ($/kWh)		0.228						Note 5
Retail energy price real growth rate (%)	3	5	7					
Useful lives (years)				Same as 'basic scenario'				Note 6
Pump life	5	10	15					
Rainwater tank life	15	25	35					
Plumbing		50						

Maintenance ($AUD2012)	0	20	54	54	104	184	Same as 'alternative scenario'	Note 7
Yield (kL/year)								Note 8
Brisbane	18	42	76				Fitted distribution	
Moreton Bay	16	43	78					
Sunshine Coast	24	48	90	Same as 'basic scenario'				
Ipswich	10	34	66					
Gold Coast	18	44	74					

Notes

1. Long period of analysis discount rate based upon (Weitzman, 2001, 2007). Water utility discount rate and period of analysis based upon (Queensland Competition Authority Accessed 25 February 2013 http://www.qca.org.au/) (PWC, 2013)
2. Considered as a single value for the period of analysis and not considered as a distribution
3. Based upon (Gurung, Sharma & Umapathi, 2012) with comparison to (Marsden Jacob, 2007)
4. Based upon (Umapathi, Chong, and Sharma, 2012) with comparison to (Beal, Hood, Gardner, Christiansen & Lane, 2008; Ferguson, 2012; Lane & Gardner, 2009; Retamal & Turner, 2010)
5. (AEMC, 2011); (IES, 2010)
6. (Gurung et al. 2012; Marsden Jacob, 2007; Stewart, 2011)
7. Costs based upon (Gurung et al. 2012) and scenario based upon (Gardiner, 2010; Marsden Jacob, 2007; Moglia et al. 2011) (Mankad et al. 2012; Walton, Gardner, Sharma, Moglia, and Tjandraatmadja, 2012)
8. Yield distributions based upon (Maheepala, Coultas & Neumann, 2013) with reference to measurements and modelling from (Beal et al. 2011; 2012; Chong et al. 2011; Umapathi et al. 2012).

SEQ were less likely to be maintained than voluntary installations (Mankad *et al.* 2012). These results followed earlier research in SEQ that indicated that 'half of the owners of required tanks reported that they never cleaned their screens and gutters or inspected inside their tanks or did so only when there were obvious problems' (Gardiner, 2010). These variables need to be clearly identified to assess the rationale for the data selected and the sensitivity of the results. For example, the cost of maintenance practices for cleaning tanks could be estimated using a typical or 'nominal' cost to provide a scenario for future practice (Domenech & Sauri, 2011; Khastagir & Jayasuriya, 2011; Marsden Jacob, 2007; Tam *et al.* 2010). However, an alternative scenario may also be justified if current maintenance practices do not meet health guidelines. This is discussed in more detail in the case study, in the next section. The uncertainty in maintenance practice and costs was captured in the following scenarios. Current maintenance is minimal and a nominal cost was assumed based on previous studies. Maintenance costs based on health guidelines was also considered.

A 'basic scenario' was developed for a long-term *government perspective* based on current practice. A 50-year period of analysis was selected with a 3% discount rate and maintenance based on current practice. The sensitivity analysis considered changes in the discount rate, period of analysis, tank maintenance as well as the effect of refining the distribution for the yield. An 'alternative scenario' was also developed to provide an upper range estimate that may reflect a *water utility* perspective. The 'alternative scenario' used the WACC for the discount rate of 6%, a period of analysis for infrastructure appraisal of 25 years as well as tank maintenance according to recommended practice. These scenarios are described in Section 12.2.1 Step 1.

12.2.2.3 Calculate the cost-effectiveness and test the sensitivity

Table 12.2 presents the levelised cost for the 'basic scenario' based on the data presented in Table 12.1. In summary, the basic scenario has a 50-year period of analysis, 3% discount rate and maintenance based on current practice. The levelised cost is presented for various locations in SEQ as well as an average for the region. The results were calculated using software called *@risk*. This program is an add-on to excel and uses distributions of input variables to simulate a distribution of the output variable. The results show the parameters for the cost-effectiveness distribution such as the mean, the 95% confidence interval and parameters describing the skew. The skew in the distribution shows that a relatively small number of tanks are very costly for the amount of water provided. The skew effects measures of central tendency such as the mean. The mean is important for capturing the effect of poorly performing tanks on the average for a region. The median provides an indication of central tendency that shows typical performance for most of the tanks and reduces the effect of the tail of the distribution, in this case the poorly performing tanks. The data in Table 12.2 can also be presented as probability distributions. Figure 12.2 shows the results for Brisbane to illustrate the shape of the distribution and the 95% confidence interval. The cost-effectiveness differed across the region in terms of the mean as well as the skew. The Sunshine Coast was the most cost-effective location for a rainwater tank, with a mean levelised cost of $7.62/kL. Ipswich was the least cost-effective location for a rainwater tank, with a mean levelised cost of $11.17/kL. The distribution for Ipswich also had the highest skew and produced a relatively high upper 95% confidence limit of $22.19/kL.

Figure 12.3 illustrates sensitivity of the cost-effectiveness to the input variables. The effect on the cost-effectiveness is calculated by increasing each input variable by one standard deviation while the other input variables remain unchanged. For example, a one standard deviation increase in the yield means a greater yield for the same cost and improves the cost-effectiveness by $2.17/kL. Conversely, a one standard deviation increase in the cost of laying a concrete slab worsens the cost-effectiveness by $0.036/kL. This illustrates that the results are sensitive to yield while they are not very sensitive to the cost of the concrete slab. Figure 12.3 shows the cost-effectiveness is sensitive to assumptions about the yield, pump life, tank life, maintenance and growth rate for electricity prices. Interestingly, the cost-effectiveness was more sensitive to assumptions about the life of a pump and tank as well as the price path for electricity rather than the actual capital costs and operation of the pump. The results are more sensitive to the uncertainty associated with projecting these input variables over the period of analysis than the variation in the collected or modelled data. In general, the results were more sensitive to operating rather than capital costs. Note that cost variables were considered as independent variables and this assumption needs to be explored further in future research. For example, a high initial capital cost for a pump may be associated with a longer life and lower energy costs.

Table 12.2 Summary of 'basic scenario' levelised cost for rainwater tanks in SEQ.

Region	Mean ($/kL)	Mode ($/kL)	Median ($/kL)	Std dev ($/kL)	Skewness	2.5% ($/kL)	97.5% ($/kL)
Brisbane	8.93	7.18	8.42	2.46	1.12	5.59	15.10
Moreton Bay	8.97	7.12	8.38	2.70	1.32	5.49	16.10
Sunshine Coast	7.62	6.19	7.28	1.91	0.95	4.90	12.23
Ipswich	11.17	8.15	10.17	4.05	1.64	6.40	22.19
Gold Coast	8.90	7.99	8.41	2.40	1.20	5.65	15.08
SEQ weighted av.	9.22	9.04	9.03	1.57	0.78	6.73	12.77

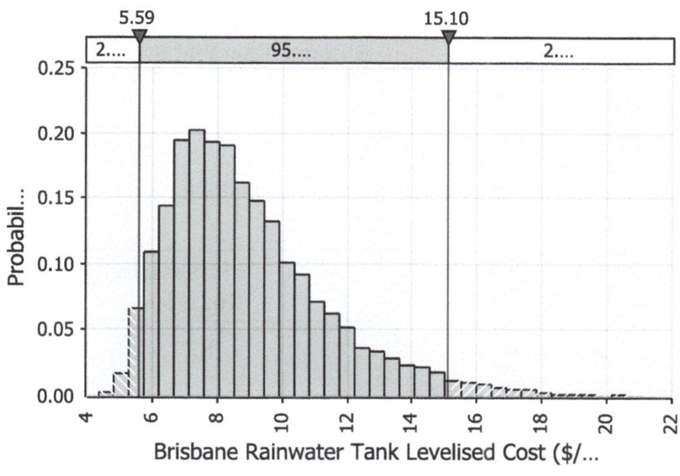

Figure 12.2 Brisbane rainwater tank cost-effectiveness probability density.

Table 12.3 further explores the sensitivity of the results by considering the uncertainty in the modelling assumptions including the discount rate, the period of analysis and the type of distribution assumed for the yield. Similar to Figure 12.3, the change in the cost-effectiveness is reported by changing each variable while the others are held constant. This enables comparison of sensitivity within Table 12.3 as well as in comparison to variables considered in Figure 12.3. For example, if maintenance is assumed to follow recommended health guidelines, it has a similar impact on the cost-effectiveness as a one standard deviation increase in the yield. Similarly, if the discount rate is changed from 3 to 6% it also has a similar effect as a one standard deviation increase in yield. Given the importance of the yield variable to the cost-effectiveness, the assumption of a triangular distribution was also reviewed by fitting a distribution to the yield data (Maheepala *et al.* 2013). The change to the yield distribution had a larger effect than a standard deviation increase in any of the variables presented in Figure 12.3 except for changing the yield itself. This illustrates that uncertainty in modelling variables is as important to the cost-effectiveness as the variation in the yield itself. Consequently, any apparent difference in cost-effectiveness requires careful scrutiny to ensure modelling assumptions are the same.

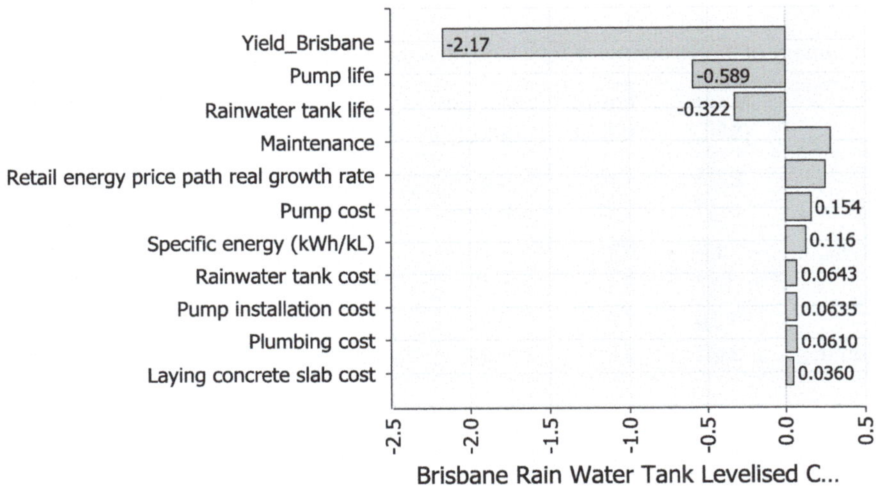

Figure 12.3 Brisbane rainwater tank cost-effectiveness Tornado chart.

Table 12.3 Sensitivity of the basic scenario to financial parameters, modelling and maintenance scenarios for Brisbane data.

	Increase in the mean levelised Cost ($/kL)
Maintenance following recommended practice	2.19
Increase discount rate from 3% to 6%	2.11
Fitted distribution*	0.92
Decrease period of analysis from 50 to 25 years	0.29

*The Johnson SU distribution was used and ranked the highest in terms of 'goodness of fit' for 65 distributions considered. The Johnson SU distribution had a Kolmogorov Smirnov statistic of 0.00606, Anderson-Darling statistic of 0.44284 and Chi Squared of 23.189.

12.2.2.4 Alternative Scenario

The 'Alternative Scenario' considers the effect of concurrent changes in a number of variables from Table 12.3 upon the cost-effectiveness. The assumptions for the Alternative Scenario are described in Table 12.1. In summary, the 'Alternative Scenario' provides a water utility perspective for the cost of capital with a

6% discount rate, a time frame for infrastructure appraisal of 25 years as well as maintenance according to recommended practice. Figure 12.4 provides a comparison of the cost-effectiveness distributions of the two scenarios. The 'Alternative Scenario' had a mean levelised cost of $13.62/kL compared to $8.93 for the 'Basic Scenario'. This is approximately a 50% increase in the levelised cost and further emphasises that modelling assumptions can produce very different cost-effectiveness results even when the yield and cost data are the same. Changes to the modelling assumptions also have an effect on the distribution of the cost-effectiveness. For example, the distribution for the 'Alternative Scenario' shifted to the right, is less concentrated about the median and has a slightly greater skew. This means that almost half of the 'Basic Scenario' has a better cost-effectiveness than the best performing 2.5% of the 'Alternative Scenario'. The dashed lines in the Figure 12.4 show the values outside of the 95% confidence interval of the 'Alternative Scenario'.

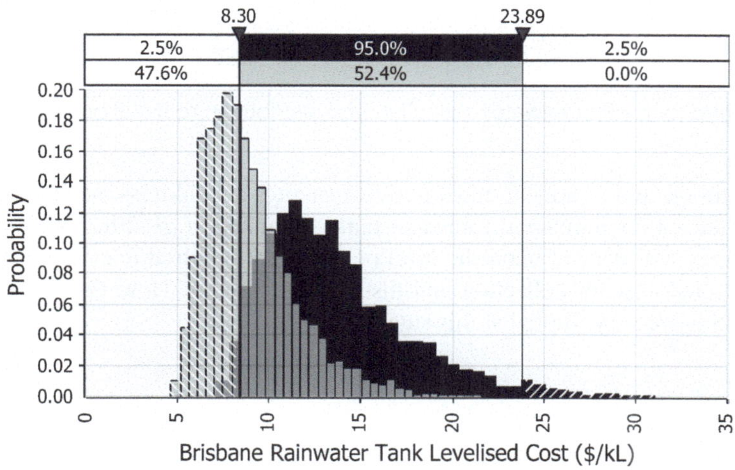

	Name	Mean ($/kL)	Mode($/ kL)	Median ($/kL)	Std Dev ($/kL)	Skewness	2.5% ($/kL)	97.5% ($/kL)
□	Basic Scenario -Brisbane Rainwater Tank Levelised Cost	8.93	7.18	8.42	2.46	1.12	5.59	15.10
■	Alternative Scenario -Brisbane Rainwater Tank Levelised Cost	13.62	11.00	12.88	3.96	1.14	8.30	23.89

Figure 12.4 Brisbane rainwater tank cost-effectiveness probability density comparison of 'basic' and 'alternative scenario'.

12.3 COST-EFFECTIVENESS OF COMMUNAL RAINWATER TANKS

12.3.1 Communal rainwater tank systems

Communal rainwater tank systems collect, store and treat rainwater across a residential development and supply treated water back to homes for either potable or non-potable use. In comparison to an individual household rainwater tank, a communal system has a single storage and treatment facility that services a cluster of houses, as shown in Figure 12.5. This type of water supply system is particularly suited for providing potable water to an off-grid community.

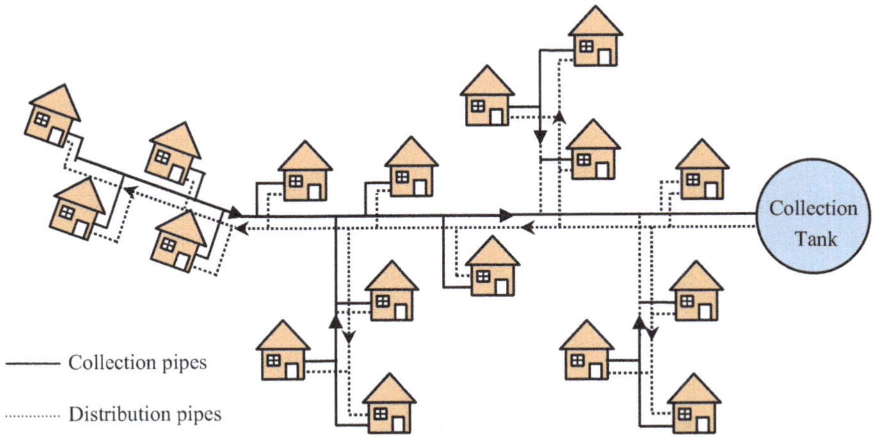

Figure 12.5 Communal rainwater tank with rainwater collection and distribution reticulation system.

A communal approach to storage and treatment may offer cost savings and address management and maintenance challenges experienced for individual rainwater tanks (Mankad *et al.* 2012; Walton *et al.* 2012). Achieving the cost savings will depend upon the level of treatment required (e.g., potable, Class A+ non-potable) and the additional cost for collection and distribution systems. However, examples of communal rainwater tank systems are very limited at this stage.

12.3.2 Method for cost-effectiveness estimation

The cost effectiveness analysis focused on the change in per household costs with changes in scale of the communal system. A conceptual design and life cycle costing was developed for a case study following the method of (Gurung & Sharma, 2014) which is described in detail in Chapter 11. Various configurations and scales were then considered to explore the effect on the system design and costing and to identify the optimum scale. The method did not explore the range of variables that may be expected to vary the cost from location to location. These factors include for example, topography of the area, end usages for rainwater supply, climatic conditions of the area (total annual rainfall and its pattern), connected roof area, tank size for the selected reliability of water supply, water pressure requirements, ongoing energy usage, and the overall density of housing development.

12.3.2.1 Case study – Results

Various housing layouts were developed for a location in South East Queensland assuming from 4 to 576 dwellings would be connected to the communal rainwater tank system, with a housing density not exceeding 20 lots per hectare and flat topography. Based on the information from current housing developments in the area, lot dimensions of 16 m by 25 m were adopted for this study. Other parameter values were based on local data such as: connected roof area of 200 m^2 (rainwater catchment), potable water demand of 82.3 L/person/day, household occupancy of 2.6 persons (OESR, 2012), historic rainfall data of the region, and sizing the communal rainwater tank based on 94% supply reliability. Cost data of the different components (e.g., pumps, pipes, tanks) required for the design of various layouts were sourced locally. An example of a layout for 24 homes is shown in Figure 12.6.

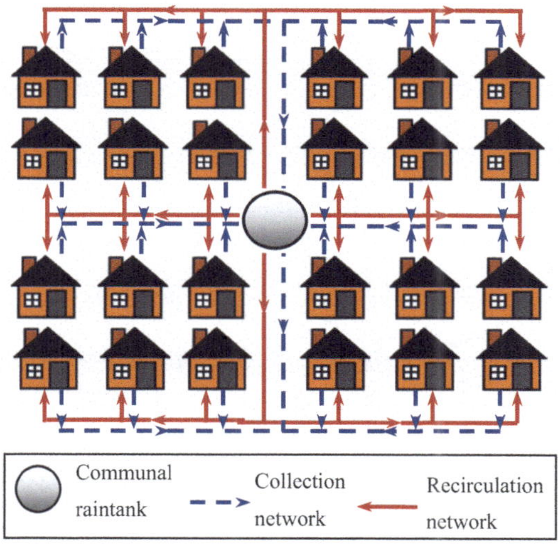

Figure 12.6 Layout of 24 homes in a communal rainwater tank system.

Life Cycle Cost (LCC) estimation using NPV method was conducted for a 50-year analysis period using a 3% discount rate. A sensitivity analysis of LCC was also conducted using a 6% discount rate. The analysis period of 50 years was selected for NPV as any longer period would make the present value of the future investment beyond 50 years insignificant (DFA, 2006) as shown in the following Equation 2 (Swamee & Sharma, 2008 and Newnan *et al.* 2002):

$$P = F(1 + i)^{-n} \tag{12.2}$$

where P is present cost, F is future cost, i is discount rate and n is the analysis period.

Table 12.4 and Figure 12.7 show the life cycle cost in NPV for the selected discount rates for each layout.

Table 12.4 Life cycle cost per household at different development sizes and discount rates.

Number of dwellings	Cost per household with discount rates	
	3%	6%
4	$46,701	$37,856
8	$28,705	$23,842
16	$20,423	$17,351
24	$17,627	$15,305
48	$15,118	$13,154
96	$12,683	$11,238
192	$11,616	$10,390
288	$11,543	$10,404
384	$11,663	$10,555
576	$11,871	$10,838

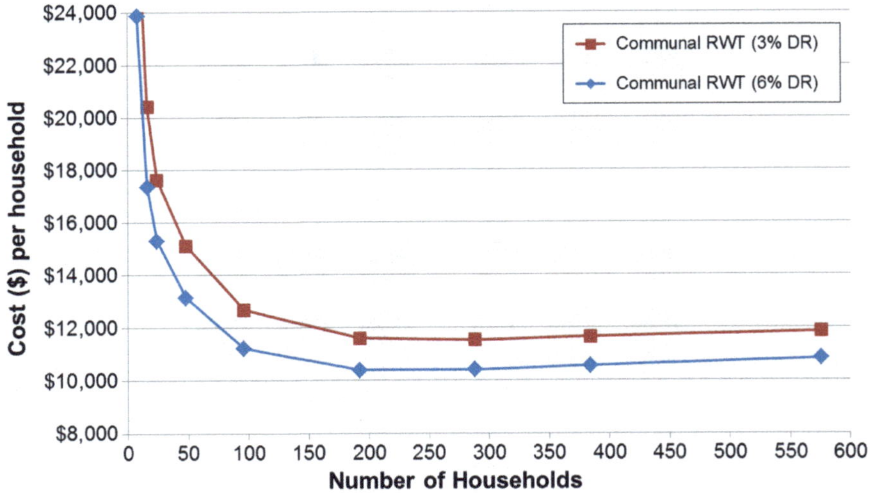

Figure 12.7 NPV costs per household in a communal rainwater harvesting system for a 3% and a 6% discount rate.

The results suggest that the optimal household scale was in the range of the 192 to 288 households, as shown in Figure 12.7. The lowest NPV was found to be $11,543 per household for a development layout of 288 households. The cost of the communal system per household starts increasing beyond this number of dwellings due to the dis-economy of scale from the increasing pipe costs not being adequately counter balanced by the economies of scale of other components within the system, as depicted in Figure 12.8. Within the optimal household layout, the cost of pipes made up 42% of overall costs, with treatment and storage units accounting for 33% and the rest made up of recurring costs.

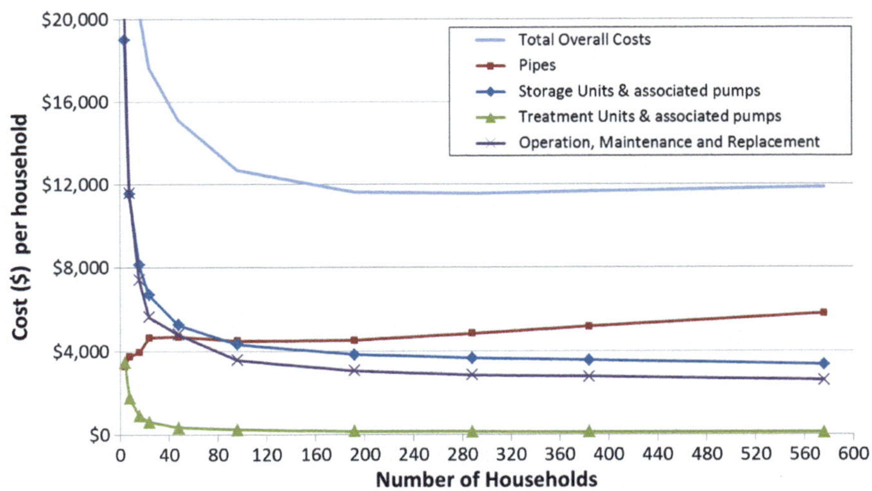

Figure 12.8 Cost per household of the various components required in a communal rainwater harvesting system.

The sensitivity analysis doubled the discount rate to 6% and explored the effect on the system cost and optimal scale and layout. The overall costs are not heavily affected by discount rates in this case as the communal system is to be built within one year, incurring major capital investment in the first year. The increased discount rate only had a small effect on the cost because of the relative importance of capital costs to operating costs. In addition, the increased discount rate caused only minor change to the optimal housing scale and layout. A similar result was also noticed by Clark (1997) for his analysis of a communal sewer model.

The levelised cost of rainwater supply from various system configurations was also estimated for both discount rates over a 50-year analysis period. Levelised costs were obtained from Equation 1 in Section 12.2.2 and have been plotted for each household layout in Figure 12.9. The levelised cost for each layout and discount rate is also listed in Table 12.5. Comparison with the NPV chart shows that there was no difference in the optimal household scale, which occurs from 192 to 288 households for both discount rates. The levelised cost of a communal system at the optimal scale was $6.11/kL for 94% supply reliability of the tank and household occupancy of 2.6 persons.

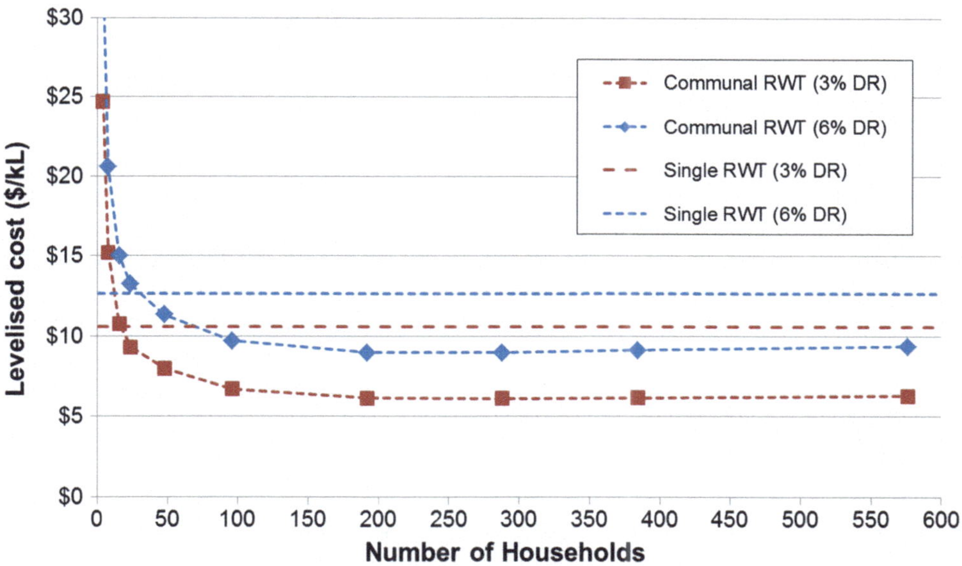

Figure 12.9 Levelised costs ($/kL) per household for a communal rainwater harvesting system and a household rainwater tank (RWT) for 3% and 6% discount rates.

As a comparison, the NPV of a single household rainwater tank (RWT) for non-potable use (at 54 L/p/day) was calculated using similar parameters (e.g., system costs, analysis period, discount rates, household occupancy) to the communal system. Note that the performance of individual rainwater tanks can have a large range due to the variability in physical performance such as yield, which was explored in the Section 12.2.2. For a 5 kL household rainwater tank, the volumetric reliability was estimated to be 90% with a NPV of $12,576 for a 3% discount rate, over the 50-year analysis period. The levelised cost was estimated as $10.60/kL and is plotted in Figure 12.9 to compare against the levelised cost of communal systems. The levelised cost was lower for a communal rainwater harvesting system with more than 24 dwellings than for a single household rainwater tank.

Table 12.5 Levelised cost ($/kL) at different development sizes.

Household numbers	Levelised cost ($/kL) with discount rates	
	3%	6%
4	$24.71	$32.69
8	$15.19	$20.59
16	$10.80	$14.98
24	$9.32	$13.22
48	$8.00	$11.36
96	$6.71	$9.71
192	$6.15	$8.97
288	$6.11	$8.98
384	$6.17	$9.11
576	$6.28	$9.36

12.4 COST BENEFIT ANALYSIS OF RAINWATER TANKS

As noted earlier in this chapter, comparing the levelised cost of water supplied by rainwater tanks with the levelised costs of alternative supply sources in a given location allows us an understanding of the relative cost-effectiveness of rainwater tanks as a water supply option.

In this section, we discuss the strengths and weaknesses of cost-effectiveness assessment, before outlining the role of cost-benefit analysis (CBA) in assisting decision-making, and presenting a CBA framework for rainwater tank assessment. We also provide a case study in the field. Benefits explored include those to the user (from increased water supply reliability), to the water supply system (from deferral of future supply augmentations), to the stormwater system (from reduced costs of stormwater management) and to the environment (from increased ecosystem services).

12.4.1 The role of cost-effectiveness assessment

Cost-effectiveness assessment is a very useful tool when project costs are well understood, but project benefits are poorly understood or difficult to quantify. In the calculation of levelised cost for water supply options, the benefit stream is typically limited to the volume of water supplied ($/kL).

However, some water supply options are recognised as having additional benefits to society beyond the volume of water produced for consumption. These could be, for example, the generation of improved environmental outcomes, diversifying the water supply risk profile, or the avoidance of other system costs. Many of these can be quantified and could therefore significantly affect the relative ranking of different water supply options.

The challenge then becomes how to incorporate these benefits into a framework for assessment that allows the full merits of each water supply option to be appropriately considered. It is possible to subtract the benefits from the cost stream within a levelised cost assessment. However, the output is not strictly a cost-effectiveness assessment and can be easily confused.

A preferred approach would be to use a framework expressly designed to incorporate all the costs and benefits of a water supply option compared with the water supply system in the absence of the option. This is a cost-benefit framework.

12.4.2 Cost-benefit analysis

CBA is an economic tool used to aid decision-making in policy contexts such as water supply planning. It is often preferred by public policy decision-makers for its rigour and transparency, and has become a standard tool of Government in recent years (Commonwealth of Australia, 2006).

CBA compares the monetary costs and benefits associated with a policy, project or option, and can be used to compare alternatives or explore the justification of an option compared to a 'base case' (Commonwealth of Australia, 2006). In the context of rainwater tanks, the relative merits (or 'net benefits') of instituting compulsory rainwater tanks on new dwellings has been assessed using CBA (MJA, 2012).

CBA builds upon cost-effectiveness assessment described above by quantifying and incorporating the range of benefits that is attributable to an option over a given time period. CBA can be applied to rainwater tanks to develop a fuller understanding of the value of rainwater tanks in the context of water supply planning.

As per cost-effectiveness assessment, the full capital and operating costs of rainwater tanks are recorded over a chosen time period, along with the yield. These are discounted to present day values using an appropriate discount rate.

Critically, a CBA assesses costs and benefits from society's point of view, rather than from the perspective of the user or the water supplier. Many of the benefit streams accrue to parties that may not bear the private costs of rainwater tanks, posing important questions about who should pay and should the investment proceed. We will discuss this further below.

By quantifying the full range of benefits and costs in dollar terms, it allows us to make an informed choice on the relative merits of rainwater tanks. By subtracting total present value costs from total present value benefits, we produce the net present value (NPV) of rainwater tanks to society. Dividing present value (PV) benefits by PV costs informs us of the ratio of benefits to costs: that is, for every dollar spent, what is the scale of benefit produced? This Benefit/Cost ratio (BCR) allows us to compare rainwater tanks against other alternatives.

Importantly, CBA is only one tool to assist decision-making. The main challenge for applying CBA to rainwater tanks is appropriately delineating and quantifying the various benefit streams that rainwater tanks produce. Not all benefits are quantifiable, and a broad range of issues need to be considered by decision-makers.

12.4.3 Proposed economic framework

An economic framework for assessing the broader range of economic costs and benefits of rainwater tanks is presented in Figure 12.10. In this framework, we conceptually commence at cost neutrality ($0), with rainwater tank costs (red bars) adding in a downwards direction, and benefit streams (in green) adding upwards.

If the combined benefit streams total more than the combined cost streams, the economic viability threshold is exceeded, and hence rainwater tanks are a justifiable economic investment for society in that context.

12.4.3.1 Cost streams

Cost streams for rainwater tanks are relatively uncontroversial, and would not differ methodologically from those outlined for cost-effectiveness assessment. Essentially, they include:

- Capital costs of the rainwater tank, including the tank itself, siting and installation costs and any associated capital (such as pumps);
- Operating costs, essentially relating to energy use of the pump; and
- Maintenance costs to keep the system in good working condition.

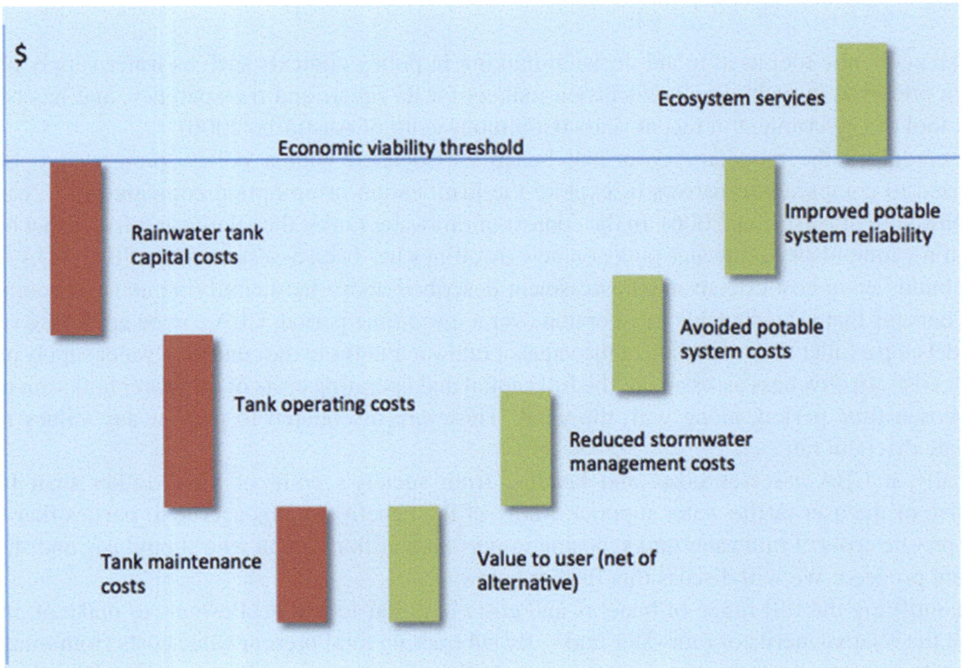

Figure 12.10 Economic framework for rainwater tank costs and benefits (adapted from Pickering, 2013).

Additional costs may be relevant, depending upon the specific circumstance. For example, potable consumption of rainwater may be found to incur health costs, which could be quantified and included.

12.4.3.2 Benefit streams

The economic benefits of rainwater tanks are significantly more complex than their costs, and accrue to different parties in different ways.

12.4.3.2.1 Value to the user

The economic value of rainwater tanks to the user refers to the additional value provided by rainwater tanks beyond which the user would have received without the tanks. In urban settings with a reliable potable water system, this economic benefit of rainwater refers predominantly to the greater water supply reliability to avoid outdoor water restrictions during times of mains water shortage.

Here we distinguish between the financial benefit to the user of avoided potable water charges, and the economic benefit of avoiding the production and delivery of potable water when substituted by rainwater tank supply. We address the latter in *avoided and deferred potable system costs* below. To include them both would double-count this benefit.

However, while values for avoiding water restrictions can be calculated, attributing these values to rainwater tanks is extremely problematic, as the reliability performance of tanks is often less than 100 per cent. That is, a 5 kL residential rainwater tank does not allow the user to overcome water restrictions 100% of the time.

12.4.3.2.2 Reduced stormwater management costs

There are two ways that rainwater tanks may reduce stormwater management costs:

(1) by reducing infrastructure required to improve stormwater quality; and
(2) by reducing costs associated with stormwater conveyance (flood management).

In many urban settings, regulations are now in place to reduce the impacts of new urban developments on stormwater quality. Impervious surfaces in new urban areas increase run-off into local waterways, increasing pollutant loads (nitrogen, phosphorus and sediment), affecting both local waterway health and that of end receiving waters. Regulatory interventions can require that minimum pollutant load reductions be achieved in new developments. These are often met with stormwater filtration areas using biofiltration or wetland creation on public land.[1]

Recent analysis (WaterByDesign, 2010 a; b) suggests that the addition of rainwater tanks in new residential developments reduces the total stormwater pollution load exiting the development, by trapping stormwater during rainfall events. This allows for the reduction in sizing of water quality management capital investments, such as bioretention and constructed wetland areas (see Figure 12.11, which shows a one third reduction in area of bioretention (0.3 m extended detention) required to meet water quality regulations with the addition of rainwater tanks on new dwellings). This saves capital, operating and replacement costs of the water quality actions and is a quantifiable economic benefit of rainwater tanks in new developments. This benefit would not be attributable to retro-fitted tanks as the WSUD devices would already be in place.

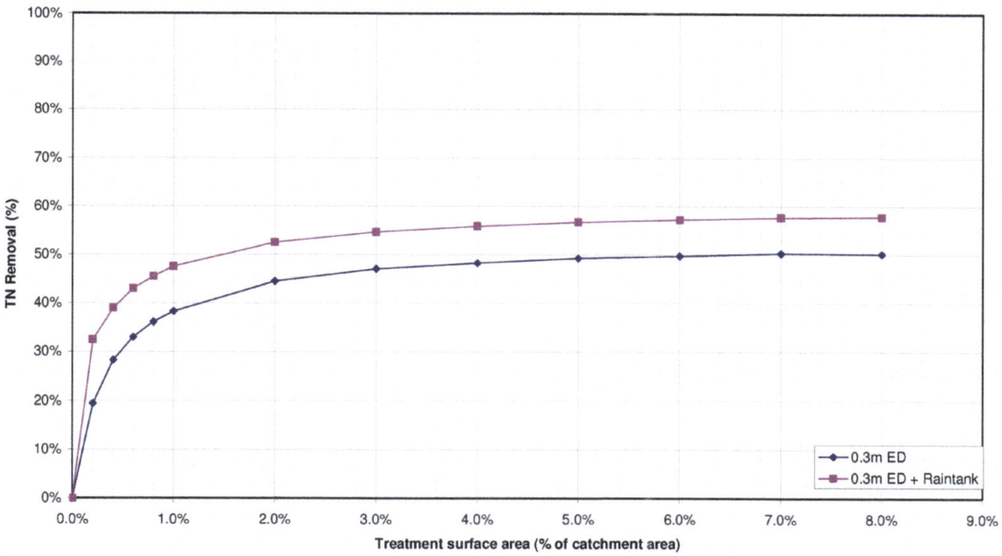

Figure 12.11 Nitrogen removal potential of bioretention with and without rainwater tanks, Weipa Australia (EDAW, 2009).

Evidence in support of savings due to a reduction in stormwater quantity is less clear, and rests on the ability of rainwater tanks to reduce the sizing of stormwater drainage infrastructure. While residential

[1]For example, the State Planning Policy for Healthy Waters in Queensland requires pollution load reductions in new residential developments.

rainwater tanks appear successful in reducing the impacts of small and medium sized rainfall events, they appear less successful in reducing peak flow events. Also, the risk of homeowners not replacing tanks or pumps when they fail may prove an unacceptable risk to stormwater managers.

No documented evidence of the resizing of drainage infrastructure due to rainwater tanks could be found for this analysis. However, where such a reduction can be identified and applied in practice, such savings are legitimate and should be included in an economic analysis.

12.4.3.2.3 Avoided and deferred potable system costs

A tangible economic benefit of rainwater tanks is the ability to substitute for water delivered through the potable system. This produces economic benefits to the potable system in three main forms:

- avoiding the variable costs of potable water supply for every unit of rainwater that substitutes for potable water;
- deferring potentially large and expensive centralised supply augmentations to the potable system (such as water sources and treatment infrastructure); and
- indefinitely avoiding localised system augmentations.

The variable costs of potable water supply include water treatment costs and the pumping costs of water delivery, for every unit of potable water replaced with rainwater. The costs of existing assets are sunk and cannot be reduced by the addition of rainwater tanks to a water supply system.

The benefit to rainwater tanks of deferring water supply augmentations can be calculated in dollar terms. In practice, this has been undertaken for the system impacts of regulatory requirements for rainwater tanks on new residential dwellings in urban areas. For example in South East Queensland, compulsory rainwater tanks in new dwellings was estimated to defer water supply augmentations by 2–3 years (MJA, 2012). By deferring large costs by several years, significant savings can be realised.

Estimating the deferral impact of a single rainwater tank can be more problematic. If the Long Run Marginal Cost (LRMC) of water supply in the relevant area is known, this can substitute for estimate of variable cost savings and deferral savings.

> *'The term LRMC is used to signify the cost effect of a change which involves some alteration in the amount or timing of future investment. Short Run Marginal Cost (SRMC), on the other hand, takes capacity as given, so relates only to changes in operating costs for example when the transport of additional water requires only additional pumping costs.'* (Turvey, 2001)

LRMC will differ by location and will reflect not just the short term delivery costs of the water supply system, but also the unit impact on the amount or timing of future investments. Ideally, the variable water charge levied by the relevant water authority will reflect LRMC, but this may not be true in all cases.

From a policy perspective, the deferral impact of mandated tanks on all new dwellings could be very significant, especially where high population growth is expected.

In addition to the above, there may be other system costs that can be avoided due to rainwater tanks. For example, it may be possible to indefinitely avoid localised infrastructure augmentations if total local potable demand remains below certain thresholds. Where sufficient penetration of rainwater tanks within these areas maintains total demand below identified thresholds, these avoided costs can be attributable to rainwater tanks.

12.4.3.2.4 Improved potable system reliability

Sufficient penetration of rainwater tanks across a water supply system may improve the reliability of the broader potable system. This improved reliability can be evidenced in the reduced incidence of outdoor water restrictions, the value of which can be estimated by calculating the community willingness to pay (WTP) for such a reduction.

Estimating the benefit to a community of avoiding water restrictions has been undertaken in a number of circumstances in Australia, where residents have experienced restrictions on outdoor use regularly over the past decade. For example:

- a study of residents in the Australian Capital Territory (ACT) using the choice modelling method was conducted by Blamey *et al.* (2001) which found that residents were willing to pay an average of only $10 per year (in 1997 dollars) to prevent a 10% reduction in water use ($0.52/kL);
- a later study conducted in the ACT by Hensher *et al.* (2006) at the peak of a ten-year drought found that water consumers were prepared to pay up to $239 annually to avoid longer and/or more severe restrictions; and
- a study conducted in South East Queensland by DBM Consulting (2007) using a choice modelling approach found that consumers were willing to pay an average of $174 per year to reduce Stage 4 water restrictions from 50% of the time to less than 1% of the time.

Critically for any analysis, the quantified impact on changes to the expected incidence of water restrictions due to the use of rainwater tanks must be estimated, before exploring the community willingness to pay for such changes.

12.4.3.2.5 Ecosystem services

Rainwater tanks are associated with improved environmental outcomes, predominantly related to improved ecosystem function in local waterways and end receiving waters. Where appropriately identified and quantified, the impact on ecosystem function is another benefit attributable to rainwater tanks.

Community willingness to pay for improvements to waterway health can be established through a range of 'stated preference' techniques, including choice modelling and contingent valuation.

Critically, there is a strong risk of double-counting benefits when estimating environmental values. For example, the water quality benefits of rainwater tanks described in the section dealing with reduced stormwater management costs (above) reflect a regulatory intervention to protect waterway health. Quantifying this benefit and estimating the waterway health improvement benefit of rainwater tanks double-counts the same benefit.

12.4.3.3 Qualitative benefits

An economic cost-benefit analysis of rainwater tanks will seek to rigorously quantify as full a range of benefits and costs as is conceivably possible. However, not all benefits are feasible to quantify in dollar terms. There exist a range of additional benefits that may not be quantified, but should be considered in qualitative terms. For example:

- Water supply risk diversification: for areas with limited supply diversification (for example, surface water supply from one geographic location), rainwater tanks in urban areas my assist in the diversification of supply risk.
- Financial risk management: large centralised water supply augmentations are expensive and can be subject to significant financial risk due to available capital and debt. In contrast, the capital and operating costs of rainwater tanks are known and relatively stable. Estimates of their cost are relatively robust.

- Water conservation message: during times of drought, governments invest considerable resources in reinforcing a water conservation message in households. A policy promoting rainwater tanks may strongly support this message, and provide households with a tangible reminder of water availability.

12.4.4 Case study – Toowoomba

A cost-benefit analysis was undertaken to assess the merits of requiring internally-plumbed rainwater tanks to be added to all new detached homes in the Toowoomba region of Queensland. Toowoomba has limited surface and groundwater storages and was significantly adversely affected by drought between 2004 and 2010.

16,000 additional rainwater tanks would be installed with new dwellings over the 40 year analysis period. In assessing a policy of compulsory rainwater tanks with new dwellings, it was accepted that replacement of tanks and pumps could not be influenced by the policy, so tank and pump replacement was excluded from the analysis. Tank and pump cost was estimated at $3,500, annual maintenance and operating costs estimated at $40 per year, and an annual tank yield of 50 kl.

The results of the analysis are summarised in Table 12.6. Quantified benefits include a significant deferral value to the water supply system of rainwater tanks, avoided potable system operating costs, and avoided stormwater management costs (reduced need for stormwater treatment through bioretention). Given high expected population growth in the region, numerous and expensive supply augmentations are planned for Toowoomba, with rainwater tanks allowing the deferral of these by several years. Some additional local augmentations can be avoided indefinitely if per-capita consumption remains below certain thresholds, which rainwater tanks are expected to critically achieve.

Table 12.6 Cost-benefit results of compulsory rainwater tanks on new houses in Toowoomba (RMCG, 2013).*

	NPV ($)	NPV ($) /tank installed
Benefits		
Deferred augmentation	38,298,434	2371
Avoided fixed operating expenditure	9,618,551	596
Avoided variable operating expenditure	3,317,062	205
Bioretention capital expenditure savings	4,797,120	297
Bioretention operating expenditure savings	540,866	33
Total benefits	56,572,032	3502
Costs		
Capital cost of tanks	56,531,715	3500
Pump replacement	0	0
Tank replacement	0	0
Operating costs	4,807,693	298
Abatement cost if tanks not replaced	354,852	22
Total costs	61,694,260	3820
Net cash flow	**–5,122,228**	**–317**
Benefit-Cost Ratio (BCR)	**0.92**	**0.92**

*40 year timeframe, 6.6% discount rate.

Benefits to the user in terms of increased supply reliability were not calculated, nor were specific ecosystem services. It was assumed that stormwater quality regulations were designed to protect ecosystem health, and as such, counting stormwater management benefits and ecosystem health benefits would represent double-counting of this benefit.

Costs were unsurprisingly dominated by the capital costs of tanks, and also included operating costs as well as other stormwater treatment actions if pumps were not replaced at their end of asset life. Pump and tank replacement costs were not included as the policy could not influence whether this was undertaken or not.

As can be seen from Table 12.6, the quantified benefits of PV \$56.6 m over 40 years fall short of the quantified costs of rainwater tanks (\$61.7 m), for a benefit-cost ratio of 0.92. As a result, the case for rainwater tanks can only be made by considering various unquantified benefits in addition to the core quantified benefits. These include risk diversification, financial risk management and supporting the public water conservation message.

12.5 DISCUSSION

The costs and benefits of a rainwater tank depends upon a number of variables, many of which are context specific. This is particularly important when applying the results from one study to another location. For example, Alam *et al.* (2012) considered the feasibility for a rainwater tank in Bangladesh which will provide 25 litres per day for drinking and cooking uses for a family. In addition, the rainwater tank provides an alternative to ground water supplies which are potentially contaminated with arsenic. In comparison, in the case of developed countries, the household water use is at least a magnitude higher, health standards may preclude potable use of rainwater and there may be a portfolio of existing alternative supplies.

Cost-benefit estimates may also differ for similar contexts because of the method and perspective of the assessment. This can be partly resolved by specifying the decision to be made and justifying the scope of costs and benefits. Specifying the variation and uncertainty in the data may also account for differing estimates.

Cost-effectiveness analysis provides a relatively simple evaluation of costs for providing water. Nonetheless, this analysis may draw upon sophisticated modelling and provide insight for policy. For example, the case study for individual rainwater tanks provided the following insights:

- Some tanks had very low yields and performed much worse than the rest of the tanks within the same rainfall region. This suggests that simple measures such as maximising the roof area connected to the tank could improve the overall performance of tanks in a region.
- Some variables may not be independent and the calculation of cost-effectiveness may require further analysis. For example, a high initial capital cost for a pump may be associated with a longer life and lower energy costs. This requires further data collection of cost and performance which may also be useful to inform consumer choice.
- The cost of maintaining a rainwater tank increased greatly if recommended health guidelines were followed. This suggests the need for further understanding of the risks to health versus the cost savings for the owner for current practice.

Communal rainwater tanks systems provide an approach to simplify and coordinate maintenance issues as well as optimise the performance of the system. Similarly, the cost-effectiveness analysis of communal rainwater tanks draws upon detailed system modelling and provides the following insights:

- communal rainwater tanks can be more cost-effective than individual rainwater tanks. However, if there are too few households connected to the system (24 houses in the case study) then the benefits of a communal system are reduced. In addition, if there are too many households connected to the system (beyond 576 houses in the case study) then the benefits from greater economies of scale also begin to reduce (Gurung & Sharma, 2014).

- the type of end-use for the water can effect system design and scaling of infrastructure. For example, a system designed for non-potable end uses would avoid costs for treatment processes such as filtration and chlorination and may justify a lower reliability (from 94 to 90%) which would result in significant cost savings for the storage tank and pipe sizes.
- the topography of the land can also effect sizing and excavation costs. For example, steeper areas require smaller pipe sizes but potentially deeper excavations for rainwater collection pipes.

Cost-effectiveness analysis allows a simple comparison to the cost of other options for providing water. As noted above, some caution is required to account for variation in performance as well as other benefits that may not be considered. Nonetheless, a comparison of the cost-effectiveness can provide insight to the variation and other benefits required to counter the difference in cost-effectiveness of two options. For example, upper range estimates for the cost of water from measures such as demand management, sea water desalination and long distance pipelines are about $1.45, $3.00 and $9.30 per kilo litre respectively (Marsden Jacobs, 2007). Although these estimates are somewhat dated, it suggests that rainwater tanks are a relatively costly method of providing water based on the upper range estimates presented in this report. This suggests that other significant benefits beyond the supply of water may be required to justify rain water tanks.

Cost-benefit analyses can be used to consider the range of other benefits from rainwater tanks. Applied cost-benefit analysis of decentralised water supply options can be a complex task and benefit valuation can be challenging. However, significant benefits to the broader water supply system of rainwater tanks can be observed, as well as to the environment through stormwater management. These benefits extend beyond those to the direct user of rainwater, to all water supply customers and broader society.

This raises important questions about who benefits and who pays for rainwater tanks. Traditionally, rainwater tanks are paid for by householders as they are considered the main beneficiary of them. Government contribution is sometimes undertaken with a 'rebate' of varying value that may or may not be linked to the broader benefit of the rainwater tank to the supply system and the environment.

Using the approach described a case could be made for Government to bear the cost of the rainwater tank, with the tank yield being charged as per other water supplies (less the value of the land lost due to situating the tank on the householder's land).

12.6 CONCLUSIONS

The economic evaluation of rainwater tanks needs to consider the decision to be made and the range of costs and benefits to be considered. This is particularly important for comparisons of alternative options.

A detailed analysis of the system design and variables can provide insight into design and policy. When coupled with cost-effectiveness analysis, a simple comparison between options for the cost of supplying water can be made. In the Australian context, rainwater tanks appear to be a relatively costly means of supplying water. However, there is large variation in the cost-effectiveness of rainwater tanks and theirperformance is dependent on a number of local factors.

Rainwater tanks can also provide a number of other benefits such as stormwater management and improved conditions for waterways. A detailed cost benefit analysis can provide insight into the benefits as well as their distribution. This can be used to compare the cost benefit ratio of water supply options as well as inform policy such as rebates to a household to account for the broader social and environmental benefits.

12.7 REFERENCES

AEMC (2011). Possible Future Retail Electricity Price Movements: 1 July 2011 to 30 June 2014, Final Report.

Alam R., Munna G., Chowdhury M. A. I., Sarkar M., Ahmed M., Rahman M. T. and Toimoor M. A. (2012). Feasibility study of rainwater harvesting system in Sylhet City. *Environmental monitoring and assessment*, **184**(1), 573–580. doi: 10.1007/s10661-011-1989-7.

Beal C. D., Hood B., Gardner T., Christiansen C. and Lane J. (2008). Energy and water metabolism of a sustainable subdivision in South East Qld: the little toe of the urban ecological footprint? Paper presented at the Proceedings of Enviro 08, 5–7 May 2008, Melbourne, Australia.

Beal C., Gardner T., Sharma A., Barton, R. and Chong M. (2011). A Desktop Analysis of Potable Water Savings from Internally Plumbed Rainwater Tanks in South East Queensland. Urban Water Security Research Alliance, Technical Report No. 26, Australia.

Beal C., Sharma A., Gardner T. and Chong M. (2012). A desktop analysis of potable water savings from internally plumbed rainwater tanks in South East Queensland. *Water Resouce Management*, **26**, 1577–1590.

Chong M. N., Mankad A., Gardner T. and Sharma A. (2011). Baseline Characteristics of Mandated Rainwater Tank Users in South East Queensland (Phase 1). Urban Water Security Research Alliance, Technical Report No. 48, Australia.

Clark R. (1997). Water Sustainability in Urban Areas: An Adelaide and Regions Case Study, Report Five: Optimum Scale for Urban Water Systems, Department of Environment and Natural Resources, Adelaide, South Australia.

Coombes P. J., Kuczera G. A. and Kalma J. D. (2003). Economic, Water Quantity And Quality Impacts From The Use Of A Rainwater Tank In The Inner City. *Australian Journal of Water Resources*, **7**, 111–120.

Domenech L. and Sauri D. (2011). A comparative appraisal of the use of rainwater harvesting in single and multi-family buildings of the Metropolitan Area of Barcelona (Spain): social experience, drinking water savings and economic costs. *Journal of Cleaner Production*, **19**(6–7), 598–608. doi: 10.1016/j.jclepro.2010.11.010.

Fane S., Robinson J. and White S. (2002). The use of levelised cost in comparing supply and demand side options. Paper presented at the 3rd World Water Congress of the International-Water-Association, Melbourne, Australia.

Farreny R., Gabarrell X. and Rieradevall J. (2011). Cost-efficiency of rainwater harvesting strategies in dense Mediterranean neighbourhood. *Resources Conservation and Recycling*, **55**(7), 686–694. doi: 10.1016/j.resconrec.2011.01.008.

Ferguson M. (2012). A 12-month rainwater tank water savings and energy use study for 52 real life installations. Paper presented at the Ozwater '12 Conference, 8–10 May 2012, Sydney, Australia.

Gardiner A. (2009). Domestic rainwater tanks: usage and maintenance patterns in south east Queensland. *Water*, **36**(1), 151–156.

Gardiner A. (2010). Do rainwater tanks herald a cultural change in household water use? *Australasian Journal of Environmental Management*, **17**(2), 100–111. Taylor and Francis Online, doi: 10.1080/14486563.2010.9725255.

Garnaut R. (2008). The Garnaut Climate Change Review – Final Report. Cambridge University Press.

Ghisi E. and Schondermark P. N. (2013). Investment Feasibility Analysis of Rainwater Use in Residences. *Water Resources Management*, **27**(7), 2555–2576. doi: 10.1007/s11269-013-0303-6.

Gurung T. R. and Sharma A. (2014). Communal rainwater tank systems design and economies of scale. *Journal of Cleaner Production*, **67**, 26–36. doi: http://dx.doi.org/10.1016/j.jclepro.2013.12.020.

Gurung T. R., Sharma A. and Umapathi S. (2012). Economics of Scale Analysis of Communal Rainwater Tanks. Urban Water Security Research Alliance, Technical Report No. 67, Australia.

IES (2010). SCENARIO MODELLING FOR ENERGY WHITE PAPER, Modelling Results. Intelligent Energy Systems report for the Australian Energy Market Operator, http://www.aemo.com.au/planning/0400-0022.pdf

IPCC (2000). Good Practice Guidance and Uncertainty Management in National Greenhouse Gas Inventories. Intergovernmental Panel on Climate Change, http://www.ipcc-nggip.iges.or.jp/public/gp/english/ (accessed 7 October 2013).

Islam M. M., Chou F. N. F. and Kabir M. R. (2011). Feasibility and acceptability study of rainwater use to the acute water shortage areas in Dhaka City, Bangladesh. *Natural Hazards*, **56**(1), 93–111. doi: 10.1007/s11069-010-9551-4.

Khastagir A. and Jayasuriya N. (2011). Investment Evaluation of Rainwater Tanks. *Water Resources Management*, **25**(14), 3769–3784. doi: 10.1007/s11269-011-9883-1.

Knights D. and Wong T. (2008). Drough proofing cities: Desalination versus rainwater tanks – a challenge to water supply managers. *Australian Journal of Water Resources*, **12**(2), 101–109.

Lane J. and Gardner T. (2009). A life-cycle comparison of decentralised and centralised water cycle approaches. Paper presented at the Ozwater'09, Melbourne, Australia.

Maheepala S., Coultas E., Neumann L. and Sharma A. (2013). Quantification of Regional Scale Water Quantity and Quality Implications of Domestic Rainwater Tanks in South East Queensland. Urban Water Security Research Alliance, Technical Report No. 104, Australia.

Mankad A. Tucker D. and Greenhill M. P. (2012). Mandated versus Retrofitted Tank Owners: Psychological Factors Predicting Maintenance and Management. Urban Water Security Research Alliance Technical Report No. 51.

Marsden Jacobs (2007). The cost-effectiveness of rainwater tanks in urban Australia: National Water Commission, Australian Government.

Moglia M., Tjandraatmadja G. and Sharma A. (2011). Initial Investigation into the Governance and Management Models for Rainwater Tank Systems. Urban Water Security Research Alliance, Technical Report No. 50, Australia.

Morales-Pinzon T., Luruena R., Rieradevall J., Gasol C. M. and Gabarrell X. (2012). Financial feasibility and environmental analysis of potential rainwater harvesting systems: A case study in Spain. *Resources Conservation and Recycling*, **69**, 130–140. doi: 10.1016/j.resconrec.2012.09.014.

Newnan D. G, Lavelle J. P. and Eschenbach T. G. (2002) Essentials of Engineering Economic Analysis, 2nd edn, Oxford University Press, USA.

OECD (2010). Projected Costs of Generating Electricity: Organisation for Economic Co-operation and Development/ International Energy Agency.

OESR (2012). Population and Dwelling Profile; South East Queensland: Office of Economic and Statistical Research, Queensland Government.

Pearce D., Atkinson G. and Mourato S. (2006). Cost-Benefit Analysis and the Environment: Recent Developments: OECD Publishing.

Pickering P. (2013). The Economic Viability of Recycled Water Projects, IBR National Water Recycling and Technology Conference, 30–31 May, Melbourne, Australia.

PWC (2013). Cost of Debt for SEQ distribution-retail water and wastewater entities – Draft: Queensland Competition Authority.

QWC (2008). South East Queensland Water Strategy – Draft March 2008. Queensland Water Commission, Queensland Government, Australia.

Retamal M. and Turner A. (2010). Unpacking the energy implications of distributed water infrastructure: how are rainwater systems performing? *Water science and technology. Water supply*, **10**(4), 546–553.

Roebuck R. M., Oltean-Dumbrava C. and Tait S. (2011). Whole life cost performance of domestic rainwater harvesting systems in the United Kingdom. *Water and Environment Journal*, **25**(3), 355–365. doi: 10.1111/ j.1747-6593.2010.00230.x.

Ruth M. (1993). Integrating Economics, Ecology and Thermodynamics: Kluwer Academic Publishers, The Netherlands.

StandardsAustralia (1999). AS/NZS 4536:1999 Life cycle costing – an application guide.

Stewart R. A. (2011). Verifying potable water savings by end use for contemporary residential water supply schemes. Waterlines Report Series No. 61, October 2011.

Swamee P. K. and Sharma A. K. (2008) Design of Water Supply Pipe Networks, John Wiley and sons, USA.

Tam V. W. Y., Tam L. and Zeng S. X. (2010). Cost-effectiveness and tradeoff on the use of rainwater tank: An empirical study in Australian residential decision-making. *Resources, Conservation and Recycling*, **54**, 178–186.

Umapathi S., Chong M. N. and Sharma A. (2012). Investigation and Monitoring of Twenty Homes to Understand Mains Water Savings from Mandated Rainwater Tanks in South East Queensland. Urban Water Security Research Alliance, Technical Report No. 63, Australia.

Walton A., Gardner J., Sharma A., Moglia M. and Tjandraatmadja G. (2012). Exploring Interventions to Encourage Rainwater Tank Maintenance: Urban Water Security Research Alliance, Technical Report No. 59, Australia.

Weitzman M. L. (2001). Gamma Discounting. *American Economic Review*, **91**(1), 260–271.

Weitzman M. L. (2007). A Review of The Stern Review on the Economics of Climate Change. *Journal of Economic Literature,* **XLV**(September 2007), 703–724.

Chapter 13

Impact of rainwater tanks on urban hydrology and stormwater quality

Matthew J. Burns, Anthony R. Ladson and Tim D. Fletcher

ABSTRACT

Urbanisation alters catchment hydrology and receiving water quality. These impacts have been identified as primary stressors to urban stream ecosystems. The protection or restoration of urban streams requires approaches to stormwater management which focus on protecting or restoring natural flow- and water quality regimes. Rainwater tanks have a major role to play in such approaches. They can reduce the frequency, magnitude and volume of urban stormwater runoff delivered to streams. They can also improve stormwater quality. Other benefits of using rainwater tanks include: urban cooling and mitigation of urban flooding. This chapter describes the changes in stormwater quality and quantity due to rainwater tank implementation in urban landscapes.

Keywords: flow-regime; harvesting; source-control; stormwater; retention waterway.

13.1 INTRODUCTION

Urbanisation alters catchment hydrology and receiving water quality. The creation of impervious surfaces (e.g., roofs, roads and pavements) and conventional stormwater drainage consistently results in a 'flashy' hydrograph and one which has increased pollutant concentrations. These changes have been identified as primary stressors to urban receiving waters and streams. Most urban streams are ecologically degraded (e.g., King *et al.* 2011; McIntosh *et al.* 2013a).

Urban stream protection or restoration requires approaches to urban stormwater management which intervene in conventional stormwater drainage and mitigates associated changes to catchment hydrology and receiving water quality. Such approaches aim to restore natural hydrology at allotment scales, ultimately to return catchment-scale flow and water quality regimes towards their pre-development condition.

A range of different stormwater source-control measures can be used to restore natural hydrology at small scales – for example, rainwater tanks, rain-gardens, infiltration trenches, and so on. Rainwater tanks, in particular, have a major role to play in mimicking natural evapotranspiration losses – a requirement to restore the ecological integrity of urban receiving waters.

This chapter provides information on the impact of urbanisation on stream hydrology and water quality; conventional approaches to urban stormwater management in Australia; and how urban stormwater can be managed for stream protection or restoration. In particular, we describe the role of rainwater tanks in restoring hydrology of urban catchments and in protecting stream health. Other benefits of using rainwater tanks are also discussed.

13.2 IMPACTS OF URBANIZATION ON STREAM HYDROLOGY AND WATER QUALITY

13.2.1 Stream hydrology

When a catchment becomes urbanised, the land surface is substantially modified and this alters catchment hydrology (Booth & Jackson, 1997). Deep-rooted vegetation such as trees – which are high water users – is cleared and replaced with lower water using vegetation (e.g., gardens and lawns). This reduces catchment evapotranspiration (Zhang *et al.* 2001). In addition to vegetation clearance, soil compaction is common, which reduces groundwater recharge (Price, 2011). Probably the most unambiguous effect of urbanisation is the covering of land with impervious surfaces (e.g., roofs, roads, and pavements). Most rain falling on impervious surfaces becomes urban stormwater runoff – little is evaporated (~0.5–1 mm/day; Boyd *et al.* 1993) and none is infiltrated. In most urban catchments, urban stormwater runoff is routed directly to streams via conventional stormwater drainage (i.e., the stormwater network).

In a recent review, Burns *et al.* (2012) summarised the impacts of urbanisation on stream hydrology which included:

- Increased frequency, magnitude and volume of storm flow;
- Increased volume of total runoff;
- Reduction in the volume of summer and winter baseflow;
- Increased frequency of low-magnitude flows; and
- Reduced storm recession time.

In general, the impacts described above change the hydrograph from stable to 'flashy' (Figure 13.1). The hydrograph of the non-urban catchment is stable, with fewer high-flow events and plentiful baseflow. In contrast, the hydrograph of the urban catchment is 'flashy' with many peaks and little baseflow. Importantly, these impacts are driven by conventional stormwater drainage which transfers water from impervious surfaces directly to streams.

13.2.2 Water quality

Urbanisation not only effects water quantity and the timing of flows, there are also effects on water quality. The degradation of the water quality of receiving waters by urban stormwater has been extensively researched (e.g., Novotny & Olem, 1994; Soranno *et al.* 1996; Hatt *et al.* 2004). In the last few decades, there have been a number of major reviews of stormwater quality which have identified pollutant types, their sources and observed range of concentrations (Torno, 1984; Terstriep *et al.* 1986; Makepeace *et al.* 1995; Duncan, 1999; Fuchs *et al.* 2004; Duncan, 2006; Sidhu *et al.* 2012).

Pollutants are generally classified into sediments, nutrients, heavy metals, organic matter (and oxygen demanding material), pathogens, as well as pesticides, herbicides and micropollutants (Table 13.1). The sources, mobilisation and treatment of pollutants will depend on their characteristics, with dissolved pollutants typically being much more difficult to remove than those attached to sediment particles (Carleton *et al.* 2000).

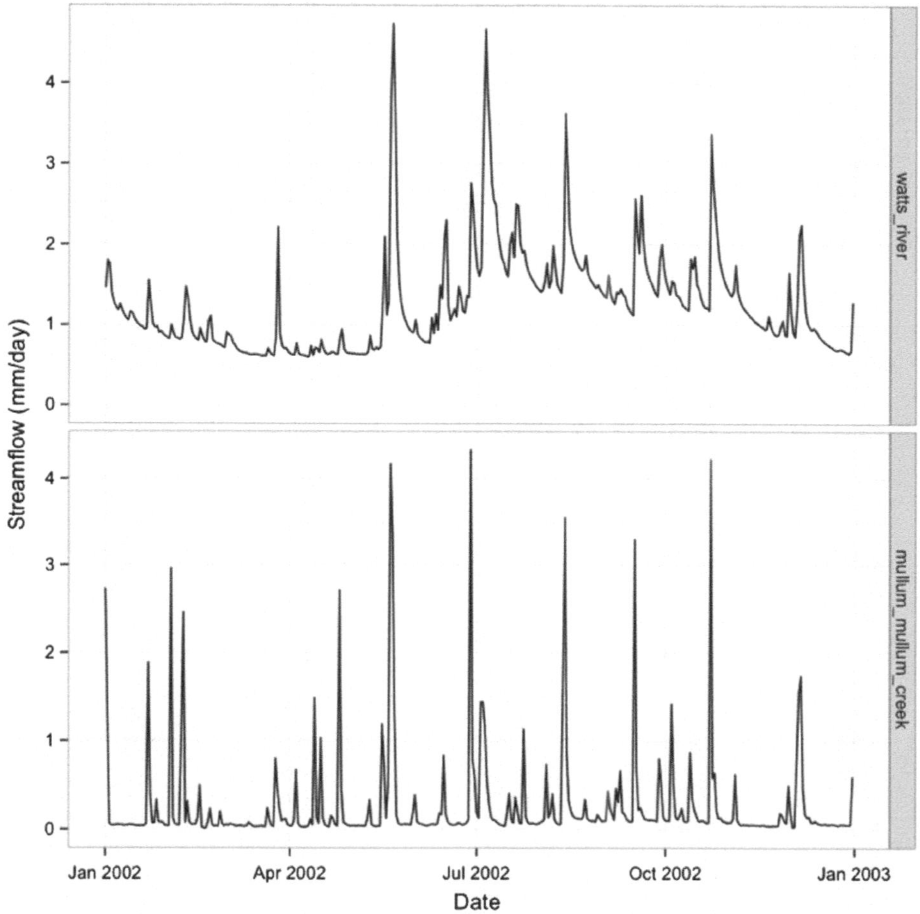

Figure 13.1 Comparison of the 2002 annual hydrograph of a non-urban (forested) catchment (top panel) with a nearby urban catchment 'flashy' (bottom panel) in Melbourne, Victoria.

Understanding the mechanisms by which urban stormwater runoff becomes polluted and which in turn, pollutes receiving waters, is critical to developing strategies to mitigate degradation. Urbanisation is a major generator of pollutants which are sourced from activities such as civil construction, house construction, transport, vegetation management and industrial processes (Ahlman *et al.* 2005). Urbaniszation also dramatically increases the efficiency with which any pollutants generated within the catchment can be mobilised and transported to receiving waters (Mason *et al.* 1999; Francey *et al.* 2011). Given this, attempts to reduce stormwater pollution are typically aimed either at reducing the mobilisation of pollutants at source (Taylor & Fletcher, 2007), treating and retaining them in appropriately designed treatment measures (Winer, 2000), or removing the runoff water altogether through rainwater and stormwater harvesting (Fletcher *et al.* 2008).

A number of reviews of roof runoff quality are available (Duncan, 1999; Duncan, 2005; Meera & Mansoor Ahammed, 2006; Lye, 2009; Abbasi & Abbasi, 2011). It is often assumed that water from roofs is relatively clean, but as can be seen from Table 13.1, the concentrations of pollutants from roofs can actually

Table 13.1 Principal sources, concentrations, impacts and treatment of pollutants in (i) general urban stormwater runoff and (ii) impervious roof runoff (data only provided where available).

Pollutant type	Principal sources	Concentrations in general urban runoff and roofs typical value (range) (mg/L or as specified)	Typical impacts	Treatment mechanisms
Suspended sediments	Erosion (construction, catchment, channel), sewer inputs, transport, industry.	General: 140 (50–500) Roofs: 20 (5–100)	Transport of other (attached) pollutants, smothering of habitat.	Sedimentation (decantation), filtration.
Turbidity		General: 60 (15–300)	Reduction in plant growth in receiving waters.	
Nutrients	Erosion (construction, catchment, channel), sewer inputs, industrial emissions, atmospheric emissions.		Eutrophication, algal blooms, change in species composition, and so on.	Filtration and sedimentation (for particulates), adsorption, bacterial and plant uptake (biofiltration), denitrification for N removal.
Total Phosphorus		General: 0.25 (0.08–0.8) Roofs: 0.13 (0.05–0.3)		
Total Nitrogen		General: 2.2 (0.7–6) Roofs: 2 (0.7–6)		
Heavy metals	Vehicle wear, industrial emissions, street furniture, roof washoff, pesticides.		Toxic effects on a range of organisms such as macroinvertebrates and plants.	Filtration and sedimentation (for particulates), adsorption.
Lead		General: 0.15 (0.04–0.6) Roofs: 0.022 (0.005–0.1)		
Zinc		General: 0.3 (0.1–1.0) Roofs: 0.16* or 4** (0.05–20)		
Copper		General: 0.08 (0.02–0.3) Roofs: 0.024 (0.007–0.08)		
Cadmium		General: 0.005 (0.001–0.02) Roofs: 0.0006 (0.0002–0.002)		
Chromium		General: 0.022 (0.006–0.11)		
Nickel		General: 0.04 (0.02–0.08)		
Iron		General: 4 (0.08–9.0)		
Manganese		General: 0.3 (0.09–0.8)		
Mercury		General: 0.25 (0.06–0.8) µg/L		

Organic pollution	Fuels, sewer inputs, leaf material, erosion.		Odours (e.g., SO2), asphyxiation of respiring organisms, release of P from sediment due to redox conditions.	Filtration, adsorption.
BOD		General: 15 Roofs: 4		
COD		General: 70 Roofs: 21		
Total organic carbon		General: 25 (15–45)		
PAHs, other hydrocarbons		General: 0.00024–0.013 (Total PAH)	Impacts on human health, aquatic organisms, and so on.	Adsorption, fine filtration, biological degradation.
Pathogens (and pathogenic indicators)	Sewer inputs, animal waste.		Impacts on health of humans and livestock. Restricts access to water for recreational purposes.	Filtration, exposure to sunlight, disinfection (UV, chemical).
Total coliforms		General: 7.0 to 1.8×10^7 CFU/100 mL		
Escherichia coli		General: 12 to 4.7×10^3 CFU/100 mL		
Faecal coliforms		General: 0.2–1.9×10^6 CFU/100 mL Roofs: 60 (6–600)		
Faecal streptococci		General: 3 to 1.4×10^6 CFU/100 mL		
Pesticides/herbicides	Landscape maintenance, agricultural land, industrial processes.		Impacts on health of humans, livestock and aquatic organisms. Possible accumulation effects through food chain.	Adsorption and fine filtration. Removal methods not yet well understood nor tested.
Glyphosate		General: <0.03–232 μg/L		
Diuron		General: 0.03–1.75 μg/L		
Dieldrin		General: <0.02–0.2 μg/L		
Isoproturon		General: <0.01–0.14 μg/L		
Metaldehyde		General: <0.02–0.58 μg/L		
AMPA		General: 0.14–9.37 μg/L		

Sources: Makepeace *et al.* (1995); Fletcher *et al.* (2005); Duncan (2006); Lawrence & Breen (2006); Zgheib *et al.* (in press).
*Non-zinc roofs
**Zinc roofs

be quite elevated, particularly for heavy metals, bacteria and nitrogen. Indeed, nitrogen concentrations from roofs are generally similar to those from stormwater runoff.

The level of atmospheric pollution can also be quite high (Gôbel *et al.* 2007), meaning that runoff even from a nominally clean roof surface can be elevated in some pollutants (Huston *et al.* 2009; Huston *et al.* 2012). For example, rainfall concentrations of nitrogen in urban areas can be around 2 mg/L (Duncan, 1999). Even particulate pollution can be elevated from roofs, due to build up and wash off of pollutants (Egodawatta *et al.* 2012; Gunawardana *et al.* 2012). Despite these observations, the concentrations of many pollutants are typically less from roofs than from other urban surfaces (roads, other paved areas), and many stormwater pollutants are not found at detectable concentrations on roofs.

The urban stormwater quality literature includes many studies that investigated variation in concentrations within a storm event (Sansalone & Buchberger, 1997; McCarthy, 2008; Davis & Birch, 2010). The concept of a 'first flush', whereby concentrations are elevated at the start of a storm due to washoff of accumulated pollutants, is regularly (although not universally) observed on roofs (Gardner *et al.* 2004; Bach *et al.* 2010; Kus *et al.* 2010; Egodawatta *et al.* 2012), leading to the use of 'first flush diverters' for rainwater harvesting systems.

While rainwater tanks (and the associated harvesting of water) are effective in reducing overall pollutant loads, due to their trapping and removal of both the water and the pollutants they contain, receiving waters may experience an increase in pollutant concentrations, as the proportion of relatively 'dirty' runoff from other impervious surfaces, will be increased (Taylor *et al.* 2005). Such an observation makes it imperative – from a stream health perspective – that rainwater and stormwater harvesting strategies be combined with effective treatment systems for runoff which is not captured and used. Given the observation that concentrations of some pollutants can be elevated even during baseflows, because of factors such as land use, leaking wastewater infrastructure and groundwater pollution (Taylor *et al.* 2005; Shepherd *et al.* 2006; Roy & Bickerton, 2012), it is important to recognise that harvesting strategies, for example, rainwater harvesting in tanks, should form just one part of an integrated approach for improving water quality in urban streams.

13.2.3 Stream ecological consequences

The impacts of urbanissation on stream hydrology and water quality have been identified as primary stressors to urban receiving waters and streams (Walsh *et al.* 2005; Wenger *et al.* 2009). In urban catchments, almost every time it rains stream biota are disturbed by poor quality urban stormwater runoff. In addition, during dry weather there is less habitat available to stream biota because of reduced baseflow.

Most attempts to restore degraded urban streams have focused on improving instream habitat or riparian zones (Walsh *et al.* 2005). Such attempts have not proved successful because they failed to address the source of stress in urban catchments – urban stormwater runoff. As discussed above, the stressors associated with urban stormwater runoff – altered flow and water quality regimes – are driven by conventional stormwater drainage. Thus to protect or restore urban streams, management should focus on intervening in conventional stormwater drainage and mitigating the associated changes to catchment hydrology and receiving water quality. Provision of rainwater tanks can be an effective mechanism to restore catchment hydrology and water quality.

13.3 CONVENTIONAL APPROACHES TO URBAN STORMWATER MANAGEMENT IN AUSTRALIA

Any discussion of the role of rainwater tanks in mitigating the effect of urbanisation on catchment hydrology needs to be considered in the context of conventional approaches to urban stormwater management. Burns *et al.* (2012) identify two dominant approaches which they describe as: 1) the drainage-efficiency approach; and 2) the load-reduction approach.

In the *drainage-efficiency approach*, the focus is on reducing flood risk. In general, runoff generated during minor storms (e.g., <5 year average recurrence interval [ARI]) is routed directly to streams via the stormwater network. For larger storms (e.g., >5 year ARI), runoff is typically conveyed to retarding basins, where peak flows are reduced to rates that equal the capacity of the downstream stormwater network (Figure 13.2). Inflows to the basin are temporarily stored and released downstream at reduced flow-rates. The amount of temporary storage controls inflow attenuation. The volume of outflow is equal to the volume of inflow. The drainage-efficiency approach alters flow and water quality regimes much to the detriment of urban streams. In most urban settings, there will always be a need for temporary storage (e.g., retarding basins) to protect the community from severe flooding (Smith *et al.* 2013). There are however, alternative ways to manage the runoff generated during minor storms (e.g., <1 year ARI) in order to protect or restore urban streams.

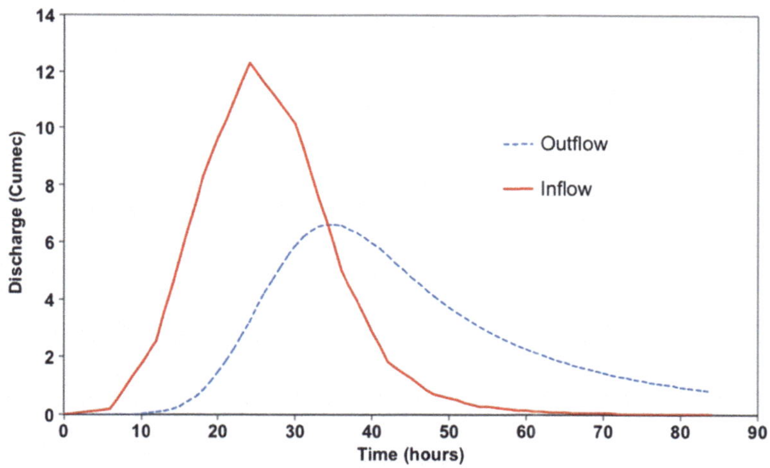

Figure 13.2 Conceptual effect of a retarding basin on catchment hydrology.

The *load-reduction approach* builds on the drainage-efficiency approach, and includes the additional objectives of reducing pollutant loads and the magnitude of geomorphically important flow events (e.g., 1.5 year ARI). This approach is commonly used in Australia, and has led to the installation of treatment strategies such as constructed wetlands in urban areas.

In the state of Victoria, the current load-reduction objectives are to reduce post-development loads of total suspended solids, nitrogen and phosphorus by 80%, 45% and 45%, respectively compared to the loads from a business as usual approach to development (Victorian Stormwater Committee, 1999). There is also an objective to ensure that the 1.5 year ARI peak runoff characteristics remain at pre-development levels, but this objective is rarely applied in practice (Wong *et al.* 2008). Loads-based targets are also common in other jurisdictions in Australia, although more recently these have been combined with flow restoration targets. For example, the state of Queensland requires new developments to retain the first 10–15 mm of impervious runoff (Queensland Government, 2010).

A focus on reducing pollutant-loads and the magnitude of large flow events fails to restore or protect streams because it does not address conventional stormwater drainage and its associated changes to hydrology and pollutant concentrations (Burns *et al.* 2012). Most running waters in the urban landscape are primarily degraded by the frequent delivery of polluted urban stormwater runoff (>100 times/year in Melbourne, Australia). Contaminant load-reduction may help protect large downstream waters (e.g., estuaries and bays), but alternative approaches designed for small streams are needed to protect or restore all running waters.

13.4 HOW URBAN STORMWATER CAN BE MANAGED FOR URBAN STREAM PROTECTION OR RESTORATION

This section describes the alternative approach to stormwater management that can protect streams. Rainwater tanks are an important aspect of this approach.

Urban stream protection or restoration requires approaches to urban stormwater management which focus on conventional stormwater drainage and its associated primary stressors (e.g., increased frequency and peak flow of runoff events accompanied by deteriorated water quality). Burns *et al.* (2012) proposed such an approach known as flow-regime management. The approach builds on the evolution of Low Impact Development and related concepts (Fletcher *et al.* 2014). Flow-regime management aims to restore natural hydrology and water quality at small scales, ultimately returning catchment-scale flow and water quality regimes towards their pre-development condition. Fundamental to flow-regime management is scale – by restoring natural hydrology from the hillslopes to the stream, all running waters are restored (even 'zero-order' streams).

To achieve flow-regime management, Burns *et al.* (2013) proposed several urban stormwater management objectives. The objectives target impervious surfaces in the landscape to achieve three things: increase volumetric losses, increase infiltration, and increase initial loss.

13.4.1 Increase volumetric losses

In the pre-development condition, most rain (>80%) falling on the landscape is evapotranspired (e.g., Benyon *et al.* 2012). In contrast, for impervious surfaces very little rainfall (<20%) is evapotranspired. This objective aims to return the amount of water lost from impervious surfaces back to the substantial volumes evapotranspired in the pre-development condition. Volumetric losses include water evapotranspired (e.g., from a rain-garden) and harvested water via rainwater tanks. Harvested water is then used for toilet flushing, clothes washing and then exported out of the catchment via the sewer network. Burns *et al.* (2014) found that achieving such levels of volumetric loss requires substantial stormwater harvesting using rainwater tanks.

13.4.2 Increase infiltration (filtered-flow)

Infiltrated rain that is not evapotranspired reaches the stream as filtered baseflow in pre-development conditions (Kirchner, 2003). For impervious surfaces, little or no rainfall is infiltrated, which is the reason baseflows are usually reduced in urban areas (Hamel *et al.* 2013). The purpose of this objective is to infiltrate some runoff from impervious areas locally, ultimately to restore baseflows. This objective can also be achieved by adding treated (filtered) impervious runoff at a low-flow rate directly to the stream (via the stormwater network). This low-flow rate (essentially a trickle) should approximate the baseflow of geographically similar, nearby undeveloped catchments, and is commonly expressed in units of L/m^2/hour.

13.4.3 Increase equivalent initial loss

For pre-developed catchments, the stream hydrograph typically begins to rise following moderate-to-large rainfall events (e.g., >15 mm). In the urban context, very small rainfall events (e.g., ~1 mm; Zaman & Ball, 1994) result in a catchment-scale response to the hydrograph. This occurs because the stormwater network connects 'hydrologically' active impervious surfaces directly to the stream. The objective is to make the 'effective ' initial loss (termed equivalent initial loss by Burns *et al.* 2013) of impervious surfaces more like natural initial loss – similar to the flow restoration target developed by the state of Queensland. Doing so means the rainfall-runoff response of urban catchments will be more like that of natural catchments as the number of small runoff events are reduced.

13.5 THE ROLE OF RAINWATER TANKS IN FLOW-REGIME MANAGEMENT

Rainwater tanks have a major role to play in restoring natural hydrologic processes at small scales for the protection or restoration of urban streams. Roofs make up approximately 40% of a developed area and contribute 50% or more of the impervious area depending on housing density (McIntosh *et al.* 2013b). Therefore, although improved treatment of roof runoff will not address all the hydrologic impacts of urbanisation, rainwater tanks provide the opportunity to address a major degrading influence. Through rainwater capture and reuse, tanks can mimic natural evapotranspiration losses, and if designed appropriately, can also deliver filtered water to the stream (either directly or through infiltration). In doing so, rainwater tanks can decrease the frequency and volume of surface runoff, thereby restoring many aspects of the natural hydrological response.

13.5.1 Increase volumetric losses

In cases where rainwater tanks supply internal end-uses (e.g., toilet flushing and washing machines), harvested water is ultimately 'lost' from the catchment via sanitary sewers. From the point of view of the catchment water balance, such losses are equivalent to evapotranspiration losses. For a parcel of undeveloped area in Melbourne, Australia, that is the same size as the impervious roof area (200 m^2) of a typical dwelling, approximately 96–116 kL/year of rainfall might be evapotranspired (Burns *et al.* 2013). If this same parcel of land were impervious, evaporation losses might only be ~24 kL/year (equivalent to 120 mm). To return the volumetric losses from this impervious area back to those levels in the pre-development condition, a typical rainwater tank would need to harvest approximately 72–92 kL/year of water (i.e., 96–116 kL/year minus 24 kL/year). Such volumes of water are substantial and represent a large proportion of total water demand in a typical domestic household in Melbourne (i.e., indoor water demand for a three person household in Melbourne, Australia can be ~131 kL/year; Roberts *et al.* 2011).

13.5.2 Increase infiltration (filtered-flow)

Rainwater tanks can be designed to help restore lost baseflows in urban catchments. A 'third' outlet pipe (i.e., not the overflow pipe or the mains water back up supply pipe) can be installed on rainwater tanks and be directed to either local soils or the stormwater network (Figure 13.3). Ideally, this pipe should have an orifice designed such that water is released at a rate which approximates the baseflow rate of nearby undeveloped catchments. Water released to local soils can be evapotranspired and infiltrated. Some infiltrated water will reach the stream as filtered subsurface flows. Alternatively, water can be released directly to the stream via the stormwater network. In this case it is suggested that water is treated using say a modular filter (Schang *et al.* 2011). An additional benefit of 'trickle release' of tank water is that it increases available tank storage which in turn decreases the frequency and magnitude of tank overflow events into the stormwater system. A simple MUSIC model of a typical allotment-scale rainwater tank (7 kL tank draining 200 m^2 of impervious roof; uniform demand of 175 L/day) and using 1 year of 6-minute rainfall data (recorded during 1959 in Melbourne, Australia), confirmed these results (Figures 13.4 and 13.5; Table 13.2). In this example, a standard design tank ('Tank_A') reduced the frequency of overflows by 64%. Configuring 'Tank_A' with a 'trickle release' pipe ('Tank_B') reduced the frequency of overflows by 88%. The diameter of the 'trickle release' pipe was sized to ensure that outflows did not exceed the baseflow rate of nearby undeveloped catchments (assumed to be 0.1 L/m^2/hour).

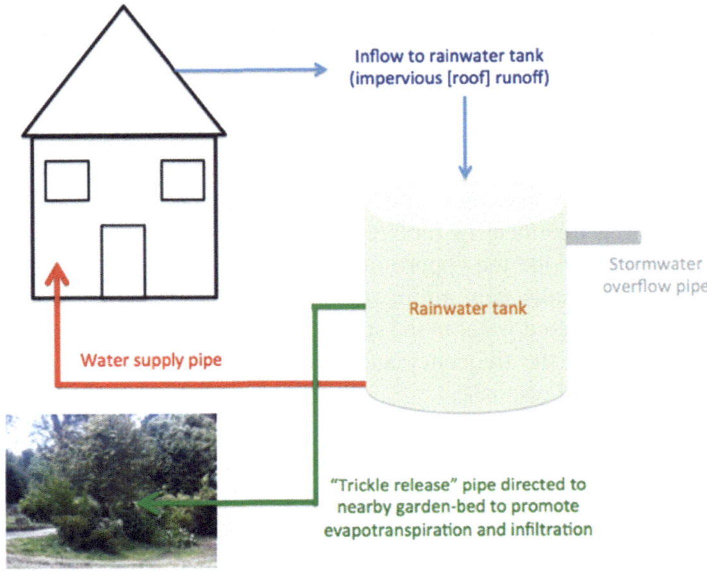

Figure 13.3 An allotment-scale rainwater tank configured with a 'trickle release' pipe.

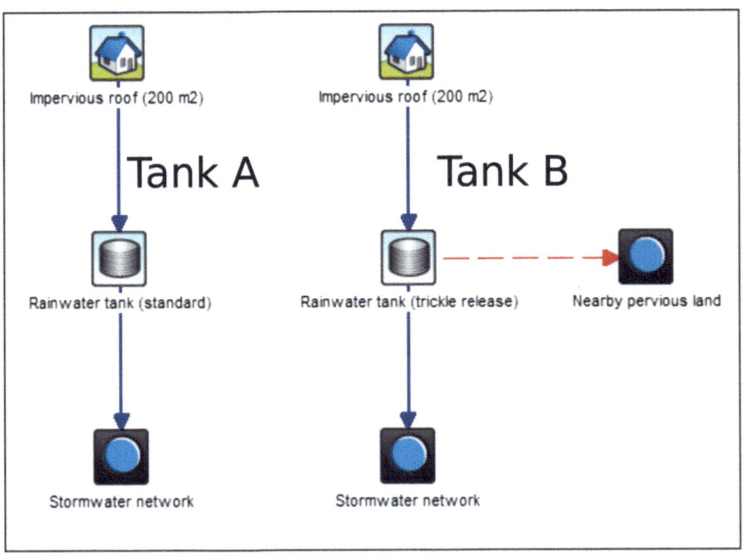

Figure 13.4 A music model for a typical allotment-scale rainwater tank. The tank configuration on the left is a 'standard design' (Tank A) whereas the one on the right is a design with a 'trickle release' pipe (Tank B).

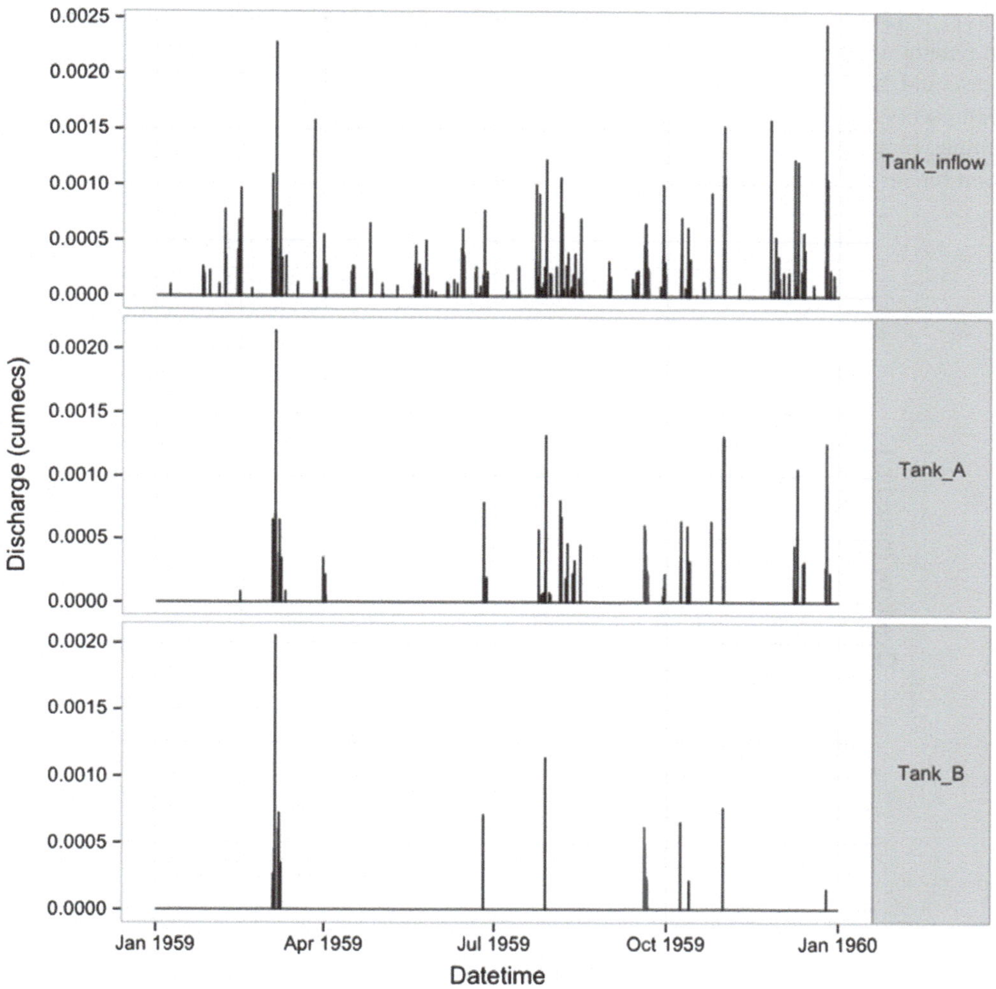

Figure 13.5 Model output from the music model shown in Figure 13.4. 'Tank_inflow' (top panel) predicted impervious (roof) runoff to the tank. The middle panel (Tank_A) shows outflow from the 'standard design' tank. The bottom panel (Tank_B) shows outflow from the tank with a 'trickle release' pipe. Tank_A did substantially reduce the frequency and magnitude of overflows; Tank_B performed even better.

Table 13.2 Stormwater overflow frequency (days/year) for each case presented in Figure 13.5.

Scenario	Frequency of stormwater delivered to streams (days/year)
Inflows to tank ('Tank_inflow')	107
'Tank_A' overflow	39
'Tank_B' overflow	13

Burns *et al.* (2012) predicted that configuring allotment-scale rainwater tanks to 'trickle' could improve median retention capacity by ~10 mm (based on the connected roof area). They also predicted that such a design could halve the frequency of untreated stormwater overflows (Figure 13.6). 'Trickled' tank water represents a virtual additional water demand. Thus, in cases where demand for rainwater is low – for example, new urban developments with reticulated recycled water – 'trickling' tanks could play an important role in the management of urban stormwater runoff.

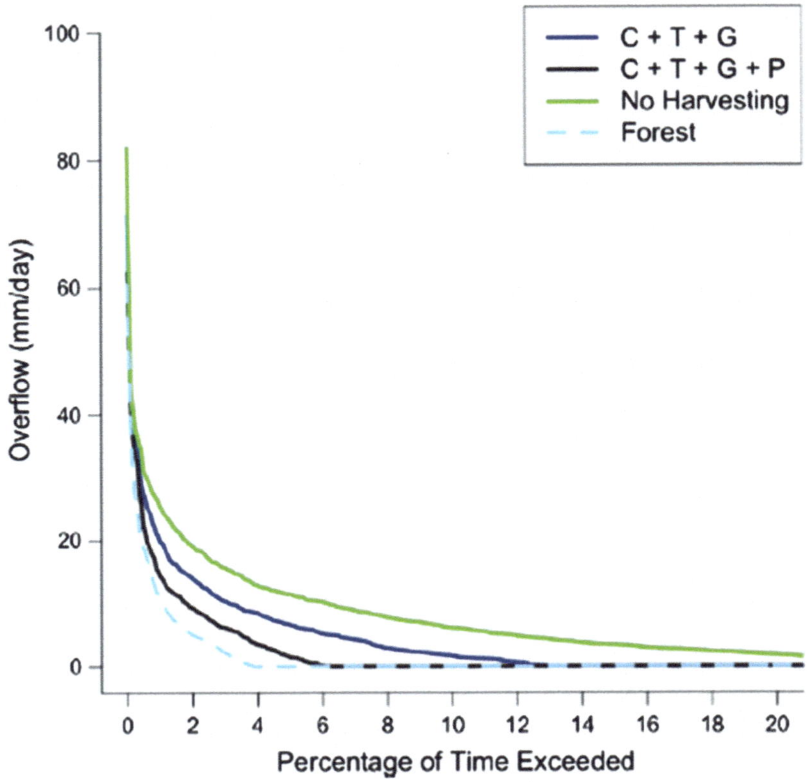

Figure 13.6 Overflow duration curves for a 10 kL rainwater tank draining a 250 m^2 impervious roof in Melbourne, Australia. The blue and black lines represent two different demand scenarios. Blue line = clothes washing (C), toilet flushing (T), and garden watering (G). Black line = same as blue, but rainwater tank is configured with a 'trickle release' pipe (P). Overflow duration curves are also shown for the conditions: 1) no rainwater tank (green line), and 2) the land-parcel was fully forested (dashed blue line). (Source: Burns *et al.* (2012).

The effectiveness of such trickle loss tanks can be improved using intelligent technologies. In Victoria, the company 'iota' (http://iota.net.au; owned and managed by the water retailer South East Water) are pioneering 'Talking Tank' systems. A 'Talking Tank' is a computer controlled rainwater tank linked via telemetry to the Bureau of Meteorology. Prior to a rain event, the computer calculates expected tank inflows based on predicted rainfall. If this inflow exceeds real-time tank storage availability, the computer actuates a solenoid valve in the tank outlet to release tank water in order to reduce stormwater overflows to the stormwater system. Ideally, tank release rates should not exceed the baseflow rate of nearby undeveloped catchments (as discussed previously).

13.5.3 Increase equivalent initial loss

If a 7 kL rainwater tank draining 200 m^2 of impervious roof (with an initial loss of 0.5 mm) is half full, then it is said to have an equivalent initial loss of 18 mm (in this case calculated as tank storage availability of 17.5 mm + impervious roof initial loss of 0.5 mm). So, >18 mm of rain would be required to result in tank overflows. Given these initial conditions, the tank is helping to make the 200 m^2 of impervious roof, generate a rainfall-runoff response more like that of a pre-development catchment. For a rainwater tank to achieve ideal equivalent initial loss performance, it must be designed appropriately – generally large tanks connected to multiple end-uses are required (e.g., Burns *et al.* 2014).

13.5.4 Water quality and tanks

Although water quality issues in rainwater tanks need to be addressed to provide a fit-for-purpose supply of water for end use (see Chapter 9), there are a range of treatment process that occur in rainwater tanks that improve the quality of water released to the environment via tank overflow.

Roof runoff can be a source of pollutants but many contaminants are removed by sedimentation in the tank, so that water from tank overflow or the trickle outlet should have lower suspended solids than inflow concentrations (Spinks *et al.* 2003; Magyar *et al.* 2011; Egodawatta *et al.* 2012). Micro-organisms are deactivated by storage, and biofilms at the water surface may remove nutrients and contaminants. Coombes *et al.* (2002) showed that concentrations of feacal coliforms, total coliforms, Ammonia, Heterotropic Plate Counts, nitrate and lead were lower in tank water than in roof runoff. In addition, the MUSIC model predicts that a typical rainwater tank can substantially reduce pollutant loads through settling, in-tank processes and diversion of water to sewer via internal end uses (Table 13.3). Further reductions were predicted when the tank was configured with a 'trickle release'.

Table 13.3 Pollutant-loads predicted by the MUSIC model (presented in Figure 13.4) for overflow from a typical rainwater tank ('Tank_A') and a tank configured with a 'trickle release' ('Tank_B').

Scenario	Total suspended solids (kg/yr)	Total phosphorus (kg/yr)	Total nitrogen (kg/yr)
Inflows to tank ('Tank_inflow')	24.4	0.05	0.34
'Tank_A' overflow	3.03	0.011	0.122
'Tank_B' overflow	1.33	0.004	0.045

13.6 OTHER BENEFITS OF RAINWATER TANKS

Much modelling research has shown that rainwater tanks can retain large rainfall events. For example, Burns *et al.* (2012) showed that the median retention capacity (analogous to catchment initial loss) of typical small-scale rainwater tank systems in Melbourne can be ~20 mm.

Coombes and Kuczera (2003) found similar results when investigating the performance of rainwater tanks across Australia, for example, a 5 kL rainwater tank in Melbourne, draining 200 m^2 of roof and supplying water to five people, could retain on average ~3.5 kL (equivalent to 18 mm of roof runoff) before overflow to the stormwater system occurred. The fact that some urban flooding is caused by high-intensity, but low total volume (<20 mm) rainfall events (http://www.bom.gov.au/water/designRainfalls/ifd), highlights the substantial potential of rainwater tanks to reduce flood risk in small urbanised catchments. This potential however, remains relatively untested by experimental monitoring and model calibration.

Burns *et al.* (2010) used the RORB model (Laurenson *et al.* 2006) to investigate the catchment-scale implications of allotment-scale rainwater tanks on flood hydrology. They found that rainwater tanks only reduced the magnitude of post-development floods by 10–20% for a hypothetical urban catchment. A significant shortcoming of this and related studies (e.g., Coombes *et al.* 2003), is a lack of hydraulic modelling, thus providing little insight on flood extents.

To explore robustly the flood benefits of rainwater tanks, hydrologic models need to be coupled with hydraulic models. In doing so, consideration must be given to the stormwater network (e.g., Qin *et al.* 2013). Such an undertaking is a topic of future research.

13.7 CONCLUSION

Rainwater tanks have the potential to mitigate the effects of urbanisation on the hydrology and water quality of urban streams. Rain tanks capturing and using rainwater for household uses can compensate for the volumetric increase in runoff that inevitably occurs with urban development. Where there is storage available at the start of a storm, rainwater tanks can increase initial loss from about 0.5 mm, typical of an impervious surface, toward values of about 20 mm which are common in undeveloped catchments. This has the potential to reduce flood risk particularly in small urban catchments as well as reducing the frequency of small runoff events and their peak discharge. Connecting tanks to a range of regular demand end-uses (such as toilets and washing machines) and/or releasing water at a 'trickle' rate into the stormwater network promotes storage availability pre-storm, and helps protect stream ecosystem health.

Protection and restoration of streams impacted by urbanisation requires addressing the source of the hydraulic stress. Rainwater tanks are an important part of the flow regime management required to reduce the number of runoff events, hydrograph flashiness whilst restoring baseflow. There is also potential for rain tanks to substantially improve runoff quality from roofs. Further research is required to demonstrate that tanks can make substantial reductions to the size of stormwater conveyance infrastructure which is designed for large storms. Part of the unknown is the lack of detailed surface hydraulic modelling coupled to tank water balances.

13.8 REFERENCES

Abbasi T. and Abbasi S. A. (2011). Sources of pollution in rooftop rainwater harvesting systems and their control. *Critical Reviews in Environmental Science and Technology*, **41**(23), 2097–2167.

Ahlman S., Malm A., Kant H., Svensson G. and Karlsson P. (2005). Modelling non-structural Best Management Practices focus on reductions in stormwater pollution. *Water Science & Technology*, **52**(5), 9–16.

Bach P. M., McCarthy D. T. and Deletic A. (2010). Redefining the stormwater first flush phenomenon. *Water Research*, **44**(8), 2487–2498.

Benyon R. G., Lane P. N. J., Theiveyanathan S., Doody T. M. and Mitchell P. J. (2012). Spatial variability in forest water use from three contrasting regions of south-eastern Australia. Acta Hort. (ISHS) 951:233–240. http://www.actahort.org/books/951/951_28.htm (accessed 5 June 2014).

Booth D. B. and Jackson C. R. (1997). Urbanization of aquatic systems: degradation thresholds, stormwater detention, and the limits of mitigation. *Journal of the American Water Resources Association*, **33**(5), 1077–1090.

Boyd M. J., Bufill M. C. and Knee R. M. (1993). Pervious and impervious runoff in urban catchments. *Hydrological Sciences*, **38**(6), 463–478.

Burns M. J., Fletcher T. D., Hatt B. E., Ladson A. R. and Walsh C. J. (2010). Can allotment-scale rainwater harvesting manage urban flood risk and protect stream health? Novatech 2010, 7th International Conference on Sustainable Techniques and Strategies in Urban Water Management. Lyon, France.

Burns M. J., Fletcher T. D., Duncan H. P., Ladson A. R. and Walsh C. J. (2012a). The stormwater retention performance of rainwater tanks at the land-parcel scale. In: 7th International Conference on Water Sensitive Urban Design. T. H. F. Wong and D. T. McCarthy (eds). Melbourne, Australia.

Burns M. J., Fletcher T. D., Walsh C. J., Ladson A. R. and Hatt B. E. (2012b). Hydrologic shortcomings of conventional urban stormwater management and opportunities for reform. *Landscape and Urban Planning*, **105**(3), 230–240.

Burns M. J., Fletcher T. D., Walsh C. J., Ladson A. R. and Hatt B. E. (2013). Setting objectives for hydrologic restoration: from site-scale to catchment-scale. Novatech 2013, 8th International Conference on Sustainable Techniques and Strategies in Urban Water Management, Lyon, France.

Burns M. J., Fletcher T. D., Duncan H. P., Hatt B. E., Ladson A. R. and Walsh C. J. (2014a). The performance of rainwater tanks for stormwater retention and water supply at the household scale: an empirical study. *Hydrological Processes*. doi: 10.1002/hyp.10142.

Burns M. J., Fletcher T. D., Ladson A. R., Walsh C. J. and Hatt B. E. (2014b). Flow-regime management at the urban land-parcel scale: a test of feasibility. *Journal of Hydrologic Engineering*. doi: 10.1061/(ASCE)HE. 1943-5584.0001002.

Carleton J. N., Grizzard T. J., Godrej A. N., Post H. E., Lampe L. and Kenel P. P. (2000). Performance of a Constructed Wetlands in Treating Urban Stormwater Runoff. *Water Environment Research*, **72**(3), 295–304.

Coombes P. J. and Kuczera G. (2003). Analysis of the performance of rainwater tanks in Australian Capital Cities. Hydrology and Water Resources Symposium, Wollongong, NSW, Engineers Australia.

Coombes P. J., Kuczera G., and Kalma J. D. (2002). Economic, water quantity and quality results from a house with a rainwater tank in the inner city. Hydrology and Water Resources Symposium. Melbourne, Australia.

Coombes P. J., Kuczera G., Frost A., O'Loughlin G. and Lees S. (2003). The Impact of Rainwater Tanks in the Upper Parramatta River Catchment. *Australian Journal of Water Resources*, **7**(2), 121–129.

Davis B. and Birch G. (2010). Comparison of heavy metal loads in stormwater runoff from major and minor urban roads using pollutant yield rating curves. *Environmental Pollution*, **158**(8), 2541–2545.

Duncan H. P. (1999). Urban Stormwater Quality: A Statistical Overview. Melbourne, Australia, Cooperative Research Centre for Catchment Hydrology: 80.

Duncan H. P. (2005). Urban stormwater pollutant characteristics. In: Australian Runoff Quality Guidelines T. H. F. Wong (ed.), Sydney, Institution of Engineers Australia.

Duncan H. P. (2006). Chapter 3 – Urban stormwater quality. In: Australian Runoff Quality T. H. F. Wong (ed.), Sydney, Australia, Institution of Engineers, Australia.

Egodawatta P., Miguntanna N. S. and Goonetilleke A. (2012). Impact of roof surface runoff on urban water quality. *Water Science & Technology*, **66**(7), 1527–1533.

Fletcher T. D., Duncan H. P., Poelsma P. and Lloyd S. D. (2005). Stormwater flow and quality, and the effectiveness of non-proprietary stormwater treatment measures – a review and gap analysis. Melbourne, Cooperative Research Centre for Catchment Hydrology (CRCCH Report 04/08), 182.

Fletcher T. D., Deletic A., Mitchell V. and Hatt B. E. (2008). Reuse of urban runoff–a review of recent Australian advances and remaining challenges. *Journal of Environmental Quality*, **37**(5), 116–127.

Fletcher T. D., Shuster W., Hunt W. F., Ashley R., Butler D., Arthur S., Trowsdale S., Barraud S., Semadeni-Davies A., Bertrand-Krajewski J.-L., Mikkelsen P. S., Rivard G., Uhl M., Dagenais D. and Viklander M. (2014). SUDS, LID, BMPs, WSUD and more–The evolution and application of terminology surrounding urban drainage. *Urban Water Journal*, 1–18.

Francey M., Fletcher T. D., Deletic A. and Duncan H. P. (2011). The testing and sensitivity of a simple method for predicting urban pollutant loads. *Journal of Environmental Engineering*, **137**(9), 782–789.

Fuchs S., Brombach H. and Weilb G. (2004). New database on urban runoff pollution. Novatech 2004, Lyon, France, Groupe de Recherche Rhone-Alpes sur les Infrastructures et l'Eau.

Gardner T., Baisden J. and Millar G. (2004). Rainwater first flush devices–are they effective? Sustainable Water in the Urban Environment. Brisbane, Australia.

Gôbel P. C., Dierkes C. and Coldewey W. (2007). Storm water runoff concentration matrix for urban areas. *Journal of Contaminant Hydrology*, **91**(1), 26–42.

Gunawardana C., Goonetilleke A., Egodawatta P., Dawes L. and Kokot S. (2012). Source characterisation of road dust based on chemical and mineralogical composition. *Chemosphere*, **87**(2), 163–170.

Hamel P., Daly E. and Fletcher T. D. (2013). Source-control stormwater management for mitigating the impacts of urbanisation on baseflow: a review. *Journal of Hydrology*, **485**(2), 201–211.

Hatt B. E., Fletcher T. D., Walsh C. J. and Taylor S. L. (2004). The influence of urban density and drainage infrastructure on the concentrations and loads of pollutants in small streams. *Environmental Management*, **34**(1), 112–124.

Huston R., Chan Y. C., Gardner T., Shaw G. and Chapman H. (2009). Characterisation of atmospheric deposition as a source of contaminants in urban rainwater tanks. *Water Research*, **43**(6), 1630–1640.

Huston R., Chan Y. C., Chapman H., Gardner T. and Shaw G. (2012). Source apportionment of heavy metals and ionic contaminants in rainwater tanks in a subtropical urban area in Australia. *Water Research*, **46**(4), 1121–1132.

King R., Baker M., Kazyak P. and Weller D. (2011). How novel is too novel? Stream community thresholds at exceptionally low levels of catchment urbanization. *Ecological Applications*, **21**, 1659–1678. doi: 10.1890/10-1357.1.

Kirchner J. W. (2003). A double paradox in catchment hydrology and geochemistry. *Hydrological Processes*, **17**(4), 871–874.

Kus B., Kandasamy J., Vigneswaran S. and Shon H. (2010). Analysis of first flush to improve the water quality in rainwater tanks. *Water Science & Technology*, **61**(2), 421–428. doi: 10.2166/wst.2010.823.

Laurenson E. M., Mein R. G. and Nathan R. J. (2006). RORB Version 6 – runoff routing program – user manual, Department of Civil Engineering, Monash University.

Lawrence I. and Breen P. (2006). Chapter 2 – Stormwater contaminants and pathways. *Australian Runoff Quality*. T. H. F. Wong. Sydney, Australia, Institution of Engineers, Australia.

Lye D. J. (2009). Rooftop runoff as a source of contamination: A review. *Science of The Total Environment*, **407**(21), 5429–5434.

Magyar M. I., Ladson A. R., Mitchell V. G. and Diaper C. (2011). The effect of rainwater tank design on sediment re-suspension and subsequent outlet water quality. *Australian Journal of Water Resources*, **15**(1), 71–84.

Makepeace D. K., Smith D. W. and Stanley S. J. (1995). Urban Stormwater Quality: Summary of Contaminant Data. *Critical Reviews in Environmental Science and Technology*, **25**(2), 93–139.

Mason Y., Ammann A. A., Ulrich A. and Sigg L. (1999). Behaviour of Heavy Metals, Nutrients, and Major Components during Roof Runoff Infiltration. *Environmental Science and Technology*, **33**(10), 1588–1597.

McCarthy D. (2008). A traditional first flush assessment of *E. coli* in urban stormwater runoff. 11th International Conference on Urban Drainage (ICUD). Edinburgh, Scotland, International Water Association.

McIntosh B. S., Aryal S., Ashbolt S., Sheldon F., Maheepala S., Gardner T., Chowdhury R., Gardiner R., Hartcher M., Pagendam D., Hodgson G., Hodgen M. and Pelzer L. (2013a). Urbanisation and stormwater management in South East Queensland–synthesis and recommendations, Urban Water Security Research Alliance, Technical Report No. 106, Brisbane, QLD, Australia.

McIntosh B. S., Hodgen M., Aryal S., Laredo L., Wolf L., Gardner T., Chowdhury R. and Maheepala S. (2013b). Ripley Valley – an application of GIS based runoff modelling to strategic stormwater harvesting assessment, Urban Water Security Research Alliance, Technical Report No. 109, Brisbane, Queensland, Australia.

Meera, V. and Mansoor Ahammed M. (2006). Water quality of rooftop rainwater harvesting systems: a review. *Journal of Water Supply: Research & Technology*, **55**(4), 257–268.

Novotny V. and Olem H. (1994). Water quality: prevention, identification and management of diffuse pollution. New York, Van Nostrand Reinhold.

Price K. (2011). Effects of watershed topography, soils, land use, and climate on baseflow hydrology in humid regions: A review. *Progress in Physical Geography*, **35**(4), 465–492. doi: 10.1177/0309133311402714.

Qin H.-p., Li Z.-x. and Fu G. (2013). The effects of low impact development on urban flooding under different rainfall characteristics. *Journal of Environmental Management*, **129**(0), 577–585.

Queensland Government (2010). Urban stormwater quality planning guidelines. Brisbane, Australia, Healthy Waters, Department of Environment and Heritage Protection.

Roberts P., Athuraliya A. and Brown A. (2011). Yarra Valley future water. Residential water use study volume 1 – winter 2010. July 2011. Available online: http://www.yvw.com.au/yvw/groups/public/documents/document/yvw1003346.pdf (accessed 15 July 2014). Yarra Valley Water.

Roy J. W. and Bickerton G. (2012). Toxic Groundwater Contaminants: An Overlooked Contributor to Urban Stream Syndrome? *Environmental Science & Technology*, **46**(2), 729–736.

Sansalone J. J. and Buchberger S. G. (1997). Partitioning and First Flush of Metals in Urban Roadway Storm Water. *Journal of Environmental Engineering*, **123**(2), 134–143.

Schang C., McCarthy D., Deletic A. and Fletcher T. (2011). Expected performances and lifespan of the enviss™ stormwater treatment technologies: results of a breakthrough analysis. Proc. 12th Int. Conf. on Urban Drainage, Porto Alegre, Brazil, 11–16.

Sidhu J., Gernjak W. and Toze S. (2012). Health Risk Assessment of Urban Stormwater, Urban Water Security Research Alliance, Technical Report No. 102, Brisbane, Queensland, Australia.

Shepherd K. A., Ellis P. A. and Rivett M. O. (2006). Integrated understanding of urban land, groundwater, baseflow and surface-water quality in The City of Birmingham, UK. *Science of The Total Environment*, **360**(1), 180–195.

Smith B. K., Smith J. A., Baeck M. L., Villarini G. and Wright D. B. (2013). Spectrum of storm event hydrologic response in urban watersheds. *Water Resources Research*, **49**(5), 2649–2663.

Soranno P. A., Hubler S. L., Carpenter S. R. and Lathrop R. C. (1996). Phosphorus loads to surface waters: A simple model to account for spatial pattern of land use. *Ecological Applications*, **6**(3), 865–878.

Spinks A. T., Coombes P. J., Dunstan R. H. and Kuczera G. (2003). Water Quality Treatment Processes in Domestic Rainwater Harvesting Systems. *28th Hydrology and Water Resources Symposium*. Wollongong, Australia, The Institution of Engineers. **2**, 227–234.

Taylor A. C. and Fletcher T. D. (2007). Non-structural measures to improve urban stormwater quality; building a knowledge base to improve their use. *Environmental Management*, **39**(5), 663–677.

Taylor G. D., Fletcher T. D., Wong T. H. F. and Breen P. F. (2005). Nitrogen composition in urban runoff – implications for stormwater management. *Water Research*, **39**(10), 1982–1989.

Terstriep M. L., Noel D. C. and Bender G. M. (1986). Sources of Urban Pollutants – Do We Know Enough? Urban Runoff Quality – Impact and Quality Enhancement Technology, Proceedings of an Engineering Foundation Conference, New Hampshire, ASCE, New York.

Torno H. C. (1984). The Nationwide Urban Runoff Program. Proceedings of the Third International Conference on Urban Storm Drainage, Goteborg, Sweden.

Victorian Stormwater Committee (1999). Urban stormwater: best practice environmental management guidelines. Melbourne, CSIRO.

Walsh C. J., Fletcher T. D. and Ladson A. R. (2005a). Stream restoration in urban catchments through re-designing stormwater systems: looking to the catchment to save the stream. *Journal of the North American Benthological Society*, **24**(3), 690–705.

Walsh C. J., Roy A. H., Feminella J. W., Cottingham P. D., Groffman P. M. and Morgan R. P. (2005b). The urban stream syndrome: current knowledge and the search for a cure. *Journal of the North American Benthological Society*, **24**(3), 706–723.

Wenger S. J., Roy A. H., Jackson C. R., Bernhardt E. S., Carter T. L., S. Filoso, Gibson, C. A., Hession W. C., Kaushal S. S., Marti E., Meyer J. L., Palmer M. A., Paul M. J., Purcell A. H., Ramirez A., Rosemond A. D., Schofield K. A., Sudduth E. B. and Walsh C. J. (2009). Twenty-six key research questions in urban stream ecology: an assessment of the state of the science. *Journal of the North American Benthological Society*, **28**(4), 1080–1098.

Winer R. (2000). National pollutant removal performance database for stormwater treatment practices, Center for Watershed Protection Ellicott City, MD.

Wong T., Knights D. and Lloyd S. (2008). Hydrologic, water quality and geomorphic indicators of catchment Effective Imperviousness. *Australian Journal of Water Resources*, **12**(2), 111.

Zaman S. and Ball J. E. (1994). Simulation of small events in an urban Catchment. International Association of Hydrogeologists. Congress (25th: 1994: Adelaide, South Australia). Barton, ACT, Institution of Engineers, Australia.

Zgheib S., Moilleron R. G. and Chebbo G. (2012). Priority pollutants in urban stormwater: Part 1 – case of separate storm sewers. *Water Research*, **46**(20), 6683–6692. doi: 10.1016/j.watres.2011.12.012.

Zhang L., Dawes W. R. and Walker G. R. (2001). Response of mean annual evapotranspiration to vegetation changes at catchment scale. *Water Resources Research*, **37**(3), 701–708.

Chapter 14

Rainwater tanks in Australia: Their social/political context, a research overview, policy implications, future research needs, and application of findings to other countries

Ted Gardner, Don Begbie, Ashok K. Sharma and Grace Tjandraatmadja

ABSTRACT

This Chapter provides an overview of the findings from the preceding 13 chapters and discusses their policy implications for Australian States, and lessons for the international community. But first, it commences with a social/technical history of the development of rainwater tank systems in Australia which has the highest adoption of any country in the world (34% of households). The major end uses are potable substitution in urban areas, motivated by the Millennium drought which occurred during most of the 2000s, and potable supply in peri-urban and rural areas. In other developed countries, the major motivation for tanks is stormwater control, an application which is relatively under researched in Australia.

The major lessons from our systematic scientific study of an essentially trial and error expansion process over 10 years are: our ability to confidently predict rain catch and its reliability over the long-term (say 100 years) provided suitable climate data is available; potable water savings from tanks can be confidently assessed using clever statistical analysis of routine water billing records; that microbiological water quality is probably not sufficient to recommend its use for drinking if traditional potable supplies are available; that lead (Pb) is likely to be a common contaminant in urban Australia that warrants treatment if rainwater is to be used for drinking. Both these microbiological and chemical water quality issues can be overcome by communal tank systems with centralised treatment and management. However, the economics are debatable compared with mains supplied water, but are more attractive than that of individual tanks.

The biggest issues are ensuring that tank systems are installed to specification (requiring a post install audit process), and that they are adequately maintained over the longer term by often disengaged homeowners. The public health implications of poor maintenance (from mosquito borne arboviruses) are likely to be substantial in tropical and semi-tropical areas, but the balance of *who pays/who gains* has yet to be fully resolved by government. The future of tanks in urban society is uncertain as the major water authorities of Australia have invested large sums in climate resilient potable water supplies (e.g., seawater desalination plants) and are often reluctant to invest in alternative water supplies such as tanks or dual reticulation of recycled water. However, recent social research has indicated that Australian water professionals are very concerned with ensuring adequate water supplies given expected climate change, and that tanks and recycled water are likely to play a major role here. There are no such doubts about the importance of tanks in supplying water of superior quality to people in developing countries, and results of this body of research has a number of direct applications, albeit with modifications to account for the different social/technological/political contexts.

Keywords: Rainwater tanks; Australia; alternative water resources; potable water substitution; social context; policy implications; implementation drivers; international leanings.

14.1 INTRODUCTION

It's not surprising that this book originated in Australia as the country has a long history of rainwater tanks for domestic supply due to highly ephemeral streams, patchy groundwater (both in quantity and quality) and seasonal rainfall patterns – all combine to require a buffering storage that leads to rainwater tanks. Tanks were initially constructed *in situ* (with solder and rivets) using galvanized corrugated iron – a ubiquitous building product in urban and rural Australia in the 19th and 20th centuries. In fact, tanks for household water supply became as Australian as *meat pies, Australian Rules football, Holden cars and kangaroos.*

As reticulated water supplies became more common in cities and larger provincial towns, tanks fell out of favour, not least because galvanized tanks commonly rusted leading to a mosquito breeding hazard. This was not only an amenity nuisance, but also was a health hazard from mosquito borne arboviruses such as Dengue fever, Ross River fever and Murray Valley encephalitis. In fact, major outbreaks of Dengue fever occurred in South East Queensland (SEQ) from 1911 to 1942, with over 185 deaths and 200 hospitalisations in the mid-1930s (Hurst, 2012). It is not surprising that local authorities banned tank installation in many urban areas (especially SEQ) in the 1950s and this ban remained in force into the 1990s.

However, tanks were a mainstay of potable water supply in rural areas and peri-urban areas without reticulated water (and sewerage). In fact, in 2013 over 34% of Australian households had tanks, with 10% using them as their main source of drinking water (ABS, 2013). In rural areas, tanks tend to be large (e.g., 40 kL+) and connected to most of the dwelling roof area. Cheap electric pumps and pressure switches then allowed pressurised water for all household uses. As power outages are common in rural areas, elevated header tanks were often installed to ensure a 1–2 day's supply of gravity pressurised water was available. About 23% of households in regional and rural Australia (3,365,000 households) currently use tank water as their major source of drinking water (ABS, 2013).

Tanks were initially galvanized corrugated iron (as previously mentioned), but over the last two decades large precast (<25 kL) and cast *in situ* (>30 kL) concrete tanks became popular in rural areas as the manufacturing technology improved. Over much the same time, rota moulded plastic tanks (≤25 kL) came onto the market in a range of shapes and colours – a technology which spilled over into the urban market. These were complemented by corrosion resistant, zincalume corrugated tanks with colour baked in at manufacture – a further adaption from the Australian roofing sector.

Most of the rural/peri-urban tank installations were rudimentary in terms of catchment protection, animal ingress and treatment before consumption. For example, gutter guards were more the exception than the rule, first flush devices were almost unheard of, lead flashing was commonly used as part of roof water proofing, and screens to prevent entry of animals were made of short lived (<10 years) corrodible brass mesh.

As technology improved over the last two decades, rain head screens on downpipes to minimize leaf contamination of tank water (which causes colour/taste issues) and stainless steel mesh screens markedly improved the physical quality of rainwater. Many households also installed under kitchen sink filters to remove suspended sediment, with some adding additional filters to reduce bacteria and remove protozoans (e.g., Giardia). A smaller number of households installed disinfection devices such as silver impregnated filters or Ultraviolet (UV) light units.

However with all such devices, maintenance is critical for successful performance, and often the filters became *sources* of contaminants (especially bacteria) rather than *sinks* for contaminants. For example, with UV units the light source needs to be replaced at least yearly, but this is rarely done. On the other hand, those solely dependent on rainwater are very aware of the operation of their collection/storage/supply system and maintenance effort is generally of a high standard. For example, pump failure means no water supply, and if collection efficiency was impaired due to overflowing gutters and/or blocked downpipes, water supply reliability is reduced and potable water might need to be purchased via water truck (at about $210 per 15 kL). This expense focuses the minds of the peri-urban residents and leads them to be highly aware of water efficient practices.

14.2 DRIVERS FOR RAINWATER TANKS IMPLEMENTATION

14.2.1 Australian context

In the mid to late 1990s, environmental sustainability emerged as a fundamental philosophy of urban living which encouraged residents to consume less energy and water, to become more self-sufficient, and to reduce their greenhouse gas footprint. Consequently, energy and water efficient buildings became popular, usually incorporating rainwater tanks for non-potable end uses, as well as solar hot water systems, solar electricity, house insulation and natural ventilation and so on. In some cases, exemplar houses and housing clusters were built and documented to demonstrate just how much reduction in grid supplied water and energy could be achieved. Examples include the Sustainable House in Sydney (Mobbs, 1998), Healthy Home at the Gold Coast (Gardner *et al.* 2003), Fig Tree Place in Newcastle (Coombes *et al.* 2000) and The Currumbin Ecovillage on the Gold Coast (Hood *et al.* 2010). For example, an 80+ percentage in self-sufficiency for water supply from 25 kL household rainwater tanks was measured for the Brisbane eco-development of Payne Road (Gardner *et al.* 2006; Beal *et al.* 2008).

This concept of self-sufficiency gained popularity in Australia's coastal urban communities and led to an explosion in the installation of rainwater tanks. For example from 1995–2013, the percentage of households in Australian cities increased from 24% to 34% (ABS, 2013). Part of this increase was due to extended drought conditions that occurred in all Australian capital cities in the so-called 'millennium drought' from 2001 to 2009 (van Dijk *et al.* 2013), which led to a mixture of self-funded, government rebated and mandated tanks in most urban areas. For example in NSW, the BASIX program required all new dwellings in the Sydney area to reduce their potable water use by 40% by installing internally plumbed rainwater tanks or other alternative non-potable sources (NSW Dept. Planning and Infrastructure, 2004). A similar type of program was introduced by the Queensland government. The Queensland Development Code MP 4.2 – Water Savings Target (QDC MP4.2) required detached dwellings in SEQ built after 1 January 2007 to achieve a mains water savings of 70 kilolitres per household per year (kL/hh/yr) (DIP, 2008). The motivation was acute as SEQ's main water storages dropped to below 17% capacity in August 2007, with about 12 months' supply of drinking water in its reservoirs at the peak of the drought (QWC, 2010), an experience not too dissimilar to that in other capital cities of Australia.

These non-subsidised, mandated programs were complemented by voluntary installations that attracted a rebate of up to $1500 from local and/or state and/or commonwealth governments. In some cases, these rebates could be claimed cumulatively by the householder. In SEQ for example, over 257,000 subsidies for rainwater tanks (worth about $250 M) were granted over the period 2006–2008 under the now defunct

Home and Garden WaterWise Rebate Scheme. The estimated total water savings was about 6,000 ML/year (Walton & Holmes, 2009). However, less than 7% of these tanks were connected inside the house for toilet flushing/washing machine end uses, as was required for the later mandated tanks. Rather, the water was used for garden irrigation and other external uses that were usually severely restricted by drought inspired water restrictions. In 2010/11 in SEQ, irrigation made up less than 5% of average total household water consumption (Beal & Stewart, 2011). That is, the externally plumbed tanks did NOT provide significant savings to potable water as potable water could not be used for external end uses in the first place under the severe water restrictions in place at the time.

Two main benefits that arose from the mandating/subsidy of rainwater tanks were, firstly, an explosion in the technology ranging from tank construction material, shape, colour and aesthetics to fit into small urban allotments (400 to 600 m^2), to collection (mainly stainless steel mesh screen and guards to prevent leaf/animal/insect ingress, and first flush devices), to mains water top-up (trickle float valve and electronic switching valve), to back flow prevention devices to stop rainwater entering the potable reticulation system, to under-sink treatment units (filters, membranes, UV disinfection units). The other major benefit was a raft of construction, installation and operational standards and guidelines (e.g., SA HB 230, 2006; Australian Government, 2004; WSAA, 2002; Chapman *et al.* 2008; QHealth, 2007; Queensland Government, 2011; NSW Health, 2007; DHS, 2003) which gave both the installing technician (i.e., the plumber) and homeowners an encyclopaedic suite of information *IF* they chose to use it.

New rainwater tank operation technologies are developing with dual objectives of potable water substitution and flood management. IOTA P/L (www.iota.net.au), a commercial arm of South East Water, one of the three water utilities in Melbourne, has developed a technology which monitors water levels in a rainwater tank and automatically releases water at a controlled rate down to a pre-defined level in the tank. The operation is governed by rainfall predictions communicated through remote link with the Bureau of Meteorology. Such a technology (which is very similar to the one in Seoul described by Han & Mun, 2011), if adopted at a large scale, can help in managing local flooding. However, successful, well-documented case studies in Australia are needed for future implementation at large scale.

14.2.2 International context

In the developed countries of Europe, the main driver for rainwater tanks is reducing stormwater inflows into the combined sewer system, thereby reducing sewer overflows, and in some cases local flooding. In Germany for example, there is a tax on impermeable areas in urban areas that increase urban runoff (both volume and peak discharge). Rainwater tanks is one method to reduce this inflow and subsidies and/or tax relief are often granted for this purpose, but do not carry over to promoting beneficial reuse of rainwater, as this would reduce the sales income of potable water utilities (Hermannn & Schmida, 2000; Nolde, 2007). Nonetheless, over 65% of new dwellings in Germany have installed rainwater tanks with an intention to augment potable water supply, adding to the existing stock of about 1.8 million systems (Schuetze, 2013). The motivation appears to be an environmental sustainability ethic, and a personal pride of partial self-sufficiency. Interestingly, Germany also has the highest penetration of domestic solar electricity systems in the world. Similar motivation occurs in some large public buildings, for example Potzdamer Platz in Berlin, as well as some large apartment buildings in Sweden (Villarreal & Dixon, 2005). Given that northern and central Europe are well watered, supplementing potable water supply with rainwater tanks does not seem to be a strong motivation for their adoption. In fact, some countries actively discourage rainwater use, with France actually banning it for even toilet and laundry uses (Vialle *et al.* 2011). This is reminiscent of that old, discredited Korean *saw* that ingesting rainwater will cause baldness (Han & Han, 2002).

In southern Europe and parts of the northern Mediterranean (e.g., Spain and Greece) where surface/ground water supplies are less reliable, rainwater tanks have been adopted for potable substitution, especially for multi residency buildings in Barcelona and Catalonia (Domenech & Sauri, 2011). However, whilst there are now some official planning regulations to mandate tank installation, there seem to be no guidelines, and certainly no subsidies. Morales-Pinzón *et al.* (2012) noted that rainwater harvesting systems are rare in Spain despite policy frameworks, such as the European Water Framework Directive, which specify measures for more sustainable use of water resources such as rainwater harvesting.

In the United Kingdom, the domestic penetration of tanks is fairly low (<6000 new tanks per year) and notwithstanding adequate guidelines for their installation, maintenance and reuse, there are no subsidies to encourage their installation, even for reducing inflow into combined sewers (Ward *et al.* 2012; Farnsworth, 2012). As with Germany, there are some public icon projects that use rainwater for toilet flushing and/or water features and/or garden irrigation, for example, London Eye and the Millennium Dome (Hills *et al.* 2001).

Across the Atlantic, the USA has developed guidelines for rainwater installation and reuse in a number of States and municipalities (USEPA, 2013; Texas Water Development Board, 2005), but it is only in the arid states of Arizona, New Mexico and Texas that tanks are encouraged and in some cases, mandated. But in some states (e.g., Colorado), their use is prohibited unless they are tagged on to a well water use license because of the often arcane water rights law of the western United States (e.g., see: www.water.state.co.us). The main use encouraged is irrigation of private lawns (Gold *et al.* 2010) which has a low public health risk (Ahmed *et al.* 2010).

In north Asia where monsoons are common and potable water supply is ample, rainwater tanks have been pioneered for flood mitigation in urban areas. In Seoul, for example, such tanks are mandated for city buildings, and their water level is centrally managed by telemetry, that is, the water level is reduced before expected storm events (Lee *et al.* 2010; Han & Mun, 2011). Reuse was not the major motivating factor, although recent innovations in private dwellings (e.g., the 1300 apartments Star Apartment complex in Seoul) are now focusing on toilet and garden irrigation, as well as flood mitigation (Han & Mun, 2011). A hydraulic analysis showed that 10 kL of storage per 100 m^2 of impermeable area could reduce the flood flow characteristics of a 1:100 year event into a 1:5 year event (Kim & Han, 2001; Han & Mun, 2011).

Similar issues motivate rainwater tank installation in Japan where over 1100 large buildings in Tokyo have tanks installed for flood mitigation. End uses such as toilet flushing and public space irrigation have gained favour particularly for large sport stadiums (Furumai & Okui, 2010). Concepts of potable substitution are slowly gaining favour along with the broader social objective of disaster preparedness if natural disasters were to cut off mains supplies (Furumai *et al.* 2008; Furumai & Okui, 2010). Most of the storage is installed in large buildings (including apartment complexes), with no subsidy available. However, more recently subsidies for residential home owners are being offered by a number of prefectures (Tjandraatmadja & Sharma, 2014).

In south Asia, rainwater tanks are important for water supply in rural areas and are also being mandated for large (>300 to 5000 m^2) buildings in water stressed states such as Uttar Pradesh in India, and cities such as Bangalore and Hyderabad. In Tamil Nadu, over 5 million private tanks have been installed (CSE, 2013). Mumbai mandates tanks for buildings >5000 m^2 but compliance has been patchy due to absence of subsidies and lax enforcement (Tjandraatmadja & Sharma, 2014). The main end use for rainwater (from large buildings) is groundwater recharge, a major source of water in India.

In other developing countries of the world, especially South Africa, Africa, Brazil and rural China, rainwater tanks are promoted by NGOs and government to supply drinking water and associated sanitation to the urban and/or rural poor. These are basic human rights under Millennium Goal # 7C (http://www.un.org/millenniumgoals/environ.shtml and http://sustainabledevelopment.un.org/focussdgs.html). In Brazil for example, over half a million domestic 16 kL tanks have been installed in the arid north east quarter of the country as part of a program to supply one million tanks servicing 5 million people under their Program for 1 Million Cisterns (P1MC – Gomes *et al.* 2014). In rural Africa (and South African shanty towns), tanks provide a supply of 'potable' water which is of a much higher quality than that collected from transient surface sources. Of course, thatched roofs in rural African areas severely limit rainfall catch. The quality of life implications are substantial in terms of reduced gastrointestinal diseases (morbidities >10% especially in the young), reduction of chronic back pain complaints, as well as freeing up time for the traditional water gatherers – the women and children (Orrico, 2003; Baguma *et al.* 2010a,b; Kahinda *et al.* 2007).

Taken overall, the use of rainwater tanks for substitution of mains water inside domestic dwellings in developed countries seems to be essentially an Australian phenomenon, with the possible exception of Germany and some Mediterranean cities (Sazakli *et al.* 2007). Developing countries where rainwater is a critical supply of potable water can learn from Australian rainwater insights and technology development in improving their rainwater tank systems for long term operation.

14.3 CURRENT AVAILABILITY OF WATER RESOURCES IN AUSTRALIA

Today, in 2014, Australia finds itself in a very different situation to that of the mid-2000s. The reservoirs of most of the capital cities (except Perth) are >80% full and the water utilities have spent around $35 Billon in providing climate resilient water supplies. For example, $7 Billion was invested in new drought resistant infrastructure in SEQ, including the SEQ Water Grid, the Western Corridor Recycled Water Scheme, the Tugun Desalination Plant and new dams and weirs, leaving the State with this debt to service. Similarly, the $3.5 Billion, 150 GL/year Wonthaggi sea water desalination plant built to service greater Melbourne, costs $657 Million/year to manage on just a care and maintenance basis (Melbourne Water, 2014). At the time of writing this book (2014), it has still not been used to supply potable water.

It is really only Perth that continues to experience a multi-decade reduction in surface water inflows and has its two sea water desalination plants (300 ML/day capacity) in full production and integrated into its water supply grid. Perth also has a strong Mediterranean-like climate (wet winters, hot, dry summers) that makes it unattractive for installing domestic rainwater tanks due to extended low rain/no rain periods each year (approximately 4+ months per year). So it's not surprising that rainwater tank penetration in Perth is about 7%, the lowest of any of Australian capital cities (ABS, 2013).

After the drought in SEQ was broken by heavy rainfall in 2008/09, with further rain filling the region's dams by 2010, the voluntary installed tanks had an uncertain future unless their amenity value (i.e., flow rate/pressure) and operating costs ($/kL) were comparable to or better than the now unrestricted mains supply (costing about $3.50/kL in SEQ). Poorly maintained tanks are a serious potential health hazard from mosquito borne arboviruses such as dengue fever (Qld Health, 2007). Moreover, the owners of these tanks believe their use and operation are at their own sole prerogative, and strongly resist any suggestion of government inspection or maintenance levy (Gardiner *et al.* 2008; Gardiner, 2009).

14.3.1 Changes in approach to alternative water supplies in Australia

The focus of water authorities in Australia is now on cost recovery of sunk infrastructure costs and minimisation of operating and maintenance costs (AWA/Deloitte, 2014). Consequently, their interest in alternative water supplies is essentially nil for domestic dwellings (especially if it competes with potable water sales) and lukewarm for public open space uses such as parks and sporting grounds. This attitude has reflected the policy position of their government shareholders who are very conscious of increasing utility charges to voters who are still recovering from the 2008 global financial crisis, and ever rising energy costs. For example, in 2013 the Queensland Government repealed QDC MP4.2 (Department of Housing and Public Works, 2013) based on a combination of lobbying by the home building industry and an economic analysis (Qld Competition Authority, 2012) that demonstrated that mandated tanks were not cost-effective for society, notwithstanding the fact it was the home owners who ultimately carried all the costs.

These economic arguments, based on both the Benefit-Cost ratio and Levelised Cost of rainwater tanks ($/kL), appear not to have influenced other states in Australia, in that NSW still retains its BASIX program, Victoria requires internally plumbed tanks or solar energy to achieve a mandated housing sustainability score, South Australia requires at least a small tank (>1 kL) connected to the hot water system, whilst the ACT has sustainable housing rules similar to NSW, that is, 40% potable water saving (see Table 3 from Tjandraatmadja *et al.* 2012). As all these states/territories have installed large climate resilient water supplies (seawater desalination plants) or increased dam reservoir capacity for inland cities (ACT), one has to wonder as to their motivation for rainwater tanks, other than for philosophical reasons of environmental sustainability. In Victoria for example, this sustainability ethic is reflected in the Office of Living Victoria whose main charter is to *encourage* water utilities, local authorities and private urban developers to install alternative urban water supplies (e.g., reticulated stormwater or recycled water) to augment traditional mains supplies (see http://www.livingvictoria.vic.gov.au/MetroFramework/MetroFramework_discussionpaper.pdf). Furthermore, stormwater management and run-off reduction are seen as critical management strategies to prevent damage from the sediment and pollutant load into the Port Phillip Bay ecosystem (Melbourne Water, 2006; Harris *et al.* 1996).

The economics of these large schemes in capital cities is far from settled (e.g., AWRCOE, 2013) with many water utilities being quite reluctant, for example, to install dual reticulation of recycled water due to its high regulatory and operational costs. For example, Sydney Water has jettisoned its plans for dual reticulation in the high-growth urban areas of north west and south west Sydney, although a number of water utilities in Victoria, such as Yarra Valley Water, are still strong supporters of such schemes. In contrast, Gold Coast City Council (located in SEQ) has mothballed its $220 M, 2500 ML/year non-potable dual reticulation scheme in Pimpama-Coomera as the water use by the customers was well below expectations, and the running costs, especially those arising from the monitoring requirements imposed by the state regulator, made the scheme uneconomic (Taylor *et al.* 2013).

A partial exception to the reluctance to invest in alternative water supplies is stormwater that is captured, stored and reused for irrigation of public open spaces. A major proponent for this concept is the arid city of Adelaide (600 mm/year rainfall with hot, dry summers), where over 12.4 GL/year of stormwater is used for this purpose, basically greening an otherwise brown, summer city scape (South Australia Strategic Plan Progress Report, 2012). Similar enthusiasm is shown in Melbourne where dry, often very hot summers, urban heat island effects, and stream water quality implications all combine to encourage official enthusiasm for these public open space irrigation schemes. However, these alternative sources of water are usually not the responsibility of the water authorities. Rather, they are initiated by an urban developer or local government authority and then often handed over to the water utilities to operate. In some cases, a state planning authority (e.g., Office of Living Victoria) might actively encourage the establishment of water sensitive suburbs, but they do not fund the construction of the water services.

Of course, when the next El Nino inspired extended drought hits eastern Australia, as is currently (2014) occurring in California (Pacific Institute, 2014 and http://gov.ca.gov/news.php?id=18379), there will be a resurgence of interest in alternative water supplies, reinvigoration of water conservation behaviour by industry and residents, and probably water restrictions on external water uses. A recent survey on the State of the Australian water sector has clearly identified that water security affected by climate change is still *top of mind* for many of the (industry) respondents, and there is strong support for a diversified urban water supply (AWA/Deloitte, 2014). Rainwater tanks would seem well suited to help resolve all these problems. But until that time, *the elephant in the room* will be the maintenance of over 1,500,000 tanks in urban Australia.

14.4 POLICY IMPLICATIONS FROM THE CHAPTERS

Water yields (kL/yr) predicted by modelling need long-term, daily climate records and demonstrated that they are responsive to tank size, connected roof area and water demand, but all 3 parameters show a plateau-type, levelling out response (Vieritz, Neumann and Cook, Chapter 2). However by 'running a tank hard' by regular water withdrawals, the yield is maximised from an otherwise small tank. On a suburb or regional basis, the yield of a population of tanks is up to 15% less than the behaviour of an average tank (Maheepala *et al.* 2013) because of variation in tank sizes, connected roof areas, household water demand and water losses. Hence, other (statistical) modelling tools are needed when considering rainwater in regional water planning studies. Somewhat surprisingly, despite the plethora of rainwater tank models, there is a dearth of model validation studies. This needs resolution if tanks are to be used with confidence in regulations (e.g., BASIX in NSW) or water planning studies. The task should not be particularly onerous as covering the variation in connected roof area, tank size, demand and, most importantly, climate is more important than a large statistical study since it is the processes that need validation in a stochastically driven model. A starting point would be the detailed rainwater tank information collected in the monitoring study of 20 homes conducted by Umapathi *et al.* (2012) and reported in Chapter 4.

Vieritz, Neumann and Cook in chapter 2 also touch briefly on predicting the overflow volume/rate from a large cohort of rainwater tanks using a continuous simulation model, but they do not relate this to the discharge characteristics of stormwater flows in urban areas, which are also affected by impervious areas other than house roofs. It is possible that one of the eWater CRC software tools such as Source Urban (http://www.ewater.com.au/products/ewater-toolkit/) may be suitable for this task. Clearly more research is needed in this area.

Models notwithstanding, we need to answer the question *do rainwater tanks really supply substantial potable water savings?* In SEQ for example, the QDC MP4.2 water savings targets were formulated on a rainwater model that predicted tanks could supply 70 kL/hh/year (Le Muth & Barry, 2006). The methodologies described by Beal *et al.* in Chapter 3 and based on studies in SEQ can be extrapolated to any urban area that has an extensive, regularly-measured water meter database, and some limited demographic information on household occupancy. At the time this modelling was undertaken (in 2006), the average per capita water use was very high, approximately 300 L/p/day (QWC, 2010), with a large component of outdoor use/irrigation. During the time the water savings studies were undertaken, the per capita water use had reduced to 150–160 L/p/day due to a combination of water restrictions, increasing adoption of water efficient appliances, demand management practices and water conservation behaviour change (Beal & Stewart, 2011). For example, simply replacing a ≤2 star rated clothes washing machines with a star rating ≥4 provided a potential savings of 8.8 kL/hh annually. Hence, it is not surprising that Beal *et al.* (2011; 2012) reported that the estimated average mains water savings in 2008 was 50 kL/household/year (kL/hh/yr) rather than the 70 kL/hh/yr predicted by modelling. The subsequent study by Chong *et al.* (2011) in 2009/10 estimated rainwater tank yield averaged 58 kL/hh/year, but ranged from 40 to 81 kL/hh/yr, with the higher yields associated with higher local rainfall and higher household water use (including garden irrigation).

A related method to measure potable water savings is detailed instrumentation of a limited number of tanks systems (say <30 households) that allows collection of fine-grained water use and energy use data which is regularly interrogated using telemetry technology (see Chapter 4). A real-time monitoring study of 20 households in SEQ was conducted by Umapathi *et al.* (2012; 2013) with smart water meters over a 12-month period to validate the mains water savings from mandated household rainwater tanks and to also measure the diurnal water demand patterns and the energy consumption. The analysis showed that the average reduction in mains water for this small population of rainwater tanks was 40 kL/hh/yr over the 12-month monitoring period, during which time the average total household water consumption was 151 kL/hh/yr.

Clearly, a reduction in water demand through a range of water efficiency, demand management and water use behaviour measures will have a significant impact on the water demand from a rainwater tank and result in less potable water savings. However, it can be confidently argued that, overall, internally plumbed rainwater tanks make a very significant contribution to potable water savings (about 30% of total household demand).

Energy use of tank systems (Chapter 6) is of particular interest on both practical and philosophical grounds as early reports of high specific energy use (e.g., >3 kWh/kL) discounts the environmental benefits accruing from self-sufficiency in water supply. Detailed studies have now established specific energies of tank systems are more often in the range 1.5 to 1.7 kWh/kL (Tjandraatmadja *et al.* 2012). Umapathi *et al.* (2012, 2013) estimated the average energy usage for rainwater supply to 20 homes in SEQ to be 71 kWh/year, with an average specific energy of 1.5 kWh/kL. Similar results were reported for 52 homes in New South Wales (Ferguson, 2011). This is still substantially more than traditional mains supplied water (circa. 0.9 kWh/kL), but considerably less than that for recycled water (2.8 kWh/kL) or desalinated seawater (3.4 kWh/kL) (Tjandraatmadja *et al.* 2013). The key to low tank specific energy use is appropriately sized pumps and high water flow rates (i.e., >10 L/minute), as energy use per unit of time for most rainwater pumps is almost constant with flow rate (Chapter 6). Hence, systems that allow high pump flow rates, such as pressure vessels and header tanks, are to be encouraged, although for single story buildings, the latter seem to be incompatible with the operating pressures required by high pressure solenoid valves in domestic water appliances.

Policy implications include encouraging installation of pressure vessels (but they must be >20 L to be effective), and promoting research activities that encourage manufactures to install low pressure solenoid valves in their water appliances. The high market penetration of low volume dual-flush toilets (>80%) indicates that the market can respond rapidly to officially endorsed water efficiency measures and water conservation messages.

Audits are very useful to ascertain if tanks are installed as per regulations, for example, recommended tank size, connected roof area, first flush devices, insect screens, and so on. An on-site assessment of mandated rainwater tanks in 223 detached dwellings in SEQ (Chapter 5) suggested good overall compliance for most criteria for mandated tanks, with tank size and connection to household appliances (toilet and cold water washing machine tap) shown to comply with requirements. However, 40% of the houses failed to meet the requirement for connected roof area – the lesser of either 100 m^2 or 50% of the roof area connected to the tank. Connected roof area has been shown to be the greatest single variable limiting tank yield in SEQ.

Voluntary installed tanks seem to be less compliant, as suggested by ACT experience where the measured yield is substantially less than modelled yield; a phenomenon the authors of Chapter 3 termed the *Functionality Factor* where tank installation quality, operational standards and behavioural characteristics are all compromised.

Maintenance is a key issue for the long–term, sustainable performance of rainwater tanks, yet there is very limited data on this for SEQ (but see Chapters 5 and 7). The issue of tank system maintenance remains a tricky problem on a number of grounds. On one hand, the regulators/water utilities require regular audit information to confirm water savings (if tanks are a critical component of a regional water strategy) and low public health risk (i.e., no mosquitos). On the other hand, Chapters 7 and 8 identified that the tank owning public are largely unaware of the need for tank maintenance, have little idea of what needs to be done, has variable self-confidence to undertake the maintenance tasks, and strongly rejects the suggestion they should pay for a mandatory tank monitoring/condition assessment program. The mandated tank owners tend to be more negative on all criteria compared with the retrofit tank owners. The authors of Chapter 7 (Moglia *et al.*) make the perceptive observation that tanks are an uncommon type of asset which are in private ownership but provide public good outcomes (i.e., reduced demand on potable water). There are very few precedents on how to manage this type of infrastructure, but whatever strategy is chosen, Moglia *et al.* counsel that it should be implemented gradually, and assessed for outcome efficacy before moving onto any legislative amendments.

Much more 'on ground' information on the state of repair of rainwater tanks is available from a recent Smart Water Fund sponsored study in Melbourne (Moglia *et al.* 2014). This survey of over 450 homes identified a 60% failure rate on criteria ranging from major (e.g., 6% had failing foundations), to suboptimal (e.g., 35% had switching valve failure and 57% had discoloured water), to minor (e.g., 31% had roof gutters full of debris). There was generally a widespread ignorance of tank system maintenance needs, especially amongst mandated tank owners. As all the tanks were new (<5 years old), more fundamental issues of screen failure were relatively rare, yet 12% reported the presence of mosquito larvae, which presents the potentially serious public health implications of arboviruses (Qld Health, 2007). Other failure issues such as overflowing gutters, blocked downpipes, failed pumps and so on impact solely on the individual householder, and are not strategically important *unless* regional water planning policy has been predicated on an assumed, consistent non-potable contribution from rainwater tanks. Of course, tree leaf discoloured rainwater in toilet bowls and so on can be an unacceptable aesthetic affront to the householder who may disconnect their tank from these internal end uses, as happened in the Payne Road development in SEQ (Beal *et al.* 2008). These types of preventable failures add to the public discourse that rainwater may not be as good as potable water for *any* internal use, although user expectations do evolve with extended rainwater use experience (Moglia *et al.* 2014).

Monitoring of tank condition and mosquito breeding is an important public health activity for government and its regulatory agencies. But this will not be without its challenges given the sense of personal tank ownership by householders, an almost uniform resistance to imposed inspection programs paid for by the householder, and resentment of what is perceived to be government interference (Chapter 7; Gardiner, 2009).

In Chapter 8, Mankad *et al.* suggest that motivational factors play a significant role in rainwater tank adoption. The effectiveness of programs that promote the installation of rainwater tanks may be increased if they focus on the benefits of rainwater tanks for householders and for the community more broadly. It makes sense to focus on the benefits of rainwater tanks for protecting the individual, their lifestyle and their property. However, it is also important to communicate that, in times of extreme drought, the chief contribution of rainwater tanks could be to reduce the pressure on drinking water supplies and thereby increase water security in urban areas.

Mankad *et al.* (2012a) also found that homeowners were rarely engaging in tank maintenance behaviours, but the higher the self-assessed competence to maintain a tank that homeowners reported, the more they engaged in tank maintenance. Most mandated tank owners had limited knowledge of their tank set-up, and participants reported only moderate perceptions of competence to be able to maintain the tank. Enabling householders by increasing their levels of competence (e.g., applied education about maintenance) could go some way to increasing rates of tank maintenance. This education should focus on 'how to' and 'why to' information that helps homeowners to the build competence in this area.

The lack of tank maintenance by mandated tank owners raises important issues for government agencies seeking to address water security through rainwater tank adoption. On the one hand, legislation that mandates installation of rainwater tanks should result in much broader uptake of these systems than measures that encourage voluntary uptake. On the other hand, the lack of choice that legislation introduces may undermine householders' willingness to maintain their systems. These findings suggest the need for agencies to pay particular attention to households with mandated tanks to encourage greater knowledge of their functioning, and increase ability to undertake tank maintenance.

Chemical water quality (Chapter 9) is always a concern if rainwater is used for drinking, as it is by about 2% of urban tank owners, and 23% of tank owners outside Australian capital cites (ABS, 2013). Results from Chapter 9 indicate that lead (Pb) is the major contaminant of health concern (health limit <10 µg/L) and after allowing for lead flashing used in roof construction, there seems to be a systemic rainfall contaminant issue in urban Australia. Maygar and Ladson (Chapter 9 authors) argue there is good experimental evidence to conclude that Pb concentration exceeds drinking water guidelines in 22% of *urban tanks* used for drinking water in Australia (27,700) increasing to 44,000 tanks if the 95 percentile figure is considered. Given that unpainted lead flashing can only explain part of this incidence, it is clear that steps need to be taken to avoid entrainment of Pb enhanced sediment at the bottom of the tank. A floating intake pipe located say 150 mm below the free water surface would seem a prudent, but under researched, mitigation practice. First flush devices are both recommended and popular, but are unlikely to have much effect on the Event Mean Concentration (see Duncan, 2006 for EMC definition) as the diverted runoff volume is usually very small (<20 L per device).

The microbiological quality of tank water is a major concern, especially if the water is used for drinking or showering. There was a general finding in SEQ that collected rainwater is contaminated with animal faecal bacteria, some of which may be pathogenic to humans (Chapter 10). There are also other zoonotic pathogens such as the protozoan parasites Giardia, and sometimes Cryptosporidium. Chapter 10 shows that natural disinfection processes can reduce faecal bacteria concentrations by 1 to 2 log10 in a matter of hours on roofs and gutters exposed to sunlight, and up to 5–8 days on the roof under shaded conditions. Once introduced to the tank, a slower inactivation process takes place over 10–15 days (Ahmed *et al.* 2014) but this is not sufficient to make the water safe to drink, based on Quantitative Microbial Risk Assessment (QMRA) methods (Ahmed *et al.* 2010; 2012). However, rigorous epidemiological studies in South Australia (Rodrigo *et al.* 2010) and anecdotal evidence from the 23% of residents in rural/peri-urban Australia that regularly drink rainwater (ABS, 2013) suggest these health risks may be overestimated.

At a policy level, the results of Chapter 10 (by Ahmed *et al.*) and the regularity of world-wide reports of zoonotic pathogens in collected rainwater (Ahmed *et al.* 2011) make it difficult to recommend rainwater for potable uses in urban areas unless it has been disinfected using chemical (e.g., chlorine, hydrogen peroxide, ozone) or mechanical methods (silver impregnated filters, UV light), or simply boiled prior to drinking. But effective operation of disinfection devices requires their regular maintenance, and the track record for this type of 'housekeeping' is not encouraging. Overall, the research supports the policy position of Australian health regulators that rainwater should *not* be used for drinking where a potable water supply is available.

Communal tanks seem the way to go in the more densely settled urban areas. Communal tanks allow centralised treatment, unlimited demand if connected to the potable mains, and a large fraction of the roof area connected to the 'catchment' area (Chapter 11). This configuration will maximise rainwater yield from connected roof areas as storage and demand limitations are essentially removed, treatment technology and quality assurance will equal that of traditional potable water, for example, as per the Australian Drinking Water Guidelines (NHMRC & NRMMC, 2014), and no extra reticulation infrastructure is required. The 45 ML/year Fitzgibbon scheme in SEQ, described in Chapter 11, is an excellent example of a communal scheme where much of the construction cost has been borne by the private sector. It is now up to the water utility/government to show that this scheme can operate safely and efficiently as per its design objectives, and hence become a precedent for similar types of schemes in eastern Australia. A problem for their wider adoption in conventional urban areas is the cost (>$4/kL) compared with traditional mains supply ($2–$3.50/kL), and the willingness of water utilities to manage a 'left field' technology on a day-by-day basis.

This raises the issue of the economics of rainwater systems. They do not seem to look financially attractive on either a Benefit Cost ratio or a Levelised Cost basis (Chapter 12). However, analyses to date have largely excluded positive externalities (e.g., stormwater quantity/quality/peak flow, reduced land take for WSUD devices) as well as some doubt on the realism of the assumed costs (White, 2014). We believe that unless tanks are mandated by some type of legislative instrument (e.g., BASIX), their future adoption will be limited to people with a high environmental sustainability ethic (as per many of the young German households). Anecdotal evidence suggests that many new home owners in Australia would prefer, say, a granite bench top in their kitchen rather than spend $5000 on an internally plumbed rainwater tank or roof mounted solar electricity panels in their fixed priced, project built homes. It's possible that steeply rising electricity and water prices in Australia and the next extended drought may moderate this attitude.

Stormwater benefits are an important driver for installing rainwater tanks in the highly urbanised European cities with a combined sewer system and a high frequency of flooding (e.g., England and Germany). This hydraulics issue is generally underexplored in Australia where dedicated sub surface stormwater conveyance infrastructure is usually built for the 1:2 year storm event, although overall flood protection of a suburb/city is based on a 1:100 year event (e.g., QUDM, 2013). The key question is whether tanks will reduce the peak discharge (Qp) such that the stormwater pipe sizing can be reduced.

In western Sydney, the Upper Parramatta River Catchment Trust (UPRCT) is responsible for stormwater management of an 11,000 ha catchment, which is administered for all other purposes by four different local authorities. The overall objective is ensure that any future urban densification (and hence increased impermeable area) since 1991 will not increase downstream flooding for any rainfall event less than or equal to the 1:100 year average return interval (ARI) event. They implement this policy by mandating distributed, on-site detention storages (OSD) which empty at a maximum controlled rate. The OSD characteristics are 455 m^3/ha of storage (about 20 kL per household), with a maximum discharge rate of 150 L/sec/ha, controlled using orifice plates. Part of this storage (the first 300 m^3/ha) is designed to discharge at the lower rate of 40 L/sec/ha to ensure that runoff rates from the 1:1.5 year ARI storm will mimic the natural stream flows, as close as is practical. The objective here is to protect creek ecosystem health.

UPRCT have subsequently modified their design guidelines (UPRCT, 2005) to account for detention storage by rainwater tanks, which is composed of a *dedicated airspace* volume (kL) which empties at the equivalent rate of 40 L/sec/ha, and a *dynamic airspace* (kL) which only empties by household water use. A formula based on tank volume, roof area and household water use is used to calculate the *dynamic airspace*, which is further discounted by a roof area formula before being credited against the mandated OSD volumes (i.e., 300 m^3 and 455 m^3/ha).

The Victorian study (see Chapter 13) confirms that rainwater tanks will reduce nitrogen and phosphorus export from roofs (by 60% to 80%) as well as reduce the frequency of those small runoff events caused by increased impervious areas. There are also clear benefits to urban creek ecosystem health, as well as reduction in the 1:1 year peak discharge (Qp) to near pre-development values. However, there is no analysis equivalent to that of Kim and Han (2008) in Seoul that showed 10 kL rainwater storage/100 m^2 roof could reduce a 1:100 year Qp event into a 1:5 year Qp behaviour, provided the tanks were automatically emptied before the storm event. This is a major research gap in Australia, as is the implication of tanks for reducing the size of WSUD devices such as bio-retention basins, which are common and often mandatory in most new urban developments (http://waterbydesign.com.au/wpcontent/uploads/drupals/seq_wsud_dos_nov_07_final_v2_0001.pdf and http://waterbydesign.com.au/techguide/).

14.5 LESSONS FOR OTHER COUNTRIES

In Australia, rainwater harvesting in urban areas has been used to substitute for mains water on a fit-for-purpose basis. The sustainability of such practice, as part of the provision of urban water services, is governed by a complex mix of social, institutional and technological factors.

Australia was also one of the first countries to try to mainstream rainwater harvesting in urban areas, often through policy trial and error, but more importantly it also conducted comprehensive scientific assessments of the systems both pre- and post-installation over the last 10 years. In other countries, especially the developing ones, the systems adopted will differ with some of the technologies used likely to be much less sophisticated. Hence, not all Australian insights may be applicable, but a number of the Australian insights have universal value.

Regions with plenty of water have historically been mostly interested in reducing stormwater inflows into the combined sewer system and stormwater management (Europe and North-Pacific Asia), but there is increasing interest in potable substitution. Germany seems a key country of interest for such purpose, given the high penetration of tanks in new domestic dwellings (about 65%). Regions subject to water scarcity, difficult water access, and stressed water resources due to overexploitation and/or climate, focus mostly on rainwater for water provision (Mediterranean Europe, parts of USA, semi-arid parts of Brazil, Africa and Asia) or recharge of traditional groundwater supply (as in India).

In developed countries such as Korea, Japan, USA and Australia, systems are designed based on the traditional service standards and often adopt telemetry and operating mechanisms that assume constant electricity supply and high labour cost. In other parts of the world, such as in developing economies, electricity may be intermittent, labour is often cheap and the mains water supply set-up differs, for example, header tanks are often adopted due to intermittent water supply, hence systems need to be designed differently, for example, more mechanical parts.

Irrespective of the system type, there is the need for models, guidelines and regulations to guide the system design, and in particular, to understand how it is possible to integrate rainwater tanks with traditional water supplies. Lessons and tools from Australia, and their impact may be useful for other countries to examine, and then modify as they develop their own standards, codes, policies, and so on suited to their needs. Examples include the impact of audits of the quality of installations in SEQ.

For those (developing) countries where rainwater replaces inferior water sources for drinking we suggest the following insights can apply:

- Rainwater is NOT pure and risk free. Hence, it is best to boil the water before drinking it. However, other end uses are probably OK without specialised treatment.
- Be careful with the roofing material used, especially lead flashing used to waterproof roofing joints, as lead leaching can be a serious contaminant in the collected rainwater. This is especially important in the young where constant exposure to lead, a neurotoxin, can cause serious development problems.
- Computer models can be used to size connected roof area/tank volume to achieve a given water supply (litres/day) for a given reliability and duration of supply. For example, it's important that tanks don't regularly run dry for extended periods of time (weeks/months) without a clear understanding of the need for a back-up plan.
- Electricity is likely to be non-existent, or intermittent at best. Hence, it's important to have a header tank to supply water by gravity 24/7. Using a petrol fuelled generator to power a cheap electric pump to fill a header tank will avoid the high specific energy use that often accompanies use of such pumps

in developed countries. This system can also mitigate the high energy use to pump (other) sources of water over long distances or high elevations (e.g., from a deep well or over hilly terrain).

- Insect and vermin control is important, increasing to essential in tropical countries where arboviruses such as dengue fever can occur. Hence, limited cash resources should be preferentially allocated to high quality stainless steel mesh screens for the inlet and overflow structures on the tank to prevent mosquito ingress.
- Sediment in the tank is a double-edged sword. On the one hand, it can immobilise heavy metals and other contaminants as well as contribute to natural die-off of pathogens (Spinks *et al.* 2005). On the other hand, if ingested in the rainwater, the contaminant dose will be concentrated. A simple solution is to have a floating off take hose inside the tank to ensure that the best available water quality is flowing through the supply tap. Sophisticated first flush devices can also contribute to better quality of collected rainwater, but simple units operated manually in low wage areas should work equally well.
- Tank material should be constructed from durable material such as concrete or corrosion resistant corrugated metal, as appropriate to climatic conditions. Whilst polyethylene tanks are relatively cheap and easily transportable, longevity in extreme weather environments of high temperatures and high solar radiation, arid conditions can be questionable. Moreover, the more the local input into tank construction, the more affordable the system is likely to be.
- It is very important to have a local education/learning group on tank operation and maintenance to ensure the learnings are well distributed though the local community. The importance of keeping up the maintenance motivation of tank owners has been verified for villagers in Uganda, where familiarity with raintanks, shared learnings through communities of practice and reinforcement of education on maintenance methods were key factors to improve the overall health of raintank users (Baguma & Loishkandl, 2010).

In some cities of developing countries, rainwater tanks are installed in large public buildings as a show case for the utility of rainwater to the wider community. Whilst this type of application has not been covered in detail in this book, a report by Cook *et al.* (2012, 2014) based on experience in a high rise building in Brisbane, SEQ, has highlighted the potential problems with poor plumbing construction. This is balanced by the opportunities to harvest other sources of water (such as air conditioning condensate and groundwater seepage into basements) once the 'mind has focused' on the task of collecting alternate water supplies for toilet flushing and amenity irrigation end uses.

The focus of rainwater in developed countries is potable water substitution, and in this application, the Australian lessons on virtually all aspects of tank systems can be more or less directly applied. Of particular relevance are the learnings from roof area/volume sizing models, pump capacity (kW), tank technologies, construction standards and maintenance guidelines, post installation and maintenance audits, and education/feedback to the householder. Assuming rainwater will not be used for drinking in well-watered European countries, the chemical and microbiological quality should not be a major issue. However, improved stormwater runoff quality may be a key advantage, especially in reducing the heavy metal export (e.g., copper and zinc) from metal roofs.

The lessons for long-term maintenance are more complex, as potable water supplies in many other developed countries are probably designed without taking into account rainwater contributions. Similarly, the health implications of arboviruses from poorly maintained tanks in northern European and northern Asian countries is likely to be much less than that in Australia and the developing countries. Nonetheless, an education program seems to be sensible, something we have yet to undertake in Australia despite a plethora of guidelines. Given the motivation of many German householders to be environmentally responsible and partially self-sufficient, it seems likely they would be very amenable to implement sensible tank maintenances information.

Finally, communal rainwater tanks would seem to have lots of potential in closely settled urban areas, especially if surface slopes are greater than 0.5% to allow installation of above ground storages (Gurung & Sharma, 2014). Centralised management of rainwater opens up a raft of reuse/treatment opportunities.

14.6 CONCLUSIONS

The body of work reported in this book should have widespread implications to other states of Australia, other developed countries, and some developing countries. However, it should be noted that the biophysical/social context of this SEQ-centric work is for tanks as suppliers of non-potable water to urban domestic dwellings. Nonetheless, most of the rainwater tank systems' analyses, assessment, modelling, monitoring and validation approaches and methods presented in this book will be useful in enhancing scientific knowledge of water professionals engaged in this area in any part of the world. The social surveys of tank owners consistently found a widespread ignorance of tank system maintenance needs, especially amongst mandated tank owners, and an almost uniform resistance to an imposed inspection program paid for by the householder. Unless self-interest (i.e., saving mains water costs) and community health (i.e., mosquitos, aka arboviruses) issues are clearly communicated, this resistance is likely to remain unchanged.

14.7 REFERENCES

Australian Bureau of Statistics (2013). Environmental issues: Water use and Conservation March 2013, 4602.0.55.003. from http://www.abs.gov.au/ausstats/abs@.nsf/Latestproducts/4602.0.55.003Main%20Features4Mar%202013?opendocument&tabname=Summary&prodno=4602.0.55.003&issue=Mar%202013&num=&view= (accessed 3 November 2014).

Ahmed W., Vieritz A., Goonetilleke A. and Gardner T. (2010). Health risks from the use of roof-harvested rainwater in Southeast Queensland, Australia, as potable or non-potable water, determined using quantitative microbial risk assessment. *Appl. Environ. Microbiol*, **76**(22), 7382–7391.

Ahmed W., Gardner T. and Toze S. (2011). Microbiological quality of roof-harvested rainwater and health risks: a review. *Journal of Environmental Quality*, **40**(1), 13–21.

Ahmed W., Hodgers L., Sidhu J., Gardner T., Richardson K., Palmer A. and Toze S. (2012). Health Risk Assessment of Roof-Captured Rainwater. Urban Water Security Research Alliance, Technical Report No. 77, Brisbane, QLD, Australia.

Ahmed W., Richardson K., Sidhu J. P. S., Jagals P. and Toze S. (2014). Inactivation of faecal indicator bacteria in a roof captured rainwater system under ambient meteorological conditions. *J. Appl. Microbiol*, **116**(1), 199–207.

Australian Government (2004). Guidance on Use of Rainwater Tanks, enHealth Council, Canberra, Australia. http://www.health.gov.au/internet/main/publishing.nsf/content/3D981B51B4FB458DCA256F1900042F6E/$File/env_rainwater.pdf (accessed 15 October 2014).

AWA/DELOITTE (2014). State of the Water Sector Report, October 2014. http://www.awa.asn.au/state_of_the_water_sector_report/ (accessed 15 October 2014).

AWRCOE (2013). Economic Viability of Recycled Water Schemes. A report of study funded by the Australian Water Recycling Centre of Excellence by Marsden Jacob Associates. November 2013. ISBN: 66 663 324 657 http://www.australianwaterrecycling.com.au (accessed 15 October 2014).

Baguma D. and Loishkandl W. (2010). Rainwater harvesting technologies and practices in rural Uganda: a case study. *Mitig Adapt Strateg Glob Change*, **15**, 355–369, doi: 10.1007/s11027-010-9223-4.

Baguma D., Loiskandl H. and Jung H. (2010). Water management, rainwater harvesting and predictive variables in rural households. *Water Resource Management*, **24**, 3333–3348.

Beal C. D. and Stewart R. A. (2011). South East Queensland Residential End Use Study: Final Report. Urban Water Security Research Alliance, Technical Report No. 47, Brisbane, QLD, Australia.

Beal C. D., Hood B. Gardner T. Lane J. and Christiansen C (2008a). Energy and water metabolism of a sustainable subdivision in South East Queensland: a reality check. *Enviro 2008*, Melbourne.

Beal C. D., Hood B., Gardner T., Christiansen C. and Lane J. (2008b). Energy and Water Metabolism of a Sustainable Subdivision in South East Qld: The Little Toe of the Urban Ecological Footprint? Paper presented at the Proceedings of Enviro 08, 5–7 May 2008, Melbourne, Australia.

Beal C. D., Stewart R. A., Huang T.-T. and Rey E. C. (2011a). SEQ residential end use study. *Australian Water Association Journal*, **38**(1), 80–84.

Beal C. D., Gardner T., Sharma A., Barton R. and Chong M. (2011b). A Desktop Analysis of Potable Water Savings from Internally Plumbed Rainwater Tanks in South East Qld. Urban Water Security Research Alliance Technical Report No. 26, Queensland, Aust.

Beal C. D., Sharma A., Gardner T. and Chong M. (2012). A desktop analysis of potable water savings from internally plumbed rainwater tanks in South-East Queensland, Australia. *Water Resouce Management*, **26**(6), 1577–1590.

Centre for Science and Environment (2013). Legislation on Rainwater Harvesting – Implementation of by Laws in different States and Cities, India. http://www.cseindia.org/node/1111 (accessed October 2014).

Chapman H., Cartwright T. and Tripodi N. (2008). Guidance Manual for the Design and Installation of Urban Roofwater Harvesting Systems in Australia (Edition 1), CRC for Water Quality and Treatment Salisbury, SA, Australia – Research Report 39.

Chong M. N., Umapathi S., Mankad A., Sharma A. and Gardner T. (2011). A Benchmark Analysis of Water Savings by Mandated Rainwater Tank Users in South East Queensland (Phase 2). Urban Water Security Research Alliance Technical Report No. 49, Queensland, Aust.

Commonwealth of Australia, Department of Health and Ageing (2010). Guidance on use of rainwater tanks. http://www.health.gov.au/internet/main/publishing.nsf/Content/DD676FA1241CDD0DCA25787000076BCD/$File/enhealth-raintank.pdf (accessed 15 October 2014).

Cook S., Sharma A., Gurung T. R., Chong M. C., Umapathi S., Gardner T., Carlin G. and Palmer A. (2012). Performance of Cluster Scale Rainwater Harvesting Systems: Analysis of Residential and Commercial Development Case Studies. Urban Water Security Research Alliance, Technical Report No. 68, Brisbane, QLD, Australia.

Cook S., Sharma A. and Gurung R. (2014). Evaluation of alternative water sources for commercial buildings: a case study in Brisbane, Australia Resources, Conservation & Recycling, 89, 86–93. http://dx.doi.org/10.1016/j.resconrec.2014.05.003.

Coombes P. J., Argue J. R. and Kuczera G. (2000a). Figtree Place: a case study in water sensitive urban development (WSUD). *Urban Water*, **1**(4), 335–343.

Department of Housing and Public Works (2013). Web site http://www.hpw.qld.gov.au/construction/BuildingPlumbing/Building/WaterSupplySystems/Pages/default.aspx (accessed 15 October 2014).

DHS (2007). Rainwater use in urban communities. Department of Human Services, State of Victoria.

DIP (2008). Water Saving Targets. For Councils, Plumbers, Builders and Developers. A Guide to the Queensland Development Code Part MP 4.2, Department of Infrastructure and Planning Effective August 2008. http://www.dlgp.qld.gov.au/resources/guideline/development-code/water-saving-targetsguidelines.pdf (accessed 15 October 2014).

DLGP (2008). Queensland Development Code MP 4.2 – Water Savings Targets. Department of Local Government and Planning, Queensland Government.

Doménech L. and Saurí D. (2011). A comparative appraisal of the use of rainwater harvesting in single and multi-family buildings of the Metropolitan Area of Barcelona (Spain): social experience, drinking water savings and economic costs. *J. Cleaner Production*, **19**, 598–608.

Duncan H. P. (1999). Urban stormwater quality: a statistical overview. Cooperative Research Centre for Catchment Hydrology, Melbourne, pp. 134.

Duncan H. P. (2006). Urban stormwater pollutant characteristics. In: *Australian Runoff Quality: A Guide to Water Sensitive Urban Design*, T. H. F. Wong (ed.), Engineers, Australia.

enHealth Council (2004). Guidance on Use of Rainwater Tanks, Australian Government, Department of Health and Ageing. http://www.health.gov.au/internet/main/publishing.nsf/content/3D981B51B4FB458DCA256F1900042F6E/$File/env_rainwater.pdf (accessed 15 October 2014).

Farnsworth L. (2012). Sustainable Water Supplies & Surface water Management. *Environment Industries Magazine*, (20), 118–120.

Ferguson M. (2011). Rainwater Tank Monitoring Report: A 12-Month One-Minute Interval Data Study of Rainwater Tank Water Savings and Energy Use for 52 Real Life Installations. Report for the State of New South Wales through Sydney Water Corporation.

Furumai H. and Okui H. (2010). Historical transition and progress of RWHM projects in Japan, 4th IWA International Rainwater Harvesting and Management Workshop as a part of IWA World Water Congress & Exhibition. 19 September, Montréal , Canada.

Furumai H., Kim J., Imbe M. and Okui H. (2008) Recent application of rainwater storage and harvesting in Japan, 3rd IWA International Rainwater Harvesting and Management Workshop as a part of IWA-Vienna World Water Congress & Exhibition, 11 September, Vienna, Austria.

Gardiner A., Skoien P. and Gardner T. (2008). Decentralised water supplies:south-east Queensland householders' experience and attitudes. *Water*, **36**(1), 53–58.

Gardiner A. (2009). Domestic rainwater tanks: usage and maintenance patterns in South East Queensland. *Water*, **36**, 151–156.

Gardiner A. (2010). Do rainwater tanks herald a cultural change in household water use? *Australasian Journal of Environmental Management*, **17**, 100–111.

Gardner T., Millar G. and Hyde R. (2003). The Healthy Home: a step towards sustainable suburbs, Published in the Proceedings of the 2nd National Water Recycling Conference, Australian Water Association and Stormwater Industry Association, Brisbane, September 2003.

Gardner E. A., Millar G. E., Christiansen C., Vieritz A. M. and Chapman H. (2006). Energy and water use at a WSUD subdivision in Brisbane, Australia. *Australian Journal of Water Resources*, **10**(3), 283–292.

Gold A., Goo R., Hair L. and Arazan N. (2010). Rainwater Harvesting: Policies, Programs, and Practices for Water Supply Sustainability, Low Impact Development 2010: Redefining Water in the City. American Society of Civil Engineers, pp. 987–1002.

Gomes U. A. F., Heller L., Cairncross S., Domenech L. and Pena J. L. (2014). Subsidizing the sustainability of rural water supply: the experience of the Brazilian rural rainwater harvesting programme. *Water International*, **39**(5), 606–619, doi:10.1080/02508060.2014.951255.

Gurung T. R. and Sharma A. (2014). Communal rainwater tank systems design and economies of scale. *Journal of Cleaner Production*, **67**, 26–36, http://dx.doi.org/10.1016/j.jclepro.2013.12.020.

Han M. and Han Y. (2002). The initiation of a rainwater harvesting campaign in Korea. *International Journal of Urban Sciences*, **6**(1), 46–51.

Han M. Y. and Mun J. S. (2011). Operational data of the star city rainwater harvesting system and its role as a climate change adaptation and a social influence. *Water Science & Technology* **63**, 2796–2801.

Harris G., Batley G., Fox D., Hall D., Jernakoff P., Molloy R., Murray A., Newell B., Parslow J., Skyring G. and Walker S. (1996). Port Phillip Bay Environmental Study Final Report, CSIRO Publishing, Melbourne, Australia.

Herrmann T. and Schmida U. (2000). Rainwater utilisation in Germany: efficiency, dimensioning, hydraulic and environmental aspects. *Urban Water*, **1**(4), 307–316.

Hills S., Birks R. and McKenzie B. (2001). The Millennium Dome "water cycle" Experiment: to evaluate water efficiency and customer perception at a recycling scheme for 6 million visitors. Proceedings of the IWA Second World Water Congress. Berlin, 15–19 October 2001.

Hood B., Gardner T., Barton R., Gardiner R., Beal C., Hyde R. and Walton C. (2010). The Ecovillage at Currumbin – a model for decentralised development, *Ozwater 2010*, Brisbane, Australia, 8–10 March 2010.

Hurst T (2012). Health risks from mosquitoes in rainwatrer tanks – a brief history of arboviruses and rainwater in SEQ. Presented at the UWSRA Decentralised Systems Project Workshop, August 2012, Brisbane. www.urbanwateralliance.org.au/publications (accessed 15 October 2014).

Ishaku H. T., Majid M. R., Johar F. (2012). Rainwater harvesting: an alternative to safe water supply in Nigerian rural communities. *Water Resources Management*, **26**, 295–305.

Kahinda J., Taigbenu A. and Boroto J. (2007). Domestic rainwater harvesting to improve water supply in rural South Africa. *Physics and Chemistry of the Earth*, **32**, 1050–1057.

Kim Y. J. and Han M. Y. (2001). Rainwater storage tank as a remedy for a local urban flood control. *Water Science and Technology: Water Supply*, **8**(1), 31–36.

Lee J., Han M. and Kim, (2010). Review on Codes and Application of Urban Rainwater Harvesting Utilization: Focused on Case Study in South Korea. *International Journal of Urban Sciences*, **14**(3), 307–319.

Le Muth N. and Barry M. (2006). Rainwater tank modelling investigation: South East Queensland Regional water supply Strategy, WBM Oceanics Australia.

Maheepala S., Coultas E., Neumann L. and Sharma A. (2013). Quantification of Regional Scale Water Quantity and Quality Implications of Rainwater Tanks in South East Queensland. Urban Water Security Research Alliance Technical Report No. 104.

Mankad A., Chong M. N., Umapathi S. and Sharma A. (2012a). The Role of Physical and Psychological Factors in Understanding Rainwater Tank Maintenance. Urban Water Security Research Alliance, Technical Report No. 64, Brisbane, QLD, Australia.

Master Plumbers and Mechanical Services Association of Australia (MPMSAA) (2008). Rainwater Tank Design and Installation Handbook. National Water Commission, Australian Government.

Melbourne Water (1996). The Port Phillip Bay Environmental Study. Publication number 1324–7905.

Melbourne Water (2014). Annual report 2013–2014. Enhancing Life and Livability. http://www.melbournewater.com.au/aboutus/reportsandpublications/Annual-Report/Documents/4977_MEL_Annual_Report_2014_LR.pd (accessed 10 October 2014).

Mobbs M. (1998). Sustainable house: Living for our future, CHOICE Books, Sydney.

Moglia M., Tjandraatmadja G., Delbridge N., Gulizia E., Sharma A. K., Butler R. and Gan K. (2014). Household Survey of savings and conditions of rainwater tanks. Melbourne, Smart Water Fund and CSIRO, Australia, 10TR4-001.

Morales-Pinzón T., Lurueña R., Rieradevall J., Gasol C. M. and Gabarrell X. (2012). Financial feasibility and environmental analysis of potential rainwater harvesting systems: a case study in Spain. *Resources, Conservation and Recycling*, **69**(0), 130–140, doi: http://dx.doi.org/10.1016/j.resconrec.2012.09.014.

NSW Health (2007). Use of Rainwater Tanks Where A Public Water Supply, Environmental Health Branch, NSW Health. http://www0.health.nsw.gov.au/policies/gl/2007/GL2007_009.html (accessed May 2013).

NSW Planning and Infrastructure (2004). Building Sustainability Index (BASIX), NSW Government, Department of Planning and Infrastructure. https://www.basix.nsw.gov.au/basixcms/ (accessed May 2013).

NHMRC and NRMMC (2014). Australian Drinking Water Guidelines. National Health and Medical Research Council and Natural Resource Management Ministerial Council, Australian Government, Canberra.

Nolde E. (2007). Possibilities of rainwater utilisation in densely populated area including precipitation runoffs from traffic surfaces. *Desalination*, **215**, 1–11.

Orrico S. R. M. (2003). Sistema associative de saneamento e seus efeitos sobre a população em comunidades do semi-arido baiano (Associative system of sanitation and its effects on the population of communities in the semiarid region of Bohia). PhD thesis, Faculty of Public Health, University of São Paulo, São Paulo, Brazil.

Pacific Institute (2014). The Californian Drought, www.californiadrought.org (accessed 10 October 2014).

Queensland Competetion Authority (2012). Report on the Assessment of Proposed Repeal of Water Saving Regulations. 35pp. http://www.qca.org.au.

Queensland Government (2011). Rainwater tanks. Department of Environment and Resource Management, Brisbane, Australia. http://www.derm.qld.gov.au/waterwise/gardening/pdf/rainwater_tanks.pdf (accessed 18 June 2011).

Queensland Health (2002). Guidelines to minimise mosquito and biting midge problems in new development areas, Queensland Health, Brisbane, Australia. http://www.health.qld.gov.au/ph/Documents/cdb/14804.pdf (accessed 15 November 2010).

Queensland Health (2007). Rainwater Tanks: A Guide to Keeping Your Tank Safe, Queensland Health, Brisbane, Australia. http://www.health.qld.gov.au/ph/Documents/ehu/32922.pdf (accessed on 11 may 2011).

QUDM (2013). Queensland Urban Drainage Manual 2013 Third Edition http://www.dews.qld.gov.au/water-supply-regulations/urban-drainage (accessed 20 October 2014).

QWC (2008). *South East Queensland Water Strategy – Draft March 2008*. Queensland Water Commission, Queensland Government, Australia.

Rodrigo S., Sinclair M., Forbes A., Cunliffe D. and Leder K. (2010). Drinking rainwater: a double blinded, randomised controlled study of water treatment filters and gastroenteritis incidence. *Amer, J of Public Health*, doi: 10.2105/AJPH.2009.185389.

Rodrigo S., Sinclair M. and Leder K. (2012). Water and sediment quality of rainwater tanks. *Journal of the Australian Water Association*, **39**(1), 83–87.

Sazakli E., Alexopoulos A. and Leotsinidis M. (2007). Rainwater harvesting, quality assessment and utilization in Kefalonia Island, Greece. *Water Research*, **41**(9), 2039–2047.

Schuetze T. (2013). Rainwater harvesting and management – policy and regulations In Germany. *Water Science & Technology, Water Supply*, **13**, 376–385.

South Australia's Strategic Plan Progress Report (2012). Audit Committee Secretariat, Department of the Premier and Cabinet, SA. http://saplan.org.au/media/BAhbBlsHOgZmSSIgMjAxMi8wOS8yNi8wMV80OV8yNV8yN19maWxlBjoGRVQ/01_49_25_27_file (accessed 20 October 2014).

Spinks A. T., Coombes P. J., Dunstan R. H. and Kuczera G. (2003). Water Quality Treatment Processes in Domestic Rainwater Harvesting Systems. 28th Hydrology and Water Resources Symposium. Wollongong, Australia. *The Institution of Engineers*, **2**, 227–234.

Spinks A. T., Dunstan R. H., Coombes P., Berghout B. and Kuczera G. (2005). Tank sludge as a sink for bacterial and heavy metal contaminants and its capacity for settlement, resuspension and flocculation enhancement. 12th International Rainwater Catchment Systems Conference: Mainsteaming Rainwater Harvesting, New Delhi, India. International Rainwater Catchment Systems Association.

Standards Australia, MPMSAA, ARID (2006). *Rainwater tank design and installation handbook*, 1edn, January 2006, Standards Australia, Master Plumbers and Mechanical Services Association of Australia, Australian Rainwater Industry Development Group. HB 230–2006, Sydney, NSW, Australia.

Talebpour M. R., Stewart R. A., Beal C., Dowling B., Sharma A. and Fane S. (2011). Rainwater energy tank end usage and energy demand: a pilot study. *Water*, March 2011, 97–101.

Taylor B., O'Halloran K., Otto A. and Geranio J. (2013). The Pros and Cons of an A+ Recycled water scheme: two years on. Presented at AWA Regional Conference Pelican Waters, Queensland, Australia.

Texas Water Development Board (2005). The Texas manual on rainwater harvesting, Austin, Texas, 3rd edn.

Tjandraatmadja G. and Sharma A. K. (2014). Policies and incentive programs for rainwater use in the urban sector. Chapter 14 in Rainwater Harvesting, Management and Use, E. Cohim, V. da Silva, Pedro Paz, L. Teixeira de Lima Brito, S. Sousa Medeiros, D. Batista dos Santos and H. Raj Gheyi, (eds), National Institute for Semi-Arid (INSA) and ABCMAC Brazilian Association of Harvesting, Management and Utilization of Rainwater (ABCMAC), Campina Grande, Brazil.

Tjandraatmadja G., Pollard C., Sharma A. and Gardner T. (2012). Optimisation of Energy Use in Household Rainwater Supply Systems. Urban Water Security Research Alliance Technical Report No. 89, http://www.urbanwateralliance.org.au/publications/UWSRA-tr89.pdf, (accessed July 2014).

Tjandraatmadja G., Pollard C., Sharma A. and Gardner T. (2013). How supply system design can reduce the energy footprint of rainwater supply in urban areas in Australia. *Water Science and Technology: Water Supply*, **13**(3), 753–760.

Umapathi S., Chong M. N. and Sharma A. (2012). Investigation and Monitoring of Twenty Homes to Understand Mains Water Savings from Mandated Rainwater Tanks in South East Queensland. Urban Water Security Research Alliance, Technical Report No. 63, Brisbane, QLD, Australia.

Umapathi S., Chong M. N. and Sharma A. K. (2013). Evaluation of plumbed rainwater tanks in households for sustainable water resource management: a real time monitoring study. *Journal of Cleaner Production*, **42**, 204–214.

UPRCT (2005). On- site stormwater detention handbook, 4th edn, Upper Parramatta River Catchment Trust http://www.uprct.nsw.gov.au/osd/osd_download_centre.htm (accessed 20 October 2014).

US EPA (2013). Rainwater harvesting: conservation, credit codes, and cost literature review and case studies, EPA-841-R-13–002, US EPA, January 2013.

van Dijk A. I. J. M., Beck H. E., Crosbie R. S., de Jeu R. A. M., Liu Y. Y., Podger G. M., Timbal B. and Viney N. R. (2013). The millennium drought in southeast Australia (2001–2009): Natural and human causes and implications for water resources, ecosystems, economy and society. *Water Resources Research*, **49**, 1040–1057.

Vialle C., Sablayrolles C., Lovera M., Huau M. C. and Montrejaud-Vignoles M. (2011). Monitoring of water quality from roof runoff: Interpretation using multivariate analysis. *Water Research*, **45**, 3765–3775.

Villarreal E. L. and Dixon A. (2005). Analysis of a rainwater collection system for domestic water supply in Ringdansen, Norrkoping, Sweden. *Building & Environment*, **40**, 1174–1184.

Walton C. and Holmes K. (2009). How much water efficiency does $321 million buy? In: *Proceedings of the 5th IWA Specialist Conference 'Efficient 2009' on Efficient Use and Management of Urban Water*, International Water Association (IWA) and Australian Water Association, 25–28 October, Sydney, Australia.

Ward S. L. (2010). Rainwater Harvesting in the UK: A Strategic Framework to Enable Transition from Novel to Mainstream, PhD thesis, University of Exeter UK.

Ward S., Barr S., Butler D. and Memon F. A. (2012). Rainwater harvesting in the UK: socio-technical theory and practice. *Technological Forecasting & Social Change*, **79**, 1354–1361.

White I. (2014). Rainwater tanks – reign no more? *Water Journal*, **41**(2), 163–169.

WSAA (2002a). WSA03–2002: Integrated Rainwater Tank Systems – a Supplement to the Water Supply Code of Australia, Melbourne, Australia, Water Services Association of Australia Inc. www.wsaa.asn.au/bookshop/Lists/eBooks/WSA03%20IRTS%20Supplement%20V1_1.pdf (accessed 20 October 2014).

WSAA (2002b). WSA03–2002: Integrated Rainwater Tank Systems – a Supplement to the Water Supply Code of Australia, Melbourne, Australia, Water Services Association of Australia Inc. www.wsaa.asn.au/bookshop/Lists/eBooks/WSA03%20IRTS%20Supplement%20V1_1.pdf (accessed 20 October 2014).

Index

A

Aesthetic, 9, 190, 202, 328
Alternative water, 181, 183, 188
Alternative water resources, 324–326
Amotivation, 185, 198, 199
Arbovirus, 151, 183, 320, 332–333
Asset Management, 152, 155, 176
Attitudinal, 193
Audit, 86, 103
Augmentation, 294–296
Australian Bureau Statistics (ABS), 51
Australian census, 59
Australian Drinking Water Guidelines (ADWG), 8, 209, 212, 222, 231, 256, 329
Autonomy, 185, 196–197
Average Return Interval, ARI, 330

B

BASIX, 49, 76, 93–95, 97, 103–105
Beliefs, 171, 188
Benefit cost ratio, 291, 296, 325, 330
Billing records, 38, 48, 59, 91
Binning, 64, 88
Bioretention, 293, 296

C

Capita water use, 57, 326
Capital cost, 10, 279, 283, 291, 297
Cisterns, 104, 110, 136, 140
Cluster, 40
Cluster-scale, 253–258, 263–264, 271
Cognitive, 171, 173, 184
Cohort(s), 52
Coliforms, 231
Communal tanks, 13, 230, 258, 329–330
Community acceptance, 2, 11, 75, 171, 186, 255
community attitudes, 182, 186
Connected (roof) area, 19, 23, 37, 39, 64, 69, 94–95, 103, 111, 115–119, 326, 329, 331
Contaminants, 8, 121, 212, 217, 269, 321
Continuing loss, 21, 29, 35, 40
Cost benefit analysis, 290–297
Cost-effectiveness, 276–285
Covenants, 50
CRC, 24, 231, 234, 326
Cryptosporidium, 234, 235, 237, 238, 329

D

Daily water use, 28
Data collection, 82, 96, 97
Data saturation, 188
Decentralised systems, 256–257, 269, 271
Demand generators, 26, 30
Demand management, 48, 51, 57
Demographics, 38, 48, 75, 195
Desalination, 6, 128, 298, 324
Detached dwelling(s), 50, 53, 104, 110, 121, 321
Detention storage, 330
Developing countries, 1, 2, 5, 14, 20, 323–324, 331, 332
Dimensionless, 28, 30, 41
Disaggregate, 25, 74, 75, 87
Disinfection, 3, 8, 247, 260, 269, 329
Diurnal water use, 13, 26, 93

Downpipes, 76, 102, 111, 115, 261, 320–321
Dual reticulation, 69, 104, 325

E
E. coli, 9, 231, 235–243
Eco-development, 187, 321
Ecologically sustainable development, 4
Economic, 11–12, 14, 103, 183, 193, 255, 279, 292
El Nino, 325–326
Encephalitis, 320
End use(s), 54, 69, 136
Energy footprint, 133
Epidemiological, 9, 230, 243, 244, 245, 249, 329
Error(s), 31, 37, 40, 42, 65, 86, 331
Evaporation, 21, 22, 23, 26, 35, 37, 309
eWater, 24, 212, 263, 326,
External use(s), 4, 104, 119, 322
Externalities, 11–12, 14, 256

F
Faecal indicator bacteria, 230–232, 238–242
Faecal pathogens, 1, 2, 14, 241
Fever, 151, 164, 177, 183, 320, 324
First flush, 109, 157, 215, 216
Flood, flooding, 5, 29, 307, 313–314, 323
Flow-regime, 309–313
Focus groups, 161, 167–168, 169, 186, 203
Functionality factor, 327

G
Giardia, 234, 235
Greenfield(s), 12, 51, 263, 264,
Greywater, 24, 69, 104, 191, 193,
Guidance, 9, 10, 146, 156, 266, 276
Guidelines, 9–10, 158, 209, 211, 231
Gutters, 2, 10, 76, 109, 212, 215, 249, 329

H
Harvesting, 48, 306, 308–309
Header tank, 137, 143–144, 145, 320, 331
Health risks, 242–248
Heavy metals, 211, 217–218, 221
Hierarchical, 199
Household, 25, 64–65, 69, 78, 91, 140

I
Impervious area, 4, 263, 309
Implementation drivers, 321–324
In situ, 50, 80, 122, 128, 137, 147, 241, 320
Initial loss, 21, 23, 29, 35–36, 308–309, 313
Installation monitoring framework, 102, 103–109
Intake pipe, 329
Integrated urban water management (IUWM), 256
Internal/external demand, 29, 38
Internalized, 185
Internally plumbed, 8, 38, 51, 104, 147, 277, 325, 330
International leanings, 322–324
Intrinsic motivation, 185, 186
IOTA, 322
Irrigation, 4, 9, 26, 34, 66, 69, 137, 195, 254, 267, 323, 332

L
Laundry, 4, 13, 25, 38, 67, 110, 156, 254, 257, 263
Lead, 209–217, 219–223
Lead flashing, 215
Levelised cost, 10, 276, 278, 282, 285, 289, 290, 325
Likert, 191, 195, 198
Local Government Area (LGA), 38, 39, 51, 52, 61, 111, 191, 195, 198
Logging, 82, 85, 90
Longitudinal, 39, 67, 247, 248
Lot size, 38, 53, 58, 64, 70
Low flow, 13, 26, 80, 134, 137, 142, 146, 308
Low-income, 20, 25, 190

M
Mains water saving, 74, 76, 87, 90, 91, 94
Mains water use, 6, 9, 13, 28, 38, 50, 56, 57, 60, 62, 76, 94
Maintenance, 156, 173, 236, 237, 282, 291, 327
Mandated, 4, 48, 50, 105, 156, 158, 194, 199
Mandated rainwater tank (MRT), 51, 195
Manual(s), 83,88, 96
Mass balance, 21, 37, 89
Mediterranean, 4, 66, 323, 324
Meter, 53, 69, 75, 80–81
Microbial quality, 8, 9, 230–231, 245–248
Millennium drought, 5, 321
Millennium Goal, 323
Model time-step, 20, 22, 24–25, 30–32, 41–42
Modelling, 7, 8, 12, 50, 55, 61, 69, 70
Monitoring, 74–76, 78–79, 82, 83–87, 94, 95–96
Monsoons, 323
Moreton Bay Regional Council, 51, 57
Mosquitos, 333
MP4.2, 61, 104, 321, 325, 326
Mplus, 193
Multiple regression, 196–197, 199
Multivariate analysis, 196
MUSIC, 24, 29, 212, 309–311, 313

N
Neurotoxin, 331
Non Government Organizations (NGOs), 323
Non potable, 5, 6, 9, 10, 19, 74, 94, 95, 103, 182, 186, 188, 190, 193, 230, 237, 245, 261, 263, 285, 289
Non-parametric, 24, 64
Normal distribution, 30
Normative comparison, 189
NVivo 8, 188

O
Occupancy, 30, 140–141
On site, 4, 39, 82, 86, 111, 122, 327
Order of calculation(s), 23, 27, 41
Orifice plates, 300
Outdoor water, 26, 29, 54, 57–58
Overflow, 33–36, 116–117
Overflow estimation, 40

P
Pair matching, 64, 70
Paradoxical effects, 202
Parramatta River Trust, 330
Peri-urban, 6, 183, 255, 320
Phone survey, 59, 60
Plastic (tanks), 320
Pluviograph, 24, 26
Policy implications, 326–331
Polyethylene, 239, 332
Potable water, 5, 51, 94, 101–102, 229, 247, 328
Potable water substitution, 322, 332
Predictions, 8, 20, 25, 27, 37
Probabilistic, 20, 24, 26, 48, 172
Probabilistic Urban Rainwater and Wastewater Reuse Simulator (PURRS), 27, 61, 62
Protection motivation, 184, 185, 190
Protection motivation theory (PM theory), 184, 190
Protozoans, 321
Proxy data, 57
Psychological, 159, 172, 191, 192, 195
Psychosocial, 182, 195–197
Public health, 2, 4, 152, 160, 171, 183, 235, 248, 249, 319, 323, 328
Pumps, 129, 130–131, 140, 146

Q
Qualitative, 186, 203, 295
Quantitative, 190, 194
Quantitative Microbial Risk Assessment (QMRA), 245–248, 329
Quantitative PCR, 237, 238, 245, 247–248
Queensland Development Code (QDC), 50, 51, 58, 61, 76, 104, 112, 114, 115, 118, 119, 121, 164, 165, 182, 321, 325, 326

R
Rain tanks, 217–223
Rain water tanks and systems, 285–290
Rainwater harvesting, 128, 145, 255–264
Rainwater tank installation, 103–109
Rainwater tank model, 23, 28, 29, 33, 39, 42
Rainwater tanks, 2–10, 12, 13, 74, 76, 77, 83, 88, 91, 128, 131, 141, 145, 146, 319–333
Rainwater yield, 48, 71
Rebate, 6, 11, 48, 63, 160, 182, 201, 276
Regression, 25, 26, 197
Reliability, 2, 7, 11, 19, 20, 25, 76, 78, 191, 254, 255, 256, 264, 267, 292, 294
Residential, 6, 20, 75, 79, 261
Respondents, 3, 171, 186–188, 191, 199
Retention waterway, 312, 313
Retrofitters, 184, 194, 200
Risk Assessment, 152, 153–154, 160, 171
Roof catchment(s), 7, 11, 66, 67, 77, 95, 101, 102, 122, 261, 263
Roof-captured rainwater, 242–248
Root square error (RSE), 64
Runoff, 2, 23, 27, 37, 212–214, 261, 306, 307

S
Scaling up, 28, 39, 90
Seasonal, 25, 30, 33, 86–87, 267
Self Determination theory (SDT), 185, 197, 199, 200
Self efficacy, 160–161
Self-determination, 185, 186, 197, 199, 200
Sewage, 4
Sewer overflow(s), 12
Shower rose, 58
SILO, 24, 87, 256
Silver, 321
Sludge, 219, 223
Smart meter, 48, 69, 74
Social, 26, 59, 182
Social context, 331–333
Socio-demographic factors, 57, 62, 69
Socio-technical analysis 168, 170, 171, 173
Solenoid, 132, 144, 146, 147, 312
Source-control, 301
Spatial averaging, 40

Spatial scale, 83
Spatially variable, 69
Specific energy, 128, 133, 134, 137, 139
Spill, 22, 29, 30–31, 41
Spreadsheet models, 24
Stochastic generator, 26, 30
Stormwater, 308–309
Stormwater quality, 3, 293, 297, 302, 306
Strategic water planning, 8
Stream ecosystem, 12, 301
Student's *t* test, 53–54
Sub-tropical, 66
Subsidies, 4, 63, 193–194, 230, 321, 323
Sweden, 4, 20, 322
Sydney Water, 76, 93–94, 325

T

TANK model, 20–24, 55
Tank rebates, 65
Telemetry, 88, 312, 323, 327, 331
Time series, 30, 67
Time step, 20, 22–33, 35, 37, 75, 87, 88, 97
Toilet flushing, 312
Toxin genes, 235–236
Trickle top up, 13, 28, 37, 77, 90, 93, 95, 110, 114, 115, 119, 121, 129, 131, 132, 196
Trigger, triggering, 26, 28, 114, 115

U

Ubiquitous, 202, 320
Ultraviolet, 3, 236, 249, 321
Urban Water Security Research Alliance (UWSRA), 186, 190, 194

V

Verification, 101
Victoria, 175, 217, 230, 231, 259, 307, 312, 325
Viruses, 151, 183, 320, 324, 332

W

Water balance, 20, 22, 27–28, 89, 262
Water grid, 183, 324
Water quality, 3, 7, 9, 12, 13, 209–217, 223
Water sensitive urban design (WSUD), 6, 73, 256, 293, 330
Water supply, 152, 155, 164, 171, 208, 213, 219, 221–222
Water Efficiency Labelling Scheme (WELS), 63, 136
Water wise, 62
Waterwise rebate, 50, 322

Y

YAS, 22–23, 27, 31, 32, 41YBS (Yield Before Spill), 22–23, 27, 31, 41
Yield, 7, 10, 11, 21–23, 33, 35–37, 42, 48, 86, 248, 277, 284, 289
Yield estimation, 31, 40

Z

Zoonotic pathogens, 237–238

IWA Publishing's authorised EU representative for General Product Safety Regulations is Diane D'Arras, 15 rue Duret, 75116 Paris, France, e-mail: safety@iwap.co.uk.

Printed and bound by CPI Group (UK) Ltd, Croydon, CR0 4YY
12/05/2026
02108869-0014